# Problems in
# MATHEMATICS
## With Hints and Solutions

V GOVOROV · P DYBOV
N MIROSHIN · S SMIRNOVA

# Problems in
# MATHEMATICS
## With Hints and Solutions

## V GOVOROV · P DYBOV
## N MIROSHIN · S SMIRNOVA

*Edited by*
Prof. A.I. Prilepko, D.Sc.

*Translated from the Russian by*
Irene Aleksanova

✳ arihant
ARIHANT PRAKASHAN (Series), MEERUT

# PREFACE

The book contains more than three thousand mathematics problems and covers each topic taught at school. The problems were contributed by 120 of the higher schools of the USSR and all the universities.

The book is divided into four parts : algebra and trigonometry, fundamentals of analysis, geometry and vector algebra, and the problems and questions set during oral examinations.

The authors considered it necessary to include some material relating to complex numbers, combinatorics, the binomial theorem, elementary trigonometric inequalities, and set theory and the method of coordinates. The authors believe that this material will help the readers systematize their knowledge in the principal divisions of mathematics.

In writing the book, the authors have used their experience of examining students in mathematics at higher schools and the preparation of television courses designed to help students revise their knowledge for the entrance examinations to higher educational establishments.

To make it easier for readers to grasp the material, some of the sections have been supplemented with explanatory text. The problems are all answered and some have additional hints or complete solutions.

The more difficult problems are marked with asterisks.

Part 4 is entitled "Oral Examination Problems and Questions" and includes samples suggested by the higher schools.

The authors hope that this book will help those who want to enter the various types of higher school, aid the teachers, and be of use to all those who want to deepen and systematize knowledge of mathematics.

**The Authors**

# CONTENTS

# Part-1

## Algebra, Trigonometry, and Elementary Functions

## 1.1. Problems on Integers. Criteria for Divisibility

**Natural numbers (N = {1, 2, 3, ...., n,......}).** Every natural number $n$ can be uniquely factored into a product of prime factors

$$n = p_1^{k_1} \, p_2^{k_2} \, ... \, p_m^{k_m},$$

where $p_1, ..., p_m$ are elementary divisors of the number $n$, and $k_1,..., k_m$ are the multiplicities of those divisors $(k_1, ...., k_m \in N)$.

To calculate the *greatest common divisor* of two natural numbers, each of their elementary common divisor must be raised to a power which is equal to the smallest of the multiplicities with which the divisor appears in the prime factorization of the given numbers and all the resulting numbers must then be multiplied together.

To calculate the *least common multiple* of two natural numbers, every prime divisor appearing in the factorization of at least one of these numbers must be raised to a power which is equal to the largest of the multiplicities with which the divisor appears in the prime factorization of the given numbers and all the resulting numbers must then be multiplied together.

**Example.** Assume $n_1 = 2^3 \cdot 5^2 \cdot 7 = 1400, n_2 = 2^2 \cdot 3^2 \cdot 5 \cdot 11 = 1980$. The greatest common divisor of $n_1$ and $n_2$ is $2^2 \cdot 5 = 20$. The least common multiplicity of $n_1$ and $n_2$ is $2^3 \cdot 3^2 \cdot 5^2 \cdot 7 \cdot 11 = 138600$.

If $a_0$ is a digit in the unit's place, $a_1$ is a digit in the ten's place, etc., of the natural number $n$, then this number can be written as

$$n = \overline{a_k a_{k-1} \, ... \, a_1 a_0} = a_k \cdot 10^k + ... + a_1 \cdot 10 + a_0.$$

**Criteria for divisibility.** The number $n$ can be divided

(1) by 2 (by 5) if and only if $a_0$ can be divided by 2 (by 5);

(2) by 4 if and only if the number $a_1 \cdot 10 + a_0$ can be divided by 4;

(3) by 3 (by 9) if and only if the sum of all the digits in that number can be divided by 3 (by 9).

1. The sum of the digits in a two-digit number is 6. If we add 18 to that number, we get a number consisting of the same digits written in the reverse order. Find the number.

2. If we multiply a certain two-digit number by the sum of its digits, we get 405. If we multiply the number consisting of the same digits written in the reverse order by the sum of the digits, we get 486. Find the number.

3. Find three numbers, the second of which is as much greater than the first as the third is greater than the second, if the product of the two smaller numbers is 85 and the product of the two larger numbers is 115.

4. The sum of two numbers is equal to 15 and their arithmetic mean is 25 per cent greater than their geometric mean. Find the numbers.

5. The difference between two numbers is 48, the difference between the arithmetic mean and the geometric mean of the numbers is 18. Find the numbers.

6. The geometric mean of two numbers exceeds by 12 the smaller of the numbers and the arithmetic mean of the same numbers is smaller by 24 than the larger of the numbers. Find the numbers.

7. Find the two-digit number if the number of its units exceeds by 2 the number of its tens and the product of the required number by the sum of its digits is equal to 144.

8. The product of the digits of a two-digit number is twice as large as the sum of its digits. If we subtract 27 from the required number, we get a number consisting of the same digits written in the reverse order. Find the number.

9. The product of the digits of a two-digit number is one-third that number. If we add 18 to the required number, we get a number consisting of the same digits written in the reverse order. Find the number.

10. The sum of the squares of the digits of a two-digit number is 13. If we subtract 9 from that number, we get a number consisting of the same digits written in the reverse order. Find the number.

11. A two-digit number is thrice as large as the sum of its digits, and the square of that sum is equal to the trebled required number. Find the number.

12. Find a two-digit number which exceeds by 12 the sum of the squares of its digits and by 16 the doubled product of its digits.

13. The sum of the squares of the digits constituting a two-digit number is 10, and the product of the required number by the number consisting of the same digits written in the reverse order is 403. Find the number.

14. If we divide a two-digit number by the sum of its digits, we get 4 as a quotient and 3 as a remainder. Now if we divide that two-digit number by the product of its digits, we get 3 as a quotient and 5 as a remainder. Find the two-digit number.

**15.** If we divide a two-digit number by a number consisting of the same digits written in the reverse order, we get 4 as a quotient and 15 as a remainder ; now if we subtract 9 from the given number, we get the sum of the squares of the digits constituting that number. Find the number.

**16.** Find the two-digit number the quotient of whose division by the product of its digits is equal to 8/3, and the difference between the required number and the number consisting of the same digits written in the reverse order is 18.

**17.** Find the two-digit number, given : the quotient of the required number and the sum of its digits is 8; the quotient of the product of its digits and that sum is 14/9.

**18.** If we divide the unknown two-digit number by the number consisting of the same digits written in the reverse order, we get 4 as a quotient and 3 as a remainder. Now if we divide the required number by the sum of its digits, we get 8 as a quotient and 7 as a remainder. Find the number.

**19.** If we divide a two-digit number by the sum of its digits, we get 6 as a quotient and 2 as a remainder. Now if we divide it by the product of its digits, we get 5 as a quotient and 2 as a remainder. Find the number.

**20.** What two-digit number is less than the sum of the squares of its digits by 11 and exceed their doubled product by 5?

**21.** Find two successive natural numbers if the square of the sum of those numbers exceeds the sum of their squares by 112.

**22.** First we increase the denominator of a positive fraction by 3 and next time we decrease it by 5. The sum of the resulting fractions proves to be equal to $\frac{2}{3}$. Find the denominator of the fraction if its numerator is 2.

**23.** The denominator of an irreducible fraction is greater than the numerator by 2. If we diminish the numerator of the inverse fraction by 3 and subtract the given fraction from the resulting one, we get $\frac{1}{15}$. Find the fraction.

**24.** Let us consider a fraction whose denominator is smaller than the square of the numerator by unity. If we add 2 to the numerator and the denominator, the fraction will exceed $\frac{1}{3}$; now if we subtract 3 from the numerator and the denominator, the fraction remains positive but smaller than $\frac{1}{10}$. Find the fraction.

**25.** There are only three positive two-digit numbers such that each number is equal to the incomplete square of the sum of its digits. Find two of them, being given that the second number exceeds the first one by 50 units.

**26.** Find the sum of all two-digit numbers which, being divided by 4, leave a remainder of 1.

**27.** Find the sum of all three-digit numbers which give a remainder of 4 when they are divided by 5.

**28.** Find the sum of all two-digit numbers which give a remainder of 3 when they are divided by 7.

**29.** Find the sum of all odd three-digit numbers which are divisible by 5.

**30.** The product of a two-digit number by a number consisting of the same digits written in the reverse order is equal to 2430. Find the number.

**31.** Find the pairs of natural numbers the difference of whose squares is 45.

**32.** There is a natural number which becomes equal to the square of a natural number when 100 is added to it, and to the square of another natural number when 168 is added to it. Find the number.

**33.** Find two natural numbers, whose sum is 85 and whose least common multiple is 102.

**34.** Find all pairs of natural numbers whose greatest common divisor is 5 and the least common multiple is 105.

**35.** Find two three-digit numbers whose sum is a multiple of 504 and the quotient is a multiple of 6.

**36.** Represent the number 19 as the difference between the cubes of natural numbers.

**37.** Find three numbers if the cube of the first number exceeds their product by 2, the cube of the second number is smaller than their product by 3, and the cube of the third number exceeds their product by 3.

**38.** Find all two-digit numbers such that the sum of the digits constituting the number is not less than 7; the sum of the squares of the digits is not greater than 30; the number consisting of the same digits written in the reverse order is not larger than half the given number.

**39.** In a four-digit number the sum of the digits in the thousands, hundreds and tens is equal to 14, and the sum of the digits in the units, tens and hundreds is equal to 15, the digit of the tens being greater by 4 than the digit of the tens. Among all the numbers satisfying these conditions, find the number the sum of the squares of whose digits is the greatest.

**40.** In a four-digit number the sum of the digits of the thousands and tens is equal to 4, the sum of the digits of the hundreds and the units is 15, and the digit of the units exceeds by 7 the digit of the thousands. Among all the numbers satisfying these conditions find the number the sum of the product of whose digit of the thousands by the digit of the units and the product of the digit of the hundreds by that of the tens assumes the least value.

**41.** Prove that if the number $a$ is equal to the sum of the squares to two unequal natural numbers, then $2a$ is also equal to the sum of the squares of two unequal natural numbers.

**42.** Find the sum of all irreducible fractions between 10 and 20 with a denominator of 3.

**43.** There are natural numbers $m$ and $n$. Find all the fractions $m / n$ whose denominator is smaller than the numerator by 16, the fraction itself is smaller than the sum of the trebled inverse and 2 , and the numerator is not greater than 30.

**44.** Given a sequence

$$u_n = \frac{(1 + (-1)^n) + 1}{5n + 6}.$$

Find the number of the terms of the sequence $(u_n)$ which satisfy the condition $u_n \in (1 / 100, \ 39 / 100)$.

## 1.2. Real Numbers. Transformation of Algebraic Expressions

**Rational numbers (Q).** Every rational number $p / q$ ($p \in \mathbf{Z}$ is an integer, $q \in \mathbf{N}$ is a natural number) can be represented as an *infinite periodic decimal fraction* (possibly with a zero period)

$$p / q = \pm a, \alpha_1 \dots \alpha_n \ (\beta_1 \beta_2 \dots \beta_m).$$

A reverse representation also holds

$$\pm a, \ \alpha_1 \dots \alpha_n \ (\beta_1 \beta_2 \dots \beta_m) = \pm a \pm 0, \alpha_1 \dots \alpha_n \ (\beta_1 \beta_2 \dots \beta_m)$$

$$= \pm a \pm \frac{\overline{\alpha_1 \dots \alpha_n \beta_1 \beta_2 \dots \beta_m} - \overline{\alpha_1 \dots \alpha_n}}{\underbrace{99 \dots 9}_{n} \ \underbrace{0 \dots 0}_{n}}$$

**Real numbers (R)** . The numbers which can be represented as various decimal fractions are called real numbers.

**Formulas for abbreviated multiplication.** The following equations hold for any numbers $a, b \in \mathbf{R}$ :

$$(a + b)^2 = a^2 + 2ab + b^2;$$

$$(a - b)^2 = a^2 - 2ab + b^2;$$

$$(a + b)^3 = a^3 + 3a^2 b + 3ab^2 + b^3;$$

$$(a - b)^3 = a^3 - 3a^2 b + 3ab^2 - b^3;$$

$$a^2 - b^2 = (a - b)(a + b);$$

$$a^3 - b^3 = (a - b)(a^2 + ab + b^2);$$

$$a^3 + b^3 = (a + b)(a^2 - ab + b^2);$$

$$a^4 - b^4 = (a - b)(a + b)(a^2 + b^2);$$

$$a^4 + b^4 = (a^2 - \sqrt{2}ab + b^2)(a^2 + \sqrt{2}ab + b^2);$$

$$a^n - b^n = (a - b)(a^{n-1} + a^{n-2}b + \dots + a^{n-k}b^k + \dots + b^{n-1}), \ n \in N;$$

$$a^{2n+1} + b^{2n+1} = (a + b)(a^{2n} - a^{2n-1}b + \dots - ab^{2n-1} + b^{2n}), \ n \in N.$$

Represent the following mixed infinite decimal periodic fractions as common fractions **(1-4)** :

**1.** 7 5 (3).

**2.** 2.1 (32).

**3.** $\dfrac{0.23(7)+\dfrac{43}{450}}{0.5(61)-\dfrac{113}{495}}$.

**4.** Find $x$ if $\dfrac{0.1\,(6)+0.\,(3)}{0.\,(3)+1.1(6)}\,x = 10.$

Calculate **(5-31).**

**5.** $\left(\dfrac{1}{3}\right)^{-10}\cdot 27^{-3}+0.2^{-4}\cdot 25^{-2}+(64^{-1/9})^{-3}.$

**6.** $(\sqrt{2+\sqrt{3}}+\sqrt{2-\sqrt{3}})^{2}.$

**7.** $3\left(\dfrac{2}{\sqrt{10}+5}+\dfrac{5}{\sqrt{10}-2}-\dfrac{7}{\sqrt{10}}\right).$

**8.** $\dfrac{(5\sqrt{3}+\sqrt{50})\,(5-\sqrt{24})}{\sqrt{75}-5\sqrt{2}}.$

**9.** $1+\sec 20°-\sqrt{3}\cot 40°.$

**10.** $\dfrac{2^{-2}\cdot 5^{3}\cdot 10^{-4}}{2^{-3}\cdot 5^{2}\cdot 10^{-5}}.$

**11.** $(6-4\,(5/\,16)^{0})^{-2}+(2/\,3)^{-1}-3/\,4.$

**12.** $\dfrac{2^{-2}+2^{0}}{(1/\,2)^{-2}-5\cdot(-2)^{-2}+(2/\,3)^{-2}}.$

**13.** $\dfrac{(0.6)^{0}-(0.1)^{-1}}{(3/\,2^{3})^{-1}\cdot(3/\,2)^{3}+(-1/\,3)^{-1}}.$

**14.** $\dfrac{3\dfrac{1}{3}\sqrt{9/\,80}-\dfrac{5}{4}\sqrt{4/\,5}+5\sqrt{1/\,5}+\sqrt{20}-10\sqrt{0.2}}{3\dfrac{1}{2}\sqrt{32}-\sqrt{4\dfrac{1}{2}}+2\sqrt{1/\,8}+6\sqrt{2/\,9}-140\sqrt{0.02}}\;\sqrt{2/\,5}.$

**15.** $\dfrac{2\cdot 4\sqrt{8\dfrac{1}{3}}-9\sqrt{1/\,3}+\sqrt{2\dfrac{1}{2}}+\dfrac{1}{2}\sqrt{1/\,3}-\dfrac{1}{3}\sqrt{27}}{1\dfrac{1}{3}\sqrt{4\,(1/\,2)}-\sqrt{0.5}+1.5\sqrt{2}+20\sqrt{1/\,50}-\sqrt{32}}\;\sqrt{2/\,3}.$

**16.** $\left(\sqrt{\left(\sqrt{2}-\dfrac{3}{2}\right)^{2}}-\sqrt[3]{(1-\sqrt{2})^{3}}\right)^{2}+2^{-3/2}\cos\dfrac{3\pi}{4}.$

**17.** $\left(\sqrt{\left(\sqrt{5}-\dfrac{5}{2}\right)^{2}}-\sqrt[3]{\left(\dfrac{3}{2}-\sqrt{5}\right)^{3}}\right)^{1/2}-\sqrt{2}\sin\dfrac{7\pi}{4}.$

**18.** $(\sqrt{\sqrt{3}+\sqrt{2}}-(\sqrt{3}-\sqrt{2})^{1/2})\,((\sqrt{3}+\sqrt{2})^{1/2}+\sqrt{\sqrt{3}-\sqrt{2}})^{-1}-\cos\dfrac{5\pi}{4}.$

**19.** $\left( \sqrt{2+\sqrt{3}} + \left( 2 - 2\cos\frac{11\pi}{6} \right) \right)^{1/2} \times (\sqrt{2 + 2\cos(\pi/6)} - \sqrt{2-\sqrt{3}})^{-1}.$

**20.** $\sqrt{25^{1/\log_8 5} + 49^{1/\log_6 7}}.$

**21.** $\cos\left( \frac{\pi}{10} \left( \log_3 \frac{1}{9} + \log_{1/9} 3 \right) \right).$

**22.** $\log_{1/2} \left( \log_3 \cos\left( \frac{\pi}{6} \right) - \log_3 \sin\left( \frac{\pi}{6} \right) \right).$

**23.** $\left[ \left( \frac{3}{2} \right)^{-0.5} \cdot 3^{\frac{1}{2}\log_3 6} + 1 \right]^{1/2} \sin\frac{7\pi}{3}.$

**24.** $\left( 7^{1/3} \cdot 3^{-\log_{27} 7} \cdot \tan\frac{5\pi}{3} + 3^{1/2} + 2 \right)^{1/2} \cdot \cos\frac{7\pi}{4}.$

**25.** $\left( (128^{3/7} \cdot 27^{1/3} \cdot 10^{-\log 48})^{-1/2} + \cot^{-1}\frac{2\pi}{3} \right)^2 + 2 \cdot 6^{1/2}.$

**26.** $\left( 3^{1/2} \cdot 8^{\log_2 3} \cdot \sin\frac{2\pi}{3} + \left( \frac{1}{3} \right)^{-4} \right)^{1/2} \cos^{-1}\frac{5\pi}{6}.$

**27.** $(x^{1/3} + y^{1/3})(x^{2/3} - x^{1/3}y^{1/3} + y^{2/3})$ for $x = 4\frac{5}{7}$, $y = 5\frac{2}{7}$.

**28.** $\frac{x-1}{x^{3/4} + x^{1/2}} \cdot \frac{x^{1/2} + x^{1/4}}{x^{1/2} + 1} \cdot x^{1/4} + 1$ for $x = 16$.

**29.** $\dfrac{a^3 - a - 2b - \dfrac{b^2}{a}}{\left( 1 - \sqrt{\dfrac{1}{a} + \dfrac{b}{a^2}} \right)(a + \sqrt{a+b})} : \left( \dfrac{a^3 + a^2 + ab + a^2 b}{a^2 - b^2} + \dfrac{b}{a-b} \right)$

for $a = 23,\ b = 22$.

**30.** $x^3 + 3x - 14$ for $x = \sqrt[3]{7 + 5\sqrt{2}} - \dfrac{1}{\sqrt[3]{7 + 5\sqrt{2}}}.$

**31.** The difference $\sqrt{|40\sqrt{2} - 57|} - \sqrt{40\sqrt{2} + 57}$ is an integer. Find that integer.

Remove the irrationality in the denominator **(32-33)**.

**32.** $\dfrac{1}{1 + \sqrt{2} + \sqrt{3}}.$

**33.** $\dfrac{1}{\sqrt{2\sqrt{3} - \sqrt{2}} \sqrt{\sqrt{2} + \sqrt[3]{3}}}$

**34.** Compare the following two numbers : $a = \dfrac{9}{\sqrt{11} - \sqrt{2}}$ and $b = \dfrac{6}{3\sqrt{3}}.$

**35.** Given : $1 < a < b + c < a + 1$, $b < c$. Prove that $a < b$.

**36.** What is larger, $\log_3 108$ or $\log_5 375$?

Arrange the following numbers in an increasing order (37-39).

**37.** $0$; $\sqrt{0.8}$; $1.2$; $11/30$; $0.91846$.

**38.** $1$; $0.37$; $65/63$; $61/59$; $\tan 33°$; $\tan(-314°)$.

**39.** $0.02$; $1$; $0.85$; $\sqrt{3}/2$; $\sqrt{0.762}$; $-\cos 571°$.

**40.** Prove that $53^{53} - 33^{33}$ is divisible by 10.

Factor (41-43).

**41.** $n^4 + 4$.        **42.** $1 + n^4 + n^8$.        **43.** $1 + x^5$.

Simplify the following expressions (44-166).

**44.** $\dfrac{a^2 - b^2}{a - b} - \dfrac{a^3 - b^3}{a^2 - b^2}$.

**45.** $\dfrac{1}{(a-b)(a-c)} + \dfrac{1}{(b-c)(b-a)} + \dfrac{1}{(c-a)(c-b)}$.

**46.** $\left(a + \dfrac{ab}{a-b}\right)\left(\dfrac{ab}{a+b} - a\right) : \dfrac{a^2 + b^2}{a^2 - b^2}$.

**47.** $\left(\dfrac{4(a+b)^2}{ab} - 16\right)\left(\dfrac{(a+b)^2 - ab}{ab}\right) : \dfrac{a^3 - b^3}{ab}$.

**48.** $\left(\dfrac{a+3b}{(a-b)^2} + \dfrac{a-3b}{a^2 - b^2}\right) : \dfrac{a^2 + 3b^2}{(a-b)^2}$.

**49.** $\left(m + n - \dfrac{4mn}{m+n}\right) : \left(\dfrac{m}{m+n} - \dfrac{n}{n-m} - \dfrac{2mn}{m^2 - n^2}\right)$.

**50.** $\left(\dfrac{1}{(m+n)^2}\left(\dfrac{1}{m^2} + \dfrac{1}{n^2}\right) + \dfrac{2}{(m+n)^3}\left(\dfrac{1}{m} + \dfrac{1}{n}\right)\right) m^2 n^2$.

**51.** $\left(\dfrac{a\sqrt{2}}{(1+a^2)^{-1}} - \dfrac{2\sqrt{2}}{a^{-1}}\right) \dfrac{a^{-3}}{1 - a^{-2}}$.

**52.** $\dfrac{\dfrac{1}{a} + \dfrac{1}{b+c}}{\dfrac{1}{a} - \dfrac{1}{b+c}}\left(1 + \dfrac{b^2 + c^2 - a^2}{2bc}\right)(a+b+c)^{-2}$.

**53.** $\dfrac{\dfrac{1}{a} - \dfrac{1}{b+c}}{\dfrac{1}{a} + \dfrac{1}{b+c}}\left(1 + \dfrac{b^2 + c^2 - a^2}{2bc}\right) : \dfrac{a-b-c}{abc}$.

**54.** $\dfrac{x^3 + y^3}{x + y} : (x^2 - y^2) + \dfrac{2y}{x+y} - \dfrac{xy}{x^2 - y^2}$.

**55.** $\left(\left(\left(\dfrac{a+1}{a-1}\right)^2 + 3\right) : \left(\left(\dfrac{a-1}{a+1}\right)^2 + 3\right)\right) : \dfrac{a^3 + 1}{a^3 - 1} - \dfrac{2a}{a-1}$.

**56.** $\left(\left(\dfrac{y}{y-x}\right)^{-2} - \dfrac{(x+y)^2 - 4xy}{x^2 - xy}\right) \dfrac{x^4}{x^2 y^2 - y^4}$.

**57.** $\left(\dfrac{a}{a^2 - 4} - \dfrac{8}{a^2 + 2a}\right) \dfrac{a^2 - 2a}{4 - a} + \dfrac{a+8}{a+2}$.

**58.** $a - \left( \dfrac{(16-a)a}{a^2-4} + \dfrac{3+2a}{2-a} + \dfrac{3a-2}{a+2} \right) : \dfrac{a-1}{a(a^2+4a+4)}.$

**59.** $\left( 6a^2 + 5a - 1 + \dfrac{a+4}{a+1} \right) : \left( 3a - 2 + \dfrac{3}{a+1} \right).$

**60.** $\left( \dfrac{\sqrt{2}}{(1-x^2)^{-1}} + \dfrac{3^{3/2}}{x^{-2}} \right) : \left( \dfrac{x^{-2}}{1+x^{-2}} \right)^{-1}.$

**61.** $\left( \dfrac{1}{a} + \dfrac{1}{b} - \dfrac{2c}{ab} \right)(a+b+2c) : \left( \dfrac{1}{a^2} + \dfrac{1}{b^2} + \dfrac{2}{ab} - \dfrac{4c^2}{a^2b^2} \right).$

**62.** $\dfrac{2b + a - \dfrac{4a^2 - b^2}{a}}{b^3 + 2ab^2 - 3a^2b} \cdot \dfrac{a^3b - 2a^2b^2 + ab^3}{a^2 - b^2}.$

**63.** $\left( \dfrac{1}{a-\sqrt{2}} - \dfrac{a^2+4}{a^3-\sqrt{8}} \right) : \left( \dfrac{a}{\sqrt{2}} + 1 + \dfrac{\sqrt{2}}{a} \right)^{-1}.$

**64.** $\dfrac{\sqrt[3]{a^5 b^{1/2}} \, \sqrt[4]{a^{-1}}}{(a^2 \sqrt[5]{ab^3})^2}.$

**65.** $\dfrac{(\sqrt[5]{a^{4/3}})^{3/2} \, (\sqrt{a^3 \sqrt{a^2 b}})^4}{(\sqrt[5]{a^4})^3} \cdot \dfrac{(\sqrt{a \sqrt{b}})^6}{(\sqrt[3]{a\sqrt{b}})^6}.$

**66.** $\left( \dfrac{1}{2+2\sqrt{a}} + \dfrac{1}{2-2\sqrt{a}} - \dfrac{a^2+1}{1-a^2} \right)\left( 1 + \dfrac{1}{a} \right).$

**67.** $\left( \dfrac{a\sqrt{a} + b\sqrt{b}}{\sqrt{a} + \sqrt{b}} - \sqrt{ab} \right)\left( \dfrac{\sqrt{a} + \sqrt{b}}{a-b} \right)^2.$

**68.** $\dfrac{x^{1/2} + x^{-1/2}}{1-x} + \dfrac{1 - x^{-1/2}}{1 + \sqrt{x}}.$

**69.** $\left( \dfrac{1}{m - \sqrt{mn}} + \dfrac{1}{m + \sqrt{mn}} \right) \cdot \dfrac{m^3 - n^3}{m^2 + mn + n^2}.$

**70.** $\dfrac{a - a^{-2}}{a^{1/2} - a^{-1/2}} - \dfrac{2}{a^{3/2}} - \dfrac{1 - a^{-2}}{a^{1/2} + a^{-1/2}}.$

**71.** $\left( \dfrac{3}{\sqrt{1+a}} + \sqrt{1-a} \right) : \left( \dfrac{3}{\sqrt{1-a^2}} + 1 \right).$

**72.** $x^{1/2} + x^{-1/2} + \dfrac{(1-x)(1-x^{-1/2})}{1+\sqrt{x}}.$

**73.** $\dfrac{\sqrt{x}+1}{1+\sqrt{x}+x} : \dfrac{1}{x^2 - \sqrt{x}}.$

**74.** $\left( \dfrac{a\sqrt{a} + b\sqrt{b}}{\sqrt{a} + \sqrt{b}} - \sqrt{ab} \right) : (a-b) + \dfrac{2\sqrt{b}}{\sqrt{a} + \sqrt{b}}.$

**75.** $\dfrac{a-b}{a+b+2\sqrt{ab}} : \dfrac{a^{1/2} - b^{-1/2}}{a^{-1/2} + b^{-1/2}}.$

**76.** $\left( \dfrac{a^{1/2}+2}{a+2a^{1/2}+1} - \dfrac{a^{1/2}-2}{a-1} \right) \cdot \dfrac{a^{1/2}+1}{a^{1/2}}.$

**77.** $(x + \sqrt{x^2-1})^2 + (x + \sqrt{x^2-1})^{-2} + 2\,(1 - 2x^2).$

**78.** $\dfrac{a\sqrt{a} + b\sqrt{b}}{(\sqrt{a}+\sqrt{b})(a-b)} + \dfrac{2\sqrt{b}}{\sqrt{a}+\sqrt{b}} - \dfrac{\sqrt{ab}}{a-b}.$

**79.** $\dfrac{x^{1/2}+1}{x + x^{1/2} + 1} : \dfrac{1}{x^{3/2} - 1}.$

**80.** $(2^{3/2} + 27y^{3/5}) : \left( \left( \frac{1}{2} \right)^{-1/2} + 3y^{1/5} \right).$

**81.** $\dfrac{\dfrac{x-y}{\sqrt{x}-\sqrt{y}} - \dfrac{x-y}{\sqrt{x}+\sqrt{y}}}{\dfrac{\sqrt{x}-\sqrt{y}}{x-y} + \dfrac{\sqrt{x}+\sqrt{y}}{x-y}} \cdot \dfrac{2\sqrt{xy}}{y-x}.$

**82.** $2(a+b)^{-1}(ab)^{1/2} \left[ 1 + \frac{1}{4} \left( \sqrt{\frac{a}{b}} - \sqrt{\frac{b}{a}} \right)^2 \right]^{1/2}.$

**83.** $\dfrac{b-x}{\sqrt{b}-\sqrt{x}} - \dfrac{b^{3/2} - x^{3/2}}{b-x}.$

**84.** $\left( \dfrac{1}{(a^{1/2}+b^{1/2})^{-2}} - \left( \dfrac{\sqrt{a}-\sqrt{b}}{a^{3/2}-b^{3/2}} \right)^{-1} \right) (ab)^{-1/2}.$

**85.** $\left( \dfrac{m+\sqrt{m^2-n^2}}{m-\sqrt{m^2-n^2}} - \dfrac{m-\sqrt{m^2-n^2}}{m+\sqrt{m^2-n^2}} \right) \dfrac{n^2}{4m\sqrt{m^2-n^2}}.$

**86.** $a \left( \dfrac{\sqrt{a}+\sqrt{b}}{2b\sqrt{a}} \right)^{-1} + b \left( \dfrac{\sqrt{a}+\sqrt{b}}{2a\sqrt{b}} \right)^{-1}.$

**87.** $\dfrac{\dfrac{1}{\sqrt{a-1}} + \sqrt{a+1}}{\dfrac{1}{\sqrt{a+1}} - \dfrac{1}{\sqrt{a-1}}} : \dfrac{\sqrt{a+1}}{(a-1)\sqrt{a+1} - (a+1)\sqrt{a-1}}.$

**88.** $\dfrac{1}{2} \left( \dfrac{\sqrt{5}+1}{1+\sqrt{5}+\sqrt{a}} + \dfrac{\sqrt{5}-1}{1-\sqrt{5}+\sqrt{a}} \right) \left( \sqrt{a} - \dfrac{4}{\sqrt{a}} + 2 \right) \sqrt{0.002}.$

**89.** $\left( \dfrac{x^{1/2}+y^{1/2}}{x^{1/2}-y^{1/2}} - \dfrac{x^{1/2}-y^{1/2}}{x^{1/2}+y^{1/2}} \right) (y^{-1/2} - x^{-1/2}).$

**90.** $\left( \dfrac{\sqrt{a}}{2} - \dfrac{1}{2\sqrt{a}} \right)^2 \left( \dfrac{\sqrt{a}-1}{\sqrt{a}+1} - \dfrac{\sqrt{a}+1}{\sqrt{a}-1} \right).$

**91.** $(a + a^{1/2}b^{1/2})(a+b)^{-1} \left[ \sqrt{a} (\sqrt{a}-\sqrt{b})^{-1} - \left( \dfrac{\sqrt{a}+\sqrt{b}}{\sqrt{b}} \right)^{-1} \right].$

**92.** $\dfrac{2\sqrt{b}}{\sqrt{a}+\sqrt{b}} + \left( \dfrac{a^{3/2}+b^{3/2}}{\sqrt{a}+\sqrt{b}} - \dfrac{1}{(ab)^{-1/2}} \right) (a-b)^{-1}.$

**93.** $\left( \dfrac{p^{3/2}+q^{3/2}}{p-q} - \dfrac{p-q}{p^{1/2}+q^{1/2}} \right) \left( \sqrt{pq} \dfrac{\sqrt{p}+\sqrt{q}}{p-q} \right)^{-1}.$

**94.** $((\sqrt[4]{a} - \sqrt{b})^{-2} + (\sqrt[4]{a} + \sqrt[4]{b})^{-2}) : \left( \dfrac{\sqrt{a}+\sqrt{b}}{a-b} \right)^2.$

**95.** $\left( \dfrac{\sqrt{a}+1}{\sqrt{ab}+1} + \dfrac{\sqrt{ab}+\sqrt{a}}{\sqrt{ab}-1} - 1 \right) : \left( \dfrac{\sqrt{a}+1}{\sqrt{ab}+1} - \dfrac{\sqrt{ab}+\sqrt{a}}{\sqrt{ab}-1} + 1 \right).$

**96.** $\left(\dfrac{a\sqrt{a}+b\sqrt{b}}{\sqrt{a}+b}\right):(a-b)+\dfrac{2\sqrt{b}}{\sqrt{a}+\sqrt{b}}.$

**97.** $\dfrac{b^{1/2}}{a^{1/2}+1}:\left(\dfrac{\sqrt{b}-\dfrac{a}{(ab)^{-0.5}}}{1-a}-\sqrt{ab}\right)+\dfrac{b}{a}\left(-3\dfrac{3}{8}\right)^{-1/3}.$

**98.** $\dfrac{\dfrac{1}{\sqrt{a}-1}+\sqrt{a-1}}{\dfrac{1}{\sqrt{a}+1}-\dfrac{1}{\sqrt{a}-1}}+\dfrac{\sqrt{a-1}}{(a-1)\sqrt{a+1}-(a+1)\sqrt{a-1}}.$

**99.** $(a^2\sqrt{b})^{-1/2}\left(\sqrt{ab}-\dfrac{ab}{a+\sqrt{ab}}\right):\dfrac{\sqrt[4]{ab}-\sqrt{b}}{a-b}.$

**100.** $\left(\dfrac{a^3-8}{a^2-5a+6}-\dfrac{(a+1)^2+3}{a-3}+\dfrac{a^2+a}{\sqrt[4]{a}}\right):\dfrac{\sqrt{ab}}{\sqrt[4]{a^{-1}b^2}}.$

**101.** $\dfrac{a^{4/3}-8a^{1/3}b}{a^{2/3}+2\sqrt[3]{ab}+4b^{2/3}}:\left(1-2\sqrt[3]{\dfrac{b}{a}}\right).$

**102.** $\left(\dfrac{a+b}{a-b}\right)^{1/2}-\dfrac{2a\sqrt{a^2-b^2}}{b^2\,(ab^{-1}+1)^2}\cdot\dfrac{1}{1+\dfrac{1-ba^{-1}}{1+ba^{-1}}}.$

**103.** $\dfrac{(a^{1/m}-a^{1/n})^2+4a^{(m+n)/mn}}{(a^{2/m}-a^{2/n})(\sqrt[m]{a^{m+1}}+\sqrt[n]{a^{m+1}})}.$

**104.** $(1-a^2):\left(\left(\dfrac{1-a^{3/2}}{1-a^{1/2}}+a^{1/2}\right)\left(\dfrac{1+a^{3/2}}{1+a^{1/2}}-a^{1/2}\right)\right)+1.$

**105.** $\dfrac{2}{a}-\left(\dfrac{a+1}{a^3-1}-\dfrac{1}{a^2+a+1}-\dfrac{2}{1-a}\right):\dfrac{a^3+a^2+2a}{a^3-1}.$

**106.** $\dfrac{a+b}{a^{2/3}-a^{1/3}b^{1/3}+b^{2/3}}-\dfrac{a-b}{a^{2/3}+a^{1/3}b^{1/3}+b^{2/3}}-\dfrac{a^{2/3}-b^{2/3}}{a^{1/3}-b^{1/3}}.$

**107.** $\left(\dfrac{x^{1/2}-y^{1/2}}{xy^{1/2}+x^{1/2}y}+\dfrac{x^{1/2}+y^{1/2}}{xy^{1/2}-x^{1/2}y}\right)\dfrac{x^{3/2}y^{1/2}}{x+y}-\dfrac{2y}{x-y}.$

**108.** $t\,\dfrac{1+2(t+4)^{-1/2}}{2-(t+4)^{1/2}}+(t+4)^{1/2}+4\,(t+4)^{-1/2}.$

**109.** $\dfrac{\sqrt{a^3}+\sqrt{b^3}}{\sqrt[3]{a^2(a-b)^2}}\cdot\dfrac{a^{2/3}\sqrt{a^3}-\sqrt{b^3}}{(a-b)^{1/3}}.$ **110.** $\left(\dfrac{2}{x^2-a^2}\right)^{-1}\left(\dfrac{x\,(x^2-a^2)^{1/2}+1}{a(x-a)^{-1/2}+(x-a)^{1/2}}\right.$

$:\left(\dfrac{x-(x^2-a^2)^{1/2}}{a^2\,(x+a)^{1/2}}\right)^{-1}\left.+\,(x^2+ax)^{-1}\right).$

**111.** $\left(\left(\sqrt{\dfrac{a}{b}} - \sqrt{\dfrac{b}{a}}\right) : \left(\sqrt{\dfrac{a}{b}} + \sqrt{\dfrac{b}{a}} - 2\right)\right) : \left(1 + \sqrt{\dfrac{b}{a}}\right),\ a > 0,\ b > 0.$

**112.** $\left(\dfrac{2a + b^{1/2}a^{1/2}}{3a}\right)^{-1} \left(\dfrac{a^{3/2} - b^{3/2}}{a - a^{1/2}b^{1/2}} - \dfrac{a - b}{\sqrt{a} + \sqrt{b}}\right).$

**113.** $\dfrac{(\sqrt{x} - 1)(\sqrt{x} + 1)}{x : \sqrt[4]{x} + \sqrt{x}} \cdot \dfrac{\sqrt[4]{x} + \sqrt{x}}{1 + \sqrt{x}} \cdot \sqrt[4]{x} - \sqrt{x} + 1.$

**114.** $\dfrac{2x^{-1/3}}{x^{2/3} - 3x^{-1/3}} - \dfrac{x^{2/3}}{x^{5/3} - x^{2/3}} - \dfrac{x + 1}{x^2 - 4x + 3}.$

**115.** $\dfrac{1}{2}\left(\sqrt{a^3 b^{-3}} - \sqrt{b^3 a^{-3}}\right) : \left(\dfrac{a^2 + b^2}{ab} + 1\right)\right) \dfrac{2(a - b)^{-1}}{(ab)^{-1/2}}.$

**116.** $a^2 (1 - a^2)^{-1/2} : \left(\dfrac{1}{1 + (a(1 - a^2)^{-1/2})^2} \cdot \dfrac{(1 - a^2)^{1/2} + a^2(1 - a^2)^{-1/2}}{1 - a^2}\right).$

**117.** $\dfrac{(\sqrt{x} - \sqrt{y})^3 + 2x^2 : \sqrt{x} + y\sqrt{y}}{x\sqrt{x} + y\sqrt{y}} + \dfrac{3\sqrt{xy} - 3y}{x - y}.$

**118.** $\left(\dfrac{1 - x^{-2}}{x^{1/2} - x^{-1/2}} - \dfrac{2}{x^2 : \sqrt{x}} + \dfrac{x^{-2} - x}{x^{1/2} - x^{-1/2}}\right)\left(1 + \dfrac{2}{x^2}\right)^{-1}.$

**119.** $\dfrac{(x^{1/3} - y^2)(x^{-1/3}y + x^{1/3}y^{-1} + 1)}{x^{-2/3}y^2 - x^{-1/3}y + x^{2/3}y^{-2} - x^{1/3}y^{-1}} : (x^{1/3}y).$

**120.** $\dfrac{1 + a\sqrt[3]{a} + a + \sqrt[3]{a^2}}{1 - \sqrt[3]{a}} + \dfrac{1}{\sqrt[3]{a^{-2}}}.$

**121.** $\left(\dfrac{(1 - x)^{1/4}}{2(1 + x)^{3/4}} + \dfrac{(1 + x)^{1/4}(1 - x)^{-3/4}}{2}\right) : (1 - x^2)^{1/4}.$

**122.** $(x^{1/4} + y^{1/4}) : \left(\left(\dfrac{\sqrt[3]{y}}{y\sqrt{x}}\right)^{3/2} + \left(\dfrac{x^{-1/2}}{\sqrt[8]{y^3}}\right)^2\right).$

**123.** $\dfrac{x^{3/2} + y^{3/2} - x^{1/2}y^{1/2}(\sqrt{x} + \sqrt{y})}{(x - y)(x^{1/2} + y^{1/2})} + \dfrac{2\sqrt{y}}{\sqrt{x} + \sqrt{y}}.$

**124.** $\dfrac{(a - b)^3 (\sqrt{a} + \sqrt{b})^{-3} + 2a\sqrt{a} + b\sqrt{b}}{a\sqrt{a} + b\sqrt{b}} - \dfrac{3(\sqrt{ab} - b)}{a - b}.$

**125.** $\left(\dfrac{\sqrt[4]{a^3} - \sqrt[4]{a}}{1 - \sqrt{a}} + \dfrac{1 + \sqrt{a}}{\sqrt[4]{a}}\right)\left(1 + \dfrac{2}{\sqrt{a}} + \dfrac{1}{a}\right)^{-1/2}.$

**126.** $\dfrac{(\sqrt[8]{a} + \sqrt[8]{b})^2 + (\sqrt[8]{a} - \sqrt[8]{b})^2}{a - \sqrt{ab}} : \dfrac{(\sqrt[4]{a} + \sqrt[8]{ab} + \sqrt[4]{b})(\sqrt[4]{a} - \sqrt[8]{ab} + \sqrt[4]{b})}{\sqrt[4]{a^3 b} - b}.$

**127.** $\left(1 - 2\sqrt[3]{\dfrac{b}{a}}\right)\left(\dfrac{a^{4/3} - 8a^{1/3}b}{a^{2/3} + 2\sqrt[3]{ab} + 4b^{2/3}}\right)^{-1}\sqrt[3]{\dfrac{1}{a^{-2}}}.$

**128.** $\left(x\sqrt[3]{\dfrac{x-1}{(x+1)^2}} + \dfrac{x-1}{\sqrt[3]{(x^2-1)^2}}\right)^{-3/5} : (x^2-1)^{4/5}.$

**129.** $\left(\dfrac{m-n}{m^{3/4} + m^{1/2}n^{1/4}} - \dfrac{m^{1/2} - n^{1/2}}{m^{1/4} + n^{1/4}}\right)\left(\dfrac{n}{m}\right)^{-1/2}.$

**130.** $\left[\dfrac{8-x}{2+\sqrt[3]{x}} : \left(2 + \dfrac{\sqrt[3]{x^2}}{2+\sqrt[3]{x}}\right) + \left(\sqrt[3]{x} + \dfrac{2\sqrt[3]{x}}{\sqrt[3]{x}-2}\right)\right]\dfrac{\sqrt[3]{x^2}-4}{\sqrt[3]{x^2}+2\sqrt[3]{x}}.$

**131.** $\left(\dfrac{\sqrt[4]{a^3}-b}{\sqrt[4]{a}-\sqrt[3]{b}} - 3\sqrt[12]{a^3b^4}\right)^{-1/2}\left(\dfrac{\sqrt[4]{a^3}+b}{\sqrt[4]{a}+\sqrt[3]{b}} - \sqrt[3]{b^2}\right), \; b > 0.$

**132.** $\dfrac{b-b^{-2}}{b^{1/2} - b^{-1/2}} - \dfrac{1-b^{-2}}{\sqrt{b} + b^{-1/2}} - b^{1/2}.$

**133.** $\left(2 - \dfrac{a}{4} - \dfrac{4}{a}\right)\left((a-4)\sqrt[3]{(a-4)^{-3}} - \dfrac{(a^2-16)^{-1/2}(a-4)^{-1/2}}{(a+4)^{-3/2}}\right)\cdot\dfrac{a+4}{a-4}.$

**134.** $\dfrac{8-x}{2+\sqrt[3]{x}} : \left(2 + \dfrac{\sqrt[3]{x^2}}{2+\sqrt[3]{x}}\right) + \left(\sqrt[3]{x} + \dfrac{2\sqrt[3]{x}}{\sqrt[3]{x}-2}\right)\dfrac{\sqrt[3]{x^2}-4}{\sqrt[3]{x^2}+2\sqrt[3]{x}}.$

**135.** $\dfrac{\left(\dfrac{\sqrt[4]{bx^3} + \sqrt[4]{a^2bx}}{\sqrt{x}+\sqrt{a}} + \sqrt[4]{bx}\right)^2 + bx + 4}{x\left(\sqrt{b} + \sqrt{4x^{-1}}\right)^2}.$

**136.** $\dfrac{a^2 + 10a + 25 + 2\sqrt{5}\left(\sqrt{a^3} + 5\sqrt{a}\right)}{(a^2 - 25)\left((\sqrt{a^3} - \sqrt{125})(a + \sqrt{5a} + 5)^{-1}\right)^{-1}}.$

**137.** $\left(\dfrac{a-2b}{\sqrt[3]{a^2} - \sqrt[3]{4b^2}} + \dfrac{\sqrt[3]{2a^2b} + \sqrt[3]{4ab^2}}{\sqrt[3]{a^2} + \sqrt[3]{4b^2} + \sqrt[3]{16ab}}\right) : \dfrac{a\sqrt[3]{a} + b\sqrt[3]{2b} + b\sqrt[3]{a} + a\sqrt[3]{2b}}{a+b}.$

**138.** $\left(\dfrac{3-\sqrt{a}}{9-a} + \dfrac{1}{3-\sqrt{a}} - 6\dfrac{a^2+162}{729-a^3}\right)^{-1} + \dfrac{a(a+9)}{54}.$

**139.** $-\left(\left(\dfrac{\sqrt{a}+\sqrt{b}}{\sqrt[4]{a}-\sqrt[4]{b}}\right)^{-1} - \dfrac{2\sqrt[4]{ab}}{b^{3/4} - a^{1/4}b^{1/2} + a^{1/2}b^{1/4} - a^{3/4}}\right)^{-1} + \sqrt{2}^{\log_4 a}.$

**140.** $\left(\dfrac{3}{\sqrt[3]{a^2} - \sqrt[3]{a}+1} - \dfrac{3}{a+1} + \dfrac{\sqrt[3]{a}-1}{\sqrt[3]{a^2}-1}\right)^{-1}\left(\dfrac{a^{-1/3}+1}{a^{1/3}}\right)^2 - a^{\frac{2}{3}\log_3\frac{1}{4}}.$

**141.** $\dfrac{\sqrt{ab}\left(\sqrt{a}-\sqrt{b}\right)}{(\sqrt[4]{b}-\sqrt[4]{a})^2\sqrt[4]{b}} - \dfrac{\sqrt[4]{16ab}\left(a + \sqrt[4]{a^3b} + \sqrt{ab}\right)}{\sqrt[4]{a^3} - \sqrt[4]{b^3}}.$

**142.** $\left[ (a^{1/4} - a^{1/8} + 1)^{-1} + (a^{1/4} + a^{1/8} + 1)^{-1} - \dfrac{2\sqrt[4]{a} - 2}{\sqrt{a} - \sqrt[4]{a} + 1} \right]^{-1} - 2^{\log_4 a - 2}.$

**143.** $\left[ \left( \dfrac{\sqrt[3]{x^2 y^2} + x\sqrt[3]{x}}{x\sqrt[3]{y} + y\sqrt[3]{x}} - 1 \right)^{-1} \left( 1 + \sqrt[3]{\dfrac{x}{y}} + \sqrt[3]{\dfrac{x^2}{y^2}} \right)^{-1} + 1 \right]^{1/3} \sqrt[3]{x - y}.$

**144.** $\dfrac{a + 10\sqrt{a} + \sqrt{20}\,(\sqrt[4]{a^3} + 5\sqrt[4]{a}) + 25}{(a - 25)(\sqrt[4]{a^3} - \sqrt{125})^{-1}\,(\sqrt{a} + \sqrt[4]{25a} + 5)}.$

**145.** $\left[ \dfrac{(\sqrt[4]{a} + \sqrt[4]{b})^2 - \sqrt[4]{16ab}}{a - b} + \dfrac{1}{\sqrt{a} + \sqrt{b}} - \left( \dfrac{a - b}{2\sqrt{b}} \right)^{-1} \right]^{-1}.$

**146.** $\left( \dfrac{2\sqrt{a} + 3\sqrt[4]{a}}{\sqrt{16a} + 12\sqrt[4]{a} + 9} - \dfrac{\sqrt[4]{a} - 3}{2\sqrt[4]{a} + 3} \right) (2 \cdot 3^{\log_{81} a} + 3).$

**147.** $2(x^2 + \sqrt{x^4 - 1}) \left( \sqrt[3]{(x^2 + 1)\sqrt{1 + \dfrac{1}{x^2}}} + \sqrt[3]{(x^2 - 1)\sqrt{1 - \dfrac{1}{x^2}}} \right)^{-2}.$

**148.** $\dfrac{2a\sqrt[3]{ab^2} - a\sqrt[6]{ab^5} - ab}{\sqrt[3]{a^2 b} - \sqrt{ab}} - 2^{1 + 2\log_8 a + \log_8 b}.$

**149.** $\left( \dfrac{a - \sqrt{a^2 - b^2}}{a + \sqrt{a^2 - b^2}} - \dfrac{a + \sqrt{a^2 - b^2}}{a - \sqrt{a^2 - b^2}} \right) : \dfrac{4\sqrt{a^4 - a^2 b^2}}{(5b)^2}.$

**150.** $\dfrac{1}{a^2} \sqrt{\left( a^6 + \dfrac{3a^4}{b^{-2}} + \dfrac{a^2 b^4}{3^{-1}} + \dfrac{1}{b^{-6}} \right)^{2/3}} + \left[ \dfrac{(b^{2/3} - a^{2/3})^3 - 2a^2 - b^2}{a^2 + (b^{2/3} - a^{2/3})^3 + 2b^2} \right]^{-3}.$

**151.** $\left( \dfrac{(a^{3/2} - \sqrt{8}\,(\sqrt{a} + \sqrt{2})}{a + \sqrt{2a} + 2} \right)^2 + \sqrt{(a^2 + 2)^2 - 8a^2}.$

**152.** $\left( \dfrac{(\sqrt[3]{x} - \sqrt[3]{a})^3 + 2x + a)}{(\sqrt[3]{x} - \sqrt[3]{a})^3 - (x + 2a)} \right)^3 + \sqrt{|\,a^3 - 3a^2 x + 3ax^2 - x^3\,|^{2/3}} : a \text{ for } x > a.$

**153.** $\dfrac{\sqrt{a^2 - 2ab + b^2}}{\sqrt{a^2 + 2ab + b^2}} + \dfrac{2a}{a + b},\ 0 < a < b.$

**154.** $\left( \dfrac{1 + \sqrt{1 - x}}{1 - x + \sqrt{1 - x}} + \dfrac{1 - \sqrt{1 + x}}{1 + x - \sqrt{1 + x}} \right)^2 \dfrac{x^2 - 1}{2} + 1,\ 0 < x < 1.$

**155.** $\dfrac{2a\sqrt{1 + x^2}}{x + \sqrt{1 + x^2}},\ x = \dfrac{1}{2} \left( \sqrt{\dfrac{a}{b}} - \sqrt{\dfrac{b}{a}} \right),\ a > 0,\ b > 0.$

**156.** $\dfrac{(m + x)^{1/2} + (m - x)^{1/2}}{(m + x)^{1/2} - (m - x)^{1/2}},\ x = \dfrac{2mn}{n^2 + 1},\ m > 0,\ 0 < n < 1.$

**157.** $\dfrac{((3b^2 + 2a^2)^2 - 24a^2b^2)^{1/2}}{3b - a^2 \cdot 2^{1-\log_2 b}} + \sqrt{a - b^2} - \sqrt{a + 2b\sqrt{a - b^2}}$,

$$a/b > \sqrt{3/2},\ b > 0,\ a \geq b^2.$$

**158.** $\left((a - b)\sqrt{\dfrac{a + b}{a - b}} + a - b\right)(a - b)\left(\sqrt{\dfrac{a + b}{a - b}} - 1\right)$,

(1) $a + b > 0,\ a - b > 0$;  (2) $a + b < 0,\ a - b < 0$.

**159.** $\dfrac{\sqrt[3]{a + \sqrt{2 - a^2}}\ \sqrt[6]{1 - a\sqrt{2 - a^2}}}{\sqrt[3]{1 - a^2}}$.

**160.** $\dfrac{\left(\dfrac{\sqrt[4]{bc^3} + \sqrt[4]{a^2bc}}{\sqrt{a} + \sqrt{c}} + \sqrt[4]{bc}\right)^2 + bc}{\sqrt{bc} + 3}$.

**161.** $\dfrac{(a^{-1} + b^{-1})(a + b)^{-1}}{\sqrt[6]{a^4 \sqrt[5]{a^{-2}}}}$.

**162.** $\dfrac{1 - a^{-1/2}}{1 + a^{1/2}} - \dfrac{a^{1/2} - a^{-1/2}}{a - 1}$ and calculate it for $a = 5$.

**163.** $\dfrac{a^{3/2} + b^{3/2}}{(a^2 - ab)^{2/3}} : \dfrac{a^{-2/3}\sqrt[3]{a - b}}{a\sqrt{a} - b\sqrt{b}}$ and calculate it for $a = 1.2$;  $b = 3/5$.

**164.** $\left(\dfrac{a^2}{a + b} - \dfrac{a^3}{a^2 + 2ab + b^2}\right) : \left(\dfrac{a}{a + b} - \dfrac{a^2}{a^2 - b^2}\right)$ and calculate it for

$a = -2.5$;  $b = 0.5$.

**165.** $f(x,\ y) = \left(\dfrac{\sqrt[4]{x^3y} - \sqrt[4]{xy^3}}{\sqrt{y} - \sqrt{x}} + \dfrac{1 + \sqrt{xy}}{\sqrt[4]{xy}}\right)^{-2} \cdot \left(1 + 2\sqrt{\dfrac{y}{x}} + \dfrac{y}{x}\right)^{1/2}$ and calculate

it for $x = 9$, $y = 0.04$.

**166.** Prove the identity $\dfrac{b^2 - 3b - (b - 1)\sqrt{b^2 - 4} + 2}{b^2 + 3b - (b + 1)\sqrt{b^2 - 4} + 2}\sqrt{\dfrac{b + 2}{b - 2}} = \dfrac{1 - b}{1 + b}$.

**167.** Indicate the domain of definition of the function and simplify the given

expression $y = x\dfrac{1 + 2(x + 4)^{-0.5}}{2 - (x + 4)^{0.5}} + (x + 4)^{0.5} + 4(x + 4)^{-0.5}$.

**168.** Prove that if $\sqrt{x^2 + \sqrt[3]{x^4y^2}} + \sqrt{y^2 + \sqrt{x^2y^4}} = a$, then $x^{2/3} + y^{2/3} = a^{2/3}$.

**169.** Find the value of the expression $\left(\dfrac{(x^2 + a^2)^{1/2} + (x^2 - a^2)^{1/2}}{(x^2 + a^2)^{1/2} - (x^2 - a^2)^{1/2}}\right)^{-2}$ for

$x = a\left(\dfrac{m^2 + n^2}{2mn}\right)^{1/2}$, $a > 0$, $m > 0$, $n > 0$, $m > n$.

**170.** Find the domain of definition and simplify the expression

$A = \dfrac{a + b}{\sqrt{a} + \sqrt{b}} : \left(\dfrac{a + b}{a - b} - \dfrac{b}{b - \sqrt{ab}} + \dfrac{a}{\sqrt{ab} + a}\right) - \dfrac{\sqrt{(\sqrt{a} - \sqrt{b})^2}}{2}$.

## 1.3. Mathematical Induction, Elements of Combinatorics. Binomial Theorem

### Mathematical Induction

The method of proving propositions, called mathematical induction (or simply induction), is based on the principle which is one of the axioms of natural number arithmetic.

*Mathematical induction.* The proposition $A(n)$ is considered to be true for all natural values of the variable if the following two conditions are fulfilled :

(1) The proposition $A(n)$ is true for $n = 1$.

(2) From the proposition that $A(n)$ is true for $n = k$ (where $k$ is *any* natural number) it follows that it is also true for the next value $n = k + 1$.

The proof by induction consists of the following two parts : in the first part we verify the truth of the proposition $A(1)$; in the second part we assume the truth of $A(n)$ for $n = k$ and prove the validity of $A(n)$ for $n = k + 1$, i.e. $A(k) \Rightarrow A(k + 1)$.

If we have proved that $A(1)$ is true, and it follows from the truth of $A(n)$ for $n = k$ that $A(n)$ is true for $n = k + 1$ (for any natural $k$), then $A(n)$ is true for all natural $n$.

**Example 1.** Prove that for $x > -1$ the inequality

$$(1 + x)^n \geq 1 + nx \tag{1}$$

(Bernoulli's inequality) holds for any natural $n$.

(1) For $n = 1$ we have $(1 + x)^1 = 1 + x$. One of the relations $>$ or $=$ holds and, therefore, $A(1)$ is true.

(2) Let us prove that the truth of $A(k)$ yields the truth of $A(k + 1)$ for any natural $k$. Suppose the inequality

$$(1 + x)^k \geq 1 + kx \tag{2}$$

is true. We multiply its both sides by $1 + x$. Since $1 + x > 0$, we have

$$(1 + x)^{k+1} \geq (1 + kx)(1 + x),$$

or $\quad (1 + x)^{k+1} \geq 1 + (k + 1)x + kx^2;$

taking into account that $kx^2 \geq 0$, we infer that

$$(1 + x)^{k+1} \geq 1 + (k + 1)x,$$

i.e. $\quad A(k) \Rightarrow A(k + 1).$

Since inequality (1) is valid for $n = 1$ and it follows from its truth at $n = k$ for any natural $k$ that it is also true for $n = k + 1$, then by induction inequality (1) is true for all natural $n$.

We can use induction to prove propositions defined for negative integer $n$ (by means of the substitution $n = -m$), and also propositions defined on the set of integers beginning with $n = m$. In the latter case, the proof is based on the following generalization of mathematical induction.

If the proposition $A(n)$, in which $n$ is an integer, is true for $n = m$ and if it follows from its truth for $n = k$, where $k$ is any integer larger than, or equal to, $m$, that it is true for $n = k + 1$, then the proposition $A(n)$ is true for any integral $n \geq m$.

**Example 2.** Prove that the inequality

$$2^n > 2n + 1 \tag{3}$$

is true for all natural $n \geq 3$.

(1) If $n = 3$, then $2^3 > 2 \cdot 3 + 1$, i.e. $A$ (3) is true.

(2) Let us prove that it follows from the truth of the inequality at $n = k$ for any natural $k \geq 3$ that it is true for $n = k + 1$. We assume that

$$2^k > 2k + 1; \tag{4}$$

then $\qquad 2^k + 2^k > 4k + 2$, i.e.

$$2^{k+1} > 2(k+1) + 1 + (2k-1). \tag{5}$$

Since $2k - 1 > 0$ for all natural $k > 1$, it follows from the validity of (5) that the inequality

$$2^{k+1} > 2(k+1) + 1$$

holds true, i.e. $A(k) \Rightarrow A(k+1)$.

We have thus carried out both parts of the proof, and, consequently, inequality (3) is valid for any $n \geq 3$.

Prove the assertions of the following problems **(1-10)**.

1. Prove that $n^3 + 3n^2 + 5n + 3$ is divisible by 3 for any natural $n$.
2. Prove that the number $4^n + 15n - 1$ is a multiple of 9 for any natural $n$.
3. Prove that $2^n > n^2$ for any natural $n \geq 5$.
4. Prove that the expression $n^3 - n$ is divisible by 24 for any odd $n$.
5. Prove that $\dfrac{1}{\sqrt{1}} + \dfrac{1}{\sqrt{2}} + \dfrac{1}{\sqrt{3}} + \ldots + \dfrac{1}{\sqrt{n}} > \sqrt{n}$ for an arbitrary natural $n \geq 2$.
6*. Prove that the equality

$$\left(1 - \frac{4}{1}\right)\left(1 - \frac{4}{9}\right)\left(1 - \frac{4}{25}\right) \cdots \left(1 - \frac{4}{(2n-1)^2}\right) = \frac{1+2n}{1-2n}$$ holds true for any natural $n$.

7. Prove that the inequality $\dfrac{1}{n+1} + \dfrac{1}{n+2} + \ldots + \dfrac{1}{n+n} > \dfrac{13}{24}$ holds true for any natural $n > 1$.
8. Prove that $|\sin m\alpha| \leq m|\sin \alpha|$ for any natural $m$ and any $\alpha \in R$.
9. Prove that $11^{n+2} + 12^{2n+1}$ is divisible by 133 for any integral nonnegative $n$.
10. Prove that $\dfrac{(2n)!}{(n!)^2} > \dfrac{4n}{n+1}$ for any natural $n > 1$.

## Combinatorics

**Permutations. Number of permutations.** An order established in a finite set is known as a *permutation* of its elements. The number of permutations in

---

\* Here and henceforth the asterisk signifies a problem of advanced difficulty.

a finite set of elements depends only on the number of elements. The number of permutations for the set of $n$ elements is designated as $P_n$. A set consisting of one element can be ordered in a unique way: the first element of the set is considered to be unique, and, therefore, $P_1 = 1$. We can prove by induction that $P_n$ is equal to the product of the first $n$ natural numbers :

$$P_n = 1 \cdot 2 \cdot 3 \dots n. \tag{1}$$

The product $1 \cdot 2 \cdot 3 \dots n$ is designated as $n!$ Therefore, $P_n = n!$. By definition it is considered that $P_0 = 0! = 1$.

**Arrangements. Number of arrangements.** A set considered together with the assigned order of its elements is said to be an *ordered* set. In an ordered set, the elements are written in parentheses in a specified order. For example, $(A, B, C)$ is an ordered set whose first element is $A$, the second element is $B$, and the third element is $C$.

Finite ordered sets are called *arrangements*. The number of arrangements of $n$ elements taken $m$ at a time is designated as $A_m^n$. It can be proved by induction that

$$A_m^n = \frac{n!}{(n-m)!}. \tag{2}$$

This formula can also be written as

$$A_m^n = n(n-1)\dots(n-m+1). \tag{3}$$

**Combinations. Number of combinations. Properties of the number of combinations.** In combinatorics finite sets are called *combinations*. The number of combinations of $n$ elements taken $m$ at a time i.e. the number of subsets, consisting of $m$ elements each, constituting a set of $n$ elements) is designated as $C_m^n$.

Calculating the number of arrangements of $n$ elements taken $m$ at a time, we can find that

$$A_m^n = C_m^n \cdot P_m, \tag{4}$$

whence it follows that

$$C_m^n = \frac{n!}{m!(n-m)!}. \tag{5}$$

This formula can also be written as

$$C_m^n = \frac{n(n-1)(n-2)\dots(n-m+1)}{1 \cdot 2 \dots m}. \tag{6}$$

The equality

$$C_m^n = C_{n-m}^n \tag{7}$$

holds true for any $n$ and $m$ $(0 \le m \le n)$. Indeed,

$$C_m^n = \frac{n!}{m!\,(n-m)!} - \frac{n!}{(n-m)!\,(n-(n-m))!} = C_{n-m}^n.$$

We can prove by induction that

$$C_0^n + C_1^n + C_2^n + \dots + C_n^n = 2^n. \tag{8}$$

We can also prove this equality if we set $a = b = 1$ in Newton's formula. The validity of the formula also follows from the fact that the sum

$$C_0^n + C_1^n + C_2^n + \dots + C_n^n$$

is a complete number of subsets of a set consisting of $n$ elements, and it is equal to $2^n$. The equality

$$C_m^n + C_{m+1}^n = C_{m+1}^{n+1} \tag{9}$$

holds true for any $n$ and $m$ such that $0 \le m \le n$. We can carry out the proof representing $C_m^n$ and $C_{m+1}^n$ by formula (5) and adding the resulting fractions together.

Solve the following equations **(11-26)**.

**11.** $\dfrac{C_{x+1}^{2x}}{C_{x-1}^{2x+1}} = \dfrac{2}{3}$, $x \in \mathbf{N}$.

**12.** $A_2^{x-1} - C_1^x = 79$, $x \in \mathbf{N}$.

**13.** $3C_2^{x+1} = 2A_2^x = x$, $x \in \mathbf{N}$.

**14.** $\dfrac{C_2^{x+1}}{C_3^x} = \dfrac{4}{5}$, $x \in \mathbf{N}$.

**15.** $12C_1^x + C_2^{x+4} = 162$, $x \in \mathbf{N}$.

**16.** $A_3^{x+1} + C_{x-1}^{x+1} = 14(x+1)$, $x \in \mathbf{N}$.

**17.** $A_3^x + C_{x-2}^x = 14x$, $x \in \mathbf{N}$.

**18.** $C_{x-4}^{x+1} = \dfrac{7}{15} A_3^{x+1}$, $x \in \mathbf{N}$.

**19.** $C_3^{x+1} : C_4^x = 6 : 5$, $x \in \mathbf{N}$.

**20.** $C_2^{x+1} \cdot A_2^x - 4x^3 = (A_{2x}^1)^2$, $x \in \mathbf{N}$.

**21.** $3C_2^{x+1} + P_2 \cdot x = 4A_x^2$, $x \in \mathbf{N}$.

**22.** $C_{m+1}^{n+1} : C_m^{n+1} : C_{m-1}^{n+1} = 5 : 5 : 3$.

**23.** $\dfrac{P_{x+3}}{A_5^x \cdot P_{x-5}} = 720$, $x \in \mathbf{N}$.

**24.** $A_2^{x+3} = C_3^{x+2} + 20$, $x \in \mathbf{N}$.

**25.** $C_3^x + C_4^x = 11 \cdot C_2^{x+1}$, $x \in \mathbf{N}$.

**26.** $11C_3^x = 24C_2^{x+1}$.

**27.** Simplify the expression
$C_0^n + 2C_1^n + 3C_2^n + \ldots + (n+1)C_n^n$, $x \in N$, eliminating $C_k^n$ ($k = 0, 1, \ldots, n$)

Solve the following inequalities **(28-37)**.

**28.** $C_m^{13} < C_{m+2}^{13}$, $m \in N$.

**29.** $C_{m-2}^{18} > C_m^{18}$, $m \in N$.

**30.** $C_6^n < C_4^n$, $n \in N$.

**31.** $5C_3^n < C_4^{n+2}$, $n \in N$.

**32.** $C_4^{x-1} - C_3^{x-1} - \dfrac{5}{4} A_2^{x-2} < 0$, $x \in N$.

**33.** $C_{x-1}^{x+1} > 3/2$, $x \in N$.

**34.** $C_{x-1}^{x+1} < 21$, $x \in N$.

**35.** $2C_5^n > 11C_3^{n-2}$, $n \in N$.

**36.** $C_{n-2}^{n+1} - C_{n-1}^{n+1} \leq 100$, $n \in N$.

**37.** $\dfrac{A_4^{n+1}}{C_{n-3}^{n-1}} > 14P_3$, $n \in N$.

**38.** How many negative terms are there in the sequence $(x_n)$, where
$x_n = C_4^{n+5} - \dfrac{143}{96} \cdot \dfrac{P_{n+5}}{P_{n+3}}$, $n \in N$.

**39.** How many positive terms are there in the sequence $(x_n)$ if
$x_n = \dfrac{195}{4P_n} - \dfrac{A_3^{n+3}}{P_{n+1}}$, $n \in N$.

**40.** Find the negative terms of the sequence $x_n = \dfrac{A_4^{n+4}}{P_{n+2}} - \dfrac{143}{4P_n}$, $n \in N$.

**41.** We take $n$ points on one side of the triangle, $m$ points on the second side, and $k$ points on the third side, none of the points being a vertex of the triangle. How many triangles are there with vertices at those points?

**42.** We must place 5 boys and 5 girls round the table so that no two boys and no two girls sit side by side. In how many ways can we do it?

**43.** Two variants of a test are suggested to twelve students. In how many ways can the students be placed in two rows so that there should be no identical variants side-by-side and that the students sitting one behind the other should have the same variant?

**44.** In a 12-storey house 9 people enter a lift cabin. It is known that they will leave the lift in groups of 2, 3, and 4 people at different storeys. In how many ways can they do so if the lift does not stop at the second storey?

**45.** There are 40 doctors in a surgical department. In how many ways can they be arranged to form the following teams : (a) a surgeon and an assistant ; (b) a surgeon and four assistants?

**46.** In how many ways can 10 identical presents be distributed among 6 children so that each child gets at least one present?

**47.** How many ways are there of distributing $n$ identical balls among $k$ boxes?

**48.** Six white and six black balls of the same size are distributed among ten urns so that there is at least one ball in each urn. What is the number of different distributions of the balls?

**49.** Five boys and three girls are sitting in a row of eight seats. In how many ways can they be seated so that not all the girls sit side-by-side?

**50\*.** Seven different objects must be divided among three people. In how many ways can it be done if one or two of them can get no objects?

**51\*.** How many natural numbers are there which are smaller than $10^4$ and are divisible by 4, whose decimal notation consists only of the digits 0, 1, 2, 3, and 5 which do not repeat in any of these numbers?

**52.** How many six-digit numbers can be formed from the digits 1, 2, 3, 4, 5, 6, and 7 so that the digits should not repeat and the terminal digits should be even?

**53.** How many four-digit numbers which are divisible by 4 can be formed from the digits 1, 2, 3, 4, and 5?

**54.** How many natural numbers smaller than $10^4$ are there, in the decimal notation of which all the digits are different?

**55.** How many five-digit numbers, which do not contain identical digits, can be written by means of the digits 1, 2, 3, 4, 5, 6, 7, 8, and 9?

**56\*.** How many four-digit numbers are there whose decimal notation contains not more than two distinct digits?

**57\*.** How many different seven-digit numbers are there the sum of whose digits is even?

**58\*.** How many different four-digit numbers can be written using the digits 1, 2, 3, 4, 5, 6, 7, and 8 so that each of them contains only one unity, if any other digit can occur several times in the notation of these numbers?

**59\*.** How many different seven-digit numbers can be written using only three digits, 1, 2, and 3, under the condition that the digit 2 occurs twice in each number?

**60\*.** How many six-digit numbers contain exactly four different digits?

**61\*.** How many different numbers, which are smaller than $2 \cdot 10^8$ and are divisible by 3, can be written by means of the digits 0, 1, and 2 (the numbers cannot begin with 0)?

**62\*.** How many four-digit numbers are there whose decimal notation contains not more than two different digits?

**63\*.** We must form a bouquet from 18 different flowers so that it should contain not less than three flowers. How many different ways are there of forming such a bouquet?

**64\*.** How many different numbers, which are smaller than $2 \cdot 10^8$, can be written by means of the digits 1 and 2?

**65\*.** How many different six-digit numbers are there whose three digits are even and three digits are odd?

**66\*.** How many different four-digit numbers can be written using each of the digits 1, 2, 3, 4, 5, 6, 7, and 8 only once so that each number should contain a unity?

**67\*.** How many different six-digit numbers are there the sum of whose digits is odd?

**68.** There were two women participating in a chess tournament. Every participant played two games with the other participants. The number of games that the men played between themselves proved to exceed by 66 the number of games that the men played with the women. How many participants were there in the tournament and how many games were played?

## Binomial Theorem

The *binomial theorem* states

$$(a+b)^n = C_0^n a^n + C_1^n a^{n-1}b + \dots + C_k^n a^{n-k}b^k + \dots + C_n^n b^n \qquad (1)$$

for any natural $n$. The coefficients $C_k^n$ appearing in (1) are called *binomial coefficients* : the $(k+1)$th term in (1) is regarded as the $k$th term of the expansion and is designated as $T_k$ :

$$T_k + C_k^n a^{n-k}b^k \ (k = 0, 1, \dots, n). \qquad (2)$$

**Properties of the binomial coefficients $C_n^k$.**

$$n! = 1 \cdot 2 \cdot \dots \cdot n, \ 0! = 1 \ \text{(by definition)},$$

$$0! = 1, \ 1! = 1, \ 2! = 2, \ 3! = 6,$$

$$4! = 24, \ 5! = 120, \ 6! = 720, \text{ and so on,}$$

$$n!(n+1) = (n+1)!$$

$$C_k^n = \frac{n!}{k!(n-k)!} = C_{n-k}^n, C_k^n + C_k^{n+1} = C_{k+1}^{n+1} \ (0 \le k \le n).$$

In particular,
$$C_0^4 = C_4^4 = 1, C_1^4 = C_3^4 = 4, C_2^4 = 6;$$
$$C_0^5 = C_5^5 = 1, \ C_1^5 = C_4^5 = 5, \ C_2^5 = C_3^5 = 10;$$
$$(a \pm b)^4 = a^4 \pm 4a^3b + 6a^2b^2 \pm 4ab^3 + b^4;$$
$$(a \pm b)^5 = a^5 \pm 5a^4b + 10a^3b^2 \pm 10a^2b^3 + 5ab^4 \pm b^5.$$

69. Find the greatest coefficient of the expansion of $(a + b)^n$ if the sum of all the coefficients is equal to 4096.

70. Find the middle term of the expansion of $\left(\sqrt{x} - \dfrac{1}{x}\right)^6$.

71. In the expansion of $\left(a\sqrt{a} + \dfrac{1}{a^4}\right)^n$, the coefficient in the second term exceeds by 44 the coefficient in the first term. Find $n$.

72. Find the term of expansion of $\left(x + \dfrac{1}{x}\right)^n$ which does not contain $x$.

73. Find the term of the expansion of $\left(\dfrac{1}{\sqrt[3]{a^2}} + \sqrt[4]{a^3}\right)^{17}$ which does not contain $a$.

74. Find the term of the expansion of $\left(\sqrt[3]{x^{-2}} + x\right)^7$ containing $x$ in the second power.

75. Find the second term of the binomial expansion of $\left(\sqrt[13]{a} + \dfrac{a}{\sqrt{a^{-1}}}\right)^m$ if $C_3^m : C_2^m = 4 : 1$.

76. Find the third term of the expansion of $\left(z^2 + \dfrac{1}{z}\sqrt[3]{z}\right)^n$ if the sum of all the binomial coefficients is equal to 2048.

77. Find $x$ in the binomial $\left(\sqrt[3]{2} + \dfrac{1}{\sqrt[3]{3}}\right)^x$ if the ratio of the seventh term from the beginning of the binomial expansion to the seventh term from its end is $1 / 6$.

78. Determine the ordinal number of the term of the expansion of $\left(\dfrac{3}{4}\sqrt[3]{a^2} + \dfrac{2}{3}\sqrt{a}\right)^{12}$ which contains $a^7$.

79. In the expansion of $\left(x\sqrt{x} + \dfrac{1}{x^4}\right)^n$, the binomial coefficient in the second term exceeds by 44 the binomial coefficient in the first term. Find the ordinal number of the term which does not contain $x$.

80. Find the term of the expansion of $\left(\sqrt[3]{x} - \dfrac{1}{\sqrt{x}}\right)^{15}$ which does not contain $x$.

81. The coefficient in $x$ in the second term of the expansion of $\left(x^2 - \dfrac{1}{4}\right)^n$ is equal to 31. Find the power $n$.

82. The sum of the coefficients in the first three terms of the expansion of $\left(x^2 - \dfrac{2}{x}\right)^m$ is equal to 97. Find the term of the expansion containing $x^4$.

83. Find $A_2^n$ if the fifth term of the expansion of $\left(\sqrt[3]{x} + \dfrac{1}{x}\right)^n$ does not depend on $x$.

84. Find the values of $x$ for which the sum of the third and the fifth term in the expansion of $\left(\sqrt{2^x} + \dfrac{1}{\sqrt{2^{x-1}}}\right)^m$ is equal to 135, and the sum of the binomial coefficients in the last three terms is equal to 22.

85. For what $x$ is the fourth term of the expansion of $(\sqrt{x}^{-1/(\log x + 1)} + \sqrt[12]{x})^6$ equal to 200?

86. In the expansion of $\left(2^x + \dfrac{1}{4^x}\right)^n$, the sum of the binomial coefficients in the first and the second term is equal to 36, and the second term of the expansion is 7 times as large as the first. Find $x$.

87. Find the values of $x$ for which the difference between the fourth and the sixth term of the expansion of $\left(\dfrac{\sqrt{2^x}}{\sqrt[16]{8}} + \dfrac{\sqrt[16]{32}}{\sqrt{2^x}}\right)^m$ is equal to 56 if it is known that the binomial power $m$ is smaller by 20 than the binomial coefficient in the third term of the expansion.

88. Find $x$ if it is known that the second term of the expansion of $(x + x^{\log x})^5$ is equal to 1000000.

89. For what value of $x$ is the sixth term of the expansion of the power of the binomial $\left(2^{\log_2 \sqrt{9^{x-1}+7}} + 2^{-\frac{1}{5}\log_2(3^{x-1}+1)}\right)^7$ equal to 84?

90*. Prove the inequality $n^{n+1} > (n+1)^n$, $n \geq 3$, $n \in \mathbf{N}$.

91. Find the power $n$ of the binomial $\left(\dfrac{x}{5} + \dfrac{2}{5}\right)^n$ if the ninth term of the expansion has the greatest coefficient.

92. Find the term of the expansion of $(a+b)^{50}$ which is the greatest in absolute value if $|a| = \sqrt{3}|b|$.

93. In the expansion of $(1 + x)^n$ by the increasing powers of $x$, the third term is four times as great as the fifth term and the ratio of the fourth term to the sixth is $40/3$. Find $n$ and $x$.

**94.** Simplify the binomial

$$\left(\frac{x+1}{x^{2/3}-x^{1/3}+1}-\frac{x-1}{x-x^{1/2}}\right)^{10}$$

and find the term of its expansion which does not contain $x$.

**95.** The sum of the coefficients in the first, second, and third terms of the expansion of $\left(x^2+\frac{1}{x}\right)^m$ is equal to 46. Find the term of the expansion which does not contain $x$.

## 1.4. Equations and Inequalities of the First and the Second Degree

**The absolute value of a real number.** The absolute value of the real number $a$ is defined by the formula.

$$|a|=\begin{cases} a \text{ if } a \ge 0, \\ -a \text{ if } a < 0. \end{cases}$$

The inequality $|x| < a$, where $a > 0$, is equivalent to the two sided inequality $-a < x < a$.

The inequality $|x| > a$, where $a > 0$, is equivalent to the collection of two inequalities $x < -a$ and $x > a$, i.e. for $a > 0$ we have

$$|x| < a \Leftrightarrow -a < x < a, |x| \le a \Leftrightarrow -a \le x \le a;$$

$$|x| > a \Leftrightarrow \begin{bmatrix} x < -a, \\ x > a, \end{bmatrix} |x| \ge a \Leftrightarrow \begin{bmatrix} x \le -a, \\ x \ge a. \end{bmatrix}$$

The inequality

$$|a+b| \le |a|+|b|, ||a|-|b|| \le |a-b|$$

is valid for any numbers $a, b \in \mathbf{R}$.

Solve the following equations **(1–3)**.

**1.** $|x+2| = 2(3-x)$.  **2.** $|3x-2| + x = 11$.
**3.** $|x| - |x-2| = 2$.
**4.** Find the least integral value of $x$ which satisfies the equation $|x-3| + 2|x+1| = 4$.
**5.** Find all $a \in \mathbf{R}$ for which the equation $a^3 + a^2|a+x| + |a^2x+1| = 1$ has not less than four different solutions which are integers.

Solve the following inequalities **(6-11)**.

**6.** $|5-2x| < 1$.
**7.** $|3-04353x-2.5| \ge 2$.

**8.** $|x - 2| \le |x + 4|$.

**9.** $|2x - 4| < x - 1$.

**10.** $2|x + 1| > x + 4$.

**11.** $|x + 2| - |x - 1| < x - \dfrac{3}{2}$.

**12.** Solve the inequality $|x + 1| + |x - 4| > 7$, indicating the least positive integer $x$ satisfying the inequality.

**13.** Find the greatest integer $x$ satisfying the inequality $\dfrac{2x + 1}{3} - \dfrac{3x - 1}{2} > 1$.

## Quadratic Equations and a Quadratic Trinomial

**An arithmetic root.** The nonnegative number $b$ such that $b^n = a$ ($n \ge 2$ is natural) is called an $n$th *arithmetic root* of the number $a$ and is designated as $\sqrt[n]{a}$. In particular, $\sqrt{c^2} = |c|$.

**An $n$th root.** The number $b$ is called an $n$th *root* of the number $a$ if $b^n = a$.

For example, the numbers 2 and –2 are fourth roots of 16; the number –3 is a cube root of –27. There is only one fourth arithmetic root of 16, namely, the number 2. The arithmetic cube root of the number –27 does not exist.

**quadratic equation.** An equation of the form

$$ax^2 + bx + c = 0 \ (a \ne 0) \qquad (*)$$

is a *quadratic equation*. The quantity $D = b^2 - 4ac$ is a *discriminant* of the quadratic equation. Then,

**(1)** if $D \ge 0$, the equation has *two real roots*

$$x_1 = \frac{-b + \sqrt{D}}{2a}, \ x_2 = \frac{-b - \sqrt{D}}{2a}$$

(for $D = 0$ the roots coincide).

**(2)** if $D < 0$, the equation *has no real roots*. The following assertions hold true :

(a) if $x_1$ and $x_2$ are roots of the quadratic equation (*), then $x_1 + x_2 = -\dfrac{b}{a}$, $x_1 \cdot x_2 = \dfrac{c}{a}$ (*Vieta's theorem*);

(b) $ax^2 + bx + c = a(x - x_1)(x - x_2)$ is the *formula for the factorization of a quadratic trinomial*.

## Solution of quadratic inequalities

$$ax^2 + bx + c > 0 \,(a \ne 0). \qquad (**)$$

**(1)** If $D > 0$, then equation (*) has two different real roots $x_1 < x_2$.

Then

$$a > 0 \Rightarrow x \in (-\infty, x_1) \cup (x_2, +\infty),$$
$$a < 0 \Rightarrow x \in (x_1, x_2).$$

(2) If $D = 0$, then equation (*) has two equal real roots $x_1 = x_2$. In that case

$$a > 0 \Rightarrow x \in (-\infty, x_1) \cup (x_1, +\infty),$$
$$a < 0 \Rightarrow x \in \phi.$$

(3) If $D < 0$, then equation (*) has no roots. In that case

$$a > 0 \Rightarrow x \in (-\infty, +\infty),$$
$$a < 0 \Rightarrow x \in \varnothing.$$

The inequalities $ax^2 + bx + c \geq 0$, $ax^2 + bx + c < 0$, and $ax^2 + bx + c \leq 0$ can be solved by analogy.

14. Find $b$ if the roots of the equation $24x^2 + bx + 25 = 0$ are positive and $x_2 = 1.5x_1$.

15. Find all solutions of the equation $(|x|+1)^2 = 4|x|+9$ belonging to the domain of definition of the function $y = \sqrt{5 - 2x}$.

16. Find all solutions of the equation $(3|x|-3)^2 = |x|+7$ belonging to the domain of definition of $y = \sqrt{x\,(x-3)}$.

17. Find all the solutions of the equation $(2|x|-1)^2 = |x|$ belonging to the domain of definition of the function $y = \log(4x-1)$.

18. Find all solutions of the equation $9x^2 - 18|x| + 5 = 0$ belonging to the domain of definition of the function $y = \ln((x+1)\times(x-2))$.

Find analytical and graphical solutions of the following equations **(19-22)**.

19. $|x^2 + 4x + 2| = (5x+16)/3$.

20. $|x^2 - 2x - 1| = (5x+1)/3$.

21. $|x^2 - 4x + 2| = (5x-4)/3$.

22. $|x^2 - 6x + 7| = (5x-9)/3$.

Solve the following equations **(23-25)**.

23. $x^2 + |x-1| = 1$.

24. $(x^2 + x + 1) + (x^2 + 2x + 3) + (x^2 + 3x + 5) + \ldots$
$\ldots + (x^2 + 20x + 39) = 4500$.

25. $x|x-4|+a = 0$.

26. For what values of $a$ does the equation $9x^2 - 2x + a = 6 - ax$ possess equal roots?

27. Find the value of $k$ for which the equation $(k-1)x^2 + (k+4)x + k + 7 = 0$ has equal roots.

28. Find the values of $a$ for which the roots of the equation $(2a-5)x^2 - 2(a-1)x + 3 = 0$ are equal.

29. For what values of $m$ does the equation $x^2 - x + m = 0$ possess no real roots?

30. For what values of $m$ does the equation $x^2 - x + m^2 = 0$ possess no real roots?

**31.** For what values of $m$ does the equation $mx^2 - (m+1)x + 2m - 1 = 0$ possess no real roots?

**32.** For what values of $c$ does the equation $(c-2)x^2 + 2(c-2)x + 2 = 0$ possess no real roots?

**33.** Find integral values of $k$ for which the equation $(k-12)x^2 + 2(k-12)x + 2 = 0$ possesses no real roots.

**34.** For what values of $a$ does the equation $x^2 + 2a\sqrt{a^2 - 3x} + 4 = 0$ possess equal roots?

**35.** Find the values of the coefficient $a$ for which the curve $y = x^2 + ax + 25$ touches the $Ox$ axis.

**36.** Find the value of $k$ for which the curve $y = x^2 + kx + 4$ touches the $Ox$ axis.

**37.** Form a quadratic equation whose roots are the numbers $\dfrac{1}{10 - \sqrt{72}}$ and $\dfrac{1}{10 + 6\sqrt{2}}$.

**38.** For what values of $k$ is the inequality $x^2 - (k-3)x - k + 6 > 0$ valid for all real $x$?

**39.** For what values of $a$ is the inequality $ax^2 + 2ax + 0.5 > 0$ valid throughout the entire number axis?

**40.** For what integral $k$ is the inequality
$$x^2 - 2(4k-1)x + 15k^2 - 2k - 7 > 0$$
valid for any real $x$?

**41.** Find the least integral value of $k$ for which the equation $x^2 - 2(k+2)x + 12 + k^2 = 0$ has two different real roots.

**42.** For what values of $a$ is the sum of the roots of the equation $x^2 + (2-a-a^2)x - a^2 = 0$ equal to zero?

**43.** For what values of $a$ does the equation $x^2 - (2^a - 1)x - 3(4^{a-1}2^{a-2}) = 0$ possess real roots?

**44.** For what values of $a$ do the graphs of the functions $y = 2ax + 1$ and $y = (a-6)x^2 - 2$ not intersect?

**45.** For what values of $p$ does the vertex of the parabola $y = x^2 + 2px + 13$ lie at a distance of 5 from the origin?

**46.** Find the value of $a$ for which one root of the equation $x^2 + (2a-1)x + a^2 + 2 = 0$ is twice as large as the other.

**47.** For what values of $a$ is the ratio of the roots of the equation $x^2 + ax + a + 2 = 0$ equal to 2?

**48.** For what values of $a$ is the ratio of the roots of the equation $ax^2 - (a+3)x + 3 = 0$ equal to 1.5?

**49.** For what values of $a$ do the roots $x_1$ and $x_2$ of the equation $x^2 - (3a+2)x + a^2 = 0$ satisfy the relation $x_1 = 9x_2$? Find the roots.

**50.** Find $a$ such that one of the roots of the equation $x^2 - \dfrac{15}{4}x + a = 0$ is the square of the other.

51. The roots $x_1$ and $x_2$ of the equation $x^2 + px + 12 = 0$ are such that $x_2 - x_1 = 1$. Find $p$.

52. Find $k$ in the equation $5x^2 - kx + 1 = 0$ such that the difference between the roots of the equation is unity.

53. For what value of $a$ is the difference between the roots of the equation $(a-2) x^2 - (a-4) x - 2 = 0$ equal to 3?

54. Find $b$ in the equation $5x^2 + bx - 28 = 0$ if the roots $x_1$ and $x_2$ of the equation are related as $5x_1 + 2x_2 = 1$ and $b$ is an integer.

55. Find $p$ in the equation $x^2 - 4x + p = 0$ if it is known that the sum of the squares of its roots is equal to 16.

56. For what values of $a$ is the difference between the roots of the equation $2x^2 - (a+1) x + (a-1) = 0$ equal to their product?

57. Find all values of $a$ for which the sum of the roots of the equation $x^2 - 2a (x-1) - 1 = 0$ is equal to the sum of the squares of its roots.

58. Find the coefficients of the equation $x^2 + px + q = 0$ such that its roots are equal to $p$ and $q$.

59. For what values of $a$ do the equations $x^2 + ax + 1 = 0$ and $x^2 + x + a$ have a root in common?

60*. Given two quadratic equations $x^2 - x + m = 0$ and $x^2 - x + 3m = 0$, $m \neq 0$. Find the value of $m$ for which one of the roots of the second equation is equal to double the root of the first equation.

61. The trinomial $ax^2 + bx + c$ has no real roots, $a + b + c < 0$. Find the sign of the number $c$.

62. Express $x_1^3 + x_2^3$ in terms of the coefficients of the equation $x^2 + px + q = 0$, where $x_1$ and $x_2$ are the roots of the equation.

63. Assume that $x_1$ and $x_2$ are roots of the equation $3x^2 - ax + 2a - 1 = 0$. Calculate $x_1^3 + x_2^3$.

64. Without solving the equation $3x^2 - 5x - 2 = 0$, find the sum of the cubes of its roots.

65. Calculate $\dfrac{1}{x_1^3} + \dfrac{1}{x_2^3}$, where $x_1$ and $x_2$ are roots of the equation $2x^2 - 3ax - 2 = 0$.

66. For what values of $a$ does the equation $(2-x)(x+1) = a$ possess real and positive roots?

67. Find all values of $p$ for which the roots of the equation $(p-3) x^2 - 2px + 5p = 0$ are real and positive.

68. Find all values of $a$ for which the inequality $(a+4) x^2 - 2ax + 2a - 6 < 0$ is satisfied for all $x \in \mathbf{R}$.

69. Find all values of $a$ for which the inequality $(a-3) x^2 - 2ax + 3a - 6 > 0$ is satisfied for all values of $x$.

70. Find all values of $a$ for which the inequality $(a-1) x^2 - (a+1) x + a + 1 > 0$ is satisfied for all real $x$.

71. For what least integral $k$ is the quadratic trinomial $(k-2) x^2 + 8x + k + 4$ positive for all values of $x$?

72. Solve the inequality $x^2 + ax + a > 0$.

**73.** Find all real values of $m$ for which the inequality $mx^2 - 4x + 3m + 1 > 0$ is satisfied for all positive $x$.

**74.** Find all values of $a$ for which both roots of the equation $x^2 - 6ax + 2 - 2a + 9a^2 = 0$ are greater than 3.

**75.** Form a quadratic equation the product of whose roots $x_1$ and $x_2$ is equal to 4, and $\dfrac{x_1}{x_1 - 1} + \dfrac{x_2}{x_2 - 1} = \dfrac{a^2 - 7}{a^2 - 4}$.

**76.** Form a quadratic equation the sum of whose roots $x_1$ and $x_2$ is equal to 2, and $\dfrac{1 - x_1}{1 + x_2} + \dfrac{1 - x_2}{1 + x_1} = 2\dfrac{4a^2 + 15}{4a^2 - 1}$.

**77.** For what values of $a$ are the roots of the equation $x^2 - 4ax + 1 = 0$ real and satisfy the conditions $x_1 \geq a$ and $x_2 \geq 0$?

**78.** For what values of $a \in \mathbf{R}$ does the equation $ax^2 + x + a - 1 = 0$ possess two distinct real roots $x_1$ and $x_2$ satisfying the inequality $\left| \dfrac{1}{x_1} - \dfrac{1}{x_2} \right| > 1$?

**79.** For what values of $a \in \mathbf{R}$ does the equation $x^2 + 1 = x / a$ possess two distinct real roots $x_1$ and $x_2$ satisfying the inequality $|x_1^2 - x_2^2| > 1 / a$?

**80.** Find all values of $a$ for which the inequality $(x - 3a) \times (x - a - 3) < 0$ is satisfied for all $x$ such that $1 \leq x \leq 3$.

**81\*.** Find all values of $k$ for which any real $x$ is a solution of at least one of the inequalities

$$x^2 + 5k^2 + 8k > 2(3kx + 2)$$

and

$$x^2 + 4x^2 \geq k(4x + 1).$$

**82\*.** For what real $a$ do the roots of the equation $x^2 - 2x - a^2 + 1 = 0$ lie between the roots of the equation $x^2 - 2(a + 1)x + a(a - 1) = 0$?

**83.** For what values of $a$ is every solution of the inequality $x^2 - x - 2 < 0$ larger than any solution of the inequality $ax^2 - 4x - 1 \geq 0$?

Solve the following inequalities **(84-109)**.

**84.** $3x^2 - 7x + 4 \leq 0$.

**85.** $3x^2 - 7x + 6 < 0$.

**86.** $3x^2 - 7x - 6 < 0$.

**87.** $x^2 - 3x + 5 > 0$.

**88.** $x^2 - 14x - 15 > 0$.

**89.** $2 - x - x^2 \geq 0$.

**90.** $x^2 - 5|x| + 6 < 0$.

**91.** $x^2 - |x| - 2 \geq 0$.

**92.** $|x^2 - 4x| < 5$.

**93.** $|x^2 + x| - 5 < 0$.

**94.** $|x^2 - 5x| < 6.$

**95.** $|x^2 - 2x| < x.$

**96.** $|x^2 - 2x - 3| < 3x - 3.$

**97.** $|x^2 - 3x| + x - 2 < 0.$

**98.** $x^2 - 7x + 12 < |x - 4|.$

**99.** $x^2 - |5x - 3| - x < 2.$

**100.** $|x - 6| > x^2 - 5x + 9.$

**101.** $|x - 6| < x^2 - 5x + 9.$

**102.** $|x - 2| \le 2x^2 - 9x + 9.$

**103.** $3x^2 - |x - 3| > 9x - 2.$

**104.** $x^2 + 4 \ge |3x + 2| - 7x.$

**105.** $x^2 - |5x + 8| > 0.$

**106.** $3|x - 1| + x^2 - 7 > 0.$

**107.** $|x - 6| > |x^2 - 5x + 9|.$

**108.** $(|x - 1| - 3)(|x + 2| - 5) < 0.$

**109.** $|x^2 - 2x - 8| > 2x.$

**110.** For what values of $k$ does the equation $kx^2 + 12x - 3 = 0$ possess a root equal to $1/5$?

**111.** For what values of the parameter $a$ is the inequality $(a^3 + (1 - \sqrt{2})a^2 - (3 + \sqrt{2})a + 3\sqrt{2})x^2 + 2(a^2 - 2)x + a > -\sqrt{2}$ satisfied for any $x > 0$?

**112.** Find all numbers $a$ for each of which the least value of the quadratic trinomial $4x^2 - 4ax + a^2 - 2a + 2$ on the interval $0 \le x \le 2$ is equal to 3.

**113.** For what value of $a$ do the roots of the equation $2x^2 + 6x + a = 0$ satisfy the condition $\dfrac{x_1}{x_2} + \dfrac{x_2}{x_1} < 2?$

## 1.5. Equations of Higher Degrees. Rational Inequalities

### Equations of Higher Degrees

If there is a polynomial $Q(x)$ such that the equality

$P(x) = D(x)Q(x)$

holds true, then we say that the polynomial $P(x)$ is divisible by the polynomial $D(x)$. In that case $P(x)$ is called a *dividend*, $D(x)$ a *divisor*. and $Q(x)$ a *quotient*.

If the polynomial $Q(x)$ does not exist, then we say that the polynomial $P(x)$ is not divisible by the polynomial $D(x)$ and we consider a division with a remainder.

Assume that $D(x)$ is a polynomial of a degree not smaller than the first. To divide the polynomial $P(x)$ by the polynomial $D(x)$ with a remainder means to represent the polynomial $P(x)$ in the form

$$P(x) = D(x) \cdot Q(x) + R(x), \qquad (1)$$

Where $Q(x)$ and $R(x)$ are polynomials, the degree of $R(x)$ being smaller than that of $D(x)$. In that case $P(x)$ is a dividend, $D(x)$ a divisor, $Q(x)$ a quotient, and $R(x)$ a remainder. *For any two polynomials $P(x)$ and $D(x)$ there are always polynomials $Q(x)$ and $R(x)$, defined uniquely, for which equation (1) holds true.*

If $D(x) = x - c$, where $c$ is a number, then equation (1) assumes the form

$$P(x) = (x - c)Q(x) + R,$$

where $R$ is a number.

**Remainder theorem.** *The remainder obtained on dividing the polynomial $P(x)$ by the binomial $x - c$ is equal to the value of the polynomial $P(x)$ for $x = c$.*

The number $c$ is called a *root* of the polynomial $P(x)$ if $P(x) = 0$.

**Corollary of the remainder theorem.** *For the polynomial $P(x)$ to be divisible by the binomial $x - c$, it is necessary and sufficient that the number $c$ should be a root of the polynomial $P(x)$.*

**Example.** $P(x) = x^3 + x^2 - 2$, $P(1) = 0$. $P(x) = (x^3 - x^2) + 2x^2 - 2$

$$= x^2(x - 1) + 2(x - 2)(x + 1) = (x - 1)(x^2 + 2x + 2),$$

i.e. $P(x) = x^3 + x^2 - 2$ is divisible by $x - 1$.

1. Solve the equation $(x - \sqrt{3})^4 - 5(x - \sqrt{3})^2 + 4 = 0$.

2. Find all solutions of the equation

$$\frac{(2|x| - 3)^2 - |x| - 6}{4x + 1} = 0$$

which belong to the domain of definition of the function $y = (2x + 1)/(x^2 - 36)$.

3. Show that if

$$z = \sqrt[3]{a + \sqrt{a^2 + b^3}} = \sqrt[3]{\sqrt{a^2 + b^3} - a},$$

then $z^3 + 3bz - 2a = 0$.

Solve the following equations (4-9).

4. $(x - 1)(x - 2)(x - 3)(x - 4) = 15$.

5. $(x + a)(x + 2a)(x + 3a)(x + 4a) = b^4$.

6. $\dfrac{2x - 2}{x^2 - 36} - \dfrac{x - 2}{x^2 - 6x} = \dfrac{x - 1}{x^2 + 6x}$.

7. $\dfrac{2x + 1}{x} + \dfrac{4x}{2x + 1} = 5$.

8. $\dfrac{x^2 - 3.5x + 1.5}{x^2 - x - 6} = 0$.

**9.** $\dfrac{2x-3}{x-1}+1=\dfrac{6x-x^2-6}{x-1}.$

**10\*.** Find all real values of $a$ for each of which the equation $\sqrt{x-a}\,(x^2+(1+2a^2)\,x+2a^2)=0$ has only two distinct roots. Write the roots.

**11.** Find all values of $n \in \mathbf{N}$ for which the equation $\dfrac{x-8}{n-10}=\dfrac{n}{x}$ has no solutions.

**12\*.** Find the values of $a$ for which the equation $x^4+(1-2a)\,x^2+a^2-1=0$ (a) has no solutions; (b) has one solution; (c) has two solutions; (d) has three solutions.

**13.** For what real values of $a$ is the sum of the roots of the equation $\dfrac{1}{x}+\dfrac{1}{a}-\dfrac{1}{a^2}=\dfrac{1}{-a^2+a+x}$ smaller than $a^3/10$?

**14.** Solve the equation
$$\dfrac{2b^2+x^2}{b^3-x^3}-\dfrac{2x}{bx+b^2+x^2}+\dfrac{1}{x-b}=0$$
For what values of $b$ is the solution of the equation unique?

**15.** Find all values of $a$ for which the inequality $\dfrac{x-2a-3}{x-a+2}<0$ is satisfied for all $x$ belonging to the interval $1 \le x \le 2$.

**16.** Find the smallest integral $x$ satisfying the inequality $\dfrac{x-5}{x^2+5x-14}>0.$

**17.** Find integral $x$'s which satisfy the inequality $x^4-3x^3-x+3<0.$

Find the largest integral $x$ which satisfies the following inequalities **(18-22)**.

**18.** $\dfrac{x-2}{x^2-9}<0.$

**19.** $\dfrac{1}{x+1}-\dfrac{2}{x^2-x+1}<\dfrac{1-2x}{x^3+1}.$

**20.** $\dfrac{x+4}{x^2-9}-\dfrac{2}{x+3}<\dfrac{4x}{3x-x^2}.$

**21.** $\dfrac{4x+19}{x+5}<\dfrac{4x-17}{x-3}.$

**22.** $(x+1)\,(x-3)^2\,(x-5)\,(x-4)^2\,(x-2)<0.$

**23.** Find integral values of $x$ satisfying the inequality $\left|\dfrac{2}{x-13}\right|>\dfrac{8}{9}.$

**24.** We call $a$ a good number if the inequality $\dfrac{2x^2 + 2x + 3}{x^2 + x + 1} \le a$ is satisfied for any real $x$.

(a) Prove that 4 is a good number.

(b) Find all good numbers.

**25.** For what values of $m$ is the inequality $\dfrac{x^2 - mx - 2}{x^2 - 3x + 4} > -1$ satisfied for all $x \in \mathbf{R}$?

**26\*.** Find all values of $k$ for which the inequality $\dfrac{x^2 + k^2}{k(6 + x)} \ge 1$ is satisfied for all $x$ such that $-1 < x < 1$.

## Rational Inequalities

*The inequality*

$$\frac{g(x)}{\varphi(x)} > 0 \qquad (1)$$

*is equivalent to the inequality*

$$g(x)\varphi(x) > 0; \qquad (2)$$

*the inequality*

$$\frac{g(x)}{\varphi(x)} < 0 \qquad (3)$$

*is equivalent to the inequality*

$$g(x)\varphi(x) < 0. \qquad (4)$$

Indeed, inequalities (1) and (2) are satisfied only at the points $x$ at which the functions $g(x)$ and $\varphi(x)$ are of the same sign, i.e. are either both positive or both negative, and inequalities (3) and (4) are satisfied only at the points $x$ at which the functions $g(x)$ and $\varphi(x)$ are of unlike signs. Therefore, we shall present arguments only for inequalities of the forms (2) and (4). When solving rational inequalities, it is useful to use the so-called *method of intervals*. Here is a brief description of the method.

Assume that

$$f(x) = (x - x_1)(x - x_2) \ldots (x - x_n),$$

where $x_1 < x_2 < \ldots < x_n$. We mark the points $x_1, x_2, \ldots, x_n$ on the $Ox$ axis. With the variation of $x$ throughout the axis the function $x - x_k$, $k = 1, 2, \ldots, n$, changes sign only when it passes through the point $x = x_n$. Therefore, throughout the $Ox$ axis the function $f(x)$ changes sign only when $x$ passes through the points $x_1, x_2, \ldots, x_n$ and on every interval

$$(-\infty, x_1), (x_1, x_2), \ldots, (x_{n-1}, x_n), (x_n, +\infty) \qquad (5)$$

the function $f(x)$ retains sign. Taking into account that $f(x) > 0$ on the interval $(x_n, +\infty)$ and changes sign when passing to each successive interval of system (5), it is easy to find all the intervals on which $f(x) > 0$ and all the intervals on which $f(x) < 0$.

**Example 1.** Solve the inequality $\dfrac{1}{x} < 1$

**Solution.** $\dfrac{1}{x} < 1 \Leftrightarrow \dfrac{1}{x} - 1 < 0 \Leftrightarrow \dfrac{1-x}{x} < 0 \Leftrightarrow \dfrac{x-1}{x} > 0.$

The function $f(x) = \dfrac{x-1}{x}$ changes sign when $x$ passes through the points $x_1 = 0$, $x_2 = 1$. We have $f(x) > 0$ on the interval $(1, +\infty)$, $f(x) < 0$ on the next interval $(0, 1)$, and $f(x) > 0$ on the next interval $(-\infty, 0)$.

*Answer :* $(-\infty, 0) \cup (1, +\infty)$.

If we solve the inequality $(f(x) \le 0$ or $f(x) \ge 0)$ nonstrictly, then the solution of the inequality at the points $x_1, x_2, \ldots, x_n$ must be verified separately.

**Example 2.** Solve the inequality $\dfrac{x}{x+2} \le \dfrac{1}{x}$.

**Solution.** $\dfrac{x}{x+2} \le \dfrac{1}{x} \Leftrightarrow \dfrac{x}{x+2} - \dfrac{1}{x} \le 0 \Leftrightarrow \dfrac{x^2 - x - 2}{x(x+2)} \le 0$

$\Leftrightarrow \dfrac{(x-2)(x+1)}{x(x+2)} \le 0.$

The function $f(x) = \dfrac{(x-2)(x+1)}{x(x+2)}$ changes sign when the variable $x$ passes through the points $x_1 = -2$, $x_2 = -1$, $x_3 = 0$, $x_4 = 2$. We have $f(x) > 0$ on the interval $(2, +\infty)$, $f(x) < 0$ on the next interval $(0, 2)$, $f(x) > 0$ on the next interval $(-1, 0)$, $f(x) < 0$ on the next interval $(-2, -1)$, and $f(x) < 0$ on the next, last, interval $(-\infty, -2)$. Taking into account that the inequality is also satisfied at the points $x_2 = -1$ and $x_4 = 2$, we get the *answer :* $(-2, -1] \cup (0, 2]$.

**Example 3.** Solve the inequality $\dfrac{x}{x-3} \le \dfrac{1}{x}$.

**Solution.** $\dfrac{x}{x-3} \le \dfrac{1}{x} \Leftrightarrow \dfrac{x}{x-3} - \dfrac{1}{x} \le 0 \Leftrightarrow \dfrac{x^2 - x + 3}{x(x-3)} \le 0.$

The function $f(x) = \dfrac{x^2 - x + 3}{x(x-3)}$ changes sign when the variable $x$ passes through the points $x_1 = 0$ and $x_2 = 3$ (we have $x^2 - x + 3 > 0$ for all $x \in R$). Since $f(x) > 0$ on the interval $(3, +\infty)$, it follows that $f(x) < 0$ on the interval $(0, 3)$ and $f(x) > 0$ on the interval $(-\infty, 0)$.

*Answer.* $(0, 3)$.

Suppose $F(x) = (x - x_1)^{k_1} (x - x_2)^{k_2} \ldots (x - x_n)^{k_n}$, where $k_1, k_2, \ldots, k_n$ are integers. If $k_j$ is an even number, then the function $(x - x_j)^{k_j}$ does not change sign when $x$ passes through the point $x_j$ and, consequently, the function $F(x)$ does not change sign. If $k_p$ is an even number, then the function $(x - x_p)^k p$ changes sign when $x$ passes through the point $x_p$ and, consequently, the function $F(x)$ also changes sign.

**Example 4.** Solve the inequality $(x - 1)^2 (x + 1)^3 (x - 4) < 0$.

*Solution.* The function $F(x) = (x - 1)^2 (x + 1)^3 (x + 4)$ changes sign only when $x$ passes through the points $x_1 = -1$, $x_2 = 4$. We have $F(x) > 0$ on the interval $(4, +\infty)$, $F(x) < 0$ on the next interval $(-1, 4)$, excluding the point $x = 1$ at which $F(x) = 0$, and $F(x) > 0$ on the last interval $(-\infty, -1)$.

　　*Answer :* $(-1, 1) \cup (1, 4)$.

**Example 5.** Solve the inequality $\dfrac{(x - 1)^2 (x + 1)^3}{x^4 (x - 2)} \leq 0$.

*Solution.* The function $F(x) = \dfrac{(x - 1)^2 (x + 1)^3}{x^4 (x - 2)^2}$ changes sign only when the variable $x$ passes through the points $x_1 = -1$, $x_2 = 2$. When $x$ passes through the points $x_3 = 0$ and $x_4 = 1$, the function $F(x)$ does not change sign. We have $F(x) > 0$ on the interval $(2, +\infty)$, $F(x) < 0$ on the next intervals $(1, 2)$, $(0, 1)$, $(-1, 0)$, and $F(x) > 0$ on the interval $(-\infty, -1)$. At the point $x_4 = 1$ the inequality is satisfied and at the point $x_3 = 0$ the function $F(x)$ is not defined.

　　*Answer :* $[-1, 0) \cup (0, 2)$.

Solve the following inequalities **(27-135)**.

27. $(x - 1)(3 - x)(x - 2)^2 > 0$.

28. $\dfrac{6x - 5}{4x + 1} < 0$.

29. $\dfrac{2x - 3}{3x - 7} > 0$.

30. $\dfrac{0.5}{x - x^2 - 1} < 0$.

31. $\dfrac{x^2 - 5x + 6}{x^2 + x + 1} < 0$.

32. $\dfrac{x^2 + 2x - 3}{x^2 + 1} < 0$.

33. $\dfrac{(x - 1)(x + 2)^2}{-1 - x} < 0$.

34. $\dfrac{x^2 + 4x + 4}{2x^2 - x - 1} > 0$.

35. $x^4 - 5x^2 + 4 < 0$.

36. $x^4 - 2x^2 - 63 \leq 0$.

**37.** $\dfrac{3}{x-2} < 1.$

**38.** $\dfrac{1}{x-1} \le 2.$

**39.** $\dfrac{4x+3}{2x-5} < 6.$

**40.** $\dfrac{5x-6}{x+6} < 1.$

**41.** $\dfrac{5x+8}{4-x} < 2.$

**42.** $\dfrac{x-1}{x+3} > 2.$

**43.** $\dfrac{7x-5}{8x+3} > 4.$

**44.** $\dfrac{x}{x-5} > \dfrac{1}{2}.$

**45.** $\dfrac{5x-1}{x^2+3} < 1.$

**46.** $\dfrac{x-2}{x^2+1} < -\dfrac{1}{2}.$

**47.** $\dfrac{x+1}{(x-1)^2} < 1.$

**48.** $\dfrac{x^2-7x+12}{2x^2+4x+5} > 0.$

**49.** $\dfrac{x^2+6x-7}{x^2+1} \le 2.$

**50.** $\dfrac{x^4+x^2+1}{x^2-4x-5} < 0.$

**51.** $\dfrac{1+3x^2}{2x^2-21x+40} < 0.$

**52.** $\dfrac{1+x^2}{x^2-5x+6} < 0.$

**53.** $\dfrac{x^4+x^2+1}{x^2-4x-5} > 0.$

**54.** $\dfrac{1-2x-3x^2}{3x-x^2-5} > 0.$

**55.** $\dfrac{x^2-5x+7}{-2x^2+3x+2} > 0.$

**56.** $\dfrac{2x^2-3x-459}{x^2+1} > 1.$

**57.** $\dfrac{x^2-1}{x^2+x+1} < 1.$

**58.** $\dfrac{x}{x^2 - 3x - 4} > 0.$

**59.** $\dfrac{x^2 + 7x + 10}{x + \dfrac{2}{3}} > 0.$

**60.** $\dfrac{3x^2 - 4x - 6}{2x - 5} < 0.$

**61.** $\dfrac{17 - 15x - 2x^2}{x + 3} < 0.$

**62.** $\dfrac{x^2 - 9}{3x - x^2 - 24} < 0.$

**63.** $\dfrac{x + 7}{x - 5} + \dfrac{3x + 1}{2} \geq 0.$

**64.** $2x^2 + \dfrac{1}{x} > 0.$

**65.** $\dfrac{x^2 - x - 6}{x^2 + 6x} \geq 0.$

**66.** $\dfrac{x^2 - 5x + 6}{x^2 - 11x + 30} < 0.$

**67.** $\dfrac{x^2 - 8x + 7}{4x^2 - 4x + 1} < 0.$

**68.** $\dfrac{x^2 - 36}{x^2 - 9x + 18} < 0.$

**69.** $\dfrac{x^2 - 6x + 9}{5 - 4x - x^2} \geq 0.$

**70.** $\dfrac{x - 1}{x + 1} < x.$

**71.** $\dfrac{1}{x + 2} < \dfrac{3}{x - 3}.$

**72.** $\dfrac{14x}{x + 1} - \dfrac{9x - 30}{x - 4} < 0.$

**73.** $\dfrac{5x^2 - 2}{4x^2 - x + 3} < 1.$

**74.** $\dfrac{x^2 - 5x + 12}{x^2 - 4x + 5} > 3.$

**75.** $\dfrac{x^2 - 3x + 24}{x^2 - 3x + 3} < 4.$

**76.** $\dfrac{x^2 - 1}{2x + 5} < 3.$

**77.** $\dfrac{x^2 + 1}{4x - 3} > 2.$

**78.** $\dfrac{x^2 + 2}{x^2 - 1} < -2.$

**79.** $\dfrac{3x-5}{x^2+4x-5} > \dfrac{1}{2}.$

**80.** $\dfrac{2x+3}{x^2+x-12} \le \dfrac{1}{2}.$

**81.** $\dfrac{5-2x}{3x^2-2x-16} < 1.$

**82.** $\dfrac{15-4x}{x^2-x-12} < 4.$

**83.** $\dfrac{1}{x^2-5x+6} \ge \dfrac{1}{2}.$

**84.** $\dfrac{(2-x^2)(x-3)^3}{(x+1)(x^2-3x-4)} \ge 0.$

**85.** $\dfrac{5-4x}{3x^2-x-4} < 4.$

**86.** $\dfrac{19-33x}{7x^2-11x+4} > 2.$

**87.** $\dfrac{0.5x+49}{10x^2+x-2} < \dfrac{1}{2}.$

**88.** $\dfrac{(x+2)(x^2-2x+1)}{4+3x-x^2} \ge 0.3.$

**89.** $\dfrac{4}{1+x} + \dfrac{2}{1-x} < 1.$

**90.** $2+\dfrac{3}{x+1} > \dfrac{2}{x}.$

**91.** $1+\dfrac{2}{x-1} > \dfrac{6}{x}.$

**92.** $\dfrac{x^4-3x^3+2x^2}{x^2-x-30} > 0.$

**93.** $\dfrac{x-1}{x} - \dfrac{x+1}{x-1} < 2.$

**94.** $\dfrac{2(x-3)}{x(x-6)} \le \dfrac{1}{x-1}.$

**95.** $\dfrac{2(x-4)}{(x-1)(x-7)} \ge \dfrac{1}{x-2}.$

**96.** $\dfrac{2x}{x^2-9} \le \dfrac{1}{x+2}.$

**97.** $\dfrac{1}{x-2} + \dfrac{1}{x-1} > \dfrac{1}{x}.$

**98.** $\dfrac{7}{(x-2)(x-3)} + \dfrac{9}{x-3} +1 < 0.$

**99.** $\dfrac{20}{(x-3)(x-4)} + \dfrac{10}{x-4} +1 > 0.$

**100.** $\dfrac{(x-2)(x-4)(x-7)}{(x+2)(x+4)(x+7)} > 1.$

**101.** $\dfrac{(x-1)(x-2)(x-3)}{(x+1)(x+2)(x+3)} > 1.$

**102.** $(x^2+3x+1)(x^2+3x-3) \geq 5.$

**103.** $(x^2-x-1)(x^2-x-7) < -5.$

**104.** $(x^2-2x)(2x-2)-9\dfrac{2x-2}{x^2-2x} \leq 0.$

**105.** $(x^2+3x)(2x+3)-16\dfrac{2x+3}{x^2+3x} \geq 0.$

**106.** $\dfrac{x^2-2x+2^{|a|}}{x^2-a^2} > 0.$

**107.** $|x^3-1| \geq 1-x.$

**108.** $\dfrac{x^2-5x+6}{|x|+7} < 0.$

**109.** $\dfrac{x^2+6x-7}{|x+4|} < 0.$

**110.** $\dfrac{|x-2|}{x-2} > 0.$

**111.** $\left|\dfrac{2}{x-4}\right| > 1.$

**112.** $\left|\dfrac{2x-1}{x-1}\right| > 2.$

**113.** $\left|\dfrac{x^2-3x-1}{x^2+x+1}\right| < 3.$

**114.** $\dfrac{x^2-7|x|+10}{x^2-6x+9} < 0.$

**115.** $\dfrac{|x+3|+x}{x+2} > 1.$

**116.** $\dfrac{|x-1|}{x+2} < 1.$

**117.** $\dfrac{|x+2|-x}{x} < 2.$

**118.** $\dfrac{1}{|x|-3} < \dfrac{1}{2}.$

**119.** $\left|\dfrac{3x}{x^2-4}\right| \leq 1.$

**120.** $\left|\dfrac{x^2-5x+4}{x^2-4}\right| \leq 1.$

**121.** $\dfrac{|x-3|}{x^2-5x+6} \geq 2.$

**122.** $\dfrac{x^2 - |x| - 12}{x - 3} \geq 2x.$

**123.** $|x| < \dfrac{a}{x}.$

**124.** $1 + \dfrac{12}{x^2} < \dfrac{7}{x}.$

**125.** $x - 17 \geq \dfrac{60}{x}.$

**126.** $\dfrac{x^2 - 4x + 5}{x^2 + 5x + 6} \geq 0.$

**127.** $\dfrac{x + 1}{x - 1} \geq \dfrac{x + 5}{x + 1}.$

**128.** $x \leq \dfrac{6}{x - 5}.$

**129.** $\dfrac{x - 1}{x^2 - x - 12} \leq 0.$

**130.** $\dfrac{30x - 9}{x - 2} \geq 25(x + 2).$

**131.** $1 < \dfrac{3x^2 - 7x + 8}{x^2 + 1} \leq 2.$

**132.** $f'(x) \geq g'(x),$ if $f(x) = 5 - 3x + \dfrac{5}{2}x^2 - \dfrac{x^3}{3},$ $g(x) = 3x - 7.$

**133.** $f'(x) \geq g'(x),$ if $f(x) = 10x^3 - 13x^2 + 7x,$ $g(x) = 11x^3 - 15x^2 - 3.$

**134.** $\dfrac{4}{x + 2} > 3 - x.$

**135.** $\dfrac{1}{x - 2} - \dfrac{1}{x} \leq \dfrac{2}{x + 2}.$

## 1.6.  Irrational Equations and Inequalities

When solving irrational equations, we must take into consideration the following theorem.

When $n$ is natural, the equation $\sqrt[2n]{f(x)} = \varphi(x)$ is equivalent to the system
$$\begin{cases} f(x) = (\varphi(x))^{2n}, \\ \varphi(x) \geq 0. \end{cases}$$

When solving irrational inequalities, we must take into consideration the following theorems.

When $n$ is natural, the inequality $\sqrt[2n]{f(x)} < \varphi(x)$ is equivalent to the system of inequalities
$$\begin{cases} f(x) < (\varphi(x))^{2n}, \\ f(x) \geq 0, \\ \varphi(x) > 0. \end{cases}$$

When $n$ is natural, the inequality $\sqrt[2n]{f(x)} > \varphi(x)$ is equivalent to the collection of two systems of inequalities

$$\begin{cases} \varphi(x) < 0, \\ f(x) \geq 0 \end{cases} \text{ and } \begin{cases} \varphi(x) \geq 0, \\ f(x) > (\varphi(x))^{2n}. \end{cases}$$

When $n$ is natural, the inequality $\dfrac{\sqrt[2n]{f(x)}}{\varphi(x)} > 1$ is equivalent to the system of inequalities

$$\begin{cases} \varphi(x) > 0, \\ f(x) > (\varphi(x))^{2n}. \end{cases}$$

When $n$ is natural, the inequality $\dfrac{\sqrt[2n]{f(x)}}{\varphi(x)} < 1$ is equivalent to the collection of two systems of inequalities

$$\begin{cases} \varphi(x) < 0, \\ f(x) \geq 0 \end{cases} \text{ and } \begin{cases} \varphi(x) > 0, \\ f(x) \geq 0, \\ f(x) < (\varphi(x))^{2n}. \end{cases}$$

Solve the following equations **(1-118).**

**1.** $(x^2 - 1)\sqrt{2x - 1} = 0.$

**2.** $(x^2 - 4)\sqrt{x + 1} = 0.$

**3.** $(9 - x^2)\sqrt{2 - x} = 0.$

**4.** $(16 - x^2)\sqrt{3 - x} = 0.$

**5.** $\sqrt{2x - 3} - \sqrt{x + 3} = 0.$

**6.** $\sqrt[3]{x} + 2\sqrt[3]{x^2} = 3.$

**7.** $\sqrt[3]{x^2} - \sqrt[3]{x} - 6 = 0.$

**8.** $\dfrac{4}{\sqrt[3]{x} + 2} + \dfrac{\sqrt[3]{x} + 3}{5} = 2.$

**9.** $\dfrac{8}{\sqrt{10 - 2x}} - \sqrt{10 - 2x} = 2.$

**10.** $\sqrt{2 - x} + \dfrac{4}{\sqrt{2 - x} + 3} = 2.$

**11.** $\dfrac{3}{\sqrt{x + 1} + 1} + 2\sqrt{x + 1} = 5.$

**12.** $\dfrac{x - 4}{\sqrt{x} + 2} = x - 8.$

**13.** $\dfrac{x\sqrt[3]{x} - 1}{\sqrt[3]{x^2} - 1} - \dfrac{\sqrt[3]{x^2} - 1}{\sqrt[3]{x} + 1} = 4.$

**14.** $\sqrt{\log_2 x} + \sqrt[3]{\log_2 x} = 2$.

**15.** $x\sqrt{x^2+15} - \sqrt{x}\,\sqrt[4]{x^2+15} = 2$.

**16.** $\sqrt{\dfrac{3-x}{2+x}} + 3\sqrt{\dfrac{2+x}{3-x}} = 4$.

**17.** $\sqrt{\dfrac{2x+1}{x-1}} - 2\sqrt{\dfrac{x-1}{2x+1}} = 1$.

**18.** $\sqrt{\dfrac{x+1}{x-1}} - \sqrt{\dfrac{x-1}{x+1}} = \dfrac{3}{2}$.

**19.** $\sqrt{12-x} = x$.

**20.** $\sqrt{7-x} = x-1$.

**21.** $x - \sqrt{x+1} = 5$.

**22.** $21 + \sqrt{2x-7} = x$.

**23.** $1 - \sqrt{1+5x} = x$.

**24.** $2\sqrt{x+5} = x+2$.

**25.** $4\sqrt{x+6} = x+1$.

**26.** $\sqrt{4+2x-x^2} = x-2$.

**27.** $\sqrt{37-x^2} + 5 = x$.

**28.** $\sqrt{6-4x-x^2} = x+4$.

**29.** $\sqrt{1+4x-x^2} = x-1$.

**30.** $\sqrt{5-x^2} = x-1$.

**31.** $\sqrt{x^2+8} = 2x+1$.

**32.** $4 + \sqrt{26-x^2} = x$.

**33.** $3x - \sqrt{18x+1} + 1 = 0$.

**34.** $\sqrt{6x-x^2} - 5 = 2x-6$.

**35.** $\dfrac{\sqrt{5-x^2}}{x+1} = 1$.

**36.** $\dfrac{1+\sqrt{2x+1}}{x} = 1$.

**37.** $\dfrac{\sqrt{13-x^2}}{x+1} = 1$.

**38.** $\dfrac{2+\sqrt{19-2x}}{x} = 1$.

**39.** $\sqrt{13-18\tan x} = 6\tan x - 3$.

**40.** $x^2 - 4x + 6 = \sqrt{2x^2 - 8x + 12}$.

**41.** $2x^2 + 3x - 5\sqrt{2x^2 + 3x + 9} + 3 = 0$.

**42.** $x^2 + \sqrt{x^2 + 2x + 8} = 12 - 2x$.

**43.** $2x^2 + \sqrt{2x^2 - 4x + 12} = 4x + 8$.

**44.** $3x^2 + 15x + 2\sqrt{x^2 + 5x + 1} = 2$.

**45.** $\sqrt[3]{16 - x^3} = 4 - x$

**46.** $\sqrt{x} - \dfrac{4}{\sqrt{2+x}} + \sqrt{2+x} = 0$.

**47.** $\sqrt{9 - 5x} = \sqrt{3 - x} + \dfrac{6}{\sqrt{3 - x}}$.

**48.** $\dfrac{4}{x + \sqrt{x^2 + x}} - \dfrac{1}{x - \sqrt{x^2 + x}} = \dfrac{3}{x}$.

**49.** $\sqrt{2x - 3} + \sqrt{4x + 1} = 4$.

**50.** $\sqrt{3x + 1} - \sqrt{x + 4} = 1$.

**51.** $\sqrt{2x + 6} - \sqrt{x + 1} = 2$.

**52.** $\sqrt{x + 5} - \sqrt{x} = 1$.

**53.** $\sqrt{2x - 4} - \sqrt{x + 5} = 1$.

**54.** $\sqrt{2x + 5} = 8 - \sqrt{x - 1}$.

**55.** $\sqrt{x + 3} + \sqrt{3x - 2} = 7$.

**56.** $\sqrt{3x + 7} - \sqrt{x + 1} = 2$.

**57.** $\sqrt{4 - x} + \sqrt{5 + x} = 3$.

**58\*.** $\sqrt{3x^2 + 6x + 7} + \sqrt{5x^2 + 10x + 14} = 4 - 2x - x^2$.

**59.** $\sqrt{3x - 5} = 3 - \sqrt{x - 2}$.

**60.** $\sqrt{x + 2} + \sqrt{3 - x} = 3$.

**61.** $\sqrt{4x + 8} - \sqrt{3x - 2} = 2$.

**62.** $\sqrt{2x + 3} + \sqrt{3x + 3} = 1$.

**63.** $\sqrt{x + 4} + \sqrt{2x + 6} = 7$.

**64.** $\sqrt{3x - 7} - \sqrt{x + 1} = 2$.

**65.** $\sqrt{15 - x} + \sqrt{3 - x} = 6$.

**66.** $\sqrt{x+5} - \sqrt{x-3} = 2$.

**67.** $2\sqrt{x-1} + \sqrt{x+3} = 2$.

**68.** $\sqrt{x + \sqrt{x+11}} + \sqrt{x - \sqrt{x+11}} = 4$.

**69\*.** $\sqrt[3]{12-x} + \sqrt[3]{14+x} = 2$.

**70.** $\dfrac{3(x-2) + 4\sqrt{2x^2 - 3x + 1}}{2(x^2 - 1)} = 1$.

**71.** $\dfrac{1}{1 - \sqrt{1-x}} - \dfrac{1}{1 + \sqrt{1-x}} = \dfrac{\sqrt{3}}{x}$.

**72.** $\dfrac{2 - \sqrt{x}}{2 - x} = \sqrt{\dfrac{2}{2-x}}$.

**73.** $\sqrt{1 + x\sqrt{x^2 + 24}} = x + 1$.

**74.** $1 + \sqrt{1 + x\sqrt{x^2 - 24}} = x$.

**75.** $\sqrt{x} + \sqrt{x - \sqrt{1-x}} = 1$.

**76.** $\sqrt{2x+1} + \sqrt{x-3} = 2\sqrt{x}$.

**77.** $\sqrt{x+1} + \sqrt{4x+13} = \sqrt{3x+12}$.

**78.** $\sqrt{x+2} - \sqrt{2x-3} = \sqrt{4x-7}$.

**79.** $\sqrt{x} + \sqrt{x-3} = \sqrt{3(x-1)}$.

**80.** $\sqrt{x-2} + \sqrt{4-x} = \sqrt{6-x}$.

**81.** $\sqrt{x+5} + \sqrt{2x-7} = 2\sqrt{x}$.

**82.** $\sqrt{x+1} + \sqrt{4x+13} = \sqrt{3x+12}$.

**83.** $\sqrt{3x+1} + \sqrt{x+4} = \sqrt{9-x}$.

**84.** $\sqrt{3x+4} + \sqrt{x-4} = 2\sqrt{x}$.

**85.** $\sqrt{2x+5} + \sqrt{5x+6} = \sqrt{12x+25}$.

**86.** $\sqrt{x+1} + \sqrt{x-1} = \sqrt{3x-1}$.

**87.** $\sqrt{x+1} - 1 = \sqrt{x - \sqrt{x+8}}$.

**88.** $\sqrt{x+3} - 1 = \sqrt{x - \sqrt{x-2}}$.

**89.** $\dfrac{\sqrt[7]{12+x}}{x} + \dfrac{\sqrt[7]{12+x}}{12} = \dfrac{64}{3}\sqrt[7]{x}$.

**90.** $\sqrt{\dfrac{20+x}{x}} + \sqrt{\dfrac{20-x}{x}} = \sqrt{6}$.

**91.** $\dfrac{(5-x)\sqrt{5-x}+(x-3)\sqrt{x-3}}{\sqrt{5-x}+\sqrt{x-3}}=2$.

**92.** $\sqrt{x^2+x+4}+\sqrt{x^2+x+1}=\sqrt{2x^2+2x+9}$.

**93.** $\sqrt{x^2-4x+3}+\sqrt{-x^2+3x-2}=\sqrt{x^2-x}$.

**94.** $\sqrt{x+3}+\sqrt{x+4}=\sqrt{x+2}+\sqrt{x+7}$.

**95.** $\sqrt{x-\sqrt{x-2}}+\sqrt{x+\sqrt{x-2}}=3$.

**96.** $\sqrt{x^2-2x+1}+\sqrt{x^2+2x+1}=2$.

**97.** $\sqrt{x^2+2x+1}-\sqrt{x^2-4x+4}=3$.

**98.** $\sqrt{x+2\sqrt{x-1}}=\sqrt{x-2\sqrt{x-1}}=2$.

**99.** $\sqrt{x+3-4\sqrt{x-1}}+\sqrt{x+8-6\sqrt{x-1}}=1$.

**100\*.** $x^3+1=2\sqrt[3]{2x-1}$.

**101.** $\dfrac{2+x}{\sqrt{2}+\sqrt{2+x}}+\dfrac{2-x}{\sqrt{2}-\sqrt{2+x}}=2\sqrt{2}$.

**102\*.** $x+\sqrt{a+\sqrt{x}}=a$.

**103\*.** $x^2-\sqrt{a-x}=a$.

**104.** $a\sqrt{x}-\sqrt{x+2ax\sqrt{x^2+7a^2}}=0$.

**105.** $x^2-4x+32=16\sqrt{x}$.

**106.** $\dfrac{a-2}{\sqrt{x+4}}=1$.

**107.** $\sqrt{\dfrac{20+x}{x}}-\sqrt{\dfrac{20-x}{x}}=\sqrt{6}$.

**108.** $\sqrt{\dfrac{2x}{x+1}}-\sqrt{\dfrac{2(x+1)}{x}}=1$.

**109.** $\dfrac{x}{x+1}-2\sqrt{\dfrac{x+1}{x}}=3$.

**110.** $\dfrac{\sqrt{x}}{\sqrt{1+x}}+\sqrt{\dfrac{1+x}{x}}=\dfrac{5}{2}$.

**111.** $\sqrt{x+8+2\sqrt{x+7}}+\sqrt{x+1-\sqrt{x+7}}=4$.

**112.** $\sqrt{3-x}+\dfrac{6}{\sqrt{3-x}}=\sqrt{9-5x}$.

**113.** $\left(\dfrac{5}{8}x\sqrt[5]{x^3}-\dfrac{10}{13}\sqrt[5]{x\sqrt{x^{11}}}\right)=56$.

**114.** $\sqrt{x+5}+\sqrt[4]{x+5}-12=0$.

**115.** $\sqrt{5x-5}+\sqrt{10x-5}=\sqrt{15x-10}$.

**116.** $x-1=\sqrt{\dfrac{x}{2}}$.

**117.** $\sqrt{x^2+8}=2x+1$.

**118.** $\dfrac{\sqrt{x+4}}{\sqrt{x-4}}-2\dfrac{\sqrt{x-4}}{\sqrt{x+4}}=\dfrac{7}{3}$.

**119\*.** How many roots does the equation $\sqrt{x^2+1}-\dfrac{1}{\sqrt{x^2-5/3}}=x$ possess? Find them.

**120.** Find the roots of the equation $\sqrt[10]{512}\sqrt{15x-21}-\sqrt{13}\times\sqrt[5]{15-6x}=0$ which can be represented by the reduced fraction $\dfrac{a}{6}$, where $a$ is an integer.

Solve the following inequalities **(121-212)**.

**121.** $(x-1)\sqrt{x^2-x-2}\geq 0$.

**122.** $(x^2-1)\sqrt{x^2-x-2}\geq 0$.

**123.** $\sqrt{\dfrac{x-2}{1-2x}}>-1$.

**124.** $\sqrt{\dfrac{3x-1}{2-x}}>1$.

**125.** $\dfrac{\sqrt{x-3}}{x-2}>0$.

**126.** $\sqrt{3x-10}>\sqrt{6-x}$.

**127.** $\sqrt{x^2+2x-3}<1$.

**128.** $\sqrt{1-\dfrac{x+2}{x^2}}<\dfrac{2}{3}$.

**129.** $\dfrac{3}{\sqrt{2-x}}-\sqrt{2-x}<2$.

**130.** $\dfrac{\sqrt{2x^2+15x-17}}{10-x}\geq 0$.

**131.** $\dfrac{x^2-13x+40}{\sqrt{19x-x^2-78}}\leq 0$.

**132.** $\sqrt{x^2}<x+1$.

**133.** $2\sqrt{x-1}<x$.

**134.** $\sqrt{x+18}<2-x$.

**135.** $x > \sqrt{24 - 5x}$.

**136.** $\sqrt{9x - 20} < x$.

**137.** $\sqrt{x + 7} < x$.

**138.** $\sqrt{2x - 1} < x - 2$.

**139.** $\sqrt{x + 78} < x + 6$.

**140.** $\sqrt{5 - 2x} < 6x - 1$.

**141.** $\sqrt{x + 61} < x + 5$.

**142.** $\sqrt{(x - 6)(1 - x)} < 3 + 2x$.

**143.** $\sqrt{2x - x^2} < 5 - x$.

**144.** $\sqrt{2x^2 - 3x - 5} < x - 1$.

**145.** $\sqrt{x^2 + 3x + 3} < 2x + 1$.

**146.** $\sqrt{x^2 - 3x - 10} < 8 - x$.

**147.** $x + 4 > 2\sqrt{4 - x^2}$.

**148.** $\sqrt{3x - x^2} < 4 - x$.

**149.** $3 - x > 3\sqrt{1 - x^2}$.

**150.** $\dfrac{1}{\sqrt{1 + x}} > \dfrac{1}{2 - x}$.

**151.** $1 - \sqrt{13 + 3x^2} > 2x$.

**152.** $x < \sqrt{2 - x}$.

**153.** $x + 3 < \sqrt{x + 33}$.

**154.** $\sqrt{x^2 - 1} > x$.

**155.** $\sqrt{2x + 14} > x + 3$.

**156.** $x - 3 < \sqrt{x - 2}$.

**157.** $x + 2 < \sqrt{x + 14}$.

**158.** $x - 1 < \sqrt{7 - x}$.

**159.** $\sqrt{9x - 20} > x$.

**160.** $\sqrt{11 - 5x} > x - 1$.

**161.** $\sqrt{x + 2} > x$.

**162.** $\sqrt{x^2 + 1} > x - 1$.

**163.** $\sqrt{(x+4)(x+3)} > 6 - x.$

**164.** $\sqrt{(x+3)(x-8)} > x + 2.$

**165.** $1 - x < \sqrt{x^2 - 2x}.$

**166.** $\sqrt{5 - x^2} > x - 1.$

**167.** $x - \sqrt{1 - |x|} < 0.$

**168.** $4 - x < \sqrt{x^2 - 2x}.$

**169.** $x < \sqrt{x^2 + x - 2}.$

**170.** $4 - x < \sqrt{2x - x^2}.$

**171.** $x - 3 < \sqrt{x^2 + 4x - 5}.$

**172.** $2x + 3 < \sqrt{x^2 + 5x + 6}.$

**173.** $\sqrt{x^2 - 3x + 2} > 2x - 5.$

**174.** $\sqrt{x^2 + x} > 1 - 2x.$

**175.** $\sqrt{8 + 2x - x^2} > 6 - 3x.$

**176.** $2x + 3 < \sqrt{-2 - 3x - x^2}.$

**177.** $x + 4 < \sqrt{-x^2 - 8x - 12}.$

**178.** $\sqrt{-x^2 + 6x - 5} > 8 - 2x.$

**179.** $\sqrt{\dfrac{1}{x^2} - \dfrac{3}{x}} < \dfrac{1}{x} - \dfrac{1}{2}.$

**180.** $x^2 \geq x\,(2 + \sqrt{12 - 2x - x^2}\,).$

**181.** $\sqrt{4 - x^2} + \dfrac{|x|}{x} \geq 0.$

**182.** $\sqrt{x} - 3 \leq \dfrac{2}{\sqrt{x} - 2}.$

**183.** $x - 3\sqrt{x - 3} - 1 > 0.$

**184.** $\dfrac{\sqrt{a + 4}}{1 - a} - 1 < 0.$

**185.** $\dfrac{\sqrt{2x - 1}}{x - 2} < 1.$

**186.** $\dfrac{\sqrt{x + 20}}{x} - 1 < 0.$

**187.** $\dfrac{\sqrt{2x^2 + 7x - 4}}{x + 4} < \dfrac{1}{2}.$

**188.** $\dfrac{1-\sqrt{21-4x-x^2}}{x+1} \geq 0.$

**189.** $\sqrt{4-\sqrt{1-x}} - \sqrt{2-x} > 0.$

**190.** $\sqrt{1-x^2} + 1 < \sqrt{3-x^2}.$

**191.** $3\sqrt{x} - \sqrt{x+3} > 1.$

**192.** $3\sqrt{x} - \sqrt{5x+5} > 1.$

**193.** $\sqrt{x+3} + \sqrt{x+15} < 6.$

**194.** $2 - \sqrt{1-x^2}\,\sqrt{4-x^2}.$

**195.** $\sqrt{x+3} + \sqrt{x+2} - \sqrt{2x+4} > 0.$

**196.** $\sqrt{x-6} - \sqrt{10-x} \geq 1.$

**197.** $\sqrt{x+3} - \sqrt{x-1} > \sqrt{2x-1}.$

**198.** $\sqrt{x+3} < \sqrt{x-1} + \sqrt{x-2}.$

**199.** $\sqrt{3x^2+5x+7} - \sqrt{3x^2+5x+2} > 1.$

**200.** $\sqrt{x+2\sqrt{x-1}} + \sqrt{x-2\sqrt{x-1}} > \dfrac{3}{2}.$

**201.** $\sqrt{5+x} - \sqrt{-x-3} < 1 + \sqrt{(x+5)(-x-3)}.$

**202.** $\dfrac{1-\sqrt{1-4x^2}}{x} < 3.$

**203.** $\sqrt{x+\dfrac{1}{x^2}} + \sqrt{x-\dfrac{1}{x^2}} > \dfrac{2}{x}.$

**204.** $a\sqrt{x+1} < 1.$

**205.** $(a+1)\sqrt{2-x} < 1.$

**206.** $\dfrac{2}{x} + 3 \leq \sqrt{41 - \dfrac{16}{x}}.$

**207.** $\sqrt{x^2-4x} > x-3.$

**208.** $\sqrt{-x^2+x+2} + 2x + 1 > 0.$

**209.** $\dfrac{\sqrt{x-5}}{\log_{\sqrt{2}}(x-4)-1} \geq 0.$

**210.** $\dfrac{|\,x+2|-|\,x\,|}{\sqrt{4-x^3}} \geq 0.$

**211.** $\left(\dfrac{1}{3}\right)^{\sqrt{x+2}} < 3^{-x}.$

**212\*.** $x + \dfrac{x}{\sqrt{x^2-1}} > \dfrac{35}{12}.$

**213.** Find the midpoint of the interval in which the inequality $2x^2 - 7\,(\sqrt{x}\,)^2 \le 4$ is satisfied.

**214.** Find the least integral positive value of $x$ which satisfies the inequality $\sqrt{x^2 + 16x + 64} > 20$.

**215.** Find the integral number which satisfies the inequality $2\sqrt{2x+1} > 3\sqrt{-x^2 - x + 6}$.

## 1.7. Systems of Equations and Inequalities

The *solution* of the system of equations (inequalities)

$$\begin{cases} f_1\,(x, y) = \varphi_1\,(x, y) \\ f_2\,(x, y) = \varphi_2\,(x, y) \end{cases} \qquad \left( \begin{cases} f_1\,(x, y) < \varphi_1\,(x, y) \\ f_2\,(x, y) < \varphi_2\,(x, y) \end{cases} \right)$$

is a pair of numbers $(x_0, y_0)$ such that the system of numerical equations (inequalities)

$$\begin{cases} f_1\,(x_0, y_0) = \varphi_1\,(x_0, y_0) \\ f_2\,(x_0, y_0) = \varphi_2\,(x_0, y_0) \end{cases} \left( \begin{cases} f_1\,(x_0, y_0) < \varphi_1\,(x_0, y_0) \\ f_2\,(x_0, y_0) < \varphi_2\,(x_0, y_0) \end{cases} \right)$$

holds true.

The *solution* of the system of equations (inequalities)

$$\begin{cases} f_1\,(x,\ y,\ z) = \varphi_1\,(x,\ y,\ z) \\ f_2\,(x,\ y,\ z) = \varphi_2\,(x,\ y,\ z) \\ f_3\,(x,\ y,\ z) = \varphi_3\,(x,\ y,\ z) \end{cases} \left( \begin{cases} f_1\,(x,\ y,\ z) < \varphi_1\,(x,\ y,\ z) \\ f_2\,(x,\ y,\ z) < \varphi_2\,(x,\ y,\ z) \\ f_3\,(x,\ y,\ z) < \varphi_3\,(x,\ y,\ z) \end{cases} \right)$$

is a triple of numbers $(x_0, y_0, z_0)$ such that the system of numerical equations (inequalities)

$$\begin{cases} f_1\,(x_0,\ y_0,\ z_0) = \varphi_1\,(x_0,\ y_0,\ z_0) \\ f_2\,(x_0,\ y_0,\ z_0) = \varphi_2\,(x_0,\ y_0,\ z_0) \\ f_3\,(x_0,\ y_0,\ z_0) = \varphi_3\,(x_0,\ y_0,\ z_0) \end{cases}$$

$$\left( \begin{cases} f_1\,(x_0,\ y_0,\ z_0) < \varphi_1\,(x_0,\ y_0,\ z_0) \\ f_2\,(x_0,\ y_0,\ z_0) < \varphi_2\,(x_0,\ y_0,\ z_0) \\ f_3\,(x_0,\ y_0,\ z_0) < \varphi_3\,(x_0,\ y_0,\ z_0) \end{cases} \right)$$

holds true.

The number of the unknowns can differ from that of the equations in the system.

To solve a system of equations (inequalities) is to find the whole set of solutions or to prove that the system has no solution.

Solve the following systems of equations **(1-77)**.

**1.** $\begin{cases} y + x - 1 = 0, \\ |\,y\,| - x - 1 = 0. \end{cases}$

**2.** $\begin{cases} x + 3\,|\,y\,| - 1 = 0, \\ x + y + 3 = 0. \end{cases}$

**3.** $\begin{cases} y - 2x + 1 = 0, \\ y - |\,x\,| - 1 = 0. \end{cases}$

4. $\begin{cases} |x-1|+y = 0, \\ 2x - y = 1. \end{cases}$

5. $\begin{cases} x + 2y - 6 = 0, \\ |x - 3| - y = 0. \end{cases}$

6. $\begin{cases} u + v = 2, \\ |3u - v| = 1. \end{cases}$

7. $\begin{cases} u + 2v = 2, \\ |2u - 3v| = 1. \end{cases}$

8. $\begin{cases} |x| + 2|y| = 3, \\ 5y + 7x = 2. \end{cases}$

9. $\begin{cases} |x - y| = 12y - 11, \\ y + 1 = 2x. \end{cases}$

10. $\begin{cases} |x - 1| + |y - 2| = 1, \\ y = 3 - |x - 1|. \end{cases}$

11. $\begin{cases} x + y - z = 2, \\ 2x - y + 4z = 1, \\ -x + 6y + z = 5. \end{cases}$

12. $\begin{cases} 2x + 3y - z = 6, \\ x - y + 7z = 8, \\ 3x - y + 2z = 7. \end{cases}$

13. $\begin{cases} x + 2y - z = 7, \\ 2x - y + z = 2, \\ 3x - 5y + 2z = -7. \end{cases}$

14. $\begin{cases} xy + x + y = 11, \\ x^2 y + xy^2 = 30. \end{cases}$

15. $\begin{cases} xy + 4 = 0, \\ x + y = 3. \end{cases}$

16. $\begin{cases} x^2 - y^2 = 16, \\ x + y = 8. \end{cases}$

17. $\begin{cases} x^2 + y^2 = 41, \\ y - x = 1. \end{cases}$

18. $\begin{cases} x^2 + y^2 = 41, \\ x + y = 9. \end{cases}$

**19.** $\begin{cases} x + y = 7, \\ y = \dfrac{6}{x}. \end{cases}$

**20.** $\begin{cases} x^2 + y^2 + 6x + 2y = 0, \\ x + y + 8 = 0. \end{cases}$

**21.** $\begin{cases} \dfrac{x}{y} + \dfrac{y}{x} = \dfrac{13}{6}, \\ x + y = 5. \end{cases}$

**22.** $\begin{cases} \dfrac{x}{y} - \dfrac{y}{x} = \dfrac{5}{6}, \\ x^2 - y^2 = 5. \end{cases}$

**23.** $\begin{cases} x^2 - xy + y^2 = 7, \\ x + y = 5. \end{cases}$

**24.** $\begin{cases} \dfrac{x+3}{y-4} - \dfrac{x-1}{y+4} + \dfrac{16}{y^2 - 16} = 0, \\ 11x - 3y = 1. \end{cases}$

**25.** $\begin{cases} \dfrac{3x + y}{x - 1} - \dfrac{x - y}{2y} = 2, \\ x - y = 4. \end{cases}$

**26.** $\begin{cases} \dfrac{x+y}{x-y} + \dfrac{x-y}{x+y} = \dfrac{13}{6}, \\ xy = 5. \end{cases}$

**27.** $\begin{cases} \dfrac{x+y}{x-y} + 6\dfrac{x-y}{x+y} = 5, \\ xy = 2. \end{cases}$

**28.** $\begin{cases} \dfrac{1}{3x} - \dfrac{1}{2y} = \dfrac{1}{3}, \\ \dfrac{1}{9x^2} - \dfrac{1}{4y^2} = \dfrac{1}{4}. \end{cases}$

**29.** $\begin{cases} \dfrac{1}{x+1} + \dfrac{1}{y} = \dfrac{1}{3}, \\ \dfrac{1}{(x+1)^2} - \dfrac{1}{y^2} = \dfrac{1}{4}. \end{cases}$

**30.** $\begin{cases} \dfrac{1}{2x - y} + y = -5, \\ \dfrac{y}{2x - y} = 6. \end{cases}$

**31.** $\begin{cases} \dfrac{1}{2x + y} + x = 3, \\ \dfrac{x}{2x + y} = -4. \end{cases}$

**32.** $\begin{cases} \dfrac{1}{x+2y} + y = 2, \\ \dfrac{y}{x+2y} = -3. \end{cases}$

**33.** $\begin{cases} \dfrac{1}{x-y} + x = 1, \\ \dfrac{x}{x-y} = -2. \end{cases}$

**34.** $\begin{cases} \dfrac{1}{y-1} - \dfrac{1}{y+1} = \dfrac{1}{x}, \\ y^2 - x - 5 = 0. \end{cases}$

**35.** $\begin{cases} \dfrac{x+y}{x-y} + \dfrac{x-y}{x+y} = \dfrac{5}{2}, \\ x^2 + y^2 = 20. \end{cases}$

**36.** $\begin{cases} 2xy - 3\dfrac{x}{y} = 15, \\ xy + \dfrac{x}{y} = 15. \end{cases}$

**37.** $\begin{cases} xy - \dfrac{x}{y} = \dfrac{16}{3}, \\ xy - \dfrac{y}{x} = \dfrac{9}{2}. \end{cases}$

**38.** $\begin{cases} x^2 + y^2 = 20, \\ xy = 8. \end{cases}$

**39.** $\begin{cases} x^2 + y^2 = 68, \\ xy = 16. \end{cases}$

**40.** $\begin{cases} x(x+y) = 9, \\ y(x+y) = 16. \end{cases}$

**41.** $\begin{cases} x^2 + xy = 15, \\ y^2 + xy = 10. \end{cases}$

**42.** $\begin{cases} y^2 + xy = 231, \\ x^2 + xy = 210. \end{cases}$

**43.** $\begin{cases} x^2 - xy = 28, \\ y^2 - xy = -12. \end{cases}$

**44.** $\begin{cases} y^2 - xy = 12, \\ x^2 - xy = 28. \end{cases}$

**45.** $\begin{cases} x^2 + y^2 = 25 - 2xy, \\ y(x+y) = 10. \end{cases}$

**46.** $\begin{cases} 5(x+y)+2xy = -19, \\ 15xy + 5(x+y) = -175. \end{cases}$

**47.** $\begin{cases} 5(x+y)+2xy = -19, \\ 3xy + x + y = -35. \end{cases}$

**48.** $\begin{cases} 4x^2 + y^2 - 2xy = 7, \\ (2x-y)y = y. \end{cases}$

**49.** $\begin{cases} x+y+xy = 5, \\ x^2 + y^2 + xy = 7. \end{cases}$

**50.** $\begin{cases} \dfrac{1}{x} + \dfrac{1}{y} = \dfrac{1}{3}, \\ x^2 + y^2 = 160. \end{cases}$

**51.** $\begin{cases} 2y^2 - 4xy + 3x^2 = 17, \\ y^2 - x^2 = 16. \end{cases}$

**52.** $\begin{cases} x^2 - xy + y^2 = 21, \\ y^2 - 2xy + 15 = 0. \end{cases}$

**53.** $\begin{cases} 2x^2 + y^2 + 3xy = 12, \\ 2(x+y)^2 - y^2 = 14. \end{cases}$

**54.** $\begin{cases} x^2 + y^2 - xy = 13, \\ x + y - \sqrt{xy} = 3. \end{cases}$

**55.** $\begin{cases} 2y^2 + xy - x^2 = 0, \\ x^2 - xy - y^2 + 3x + 7y + 3 = 0. \end{cases}$

**56.** $\begin{cases} xy + 3y^2 - x + 4y - 7 = 0, \\ 2xy + y^2 - 2x - 2y + 1 = 0. \end{cases}$

**57.** $\begin{cases} 2xy + y^2 - 4x - 3y + 2 = 0, \\ xy + 3y^2 - 2x - 14y + 16 = 0. \end{cases}$

**58.** $\begin{cases} 3x^2 + xy - 2x + y - 5 = 0, \\ 2x^2 - xy - 3x - y - 5 = 0. \end{cases}$

**59.** $\begin{cases} x^3 + y^3 = 35, \\ x + y = 5. \end{cases}$

**60.** $\begin{cases} x - y = 1, \\ x^3 - y^3 = 7. \end{cases}$

**61.** $\begin{cases} x^3 + y^3 = 7, \\ xy(x+y) = -2. \end{cases}$

**62.** $\begin{cases} \dfrac{x^2}{y} + \dfrac{y^2}{x} = 18, \\ x + y = 12. \end{cases}$

**63.** $\begin{cases} x^4 + y^4 = 82, \\ xy = 3. \end{cases}$

**64.** $\begin{cases} x^3 + y^3 = 7, \\ x^3 y^3 = -8. \end{cases}$

**65.** $\begin{cases} (x^2 + y^2)xy = 78, \\ x^4 + y^4 = 97. \end{cases}$

**66.** $\begin{cases} 3xy - x^2 - y^2 = 5, \\ 7x^2 y^2 - x^4 - y^4 = 155. \end{cases}$

**67.** $\begin{cases} x + y + z = 13, \\ x^2 + y^2 + z^2 = 91, \\ y^2 = xz. \end{cases}$

**68.** $\begin{cases} \dfrac{xy}{x + y} = 1, \\ \dfrac{xz}{x + z} = 2, \\ \dfrac{yz}{y + z} = 3. \end{cases}$

**69.** $\begin{cases} y^3 - 9x^2 + 27x - 27 = 0, \\ z^3 - 9y^2 + 27y - 27 = 0, \\ x^3 - 9z^2 + 27z - 27 = 0. \end{cases}$

**70.** $\begin{cases} 2y^3 + 2x^2 + 3x + 3 = 0, \\ 2z^3 + 2y^2 + 3y + 3 = 0, \\ 2x^3 + 2z^2 + 3z + 3 = 0. \end{cases}$

**71.** $\begin{cases} y^3 - 6x^2 + 12x - 8 = 0, \\ z^3 - 6y^2 + 12y - 8 = 0, \\ x^3 - 6z^2 + 12z - 8 = 8. \end{cases}$

**72.** $\begin{cases} 2x + 3y = 4, \\ 5x + 4y = 3. \end{cases}$

**73.** $\begin{cases} 4x - y + 4z = 0, \\ x + 5y - 2z = 3, \\ -x + 8y - 2z = 1. \end{cases}$

**74.** $\begin{cases} x^2 y^2 - 2x + y^2 = 0, \\ 2x^2 - 4x + 3 + y^3 = 0. \end{cases}$

**75.** $\begin{cases} 6x + y^2 - z^2 = 6, \\ x^2 - y - 4z = -4, \\ 21x^2 - 2y^2 + 3y = 22z^2. \end{cases}$

**76.** $\begin{cases} x^2 = (x-a)y, \\ y^2 - xy = 9ax. \end{cases}$

**77.** $\begin{cases} x^2 + 2xy + y^2 - x - y = 6, \\ x - 2y = 3. \end{cases}$

**78.** Find all values of $a$ for each of which the system of equations
$$\begin{cases} 2x + 2(a-1)y = a-4, \\ 2|x+1| + ay = 2. \end{cases}$$
has a unique solution. Find that solution.

**79\*.** Find all values of $a$ for each of which the system of equations
$$\begin{cases} ax + (a-1)y = 2+4a, \\ 3|x| + 2y = a-5 \end{cases}$$
has a unique solution. Find that solution.

For what values of $a$ does the following systems of equations have solutions? Find those solutions **(80-83).**

**80.** $\begin{cases} x + ay = 1, \\ ax + y = 2a. \end{cases}$

**81.** $\begin{cases} (a+1)x - y = a+1, \\ x + (a-1)y = 2. \end{cases}$

**82.** $\begin{cases} ax + y = a, \\ x + ay = a^2. \end{cases}$

**83.** $\begin{cases} x + ay = 1, \\ ax + y = a^2. \end{cases}$

For what values of $a$ does each of the following systems of equations **(84-85)** have an infinite number of solutions?

**84.** $\begin{cases} 3x + ay = 3, \\ ax + 3y = 3. \end{cases}$

**85.** $\begin{cases} (a-2)x + 27y = 4.5, \\ 2x + (a+1)y = -1. \end{cases}$

**86.** For what value of $a$ is the sum of the squares of the numbers, constituting the solution of the system of equations

$$\begin{cases} 3x - y = 2 - a, \\ x + 2y = a + 1, \end{cases}$$

the least?

Solve the following systems of equations and investigate them with respect to $a$ (87–90).

**87.** $\begin{cases} 2x + 3y = 5, \\ x - y = 2, \\ x + 4y = a. \end{cases}$

**88.** $\begin{cases} 2\sqrt{x} - 2 \arccos y + z = 1, \\ 5\sqrt{x} + \arccos y + z = 6a - 14, \\ \sqrt{x} + \arccos y + 2z = 2a + 1. \end{cases}$

**89.** $\begin{cases} 3 \cdot 2^x + 2y - 3 \arcsin z = 7, \\ 2^x - y - \arcsin z = -6, \\ 5 \cdot 2^x - y + \arcsin z = 6a + 2. \end{cases}$

**90.** $\begin{cases} 2\sqrt{\tan x} + 2y^2 + z = 14, \\ 3\sqrt{\tan x} - y^2 + 2z = 20 - 4a, \\ \sqrt{\tan x} + y^2 + z = 10. \end{cases}$

For what values of $a$ is there at least one $c$, for any $b$, such that each of the following systems (91-94) has at least one solution?

**91\*.** $\begin{cases} 2x + by = ac^2 + c, \\ bx + 2y = c - 1. \end{cases}$

**92\*.** $\begin{cases} x + by = ac^2 + c, \\ bx + 2y = c - 1. \end{cases}$

**93\*.** $\begin{cases} 2x + by = c^2, \\ bx + 2y = ac - 1. \end{cases}$

**94\*.** $\begin{cases} bx + y = ac^2, \\ x + by = ac + 1. \end{cases}$

**95.** For what values of $k$ do all the solutions of the system of equations

$$\begin{cases} x + ky = 3, \\ kx + 4y = 6 \end{cases}$$

satisfy the conditions $x > 1, \ y > 0$?

**96.** Find all values of $b$ for which $x$ and $y$ satisfying the system of equations

$$\begin{cases} 2x - y = 5b, \\ 4x + y = 3b^2 - 10b \end{cases}$$

also satisfy the inequality $x - y > 3.$

**97.** For what real values of $n$ do the solutions of the system of equations

$$\begin{cases} x - 2y = n, \\ 2nx - 9y = -2 \end{cases}$$

satisfy the conditions $x > 1 / 2n, \;\; y > 0$?

**98.** Find the values of $a$ for which the numbers 1 and $a$ lie between the numbers $x_0$ and $y_0$, where $(x_0, y_0)$ is a solution of the system of equations

$$\begin{cases} x + y - 1 = 2a, \\ 2xy = a^2 - a. \end{cases}$$

**99.** Solve the system of equations

$$\begin{cases} \dfrac{9a}{y} - \dfrac{72a}{x} = \dfrac{a-12}{a-6}, \\ 9x + 6ay = a^2 y; \end{cases}$$

and find the values of $a$ for which all the solutions $(x, y)$ satisfy the inequality $x + y > 0.$

**100.** Find the greatest value of $x$ which satisfies the system of equations

$$\begin{cases} x^3 + y^3 = 35, \\ x^2 y + xy^2 = 30. \end{cases}$$

**101.** We call a table of the form

$$\begin{pmatrix} a & b \\ c & d \end{pmatrix}$$

a matrix, where $a, b, c,$ and $d$ are real numbers. If

$$A = \begin{pmatrix} a & b \\ c & d \end{pmatrix}, \quad B = \begin{pmatrix} l & m \\ n & p \end{pmatrix},$$

then $A = B$ means that $a = l, \;\; b = m, \;\; c = n,$ and $d = p.$ The product $A \cdot B$ of the matrices $A$ and $B$ is a matrix

$$\begin{pmatrix} al + bn & am + bp \\ cl + dn & cm + dp \end{pmatrix}$$

(a) Find the product of the matrices $\begin{pmatrix} 2 & 1 \\ 1 & 3 \end{pmatrix}$ and $\begin{pmatrix} 1 & 2 \\ 2 & 3 \end{pmatrix}$.

(b) Find the matrix $X$ such that $\begin{pmatrix} 1 & 2 \\ 2 & 3 \end{pmatrix} \cdot X = \begin{pmatrix} 4 & 7 \\ 7 & 11 \end{pmatrix}$.

## Systems of Irrational Equations

Solve the following systems of equations (**102-120**).

**102.** $\begin{cases} x + y + \sqrt{x+y} = 20, \\ x^2 + y^2 = 136. \end{cases}$

**103.** $\begin{cases} \sqrt{\dfrac{x}{y}} + \sqrt{\dfrac{y}{x}} = \dfrac{5}{2}, \\ x + y = 5. \end{cases}$

**104.** $\begin{cases} \sqrt{\dfrac{2x-1}{y+2}} + \sqrt{\dfrac{y+2}{2x-1}} = 2, \\ x + y = 2. \end{cases}$

**105.** $\begin{cases} \sqrt{\dfrac{x}{y}} + \sqrt{\dfrac{y}{x}} = \dfrac{5}{2}, \\ x^2 + y^2 = 15. \end{cases}$

**106.** $\begin{cases} \sqrt{\dfrac{x}{y}} - \sqrt{\dfrac{y}{x}} = \dfrac{3}{2}, \\ x + y + xy = 9. \end{cases}$

**107.** $\begin{cases} \sqrt{x^2} + y = 5, \\ y^2 - x = 7. \end{cases}$

**108.** $\begin{cases} 4^{\frac{x}{y} - \frac{3y}{x}} = 16, \\ \sqrt{x} - \sqrt{2y} = \sqrt{12} - \sqrt{8}. \end{cases}$

**109.** $\begin{cases} \sqrt[3]{x} - \sqrt[3]{y} = 2, \\ xy = 27. \end{cases}$

**110.** $\begin{cases} x + y = 9, \\ x^{\frac{1}{3}} + y^{\frac{1}{3}} = 3. \end{cases}$

**111.** $\begin{cases} \sqrt[3]{x} + \sqrt[3]{y} = 4, \\ x + y = 28. \end{cases}$

**112.** $\begin{cases} \sqrt{x} + \sqrt{y} = 10, \\ \sqrt[4]{x} + \sqrt[4]{y} = 4. \end{cases}$

**113.** $\begin{cases} \sqrt{x+y} + \sqrt{2x+y+2} = 7, \\ 3x + 2y = 23. \end{cases}$

**114.** $\begin{cases} x\sqrt{y} + y\sqrt{x} = 6, \\ x^2 y + y^2 x = 20. \end{cases}$

**115.** $\begin{cases} \sqrt{\dfrac{20y}{x}} = \sqrt{x+y} + \sqrt{x-y}, \\ \sqrt{\dfrac{16x}{5y}} = \sqrt{x+y} - \sqrt{x-y}. \end{cases}$

**116.** $\begin{cases} x\sqrt{(x+y)^2} = 3x, \\ x\left(\sqrt{(x-y)^2} - 1\right)^2 = 0. \end{cases}$

**117.** $\begin{cases} (x-y) + \sqrt{\dfrac{x-y}{2(x+y)}} = \dfrac{10}{x+y}, \\ x^2 + y^2 = 17. \end{cases}$

**118.** $\begin{cases} \sqrt[3]{x} + \sqrt[3]{y} = 3, \\ \sqrt[3]{x^2} - \sqrt[3]{xy} + \sqrt[3]{y^2} = 3. \end{cases}$

**119.** $\begin{cases} \sqrt{7x+y} + \sqrt{x+y} = 6, \\ \sqrt{x+y} - y + x = 2. \end{cases}$

**120.** $\begin{cases} x - y + \sqrt{x^2 - 4y^2} = 2, \\ x^5 \sqrt{x^2 - 4y^2} = 0. \end{cases}$

**121.** For what values of $a$ do the curves $y = 1 + \dfrac{x^2}{a^3}$ and $y = 4\sqrt{x}$ possess only one point in common?

## Systems of Inequalities in One Unknown

**122.** Solve the system of inequalities
$$\begin{cases} 2(3x-1) < 3(4x+1)+16, \\ 4(2+x) < 3x+8. \end{cases}$$

**123.** Find the greatest integral $x$ which satisfies the system of inequalities
$$\begin{cases} 0.5(2x-5) > \dfrac{2-x}{2} + 1, \\ 0.2(3x-2) + 3 > \dfrac{4x}{3} - 0.5(x-1). \end{cases}$$

**124.** Find natural $x$ which satisfy the system of inequalities

$$\begin{cases} x + 3 < 4 + 2x, \\ 5x - 3 < 4x - 1. \end{cases}$$

**125.** Indicate the integral part of the numbers $x$ which satisfy the system of inequalities

$$\begin{cases} \dfrac{x-1}{2} - \dfrac{2x+3}{3} + \dfrac{x}{6} < 2 - \dfrac{x+5}{2}, \\ 1 - \dfrac{x+5}{8} + \dfrac{4-x}{2} < 3x - \dfrac{x+1}{4}. \end{cases}$$

**126.** Find all values of $a$ for which the equation

$$\cos x = \dfrac{a - 1.5}{2 - 0.5a}$$

possesses solutions.

**127.** Find all values of $\alpha$ for which the system of inequalities

$$\begin{cases} x^2 + 2x + \alpha \le 0, \\ x^2 - 4x - 6\alpha \le 0 \end{cases}$$

has a unique solution.

**128\*.** Find $a < 0$ for which the inequalities $2\sqrt{ax} < 3a - x$ and $x - \sqrt{\dfrac{x}{a}} > \dfrac{6}{a}$ have solutions in common.

**129.** Find all values of $k$ for which there is at least one common solution of the inequalities $x^2 + 4kx + 3k^2 > 1 + 2k$ and $x^2 + 2kx \le 3k^2 - 8k + 4$.

**130.** Find all values of $k$ for which every solution of the inequality $x^2 + 3k^2 - 1 \ge 2k\,(2x - 1)$ is a solution of the inequality $x^2 - (2x - 1)\,k + k^2 \ge 0$.

Solve the following two-sided inequalities **(131-133)**.

**131.** $0 < \dfrac{3x - 1}{2x + 5} < 1$.

**132.** $1 \le \dfrac{2 - x}{x + 1} \le 2$.

**133.** $1 < \dfrac{3x - 1}{2x + 1} < 2$.

Solve the following systems of inequalities **(134-141)**.

**134.** $\begin{cases} x^2 - x - 6 \ge 0, \\ x^2 - 4x < 0. \end{cases}$

**135.** $\begin{cases} x^2 - 4 < 0, \\ x + 1 > 0, \\ \dfrac{1}{2} - x > 0. \end{cases}$

**136.** $\begin{cases} \dfrac{x}{3} - \dfrac{4}{3} \le \dfrac{4}{x}, \\ \dfrac{1}{x} > -1, \\ x^2 + 3x + 1 > 0. \end{cases}$

**137.** $\begin{cases} \dfrac{1}{3x} < 1, \\ x + \dfrac{4}{3} \ge \dfrac{4}{3x}, \\ 9x^2 - 9x + 1 < 0. \end{cases}$

**138.** $\begin{cases} |x| \ge 1, \\ |x-1| < 3. \end{cases}$

**139.** $\begin{cases} |x^2 - 4x| < 5, \\ |x+1| < 3. \end{cases}$

**140.** $\begin{cases} |x^2 + 5x| < 6, \\ |x+1| \le 1. \end{cases}$

**141.** $\begin{cases} |x^2 + 5x| < 6, \\ |x+1| < 2. \end{cases}$

## Systems of Inequalities in Two Unknowns

**142.** Given a system of inequalities
$$\begin{cases} 2x - y > a, \\ 3x + 2y > 3a. \end{cases}$$
(a) Indicate at least one solution of the system for $a = 0$.
(b) Is it true that all solutions of the system satisfy the inequality $5x + y > 4a$?
(c) Is it true that all solutions of the system satisfy the inequality $x + 3y > 2a$?

**143.** For what $a \in R$ does the point $(a, a^2)$ is inside the triangle bounded by the straight lines $y = x + 1$, $y = 3 - x$ and $y = -2x$?

**144.** Find all values of $a$ for which the system of inequalities
$$\begin{cases} x^2 + 2xy - 7y^2 \ge \dfrac{1-a}{a+1}, \\ 3x^2 + 10xy - 5y^2 \le -2 \end{cases}$$
possesses a solution.

**145.** Depict on a plane a set of points whose coordinates satisfy the equation $|x| + |y| = 1$.

On the coordinate plane $xOy$ hatch a set of points whose coordinates satisfy the following inequalities in two variables $x$ and $y$ (show the boundary belonging to the set by a solid line and that not belonging to the set by a dash line) **(146-185).**

**146.** $x^2 - x < y - xy$.    **147.** $|xy| < 1$.    **148.** $|y| + \dfrac{1}{2} \le e^{-|x|}$.

**149.** $|x^2 + y^2 - 2| \le 2(x + y)$.      **150.** $|y| + 2|x| \le x^2 + 1$.

**151.** $x^2 + y^2 \le 2(|x| + |y|)$.      **152.** $4 \le x^2 + y^2 \le 2(|x| + |y|)$.

**153.** $\dfrac{y - |x|}{xy^2} \ge 0$.

**154.** $\dfrac{y - 1 + |x - 1|}{y - x^2 + 2x} \le 0$.

**155.** $x + y - 2 \le 0$,   $2y + 5x \ge 10$,   $5x - 2y - 10 \le 0$.

**156.** $3x + 2y + 1 \ge 0$,   $3x + 2y - 3 \le 0$.

**157.** $x + y \le 1$,   $-x - y \le 1$.

**158.** $x - y \le 1$,   $x + y \le 2$,   $y \le 2$.

**159.** $2y - x \le 6$,   $9x + 4y \le 56$,   $3x + 5y \ge 4$.

**160.** $x - 3y + 13 \le 0$,   $y + 5 \le 5x$,   $4y + 28 \ge 7x$.

**161.** $\log_{\frac{1}{3}} (2x + y - 2) > \log_{\frac{1}{3}} (y + 1)$,   $\sqrt{y - 2x - 3} < \sqrt{3 - 2x}$.

**162.** $\log_{\frac{1}{2}} (x + y - 1) > \log_{\frac{1}{2}} y$,   $\sqrt{y - x - 1} < \sqrt{2 - x}$.

**163.** $x^2 + y^2 \ge 1$,   $x^2 + y^2 \le 16$.      **164.** $x^2 + y^2 \le 9$,   $x + y \ge 0$.

**165.** $y^2 \le 4 - x^2$,   $3x + 2y - 3 \le 0$.

**166.** $x^2 - 4x - y + 3 \le 0$,   $2x - y - 2 \ge 0$.

**167.** $y \ge x^2 - 2x - 3$,   $y < x + 1$.      **168.** $y \ge \sqrt{1 - x^2}$,   $y + |x| \le 4$.

**169.** $2y - x^2 < 0$,   $x^2 - y^2 \ge 0$.      **170.** $y \ge x^2$,   $y \le 4 - x^2$.

**171.** $y \ge x^2 - 4x + 3$,   $y < x^2 + 4x + 3$.      **172.** $2y \ge x^2$,   $y \le -2x^2 + 3x$.

**173.** $y - |\log_2 x| > 0$,   $y - 2 \ge 0$.      **174.** $y \le \log_2 x$,   $4x - 3y - 12 \le 0$.

**175.** $\sqrt{x^2 - 3y^2 + 4x + 4} \le 2x + 1$,   $x^2 + y^2 \le 4$.

**176.** $\sqrt{3x^2 + 3y^2 - 3} \ge 2y + 1$,   $y + 4 \ge 2\sqrt{3}|x|$.

**177.** $\sqrt{x + y} \ge x$,   $y \le 2$.

**178.** $y + x^2 \le 0$,   $y - 2x + 3 \ge 0$,   $y + 1 \le 0$.

**179.** $2x + y^2 \le 0$,   $x + 2 \ge 0$,   $y^2 - 1 \le 0$.

**180.** $xy \le 1$,   $x + y \ge 0$,   $y - x \le 0$.

**181.** $x^2 + y^2 - 4 \le 0$, $y - x^2 + 3 < 0$, $x \ge 0$.

**182.** $x^2 + y^2 \le 9$, $y \ge x^2 - 3$, $|x| \le 1$.

**183.** $\log_2 (2y - x^2 + 1) > \log_2 y$, $\sqrt{y - 2x + 4} < \sqrt{8 - x}$.

**184.** $|x| + |y| < 4$, $\log_2 (2y - x^2 + 4) > \log_2 (y + 1)$.

**185.** $|x + y| + |x - y| \le 4$, $|x| \le 1$, $y \ge \sqrt{x^2 - 2x + 1}$.

## 1.8. The Domain of Definition and the Range of a Function

The number *function f* is a mapping of the subset $D$ of the set $\boldsymbol{R}$ onto the subset $E$ of the set $\boldsymbol{R}$. The set $D$ is the *domain of definition* and $E$ is the *range* of the function $f$. The domain of definition of the function $f$ is designated as $D(f)$ and the range as $E(f)$. The value of $f$ at the point $x$ is designated as $f(x)$. A number belonging to $D(f)$ is usually denoted by a letter, say, $x$ and is called an *independent variable* or an *argument*. The dependent variable or value of a function is usually also denoted by a letter, say, $y$. Every value of the independent variable $x$ from the domain $D(f)$ of the function $f$ is associated with a definite value $y = f(x)$ of the dependent variable $y$ from the range $E(f)$ of the function $f$.

**Example 1.** The domain of the function $y = \sqrt{4 - x^2}$ is the set $D = [-2, 2]$, and the range is the set $E = [0, 2]$.

**Example 2.** The domain of the function $y = \arccos (2 - x)$ is the set $D = [1, 3]$, i.e. a set of the values of $x$ on which $-1 \le 2 - x \le 1$, and the range is the set $E = [0, \pi]$.

**Example 3.** The domain of the function $y = \sqrt{x(2 - x)} \log (x - 1)$ is the set $E = (1, 2]$, i.e. the greatest set on which both functions, $\sqrt{x(2 - x)}$ and $\log (x - 1)$, are defined.

**Example 4.** The domain of the function $y = 4 \sin x - 3 \cos x$ is the set $\boldsymbol{R}$, and the range is the set $E = [-5, 5]$. Indeed, setting $\sin \alpha = \dfrac{3}{5}$, $\cos \alpha = \dfrac{4}{5}$, where $\alpha = \arctan \dfrac{3}{4}$, we reduce the given function to the form.

$$y = 5 \left( \frac{4}{5} \sin x - \frac{3}{5} \cos x \right) = 5 \sin (x - \alpha),$$

i.e. $y = 5 \sin \left( x - \arctan \dfrac{3}{4} \right)$, and this function assumes all the values belonging to the interval $[-5, 5]$.

Find the domains of definition of the following functions **(1-112)**.

**1.** $y = \sqrt{2x - x^2}$.

**2.** $y = \sqrt{x - 1} \sqrt{x + 1}$.

**3.** $y = \sqrt{x - 1} + \sqrt{6 - x}$.

**4.** $y = \sqrt{x^2 - 5x + 6}$.

**5.** $y = \sqrt{\dfrac{x+3}{5-x}}$.

**6.** $f(x) = \sqrt{2-x} + \sqrt{1+x}$.

**7.** $y = \sqrt{-4x^2 + 4x + 3}$.

**8.** $y = \sqrt{6 + 7x - 3x^2}$.

**9.** $y = \dfrac{1}{x-1} + \sqrt{2+x}$.

**10.** $y = \sqrt{\dfrac{1}{2x^2 - 5x - 3}}$.

**11.** $f(x) = \sqrt{4x - x^3}$.

**12.** $f(x) = \sqrt{3x - x^3}$.

**13.** $y = \dfrac{1}{x^3 + x - 2}$.

**14.** $y = \dfrac{\sqrt{4 - 3x - x^2}}{x+4}$.

**15.** $y = \dfrac{\sqrt{3x-7}}{\sqrt[6]{x+1}-2}$.

**16.** $f(x) = \dfrac{\sqrt{12 + x - x^2}}{x(x-2)}$.

**17.** $y = \sqrt{5 - x - \dfrac{6}{x}}$.

**18.** $f(x) = \sqrt{x^2 - x - 20} + \sqrt{6 - x}$.

**19.** $f(x) = \dfrac{\sqrt{x^2 + x - 6}}{x^2 - 4}$.

**20.** $y = \dfrac{\sqrt{x + 12 - x^2}}{x^2 - 9}$.

**21.** $y = \left(\dfrac{1}{2}\right)^{\sqrt{4-x^2}} + \dfrac{1}{x-1}$.

**22.** $y = \sqrt{\dfrac{\sqrt{17 - 15x - 2x^2}}{x+3}}$.

**23.** $y = \sqrt{\dfrac{7-x}{\sqrt{4x^2 - 19x + 12}}}$.

**24.** $y = \sqrt{\dfrac{x^2 - 7x + 12}{x^2 - 2x - 3}}$.

**25.** $y = \sqrt{\dfrac{x^2 - 5x + 6}{x^2 + 6x + 8}}$.

**26.** $y = \sqrt{x - x^2} + \sqrt{3x - x^2 - 2}$.

**27.** $y = \sqrt{x^2 - x - 20} + \dfrac{1}{\sqrt{x^2 - 5x - 14}}.$

**28.** $y = \dfrac{1}{\sqrt{14 + 5x - x^2}} + \sqrt{x^2 - x - 20}.$

**29.** $y = \sqrt{\dfrac{x^4 - 3x^2 + x + 7}{x^4 - 2x^2 + 1} - 1}.$

**30.** $f(x) = \dfrac{1}{\sin^4 x + \cos^4 x}.$

**31.** $f(x) = \arcsin 3^x.$

**32.** $f(x) = \sqrt{(\sin x + \cos x)^2 - 1}.$

**33.** $y = \dfrac{\sqrt{\cos x - \dfrac{1}{2}}}{\sqrt{6 + 35x - 6x^2}}.$

**34.** $y = \dfrac{\log_3 (x^2 + 1)}{\sin^2 x - \sin x + 0.25}.$

**35.** $y = \dfrac{1}{3 - \log_3 (x - 3)}.$

**36.** $y = \dfrac{\sqrt{x + 5}}{\log (9 - x)}.$

**37.** $f(x) = \dfrac{\sqrt{3 \log_{64} x - 1}}{\sqrt[3]{2x - 11}}.$

**38.** $y = \log_2 \dfrac{x - 2}{x + 2}.$

**39.** $f(x) = \log \dfrac{x + 3}{x + 1}.$

**40.** $y = \sqrt{\log (x + 1)}.$

**41.** $y = \log \dfrac{x^2 + 8x + 7}{x^2 + 7}.$

**42.** $y = \sqrt{1 - x} + \log (x + 1).$

**43.** $y = \sqrt{x + 1} + \log (1 - x).$

**44.** $y = \log ((x^2 - 3x)(x + 5)).$

**45.** $y = \sqrt{4x - x^2} - \log_3 (x - 2).$

**46.** $y = \sqrt{x^2 + 4x - 5} \cdot \log (x + 1).$

**47.** $f(x) = \log (5x^2 - 8x - 4) + \sqrt{x - 1}.$

**48.** $y = \sqrt{x^2 + 4x - 5} \cdot \log (x + 5).$

**49.** $y = \dfrac{\log (3 - 2x - x^2)}{\sqrt{x}}.$

**50.** $y = \sqrt{\log \dfrac{3 - x}{x}}.$

**51.** $y = \sqrt{\log \dfrac{1-2x}{x+3}}$.

**52.** $f(x) = \sqrt[4]{x - |x|} + \log(x+2)$.

**53.** $y = \dfrac{\sqrt{x^2 - 5x + 6}}{\log(x+10)^2}$.

**54.** $y = \dfrac{\log x}{\sqrt{x^2 - 2x - 63}}$.

**55.** $y = \sqrt{\log \dfrac{5x - x^2}{4}}$.

**56.** $y = \sqrt{(x^2 - 3x - 10)\log^2(x-3)}$.

**57.** $f(x) = \log(1 - \sqrt{4 - x^2})$.

**58.** $y = \log(5x^2 - 8x - 4) + (x+3)^{-0.5}$.

**59.** $y = \sqrt{\dfrac{1 - 5^x}{7^{-x} - 7}}$.

**60.** $y = \sqrt{4x - x^2} + \log(x^2 - 1)$.

**61.** $y = \sqrt{1 - \log(x-1)} + \sqrt{\dfrac{4-x}{x+2}}$.

**62.** $y = \sqrt{\log_{0.3}\dfrac{x-1}{x+5}}$.

**63.** $y = \sqrt{\log_{0.4}(x - x^2)}$.

**64.** $y = \sqrt{\log_{0.3}(x^2 - 5x + 7)}$.

**65.** $y = \sqrt{\log_{0.5}(x^2 - 9)} + 4$.

**66.** $y = \sqrt{\log_{0.4}\dfrac{x-1}{x+5} \cdot \dfrac{1}{x^2 - 36}}$.

**67.** $f(x) = \sqrt{\log_{0.5}(-x^2 + x + 6)} + \dfrac{1}{x^2 + 2x}$.

**68.** $y = \sqrt{\dfrac{-\log_{0.3}(x-1)}{\sqrt{-x^2 + 2x + 8}}}$.

**69.** $f(x) = \sqrt{16x - x^5} + \log_{\frac{1}{2}}(x^2 - 1)$.

**70.** $y = \sqrt{\log_{\frac{1}{2}}\dfrac{x}{x^2 - 1}}$.

**71.** $f(x) = \sqrt{4^{\sqrt{\frac{3x^2 + 18x + 29}{x+3}}} - 2^{6x+17}}$.

**72.** $y = \sqrt{\log_{0.5}(3x - 8) - \log_{0.5}(x^2 + 4)}$.

**73.** $f(x) = \sqrt{4x - x^3} + \log(x^2 - 1)$.

**74.** $y = \sqrt[4]{\dfrac{1}{2}\log_4 16 - \log_8(x^2 - 4x + 3)}$.

**75.** $f(x) = \log_4\left(2 - \sqrt[4]{x} - \dfrac{2\sqrt{x}+1}{\sqrt{x}+2}\right).$

**76.** $y = \sqrt{\dfrac{3^x - 4^x}{2x^2 - x - 8}}.$

**77.** $f(x) = \log_2\left(-\log_{\frac{1}{2}}\left(1 + \dfrac{6}{\sqrt[4]{x}}\right) - 2\right).$

**78.** $y = \dfrac{\sqrt{6x - x^2 - 5}}{5^{x-2} - 1}.$

**79.** $y = \dfrac{x}{\sqrt{x^2 - 5x + 6}}.$

**80.** $y = \sqrt{-x^2 + 2x + 3} + \log_3(x - 1).$

**81.** $y = \log\dfrac{x}{x-2} - \sqrt{x - 3}.$

**82.** $f(x) = \dfrac{\sqrt{x^2 - 2x}}{\log_5(x - 1)}.$

**83.** $f(x) = \log_{2x-5}(x^2 - 3x - 10).$

**84.** $f(x) = \sqrt[6]{4^x + 8^{\frac{2}{3}(x-2)} - 52 - 2^{2(x-1)}}.$

**85.** $y = \log_{1.7}\left(\dfrac{2 - f'(x)}{x + 1}\right)^{\frac{1}{2}},$

  where $f(x) = \dfrac{1}{3}x^3 - \dfrac{3}{2}x^2 - 2x + \dfrac{3}{2}.$

**86.** $y = \sqrt{\dfrac{\log_{0.3}|x - 2|}{|x|}}.$

**87.** $y = \sqrt[6]{x + x^2 - 2x^3}.$

**88.** $y = \sqrt{x - 4} - \dfrac{x}{x - 5} + \log(39 - x).$

**89.** $y = \log(1 - \log(x^2 - 5x + 16)).$

**90.** $y = \log_{0.5}\left(-\log_2\dfrac{3x - 1}{3x + 2}\right).$

**91.** $y = \sqrt{\log\log x - \log(4 - \log x) - \log 3}.$

**92.** $y = \sqrt{\log_{x-2}(x^2 - 8x + 15)}.$

**93.** $y = \log\left(\sqrt{8^{-2+\log x}} - \sqrt[3]{4^{2-\log x}}\right).$

**94.** $y = \log_{100x}\dfrac{2\log x + 1}{-x}.$

**95.** $y = \log_2\left(-\log_{\frac{1}{2}}\left(1 + \dfrac{1}{\sqrt[4]{x}}\right) - 1\right).$

**96.** $y = \log_{|x|-4}2.$

**97.** $y = \sqrt{\sin x} + \sqrt{16 - x^2}.$

**98.** $y = \log (\log^2 x - 5 \log x + 6)$.

**99.** $y = \sqrt{\log_{\frac{1}{2}} \dfrac{x-1}{3x+5}}$.

**100.** $y = \log \sin (x-3) - \sqrt{16 - x^2}$.

**101.** $y = \dfrac{\log x}{\sqrt{x^2 - 2x - 63}}$.

**102.** $y = \arcsin \dfrac{x-3}{2} - \log (4-x)$.

**103.** $y = \sqrt{3-x} + \arcsin \dfrac{3-2x}{5}$.

**104.** $y = (x+0.5)^{\log_{(0.5+x)} \frac{x^2+2x-3}{4x^2-4x-3}}$.

**105.** $y = \log_{100x} \dfrac{2\log x + 2}{-x}$.

**106.** $y = \arccos \dfrac{2x+1}{2\sqrt{2x}}$.

**107.** $y = \arccos \dfrac{2}{2+\sin x}$.

**108.** $y = \sqrt{3\sin x - 1}$.

**109.** $y = \sqrt{2\sin \dfrac{x}{2}}$.

**110.** $y = \dfrac{1}{\sqrt{4\cos x + 1}}$.

**111.** $y = \sqrt{-2\cos^2 x + 3\cos x - 1}$.

**112.** $y = \sqrt{\sin^2 x - \sin x}$.

Find the domains of definition and the ranges of the following functions (113-120).

**113.** $y = \dfrac{x}{|x|}$.

**114.** $f(x) = \sqrt{x - x^2}$.

**115.** $y = \sqrt{3x^2 - 4x + 5}$.

**116.** $y = \log (3x^2 - 4x + 5)$.

**117.** $y = \log (5x^2 - 8x + 4)$.

**118.** $f(x) = \sqrt{x-1} + 2\sqrt{3-x}$.

**119.** $f(x) = \log_2 \dfrac{\sin x - \cos x + 3\sqrt{2}}{\sqrt{2}}$.

**120.** $f(x) = \sqrt{2-x} + \sqrt{1+x}$.

**121.** Find the set of all $x$ for which there are no functions $f(x) = \log_{x+3}^{x-2} 2$ and $g(x) = \sqrt{x^2 - 9}$.

**122*.** For what real values of $a$ does the range of the function $y = \dfrac{x-1}{a - x^2 + 1}$ not contain any values belonging to the interval $[-1, -1/3]$?

**123*.** For what real values of $a$ does the range of the function $y = \dfrac{x+1}{a + x^2}$ contain the interval $[0, 1]$?

**124*.** For what real values of $a$ does the range of the function $y = \dfrac{x-1}{1 - x^2 - a}$ not contain any value from the interval $[-1, 1]$?

## 1.9. Exponential and Logarithmic Equations and Inequalities

*The logarithm of a positive number $x$ to the base $a (a > 0, a \ne 1)$ is a number equal to the power to which a must be raised in order to obtain $x$.*

**1°. Fundamental identities.** The identities used in solutions of exponential and logarithmic equations and inequalities ($a > 0, a \ne 1$) are as follows :

$$a^x = a^y \Leftrightarrow x = y. \tag{1}$$

$$a^x \cdot a^y = a^{x+y}. \tag{2}$$

$$a^0 = 1. \tag{3}$$

$$a^x \cdot b^x = (a \cdot b)^x, b > 0. \tag{4}$$

$$(a^x)^y = a^{xy}. \tag{5}$$

$$a^{\log_a x} = x, \quad x > 0. \tag{6}$$

$$\log_a xy = \log_a |x| + \log_a |y|, \quad xy > 0. \tag{7}$$

$$\log_a \frac{x}{y} = \log_a |x| - \log_a |y|, xy > 0. \tag{8}$$

$$\log_a x^\alpha = \alpha \cdot \log_a x, x > 0. \tag{9}$$

$$\log_a x^{2m} = 2m \log_a |x|, x\, 0, m \in \mathbf{N}. \tag{10}$$

$$\log_a b = \frac{\log_x b}{\log_x a}, x > 0, x \ne 1, \ b > 0. \tag{11}$$

$$\log_a b = \frac{1}{\log_b a}, \quad b > 0, \ b \ne 1. \tag{12}$$

$$\log_a x = \log_{a^k} x^k, \quad x > 0, \ k \in \mathbf{R}, \ k \ne 0. \tag{13}$$

$$\log_{a^k} x^m = \frac{m}{k} \log_a x, x > 0; m, k \in \mathbf{R}, k \ne 0. \tag{14}$$

**2°. Exponential equations.** If we have an equation of the form $a^{f(x)} = b \, (a > 0)$, then

(1) $x \in \varnothing$ if $b \le 0$;

(2) $f(x) = \log_a b$ if $b > 0$, $a \ne 1$;

(3) the equation is satisfied for all $x \in D(f)$ if $a = 1$, $b = 1$ (since $1^{f(x)} = 1$ for all $x \in D(f)$);

(4) $x \in \varnothing$ if $a = 1$, $b \ne 1$.

Here are some examples of solving exponential equations.

**Example 1.**   Solve the equation $3^{x^2 - 5x + 6} = 1$.

*Solution.*   Applying identity (3), we get $x^2 - 5x + 6 = 0$.

Hence $x_1 = 2$, $x_2 = 3$.

*Answer .*   $x = 2$, $x = 3$.

**Example 2.**   $\left(\dfrac{3}{7}\right)^{3x-7} = \left(\dfrac{7}{3}\right)^{7x-3}$.

*Solution.*   Note that $\dfrac{3}{7} = \left(\dfrac{7}{3}\right)^{-1}$ and write the equation as

$$\left(\frac{3}{7}\right)^{3x-7} = \left(\frac{3}{7}\right)^{-7x+3}.$$

Using identity (1), we get
$3x - 7 = -7x + 3$, $x = 1$.

*Answer.*   1.

**Example 3.**   $0.125 \cdot 4^{2x-8} = \left(\dfrac{0.25}{\sqrt{2}}\right)^{-x}$.

*Solution.*   Passing to the base of the power 2, we get

$$\frac{1}{8} \cdot 2^{2(2x-8)} = \left(\frac{1}{4} \cdot 2^{-\frac{1}{2}}\right)^{-x}$$

or   $2^{-3} \cdot 2^{2(2x-8)} = (2^{-2} \cdot 2^{-\frac{1}{2}})^{-x}$.

According to equation(2), we have
$$2^{-3+2(2x-8)} = (2^{-2-0.5})^{-x}.$$

Using identity (5), we write
$$2^{-3+4x-16} = 2^{2.5x}$$

According to equality (1), the last equation is equivalent to the equation.

$$-3 + 4x - 16 = 2.5x,$$

whence $x = 38 / 3$.

*Answer.*     38 / 3.

Let us consider some examples of solving exponential equations which can be reduced to quadratic equations.

**Example 4.**   $5^{2x-1} + 5^{x+1} = 250.$

*Solution.*     We use identity (2) and write the initial equation in the form $5^{2x} \cdot 5^{-1} + 5^x \cdot 5 - 250 = 0$. Substituting $5^x = t > 0$ into the last equation, we obtain

$$\frac{1}{5}t^2 + 5t - 250 = 0.$$

Hence $t_1 = -50, t_2 = 25$. The value $t_1 = -50$ does not fulfil the condition $t > 0$, and so $5^x = 25, x = 2.$

*Answer.* 2.

**Example 5.**   $9^x + 6^x = 2 \cdot 4^x.$

*Solution.* We divide both sides of the equation by $4^x$ : $\left(\dfrac{9}{4}\right)^x + \left(\dfrac{6}{4}\right)^x - 2 = 0$, or

$$\left(\frac{3}{2}\right)^{2x} + \left(\frac{3}{2}\right)^x - 2 = 0.$$

We introduce the designation $\left(\dfrac{3}{2}\right)^x = t \ (t > 0)$

and then the last equation looks like
$$t^2 + t - 2 = 0, \ t_1 = -2, \ t_2 = 1.$$

The value of $t_1$ does not fulfil the condition $t > 0$. Consequently, $\left(\dfrac{3}{2}\right)^x = 1, \ x = 0.$

*Answer.* 0.

The following examples illustrate the solution of equations by putting the common factor before the brackets.

**Example 6.**   $5^{x+1} - 5^{x-1} = 24.$

*Solution.* In the left-hand side of the equation we put the common factor $5^{x-1}$ before the brackets : $5^{x-1}(5^2 - 1) = 24.$

We obtain $5^{x-1} = 1$, whence $x - 1 = 0, \ x = 1.$

*Answer.* 1.

**Example 7.**   $6^x + 6^{x+1} = 2^x + 2^{x+1} + 2^{x+2}.$

*Solution.*     After factoring out $6^x$ o the left-hand side and $2^x$ on the right-hand side, we obtain $6x(1 + 6) = 2(1 + 2 + 4)$, or $6^x = 2^x$. We divide both sides of the equation by $2^x \neq 0 : 3^x = 1, x = 0.$

*Answer.* 0.

**3°. Exponential Inequalities.** When we solve exponential inequalities $a^{f(x)} > b \ (a > 0)$, we have

(1) $x \in D(f)$ if $b \leq 0,$

(2) if $b > 0$, then we have

$f(x) > \log_a b$ for $a > 1$,

$f(x) < \log_a b$ for $0 < a < 1$;

for $a = 1$ the inequality is equivalent to the numerical inequality $1 > b$.

Let us consider solutions of some exponential inequalities.

**Example 8.** $2^{x+2} > \left(\dfrac{1}{4}\right)^{\frac{1}{x}}$ or $2^{x+2} > (2^{-2})^{\frac{1}{x}}$.

*Solution.* According to identity (5) we have $2^{x+2} > 2^{-\frac{2}{x}}$. Since the base $2 > 1$, we have $x + 2 > -\dfrac{2}{x}$ (the sign of the inequality is retained). Solving the last inequality, we obtain $x \in (0, +\infty)$.

**Example 9.** $(1.25)^{1-x} < (0.64)^{2(1+\sqrt{x})}$.

*Solution.* We write the initial inequality in the form

$$\left(\frac{5}{4}\right)^{1-x} < \left(\frac{16}{25}\right)^{2(1+\sqrt{x})},$$

or

$$\left(\frac{4}{5}\right)^{x-1} < \left(\frac{4}{5}\right)^{2\cdot 2(1+\sqrt{x})}.$$

Since the base $0 < \dfrac{4}{5} < 1$, the last inequality is equivalent to the inequality $x - 1 > 4(1 + \sqrt{x})$ (the sign of the inequality has changed to the opposite.) Next we have $x - 4\sqrt{x} - 5 > 0$, whence $(\sqrt{x} - 5)(\sqrt{x} + 1) > 0$, i.e. $\sqrt{x} > 5$. The final result is $x > 25$.

*Answer.* $x \in (25, +\infty)$.

**4°. Logarithmic equations.** Let us consider the methods of solving logarithmic equations which are most often used.

1. *Solution of equations based on the definition of a logarithm.*

**Example 10.** $\log_3 (5 + 4\log_3 (x-1)) = 2$.

*Solution.* The logarithm of the number $b$ to the base $a$ ($a > 0, a \neq 1$) is the power to which the number $a$ must be raised in order to obtain the number $b$. Consequently,

$$5 + 4\log_3 (x-1) = 3^2,$$

or

$$4\log_3 (x-1) = 9 - 5, \quad \log_3 (x-1) = 1.$$

And again by the definition of a logarithm, we have

$$x - 1 = 3^1, \quad x = 4.$$

The verification, which is a part of the solution, confirms the truth of the result obtained.

*Answer.* $x = 4$.

2. *Solving equations by taking antilogarithms.*

**Example 11.** $\log_2 (3 - x) + \log_2 (1 - x) = 3$.

*Solution.* To find the domain of the function appearing on the left-hand side, we form a system of inequalities
$$\begin{cases} 3 - x > 0, \\ 1 - x > 0, \end{cases} \Rightarrow x < 1.$$

Using identity (7), we can write
$$\log_2 ((3 - x)(1 - x)) = 3,$$
and by the definition of a logarithm we have
$$(3 - x)(1 - x) = 2^3, \quad \text{or} \quad x^2 - 4x - 5 = 0.$$

Then $x_1 = 5$, $x_2 = -1$. Since the first value of the unknown does not belong to the domain of definition, the final result is $x = -1$.

*Answer.* $-1$.

3. *Using the fundamental logarithmic identity.*

**Example 12.** $\log_2 (9 - 2^x) = 10^{\log(3-x)}$.

*Solution.* The domain of definition is
$$\begin{cases} 9 - 2^x > 0, \\ 3 - x > 0, \end{cases} \Rightarrow \begin{cases} 2^x < 9, \\ x < 3, \end{cases} \Rightarrow \begin{cases} x < \log_2 9, \\ x < 3, \end{cases} \Rightarrow x < 3.$$

Using identity (6) on the right-hand side, we get
$$\log_2 (9 - 2^x) = 3 - x.$$

By the definition of a logarithm,
$$2^{3-x} = 9 - 2^x, \quad \frac{2^3}{2^x} = 9 - 2^x,$$
$$2^{2x} - 9 \cdot 2^x + 8 = 0,$$
$$2^x = 1, \quad x = 0; \quad 2^x = 8, \quad x = 3 \text{ is an extraneous root.}$$

*Answer.* $x = 0$.

4. *Taking logarithms.*

**Example 13.** $(x + 1)^{\log(x+1)} = 100(x + 1)$.

*Solution.* The domain of definition is $x + 1 > 0 \Rightarrow x > -1$.

We take logarithms of both sides of the equation to the base 10 :
$$\log (x+1) \cdot \log (x+1) = \log 100 + \log (x + 1).$$

We designate $\log (x + 1) = t$ and then the equation assumes the form $t^2 - t - 2 = 0$. Its solutions are $t_1 = -1$, $t_2 = 2$, i.e.
$$\log (x+1) = -1, \quad x + 1 = \frac{1}{10}, \quad x = -0.9;$$
$$\log (x+1) = 2, \quad x + 1 = 100, \quad x = 99.$$

*Answer.* $-0.9$, 99.

**5. A change of variable.**

**Example 14.** Solve the equation $(\log x)^2 - \log x^3 + 2 = 0$.

*Solution.* We introduce the variable $t = \log x$, $x > 0$, and the initial equation assumes the form $t^2 - 3t + 2 = 0$. Its solutions are $t_1 = 1$, $t_2 = 2$. Whence we have

$$\log x = 1, \quad x = 10;$$
$$\log x = 2, \quad x = 100.$$

*Answer.* 10, 100.

**6. Passing to a new base.**

**Example 15.** $1 + \log_2 (x - 1) = \log_{(x-1)} 4$.

*Solution.* The domain of definition is $x > 1$, $x \ne 2$.

By property (11),

$$\log_{(x-1)} 4 = \frac{\log_2 4}{\log_2 (x-1)} = \frac{2}{\log_2 (x-1)}.$$

We designate $\log_2 (x - 1) = y$ and then the initial equation assumes the form $1 + y = \dfrac{2}{y}$, or $y^2 + y - 2 = 0$, whence $y_1 = -2$, $y_2 = 1$.

$$\log_2 (x-1) = -2, \quad x - 1 = \frac{1}{4}, \quad x_1 = \frac{5}{4},$$
$$\log_2 (x-1) = 1, \quad x - 1 = 2, \quad x_2 = 3.$$

*Answer.* $\dfrac{5}{4}$, 3.

**5°. Logarithmic inequalities.** When solving logarithmic inequalities, it is necessary to remember that the function $y = \log_a x (a > 0,\ a \ne 1,\ x > 0)$ is decreasing if $0 < a < 1$ and increasing if $a > 1$. Therefore, an inequality of the form

$$\log_a f(x) < \log_a \varphi(x)$$

is equivalent to the system

$$\begin{cases} f(x) > \varphi(x), \\ \varphi(x) > 0 \end{cases}$$

for $0 < a < 1$, and to the system

$$\begin{cases} f(x) < \varphi(x), \\ f(x) > 0 \end{cases}$$

for $a > 1$.

Let us consider solutions of some typical logarithmic inequalities.

**Example 16.** $\log_{\frac{1}{5}} \dfrac{4x + 6}{x} \ge 0$.

*Solution.* Since $\log_{\frac{1}{5}} 1 = 0$, the given inequality can be written as

$$\log_{\frac{1}{5}} \frac{4x+6}{x} \geq \log_{\frac{1}{5}} 1.$$

When the domain of the function is taken into account, the inequality is equivalent to the system of inequalities

$$\begin{cases} \dfrac{4x+6}{x} > 0, \\ \dfrac{4x+6}{x} \leq 1, \end{cases} \text{ or } \begin{cases} \dfrac{4x+6}{x} > 0, \\ \dfrac{4x+6}{x-1} \leq 0. \end{cases}$$

We solve the inequalities by the method of intervals

$$\begin{cases} 4\left(x+\dfrac{3}{2}\right)\cdot x > 0, \\ 3x\,(x+2) \leq 0, \end{cases} \Rightarrow \begin{cases} x\left(x+\dfrac{3}{2}\right) > 0, \\ x\,(x+2) \leq 0, \end{cases} \Rightarrow x \in \left[-2,\ -\dfrac{3}{2}\right).$$

*Answer.* $x \in \left[-2,\ -\dfrac{3}{2}\right).$

**Example 17.** $\log_{2x+3} x^2 < \log_{2x+3}(2x+3).$

*Solution.*    $\log_{2x+3} x^2 < \log_{2x+3}(2x+3).$ The given inequality is equivalent to the collection of the systems

$$\begin{bmatrix} \begin{cases} 0 < 2x+3 < 1, \\ x^2 > 2x+3, \end{cases} \\ \begin{cases} 2x+3 > 1, \\ 0 < x^2 < 2x+3. \end{cases} \end{bmatrix} \qquad \begin{matrix} (1) \\ \\ (2) \end{matrix}$$

Solving system (1), we obtain

$$\begin{cases} -\dfrac{3}{2} < x < -1, \\ (x-3)(x+1) > 0. \end{cases} \qquad (3)$$

System (3) is equivalent to the collection of two systems

$$\begin{bmatrix} \begin{cases} -\dfrac{3}{2} < x < -1, \\ x > 3; \end{cases} \\ \begin{cases} -\dfrac{3}{2} < x < -1, \\ x < -1. \end{cases} \end{bmatrix} \qquad \begin{matrix} (4) \\ \\ (5) \end{matrix}$$

System (4) has no solutions. The solution of system (5) is $x \in \left(-\dfrac{3}{2},\ -1\right).$

Solving system (2), We obtain

$$\begin{cases} x > -1. \\ (x-3)(x+1) < 0, \end{cases} \text{ or } \begin{cases} x > -1, \\ -1 < x < 3, \end{cases} \text{ i.e. } x \in (-1,\, 3).$$

*Answer.* $x \in \left( -\dfrac{3}{2},\, -1 \right) \cup (-1,\, 3).$

**Example 18.** $\log_{\frac{x+4}{2}} \log_2 \dfrac{2x-1}{3+x} < 0.$

*Solution.* The solution of the given inequality is the union of the sets of solutions of the systems.

$$\begin{cases} 0 < \dfrac{x+4}{2} < 1, \\[2mm] \log_2 \dfrac{2x-1}{3+x} > 0, \\[2mm] \dfrac{2x-1}{3+x} > 0, \\[2mm] \log_2 \dfrac{2x-1}{3+x} > 1; \end{cases} \tag{1}$$

$$\begin{cases} \dfrac{x+4}{2} > 1, \\[2mm] \log_2 \dfrac{2x-1}{3+x} > 0, \\[2mm] \dfrac{2x-1}{3+x} > 0, \\[2mm] \log_2 \dfrac{2x-1}{3+x} < 1. \end{cases} \tag{2}$$

Solving system (1), we obtain

$$\begin{cases} 0 < x+4 < 2, \\[2mm] \dfrac{2x-1}{3+x} > 1, \\[2mm] \left( x - \dfrac{1}{2} \right)(x+3) > 0, \\[2mm] \dfrac{2x-1}{x+3} > 2; \end{cases} \Rightarrow \begin{cases} -4 < x < -2, \\[2mm] \left( x - \dfrac{1}{2} \right)(x+3) > 0, \\[2mm] \dfrac{2x-1-2x-6}{3+x} > 0; \end{cases}$$

$$\Rightarrow \begin{cases} -4 > x < -2, \\[2mm] x+3 < 0, \\[2mm] x - \dfrac{1}{2} < 0; \end{cases} \Rightarrow x \in (-4,\, -3).$$

Solving system (2), we obtain

$$\begin{cases} x > -2, \\ \dfrac{2x-1}{x+3} > 1, \\ 2\left(x - \dfrac{1}{2}\right)(x+3) > 0, \\ \dfrac{2x-1}{3+x} < 2; \end{cases} \Rightarrow \begin{cases} x > -2, \\ \dfrac{2x-1-x-3}{x+3} > 0, \\ \left(x - \dfrac{1}{2}\right)(x+3) > 0, \\ \dfrac{2x-1-2x-6}{3+x} < 0; \end{cases}$$

$$\Rightarrow \begin{cases} x > -2, \\ x > 4, \\ x > \dfrac{1}{2}; \end{cases} \Rightarrow x \in (4, +\infty).$$

*Answer.* $(-4, -3) \cup (4, +\infty)$.

**6°. Systems of exponential and logarithmic equations.** Let us consider some examples of solving systems of this kind.

**Example 19.** $\begin{cases} 2^{y-x}(x+y) = 1, \\ (x+y)^{x-y} = 2. \end{cases}$

*Solution.* The domain of definition is $x + y > 0$.

$$\begin{cases} x + y = \dfrac{1}{2^{y-x}}, \\ (x+y)^{x-y} = 2. \end{cases}$$

From the first equation we find $x + y = 2^{x-y}$ and substitute it into the second equation. Then

$$(2^{x-y})^{x-y} = 2 \Rightarrow 2^{(x-y)2} = 2 \Rightarrow (x-y)^2 = 1.$$

The solution of the given system is the solution of the collection of systems

$$\begin{cases} x - y = 1, \\ x + y = 2, \end{cases} \text{ and } \begin{cases} x - y = -1, \\ x + y = \dfrac{1}{2}. \end{cases}$$

*Answer.* $\left(\dfrac{3}{2}, \dfrac{1}{2}\right), \left(-\dfrac{1}{4}, \dfrac{3}{4}\right).$

**Example 20.** $\begin{cases} \log_3 x + \log_3 y = 2 + \log_3 2, \\ \log_{27}(x+y) = \dfrac{2}{3}. \end{cases}$

*Solution.* The domain of definition is $x > 0, \ y > 0$.

$$\begin{cases} \log_3 x + \log_3 y = \log_3 9 + \log_3 2, \\ \log_{27}(x+y) = \dfrac{2}{3}. \end{cases}$$

In the first equation we use identity (7) and in the second equation we use the definition of a logarithm. We obtain

$$\begin{cases} \log_3(xy) = \log_3 18, \\ x+y = (3^3)^{2/3}, \end{cases} \Rightarrow \begin{cases} xy = 18, \\ x+y = 9. \end{cases}$$

Solving the last system, we also get a solution of the given system.
*Answer.* {(6, 3), (3, 6)} .

## Exponential Equations

Solve the following equations **(1-82)**.

**1.** $\dfrac{2^{x-1} \cdot 4^{x+1}}{8^{x-1}} = 64.$

**2.** $\dfrac{0.2^{x+0.5}}{\sqrt{5}} = \dfrac{(0.04)^x}{25}.$

**3.** $\left(\dfrac{5}{3}\right)^{x+1} \cdot \left(\dfrac{9}{25}\right)^{x^2+2x-11} = \left(\dfrac{5}{3}\right)^9.$

**4.** $9^{|3x-1|} = 3^{8x-2}.$

**5.** $2^{\frac{3}{\log_2 x}} = \dfrac{1}{64}.$

**6.** $4^{\frac{2}{x}} - 5 \cdot 4^{\frac{1}{x}} + 4 = 0.$

**7.** $5^x - 24 = \dfrac{25}{5^x}.$

**8.** $4^x - 10 \cdot 2^{x-1} = 24.$

**9.** $4^{x^2+2} - 0.2^{x^2+2} + 8 = 0.$

**10.** $4^{x-\sqrt{x^2-5}} - 12 \cdot 2^{x-1-\sqrt{x^2-5}} + 8 = 0.$

**11.** $64 \cdot 9^x - 84 \cdot 12^x + 27 \cdot 16^x = 0.$

**12.** $4 \cdot 2^{2x} - 6^x = 18 \cdot 3^{2x}.$

**13.** $7^{x+2} - \dfrac{1}{7} \cdot 7^{x+1} - 14 \cdot 7^{x-1} + 2 \cdot 7^x = 48.$

**14.** $3^{2x-3} - 9^{x-1} + 27^{\frac{2x}{3}} = 675.$

**15.** $5^{2x} - 7^x - 5^{2x} \cdot 35 + 7^x \cdot 35 = 0.$

**16.** $5^{x-1} = 10^x \cdot 2^{-x} \cdot 5^{x+1}.$

**17.** $2^{x+1} \cdot 5^x = 200.$

**18.** $3 \cdot 5^{2x-1} - 2 \cdot 5^{x-1} = 0.2.$

**19.** $3^{2x+1} + 10 \cdot 3^x + 3 = 0$.

**20.** $2^{2x+1} - 33 \cdot 2^{x-1} + 4 = 0$.

**21.** $4^{-\frac{1}{x}} + 6^{-\frac{1}{x}} = 9^{-\frac{1}{x}}$.

**22.** $[(\sqrt[5]{27})^{\frac{x}{4}}\sqrt{\frac{x}{3}}]^{\frac{x}{4}+\sqrt{\frac{x}{3}}} = \sqrt[4]{3^7}$.

**23.** $2^{2x-3} = 4^{x^2-3x-1}$.

**24.** $4^x - 3^{x-\frac{1}{2}} = 3^{x+\frac{1}{2}} - 2^{2x-1}$.

**25.** $7 \cdot 3^{x+1} - 5^{x+2} = 3^{x+4} - 5^{x+3}$.

**26.** $3 \cdot 16^x + 2 \cdot 81^x = 5 \cdot 36^x$.

**27.** $64^{\frac{1}{x}} - 2^{3+\frac{3}{x}} + 12 = 0$.

**28.** $2^{2+x} - 2^{2-x} = 15$.

**29.** $9^x - 2^{x+0.5} = 2^{x+3.5} - 3^{2x-1}$.

**30.** $(3^{x^2-7.2x+3.9} - 9\sqrt{3}) \log (7 - x) = 0$.

**31.** $3^{12x-1} - 9^{6x-1} - 27^{4x-1} + 81^{3x+1} = 2192$.

**32.** $5^{2x} = 3^{2x} + 2 \cdot 5^x + 2 \cdot 3^x$.

**33.** $x^{5 \sin 3x+2} = \dfrac{1}{\sqrt{x}}$.

**34.** $2^{2 \log 4x-1} - 7^{\log 4x} = 7^{\log 4x-1} - 3 \cdot 4^{\log 4x}$.

**35.** $7^{\log x} - 5^{\log x+1} = 3 \cdot 5^{\log x-1} - 13 \cdot 7^{\log x-1}$.

**36.** $15 \cdot 2^{x+1} + 15 \cdot 2^{-x+2} = 135$.

**37.** $4^{x+\sqrt{x^2-2}} - 5 \cdot 2^{x-1+\sqrt{x^2-2}} = 6$.

**38.** $((2\sqrt{x} + 3)^{\frac{1}{2\sqrt{x}}})^{\frac{2}{\sqrt{x}-1}} = 4$.

**39.** $3^{2x+5} = 3^{x+2} + 2$.

**40.** $4^{\frac{1}{x}-2} = \dfrac{\log \sqrt{10}}{2}$.

**41.** $\sqrt{7}^{2x^2-5x-6} = (\sqrt{2})^{3 \log_2 49}$.

**42.** $2^{x^2-1} - 3^{x^2} = 3^{x^2-1} - 2^{x^2+2}$.

**43.** $9^{x^2-1} - 36 \cdot 3^{x^2-3} + 3 = 0$.

**44.** $3^{4x+8} - 4 \cdot 3^{2x+5} + 28 = 2\log_2 \sqrt{2}$.

**45.** $2 \cdot 3^{x+1} - 6 \cdot 3^{x-1} - 3^x = 9$.

**46.** $3^{\log \tan x} - 2 \cdot 3^{\log \cot x+1} = 1$.

**47.** $3^{2x+1} = 3^{x+2} + \sqrt{1 - 6 \cdot 3^x + 3^{2(x+1)}}$.

**48.** $\left(\dfrac{5}{12}\right)^x \cdot \left(\dfrac{6}{5}\right)^{x-1} = (0.3)^{-1}$.

**49.** $x^2 \cdot 2^{x+1} + 2^{|x-3|+2} = x^2 \cdot 2^{|x-3|+4} + 2^{x-1}$.

**50.** $5^x \cdot \sqrt[x]{8^{x-1}} = 500$.

**51.** $x^{\frac{5}{4}-2\cos 3x} = \sqrt[4]{x}.$

**52.** $32^{\frac{x+5}{x-7}} = 0.25 \cdot 128^{\frac{x+17}{x-3}}.$

**53.** $\left(\dfrac{4}{9}\right)^{\sqrt{x}} = (2.25)^{\sqrt{x}-4}.$

**54.** $2^x \cdot 5^{x-1} = 0.2 \cdot 10^{2-x}.$

**55.** $4^{3+2\cos 2x} - 7 \cdot 4^{1+\cos 2x} - 4^{1/2} = 0.$

**56.** $(\sqrt{5+\sqrt{24}})^x + (\sqrt{5-\sqrt{24}})^x = 10.$

**57.** $2^{x^2-6} \cdot 3^{x^2-6} = \dfrac{1}{6^5}(6^{x-1})^4.$

**58.** $3^{\log_3^2 x} + x^{\log_3 x} = 162.$

**59.** $0.5^{x^2} \cdot 2^{2x+2} = 64^{-1}.$

**60.** $9^{2x+4} = 26 \cdot 3^{2x+3} + 3.$

**61.** $\left(\dfrac{4}{9}\right)^x \cdot \left(\dfrac{27}{8}\right)^{x-1} = \dfrac{\log 4}{\log 8}.$

**62.** $0.125 \cdot 4^{2x-3} = \left(\dfrac{\sqrt{2}}{8}\right)^{-x}.$

**63.** $2^{\sin^2 x} + 4 \cdot 2^{\cos^2 x} = 6.$

**64.** $\sqrt[x-4]{5^{\frac{x}{\sqrt{x}+2}} \cdot (0.2)^{\frac{4}{\sqrt{x}+2}}} = 125 \cdot (0.04)^{\frac{x-2}{x-4}}.$

**65.** $5^{1+\log_4 x} + 5^{\log_{1/4} x-1} = \dfrac{26}{5}.$

**66.** $3 \cdot 4^x + \dfrac{1}{3} \cdot 9^{x+2} = 6 \cdot 4^{x+1} - \dfrac{1}{2} \cdot 9^{x+1}.$

**67.** $\sqrt[x]{64} - \sqrt{2^{3x+3}} + 12 = 0.$

**68.** $4^{\log x+1} - 6^{\log x} - 2 \cdot 3^{\log x^2+2} = 0.$

**69.** $4^x \cdot 5^{x+1} = 5 \cdot 20^{2-x}.$

**70.** $3^{x+1} + 18 \cdot 3^{-x} = 29.$

**71.** $5^{\log x} - 3^{\log x-1} = 3^{\log x+1} - 5^{\log x-1}.$

**72.** $5^{x-1} + 5 \cdot 0.2^{x-2} = 26.$

**73.** $3 \cdot 2^{x/2} - 7 \cdot 2^{x/4} = 20.$

**74.** $x^{1-\frac{1}{3}\log x^2} = \dfrac{1}{\sqrt[3]{100}}.$

**75.** $(\sqrt{x})^{\log_3 x-1} = 3.$

**76.** $x^{\log x} = 1000x^2.$

**77.** $(0.4)^{\log^2 x+1} = (6.25)^{2-\log x^3}.$

**78.** $(0.6)^x \left(\dfrac{25}{9}\right)^{x^2-12} = \left(\dfrac{27}{125}\right)^3.$

**79.** $6 \cdot 9^{1/x} - 13 \cdot 6^{1/x} + 6 \cdot 4^{1/x} = 0.$

**80.** $x^{\frac{\log x + 5}{3}} = 10^{5 + \log x}$.

**81.** $2^{\log_3 x} \cdot 5^{\log_3 x} = 400$.

**82.** $9^{1 + \log_3 x} - 3^{1 + \log_3 x} - 210 = 0$.

**83.** For what value of $p$ does the equation $p \cdot 2^x + 2^{-x} = 5$ possess a unique solution?

## Exponential Inequalities

Solve the following inequalities **(84-124)**.

**84.** $2^{3-6x} > 1$.

**85.** $16^x > 0.125$.

**86.** $(0.3)^{2x^2 - 3x + 6} < 0.00243$.

**87.** $\left(\dfrac{1}{3}\right)^{\sqrt{x+2}} > 3^{-x}$.

**88.** $0.1^{4x^2 - 2x - 2} \le 0.1^{2x - 3}$.

**89.** $\sqrt[x+1]{3} > 9$.

**90.** $8^{\sqrt{8^x}} > 4096$.

**91.** $\left(\dfrac{2}{5}\right)^{\frac{6-5x}{2+5x}} < \dfrac{25}{4}$.

**92.** $2^x + 2^{-x+1} - 3 < 0$.

**93.** $\dfrac{4^x + 2x - 4}{x - 1} \le 2$.

**94.** $4^{-x+0.5} - 7 \cdot 2^{-x} - 4 < 0$.

**95.** $x^2 \cdot 5^x - 5^{2+x} < 0$.

**96.** $4^{x+1} - 16^x < 2\log_4 8$.

**97.** $\dfrac{2^{x-1} - 1}{2^{x+1} + 1} < 2$.

**98.** $\dfrac{2^{1-x} - 2^x + 1}{2^x - 1} \le 0$.

**99.** $4^x - 2^{2(x-1)} + 8^{\frac{2(x-2)}{3}} > 52$.

**100.** $(0.3)^{2 + 4 + 6 + \dots + 2x} > (0.3)^{72}$, $x \in N$.

**101.** $2^{2x+1} - 21 \cdot \left(\dfrac{1}{2}\right)^{2x+3} + 2 \ge 0$.

**102.** $3^{4-3x} - 35\left(\dfrac{1}{3}\right)^{2-3x} + 6 \ge 0$.

**103.** $x^{\log^2 x - 3\log x + 1} > 1000$.

**104.** $2^{x+2} - 2^{x+1} + 2^{x-1} - 2^{x-2} \le 9$.

**105.** $5^{2x+1} > 5^x + 4.$

**106.** $2^{x+2} - 2^{x+3} - 2^{x+4} > 5^{x+1} - 5^{x+2}.$

**107.** $3^{72} \cdot \left(\dfrac{1}{3}\right)^x \cdot \left(\dfrac{1}{3}\right)^{\sqrt{x}} > 1.$

**108.** $4^x - 2 \cdot 5^{2x} - 10^x > 0.$

**109.** $25^{-x} - 5^{-x+1} \geq 50.$

**110.** $3^{\log x + 2} < 3^{\log x^2 + 5} - 2.$

**111.** $\sqrt{9^x - 3^{x+2}} > 3^x - 9.$

**112.** $9 \cdot 4^{-1/x} + 5 \cdot 6^{-1/x} < 4 \cdot 9^{-1/x}.$

**113.** $8 \cdot \dfrac{3^{x-2}}{3^x - 2^x} > 1 + \left(\dfrac{2}{3}\right)^x.$

**114.** $\left(\dfrac{1}{2}\right)^{\log_2(x^2-1)} > 1.$

**115.** $\left(\dfrac{1}{3}\right)^{\frac{|x+2|}{2-|x|}} > 9.$

**116.** $x^2 2^{2x} + 9(x+2)2^x + 8x^2 \leq (x+2)2^{2x} + 9x^2 2^x + 8x + 16.$

**117.** $\left(\dfrac{1}{3}\right)^{\log_{1/3}(3 \cdot (1/2)^x + 5)} < 2^{1+x}.$

**118.** $4^x - 3 \leq 2^{x+1}.$

**119.** $\left(\dfrac{1}{3}\right)^{x+\frac{1}{2}-\frac{2}{x}} > \dfrac{1}{\sqrt{27}}.$

**120.** $\sqrt{9^x + 3^x - 2} \geq 9 - 3^x.$

**121.** $3^{\sqrt{x}} > 2^a.$

**122.** $2^x + 2^{|x|} \geq 2\sqrt{2}.$

**123.** $(0.2)^{\frac{2x-3}{x-2}} > 5.$

**124.** $\left(\dfrac{1}{5}\right)^{\frac{2x+1}{1-x}} > \left(\dfrac{1}{5}\right)^{-3}.$

## Logarithmic Equations

Solve the following equations **(125-261).**

**125.** $\log_{x-1} 3 = 2.$

**126.** $\log_4 (2\log_3 (1 + \log_2(1 + 3\log_3 x))) = \dfrac{1}{2}.$

**127.** $\log_3 (1 + \log_3(2^x - 7)) = 1.$

**128.** $\log_3 (3^x - 8) = 2 - x.$

**129.** $\dfrac{\log_2 (9-2^x)}{3-x} = 1.$

**130.** $\log_{5-x} (x^2 - 2x + 65) = 2.$

**131.** $\log_3 \left( \log_9 x + \dfrac{1}{2} + 9^x \right) = 2x.$

**132.** $\log_3 (x+1) + \log_3 (x+3) = 1.$

**133.** $\log_7 (2^x - 1) + \log_7 (2^x - 7) = 1.$

**134.** $\log 5 + \log (x+10) - 1 = \log (21x - 20) - \log (2x-1).$

**135.** $1 - \log 5 = \dfrac{1}{3} \left( \log \dfrac{1}{2} + \log x + \dfrac{1}{3} \log 5 \right).$

**136.** $\log x - \dfrac{1}{2} \log \left( x - \dfrac{1}{2} \right) = \log \left( x + \dfrac{1}{2} \right) - \dfrac{1}{2} \log \left( x + \dfrac{1}{8} \right).$

**137.** $3^{\log_3 \log \sqrt{x}} - \log x + \log^2 x - 3 = 0.$

**138.** $(x-2)^{\log^2(x-2) + \log (x-2)^5 - 12} = 10^2 \log (x-2).$

**139.** $9^{\log_3(1-2x)} = 5x^2 - 5.$

**140.** $x^{1+\log x} = 10x.$

**141.** $x^{2 \log x} = 10x^2.$

**142.** $x^{\frac{\log x + 5}{3}} = 10^{5 + \log x}.$

**143.** $x^{\log_3 x} = 9.$

**144.** $(\sqrt{x})^{\log_5 x - 1} = 5.$

**145.** $x^{\log x + 1} = 10^6.$

**146.** $x^{\frac{\log x + 7}{4}} = 10^{\log x + 1}.$

**147.** $x^{\log_{\sqrt{x}} (x-2)} = 9.$

**148.** $\left( \dfrac{\log x}{2} \right)^{\log^2 x + \log x^2 - 2} = \log \sqrt{x}.$

**149.** $3 \sqrt{\log_2 x} - \log_2 8x + 1 = 0.$

**150.** $\log^2 x - 3 \log x = \log (x^2) - 4.$

**151.** $\log_{1/3} x - 3 \sqrt{\log_{1/3} x} + 2 = 0.$

**152.** $2 (\log_x \sqrt{5})^2 - 3 \log_x \sqrt{5} + 1 = 0.$

**153.** $\log_2^2 x + 2 \log_2 \sqrt{x} - 2 = 0.$

**154.** $(a^{\log_b x})^2 - 5y^{\log_b a} + 6 = 0.$

**155.** $\log^2 (100x) + \log^2 (10x) = 14 + \log \left( \dfrac{1}{x} \right).$

**156.** $\log_4 (x+3) - \log_4 (x-1) = 2 - \log_4 8.$

**157.** $\log_4 (x^2 - 1) - \log_4 (x-1)^2 = \log_4 \sqrt{(4-x)^2}.$

**158.** $2 \log_3 \dfrac{x-3}{x-7} + 1 = \log_3 \dfrac{x-3}{x-1}.$

**159.** $2\log_4 (4-x) = 4 - \log_2 (-2-x).$

**160.** $3 + 2\log_{x+1} 3 = 2\log_3 (x+1).$

**161.** $\log_x (9x^2) \cdot \log_3^2 x = 4.$

**162.** $\log_{1/2}^2 (4x) + \log_2 \left(\dfrac{x^2}{8}\right) = 8.$

**163.** $\log_{0.5x} x^2 - 14\log_{16x} x^3 + 40\log_{4x} \sqrt{x} = 0.$

**164.** $6 - (1 + 4 \cdot 9^{4 - 2\log_{\sqrt{3}} 3}) \cdot \log_7 x = \log_x 7, \; x \in Q.$

**165.** $\log_3 (4 \cdot 3^x - 1) = 2x + 1.$

**166.** $\log_3 (3^x - 6) = x - 1.$

**167.** $\log_3 (4^x - 3) + \log_3 (4^x - 1) = 1.$

**168.** $\log_3 (\log_{1/2}^2 x - 3\log_{1/2} x + 5) = 2.$

**169.** $\log_5 \left(\dfrac{2+x}{10}\right) = \log_5 \left(\dfrac{2}{x+1}\right).$

**170.** $1 + 2\log_{(x+2)} 5 = \log_5 (x+2).$

**171.** $\log_4 2^{4x} = 2^{\log_2 4}.$

**172.** $\log_2 \left(\dfrac{x}{4}\right) = \dfrac{15}{\log_2 \dfrac{x}{8} - 1}.$

**173.** $\dfrac{1 - 2(\log x^2)^2}{\log x - 2(\log x)^2} = 1.$

**174.** $\log_2 (4 \cdot 3^x - 6) - \log_2 (9^x - 6) = 1.$

**175.** $\dfrac{1}{2}\log (5x - 4) + \log \sqrt{x+1} = 2 + \log 0.18.$

**176.** $\log_4 (2 \cdot 4^{x-2} - 1) + 4 = 2x.$

**177.** $\log_x \sqrt{5} + \log_x (5x) - 2 \cdot 25 = (\log_x \sqrt{5})^2.$

**178.** $\log (\log x) + \log (\log x^4 - 3) = 0.$

**179.** $\log_3 x - 2\log_{1/3} x = 6.$

**180.** $\dfrac{2\log x}{\log (5x - 4)} = 1.$

**181.** $2\log_8 (2x) + \log_8 (x^2 + 1 - 2x) = \dfrac{4}{3}.$

**182.** $\dfrac{1}{6}\log_2 (x - 2) - \dfrac{1}{3} = \log_{1/8} \sqrt{3x - 5}.$

**183.** $2\log_3 (x - 2) + \log_3 (x - 4)^2 = 0.$

**184.** $\sqrt{\log_2 (2x^2) \cdot \log_4 (16x)} = \log_4 x^3.$

**185.** $\dfrac{3\log x + 19}{3\log x - 1} = 2\log x + 1.$

**186.** $\dfrac{\log(\sqrt{x+1} + 1)}{\log \sqrt[3]{x} - 40} = 3.$

**187.** $\log_3^2 6 - \log_3^2 2 = (\log^2 x - 2)\log_3 12.$

**188.** $1 - \dfrac{1}{2}\log\,(2x - 1) = \dfrac{1}{2}\log\,(x - 9).$

**189.** $\dfrac{\log\sqrt{x + 7} - \log 2}{\log 8 - \log\,(x - 5)} = -1.$

**190.** $\log\,(3x^2 + 7) = \log\,(3x - 2) = 1.$

**191.** $\dfrac{1}{3}\log\,(x^2 - 16x + 20) - \log\sqrt[3]{7} = \dfrac{1}{3}\log\,(8 - x).$

**192.** $\log_6 2^{x+3} - \log_6\,(3^x - 2) = x.$

**193.** $\left(1 + \dfrac{1}{2x}\right)\log 3 + \log 2 = \log\,(27 - \sqrt[x]{3}).$

**194.** $\log\,(5^{x-2} + 1) = x + \log 13 - 2\log 5 + (1 - x)\log 2.$

**195.** $\dfrac{1}{2}\log x + 3\log\sqrt{2 + x} = \log\sqrt{x\,(x + 2)} + 2.$

**196.** $\log_2\,(4^x + 1) = x + \log_2\,(2^{x+3} - 6).$

**197.** $\dfrac{1}{\log_6\,(x + 3)} + \dfrac{2\log_{0.25}(4 - x)}{\log\,(3 + x)} = 1.$

**198.** $\log_3\,(9^x + 9) = x + \log_3\,(28 - 2\cdot 3^x).$

**199.** $\log\,(\log x) + \log\,(\log x^3 - 2) = 0.$

**200.** $\log_{\sqrt 5}\,(4^x - 6) - \log_{\sqrt 5}\,(2^x - 2) = 2.$

**201.** $\log_2\,(4^x + 4) = \log_2 2^x + \log_2\,(2^{x+1} - 3).$

**202.** $\log_{\sqrt 5} x \cdot \sqrt{\log_x 5\sqrt 5 + \log_{\sqrt 5} 5\sqrt 5} = -\sqrt 6.$

**203.** $\log\,(3^x - 2^{4-x}) = 2 + \dfrac{1}{4}\log 16 - \dfrac{x\log 4}{2}.$

**204.** $\left(\dfrac{1}{5}\right)^{\log^2 x - \log x} = \dfrac{1}{125}\cdot 5^{\log x - 1}.$

**205.** $x^{3\log^3 x - \frac{2}{3}\log x} = 100\sqrt[3]{10}.$

**206.** $\log_2\,(25^{x+3} - 1) = 2 + \log_2\,(5^{x+3} + 1).$

**207.** $\log_3\sqrt{130 - 7^{\log_x\,(6-x)}} = 2.$

**208.** $\log 2 + \log\,(4^{x-2} + 9) = 1 + \log\,(2^{x-2} + 1).$

**209.** $\log_2\,(4^{x+1} + 4)\cdot \log_2\,(4^x + 1) = \log_{1/\sqrt 2}\sqrt{\dfrac{1}{8}}.$

**210.** $\log_2\,(2x^2)\cdot \log_2\,(16x) = \dfrac{9}{2}\log_2^2 x.$

**211.** $\log_4 + \left(1 + \dfrac{1}{2x}\right)\log 3 = \log\,(\sqrt[x]{3} + 27).$

**212.** $\log\,(x^3 + 27) - 0.5\log\,(x^2 + 6x + 9) = 3\log\sqrt[3]{7}.$

**213.** $5^{\log x} = 50 - x^{\log 5}.$

**214.** $|\,x - 1\,|^{\log^2 x - \log x^2} = |\,x - 1\,|^3.$

**215.** $x^{3\log x - \frac{1}{\log x}} = \sqrt[3]{10}.$

**216.** $|x - 10|\log_2 (x - 3) = 2(x - 10).$

**217.** $\log_4 \log_2 x + \log_2 \log_4 x = 2.$

**218.** $(6x - 5)|\ln (2x + 2 \cdot 3)| = 8 \ln (2x + 2 \cdot 3).$

**219.** $\sqrt{\log_9 (9x^8)} \cdot \log_3 (3x) = \log_3 x^3.$

**220.** $\log^2 (100x) - \log^2 (10x) + \log^2 x = 6.$

**221.** $9^{\log_{1/3}(x+1)} = 5^{\log_{1/5}(2x^2+1)}.$

**222.** $\log^2 (4 - x) + \log (4 - x) \cdot \log \left(x + \frac{1}{2}\right) = 2 \log^2 \left(x + \frac{1}{2}\right).$

**223.** $2 \log_2 \dfrac{x - 7}{x - 1} + \log_2 \dfrac{x - 1}{x + 1} = 1.$

**224.** $\log_{3x+7} (9 + 12x + 4x^2) + \log_{2x+3}(6x^2 + 23x + 21) = 4.$

**225.** $\log \sqrt{1 + x} + 3 \log \sqrt{1 - x} = \log \sqrt{1 - x^2}.$

**226.** $\dfrac{\log (35 - x^3)}{\log (5 - x)} = 3.$

**227.** $\log_x 2 - \log_4 x + \dfrac{7}{6} = 0.$

**228.** $\log (6 \cdot 5^x + 25 \cdot 20^x) = x + \log 25.$

**229.** $\sqrt{\log_2 x} - 0.5 = \log_2 \sqrt{x}.$

**230.** $\log_{1/3} \left[2\left(\dfrac{1}{2}\right)^x - 1\right] = \log_{1/3}\left(\left(\dfrac{1}{4}\right)^x - 4\right).$

**231.** $\log_6 \sqrt{x - 2} + \dfrac{1}{2} \log_6 (x - 11) = 1.$

**232.** $\log_3 (3^{x^2-13x+28} + \dfrac{2}{9}) = \log_5 0.2.$

**233.** $\log_7 2 + \log_{49} x = \log_{1/7} \sqrt{3}.$

**234.** $(\log_{\sin x} 2) \cdot (\log_{\sin^2 x} a) + 1 = 0.$

**235.** $\log_{\sqrt{x}} a \cdot \log_{a^2} \left(\dfrac{a^2 - 4}{2a - x}\right) = 1 \quad .$

**236.** $\sqrt{(\log_{0.04} x) + 1} + \sqrt{(\log_{0.2} x) + 3} = 1.$

**237.** $\log_{3x} \left(\dfrac{3}{x}\right) + \log_3^2 x = 1.$

**238.** $\dfrac{\log_x (2a - x)}{\log_x 2} + \dfrac{\log_a x}{\log_a 2} = \dfrac{1}{\log_{(a^2-1)} 2}.$

**239.** $\dfrac{\log_{a^2\sqrt{x}} a}{\log_{2x} a} + \log_{ax} a \cdot \log_{1/a} 2x = 0.$

**240.** $\log_4 (x + 12) \log_x 2 = 1.$

**241.** $2^{\sqrt{\log_2 x}-2} + 2^{-\sqrt{\log_2 x}-2} = 1$.

**242.** $\log_{0.1} \sin 2x + \log \cos x - \log 7 = 0$.

**243.** $\log_2 (3-x) - \log_2 \dfrac{\sin \frac{3\pi}{4}}{5-x} = \dfrac{1}{2} + \log_2 (x+7)$.

**244.** $\log_4 (2+\sqrt{x+3}) = 2\cos \dfrac{5\pi}{3}$.

**245.** $\log_{1/5} (2x+5) = \log_{1/5} (16-x^2) + \tan \dfrac{5\pi}{4}$.

**246.** $\log_{1/3} \sqrt{x^2 - 2x} = \sin \dfrac{11\pi}{6}$.

**247.** $\log_{16} x + \log_4 x + \log_2 x = 7$.

**248.** $\log 10 + \dfrac{1}{3} \log (3^{2\sqrt{x}} + 271) = 2$.

**249.** $2\log_3 x + \log_3 (x^2 - 3) = \log_3 0.5 + 5^{\log_5 (\log_3 8)}$.

**250.** $\log_5 (3^x + 10) + 7.1^0 = \log_5 (9^x + 56)$.

**251.** $(\log_4 x - 2) \cdot \log_4 x = \dfrac{3}{2} (\log_4 x - 1)$.

**252.** $\log_5 x + \log_{25} x = \log_{1/5} \sqrt{3}$.

**253.** $\log (64 \sqrt[24]{2^{x^2-40x}}) = 0$.

**254.** $\log_2 (9 - 2^x) = 3 - x$.

**255.** $2(\log 2 - 1) + \log (5^{\sqrt{x}} + 1) = \log (5^{1-\sqrt{x}} + 5)$.

**256.** $\log_x 3 + \log_3 x = \log_{\sqrt{x}} 3 + \log_3 \sqrt{x} + \dfrac{1}{2}$.

**257.** $3\sqrt{\log x} + 2\log \sqrt{x^{-1}} = 2$.

**258.** $x + \log (1 + 2^x) = x \log 5 + \log 6$.

**259.** $\log_x (125x) \cdot \log_{25}^2 x = 1$.

**260.** $\log_{2-2x^2} (2 - x^2 - x^4) = 2 - \dfrac{1}{\log_{4/3} (2-2x^2)}$.

**261.** $\log_{3-4x^2} (9 - 16x^4) = 2 + \dfrac{1}{\log_2 (3-4x^2)}$.

**262.** For what values of $a$ does the equation $2\log_3^2 x - |\log_3 x| + a = 0$ possess four solutions?

## Logarithmic Inequalities

Solve the following inequalities (263-384).

**263.** $\log_{1/3}(5x-1) > 0$.

**264.** $\log_5(3x-1) < 1$.

**265.** $\log_{0.5}(1+2x) > -1$.

**266.** $\log_{0.5}(x^2-5x+6) > -1$.

**267.** $\log_8(x^2-4x+3) \leq 1$.

**268.** $\log(x^2-5x+7) < 0$.

**269.** $\log_7 \dfrac{2x-6}{2x-1} > 0$.

**270.** $\log_{1.5} \dfrac{2x-8}{x-2} < 0$.

**271.** $\log_3 \dfrac{1-2x}{x} \leq 0$.

**272.** $\log_{1/3} \dfrac{2-3x}{x} \geq -1$.

**273.** $\log_{1/4} \left( \dfrac{35-x^2}{x} \right) \geq -\dfrac{1}{2}$.

**274.** $\log_3 |3-4x| > 2$.

**275.** $\log_{0.5}^2 x + \log_{0.5} x - 2 \leq 0$.

**276.** $|\log_3 x| - \log_3 x - 3 < 0$.

**277.** $\log_2 x \leq \dfrac{2}{\log_2 x - 1}$.

**278.** $\dfrac{\log^2 x - 3\log x + 3}{\log x - 1} < 1$.

**279.** $\dfrac{1}{1+\log x} + \dfrac{1}{1-\log x} > 2$.

**280.** $\log_{1/4}(2-x) > \log_{1/4}\left(\dfrac{2}{x+1}\right)$.

**281.** $\dfrac{x-1}{\log_3(9-3^x)-3} \leq 1$.

**282.** $\dfrac{1}{\log_4 \dfrac{x+1}{x+2}} < \dfrac{1}{\log_4(x+3)}$.

**283.** $\log_{3x+5}(9x^2+8x+8) > 2$.

**284\*.** $\dfrac{\log_5(x^2-4x+11)^2 - \log_{11}(x^2-4x-11)^3}{\sqrt{2-5x-3x^2}} \geq 0$.

**285.** $\log_3 \dfrac{1+2x}{1+x} < 1$.

**286.** $3^{\log_3 \sqrt{x-1}} < 3^{\log_3 (x-6)} + 3.$

**287.** $\log_{0.2} (x^2 - x - 2) > \log_{0.2} (-x^2 + 2x + 3).$

**288.** $\log_{1/6} (x^2 - 3x + 2) + 1 < 0.$

**289.** $\log_2 (x^2 - 2x) - 3 > 0.$

**290.** $\log_{1/3} (3x + 4) > \log_{1/3} (x^2 + 2).$

**291.** $\left(\dfrac{1}{2}\right)^{\log_2 (x^2-1)} > 1.$

**292.** $\dfrac{\log_2 (x+1)}{x-1} > 0.$

**293.** $\log_{0.5} \sqrt{\dfrac{x-4}{x+3}} < \log_{0.5} 2.$

**294.** $\log_{0.5} (x^2 - 3x + 4) - \log_{0.5} (x-1) < -1.$

**295.** $\log_{0.5} (3x - 4) < \log_{0.5} (x-2).$

**296.** $\log_{0.5} (4 - x) \geq \log_{0.5} 2 - \log_{0.5} x - 1).$

**297.** $\log_{0.1} (x^2 + x - 2) > \log_{0.1} (x + 3).$

**298.** $1 + \log_2 (x - 2) > \log_2 (x^2 - 3x + 2).$

**299.** $\log_{1/3} (2^{x+2} - 4^x) \geq -2.$

**300.** $\log_{1/\sqrt{5}} (6^{x+1} - 36^x) \geq -2.$

**301.** $\log_{1/3} (\log_4 (x^2 - 5)) > 0.$

**302.** $\log_{0.5} \left(\log_6 \dfrac{x^2 + x}{x + 4}\right) < 0.$

**303.** $\log_{0.1} \left(\log_2 \dfrac{x^2 + 1}{|x - 1|}\right) < 0.$

**304.** $\log_x (\log_9 (3^x - 9)) < 1.$

**305.** $\log_{0.5} \left(\log_8 \dfrac{x^2 - 2x}{x - 3}\right) < 0.$

**306.** $\dfrac{1 - \log_{0.5}(-x)}{\sqrt{2 - 6x}} < 0.$

**307.** $\dfrac{1 - \log_{0.5} (-x)}{\sqrt{-2 - 6x}} < 0.$

**308.** $\dfrac{(x - 0.5)(3 - x)}{\log_2 |x - 1|} > 0.$

**309.** $\dfrac{x^2 - 4}{\log_{1/2} (x^2 - 1)} < 0.$

**310.** $\log_3 \dfrac{|x^2 - 4x| + 3}{x^2 + |x - 5|} \geq 0.$

**311.** $\log_{x^2} \dfrac{4x - 5}{|x - 2|} \geq \dfrac{1}{2}.$

**312.** $\dfrac{\log_2 (3 \cdot 2^{x-1} - 1)}{x} \geq 1.$

**313.** $\log_{1/3} (x + 1) > \log_3 (2 - x).$

**314.** $\log_{1/5} (x^2 - 6x + 18) + 2\log_5 (x - 4) < 0.$

**315.** $\log_{1/3} x > \log_x 3 - \dfrac{5}{2}.$

**316.** $\log_2 (2^x - 1) \cdot \log_{1/2}(2^{x+1} - 2) > -2.$

**317.** $\log_4 (18 - 2^x) \cdot \log_2 \dfrac{18 - 2^x}{8} \leq -1.$

**318.** $\log_5 (x - 3) + \dfrac{1}{2}\log_5 3 < \dfrac{1}{2}\log_5 (2x^2 - 6x + 7).$

**319.** $\log_2 (2x - 1) > \log_{1/\sqrt{2}} 2.$

**320.** $\log_{4/3}(\sqrt{x + 3} - \sqrt{x}) + \log_{4/9}\left(\dfrac{2}{3}\right) \geq 0.$

**321.** $2\log_5 x - \log_x 125 < 1.$

**322.** $\log_5 \sqrt{3x + 4} \cdot \log_x 5 > 1.$

**323.** $\log_3 x + \log_{\sqrt{3}} x + \log_{1/3} x < 6.$

**324.** $\left(\dfrac{1}{3}\right)^{\log_{1/9}\left(x^2 - \frac{10}{3}x + 1\right)} \leq 1.$

**325.** $\log_{0.5}^2 x + 6 \geq 5\log_{0.5} x.$

**326.** $\left(\dfrac{1}{10}\right)^{\log_{x-3}(x^2 - 4x + 3)} \geq 1.$

**327.** $\log_2^2 \dfrac{4x - 3}{4 - 3x} > -\dfrac{1}{2}.$

**328.** $\log_{2x}(x^2 - 5x + 6) < 1.$

**329.** $\log_x (x^3 - x^2 - 2x) < 3.$

**330.** $\log_{2x+3} x^2 < 1.$

**331.** $\log_{(x-3)} (2 (x^2 - 10x + 24)) \geq \log_{(x-3)} (x^2 - 9).$

**332.** $\log_{(x-\sqrt{2})} \dfrac{x + 7}{x - 2} \leq \log_{(x-\sqrt{2})} 2x.$

**333.** $\log_{(x-4.5)} \dfrac{x + 4}{2x - 6} \leq \log_{(x-4.5)} (x - 5).$

**334.** $\log_{|x|} (\sqrt{9 - x^2} - x - 1) \geq 1.$

**335.** $\log_x 2x \leq \sqrt{\log_x (2x^3)}.$

**336.** $\log_{\log_2 (0.5x)} (x^2 - 10x + 22) > 0.$

**337.** $\log_{\frac{x+6}{3}}\left(\log_2 \dfrac{x-1}{2+x}\right) > 0.$

**338.** $\log_{\frac{3x}{x^2+1}}(x^2 - 2.5x + 1) \geq 0.$

**339.** $\log_{x^2}(2+x) < 1.$

**340.** $\log_{9x^2}(6 + 2x - x^2) \leq \dfrac{1}{2}.$

**341.** $\log_{1/x}(2.5x - 1) \geq -2$

**342.** $\log_\pi(x + 27) - \log_\pi(16 - 2x) < \log_\pi x.$

**343.** $\left(\dfrac{1}{2}\right)^{\log_2(x^2-1)} > 1.$

**344.** $\left(\dfrac{2}{5}\right)^{\log_{0.25}(x^2+5x+8)} \leq 2.5.$

**345.** $2^{\log_{2-x}(x^2+8x+15)} < 1.$

**346.** $(0.5)^{\log_5 \log_{0.3}(x-0.7)} < 1.$

**347.** $(0.5)^{\log_{1/3}\frac{x+5}{x^2+3}} > 1.$

**348.** $(0.5)^{\log_3 \log_{1/5}\left(x^2-\frac{4}{5}\right)} < 1.$

**349.** $\log_2\left(\sin\dfrac{x}{2}\right) < -1.$

**350.** $\log_4(3^x - 1)\log_{1/4}\dfrac{3^x-1}{16} \leq \dfrac{3}{4}.$

**351.** $\log_5 x + \log_x \dfrac{x}{3} < \dfrac{\log_5 x\,(2-\log_3 x)}{\log_3 x}.$

**352.** $\log_{1/2} x + \log_3 x > 1.$

**353.** $\dfrac{1-\log_4 x}{1+\log_2 x} \leq \dfrac{1}{2}.$

**354.** $\sqrt{\log_{1/2}^2 x + 4\log_3 \sqrt{x}} < \sqrt{2}\,(4 - \log_{16} x^4).$

**355.** $\log_{\frac{x+3}{x-3}} 4 < 2\,(\log_{1/2}(x-3) - \log_{\frac{\sqrt{2}}{2}}\sqrt{x+3}).$

**356.** $\sqrt{\log_2 \dfrac{3-2x}{1-x}} < 1.$

**357.** $\sqrt{x^{\log_2 \sqrt{x}}} \geq 2.$

**358.** $\log_{\sqrt{2}}(5^x - 1)\log_{\sqrt{2}}\dfrac{2\sqrt{2}}{5^x - 1} > 2.$

**359.** $\log_3 \dfrac{2^x - 5}{27} \cdot \log_{1/3} (2^x - 5) < 2$

**360.** $\left(\dfrac{x}{10}\right)^{\log x - 2} < 100.$

**361.** $\dfrac{1}{\log_3 (x+1)} < \dfrac{1}{2 \log_9 \sqrt{x^2 + 6x + 9}}.$

**362.** $\log 10^{\log (x^2 + 21)} > 1 + \log x.$

**363.** $\log_a (x-1) + \log_a x > 2$

**364.** $\dfrac{3\log_a x + 6}{\log_a^2 x + 2} > 1.$

**365.** $2\log_5 (x - 3d + 2) - \log_{\sqrt{5}} (x + 2d - 8) \le 4.$

**366.** $\dfrac{1}{5 - \log_a x} + \dfrac{2}{1 + \log_a x} < 1.$

**367.** $\log^4 x - 13\log^2 x + 36 > 0.$

**368.** $\log_{1/2} \left( \log_2 \dfrac{x}{1+x} \right) > 0.$

**369.** $\log_2 x^2 + \log_{\sqrt{2}} (x-1) < \log_{\sqrt{2}} \log_{\sqrt{2}} 2.$

**370.** $\log^2 x \ge \log x + 2.$

**371.** $\log_{1/3} x + 2\log_{19} (x-1) \le \log_{13} 6.$

**372.** $\dfrac{1 + \log_{x+1}(x-3)}{\log_{x+1} 3} < \log_3 (2x - 3).$

**373.** $\log_a (1 - 8a^{-x}) \ge 2(1 - x) \quad (a \in R).$

**374.** $2\log_{1/2} (x-1) \le \dfrac{1}{3} - \dfrac{1}{\log (x^2 - x)^8}.$

**375.** $\log_2 (4^x - 5 \cdot 2^x + 2) > 2.$

**376.** $\log_{x^2 + 2x - 3} \dfrac{|x + 4| - |x|}{x - 1} > 0.$

**377.** $(\log_{1/2} x)^2 - \log_{1/2} x^2 > (\log_{1/2} 3)^2 - 1.$

**378.** $\log_2 (x + 14) + 2\log_4 (x + 2) < 2\log_{1/2} \left(\dfrac{1}{8}\right).$

**379.** $\log_{1/4} (x + 1) \ge -2\log_{1/16} 2 + \log_{1/4}(x^2 + 3x + 8).$

**380.** $\log_{1/2} \log_3 \dfrac{x+1}{x-1} \ge 0.$

**381.** $\log_{0.5} \log_5 (x^2 - 4) > \log_{0.5} 1.$

**382.** $\log_{1/4} x^2 + \dfrac{1}{\log_{(x-1)} \frac{1}{2}} \ge \log_{1/2} 2$

**383.** $\log_2 \dfrac{3x - 1}{2 - x} < 1.$

**384.** $\log_{0.5} (x + 5)^2 > \log_{1/2}(3x - 1)^2.$

**385.** The inequality $\log_a (x^2 - x - 2) > \log_a(-x^2 + 2x + 3)$ is known to be satisfied for $x = \dfrac{4}{9}$. Find all solutions of this inequality.

## Systems of Exponential and Logarithmic Equations

Solve the following systems of equations **(386-427)**.

**386.** $\begin{cases} \log \sqrt{5-x} + \log 2 = \log(x+3), \\ x^2 + 7x - 8 = 0. \end{cases}$

**387.** $\begin{cases} 4^{x+y} = 128, \\ 5^{3x-2y-3} = 1. \end{cases}$

**388.** $\begin{cases} 4^{(x-y)^2-1} = 1, \\ 5^{x+y} = 125. \end{cases}$

**389.** $\begin{cases} 3^{2x} - 2^y = 77, \\ 3^x - 2^{y/2} = 7. \end{cases}$

**390.** $\begin{cases} 2^x + 2^y = 12, \\ x + y = 5. \end{cases}$

**391.** $\begin{cases} \log_4 x - \log_2 y = 0, \\ x^2 - 5y^2 + 4 = 0. \end{cases}$

**392.** $\begin{cases} 3 \cdot 2^x + 2 \cdot 3^y = 2.75, \\ 2^x - 3^y = -0.75. \end{cases}$

**393.** $\begin{cases} 7^x - 16y = 0, \\ 4^x - 49y = 0. \end{cases}$

**394.** $\begin{cases} 4^{\frac{x}{y}} - 3 \cdot 4^{\frac{5y-x}{y}} = 16, \\ \sqrt{x} - \sqrt{2y} = \sqrt{12} - \sqrt{8}. \end{cases}$

**395.** $\begin{cases} 3^x \cdot 2^y = 972, \\ \log_{\sqrt{3}}(x-y) = 2. \end{cases}$

**396.** $\begin{cases} \log_4 x - \log_x y = \dfrac{7}{6}, \\ xy = 16. \end{cases}$

**397.** $\begin{cases} y - \log_3 x = 1, \\ x^y = 3^{12}. \end{cases}$

**398.** $\begin{cases} x^{\log y} = 2, \\ xy = 20. \end{cases}$

**399.** $\begin{cases} \log_2 x + 2\log_2 y = 3, \\ \qquad x^2 + y^4 = 16. \end{cases}$

**400.** $\begin{cases} \dfrac{1}{x} - \dfrac{1}{y} = \dfrac{2}{15}, \\ \log_3 x + \log_3 y = 1 + \log_3 5. \end{cases}$

**401.** $\begin{cases} x + y = 4 + \sqrt{y^2 + 2}, \\ \log x - 2\log 2 = \log\left(1 + \dfrac{1}{2}y\right). \end{cases}$

**402.** $\begin{cases} 8^{\log_9(x-4y)} = 1, \\ 4^{x-2y} - 7\cdot 2^{x-2y} = 8. \end{cases}$

**403.** $\begin{cases} 5\log_2 x = \log_2 y^3 - \log_{\sqrt{2}} 2, \\ \quad \log_2 y = 8 - \log_{\sqrt{2}} x. \end{cases}$

**404.** $\begin{cases} \log(x^2 + y^2) - 1 = \log 13, \\ \log(x+y) - \log(x-y) = 3\log 2. \end{cases}$

**405.** $\begin{cases} \dfrac{x}{y^2} + \dfrac{y}{x^2} = \dfrac{9}{8}, \\ \log_2 x + \log_{\sqrt{2}} \sqrt{y} = 3. \end{cases}$

**406.** $\begin{cases} \log\sqrt{(x+y)^2} = 1, \\ \log y - \log|x| = \log 2. \end{cases}$

**407.** $\begin{cases} x^2 + y^2 = 4 + \dfrac{1}{2}y, \\ \log_3(x+2y) + \log_{1/3}(x-2y) = 1. \end{cases}$

**408.** $\begin{cases} 3\cdot\left(\dfrac{2}{3}\right)^{2x-y} + 7\cdot\left(\dfrac{2}{3}\right)^{\frac{2x-y}{2}} - 6 = 0, \\ \log(3x-y) + \log(y+x) = 4\log 2. \end{cases}$

**409.** $\begin{cases} \log_5 x + 3^{\log_3 y} = 7, \\ \qquad x^y = 625^3. \end{cases}$

**410.** $\begin{cases} 3^{\log_x 2} = y^{\log_5 y}, \\ 2^{\log_y 3} = x^{\log_7 x}. \end{cases}$

**411.** $\begin{cases} xy = a^2, \\ (\log x)^2 + (\log y)^2 = \dfrac{5}{2}(\log a^2)^2. \end{cases}$

**412.** $\begin{cases} 2^x \cdot 8^{-y} = 2\sqrt{2}, \\ \log_9 \dfrac{1}{x} + 0.5 = \dfrac{1}{2} \log_3 (9y). \end{cases}$

**413.** $\begin{cases} 10^{3-\log (x-y)} = 250, \\ \sqrt{x-y} + \dfrac{1}{2}\sqrt{x+y} = \dfrac{26-y}{\sqrt{x-y}}. \end{cases}$

**414.** $\begin{cases} \log_4 x - \log_2 y = 0, \\ x^2 - 2y^2 = 8. \end{cases}$

**415.** $\begin{cases} \log_5 x + \log_5 7 \cdot \log_7 y = 1 + \log_5 2, \\ 3 + 2\log_2 y = \log_2 5 (2 + 3\log_5 x). \end{cases}$

**416.** $\begin{cases} \log_2 (x-y) = 5 - \log_2 (x+y), \\ \dfrac{\log x - \log 4}{\log y - \log 3} = -1. \end{cases}$

**417.** $\begin{cases} \log_2 x + \log_4 y = -2\log_{1/2} 4, \\ \log_4 x + \log_2 y = 5\log 10. \end{cases}$

**418.** $\begin{cases} 2\,(\log_y x + \log_x y) = 5, \\ xy = 8. \end{cases}$

**419.** $\begin{cases} 2\log_2 x - 3^y = 15, \\ 3^y \cdot \log_2 x = 2\log_2 x + 3^{y+1}. \end{cases}$

**420.** $\begin{cases} \log_y x + \log_x y = 2 \cdot 5, \\ xy = 27. \end{cases}$

**421.** $\begin{cases} \log_2 x - \log_2 y + \log_2 (y+1) = 1 + \log_2 3, \\ 2^{y^2} - 2^x \cdot 16^y = 0. \end{cases}$

**422.** $\begin{cases} \log_3 (\log_2 x) + \log_{1/3} (\log_{1/2} y) = 1, \\ xy^2 = 4. \end{cases}$

**423.** $\begin{cases} \log_x (xy) = \log_y x^2, \\ y^{2\log_y x} = 4y + 3. \end{cases}$

**424.** $\begin{cases} 2\log_4 (y+1) + \log_2 y = \log_2 \left( \dfrac{x}{y} - 2 \right), \\ 5 + \log_2 \dfrac{y}{x} = \dfrac{6}{\log_2 \dfrac{x}{y}}. \end{cases}$

**425.** $\begin{cases} 4^{\frac{x}{y} + \frac{y}{x}} = 32, \\ \log_3 (x-y) = 1 - \log_3 (x+y). \end{cases}$

**426.** $\begin{cases} 0.5 \cdot \log_2 x - \log_4 y = 0, \\ \quad\quad x^2 - 5y^2 + 4 = 0. \end{cases}$

**427.** $\begin{cases} \quad\quad 3^x \cdot 2^y = 576, \\ \log_{\sqrt{2}} (y - x) = 4. \end{cases}$

## 1.10. Transformations of Trigonometric Expressions. Inverse Trigonometric Functions

**The Fundamental Formulas**

$$\sin^2 \alpha + \cos^2 \alpha = 1. \tag{1}$$

$$\tan^2 \alpha + 1 = \frac{1}{\cos^2 \alpha}, \alpha \neq \frac{\pi}{2} + \pi n, n \in \mathbf{Z}. \tag{2}$$

$$\cot^2 \alpha + 1 = \frac{1}{\sin^2 \alpha}, \alpha \neq \pi n \; n \in \mathbf{Z}. \tag{3}$$

$$\sin (\alpha + \beta) = \sin \alpha \cos \beta + \cos \alpha \sin \beta. \tag{4}$$

$$\sin (\alpha - \beta) = \sin \alpha \cos \beta - \cos \alpha \sin \beta. \tag{5}$$

$$\cos (\alpha + \beta) = \cos \alpha \cos \beta - \sin \alpha \sin \beta. \tag{6}$$

$$\cos (\alpha - \beta) = \cos \alpha \cos \beta + \sin \alpha \sin \beta. \tag{7}$$

$$\tan(\alpha + \beta) = \frac{\tan \alpha + \tan \beta}{1 - \tan \alpha \tan \beta}, \alpha, \beta, \alpha + \beta \neq \frac{\pi}{2} + \pi n, n \in \mathbf{Z}. \tag{8}$$

$$\tan (\alpha - \beta) = \frac{\tan \alpha - \tan \beta}{1 + \tan \alpha \tan \beta}, \alpha, \beta, \alpha - \beta \neq \frac{\pi}{2} + \pi n, n \in \mathbf{Z}. \tag{9}$$

$$\sin 2\alpha = 2 \sin \alpha \cos \alpha. \tag{10}$$

$$\cos 2\alpha = \cos^2 \alpha - \sin^2 \alpha = 2 \cos^2 \alpha - 1 = 1 - 2 \sin^2 \alpha \tag{11}$$

$$\tan 2\alpha = \frac{2 \tan \alpha}{1 - \tan^2 \alpha}, \alpha, 2\alpha \neq \frac{\pi}{2} + \pi n, n \in \mathbf{Z}. \tag{12}$$

$$\sin 3\alpha = 3 \sin \alpha - 4 \sin^3 \alpha. \tag{13}$$

Here is a brief derivation of formula (13) : $\sin 3\alpha = \sin (2\alpha + \alpha)$ $= \sin 2\alpha \cos \alpha + \cos 2\alpha \sin \alpha = 2 \sin \alpha \cos^2 \alpha + (\cos^2 \alpha - \sin^2 \alpha) \sin \alpha = 2 \sin \alpha$ $(1 - \sin^2 \alpha) + (1 - 2 \sin^2 \alpha) \sin \alpha = 3 \sin \alpha - 4 \sin^3 \alpha.$

$$\cos 3\alpha = 4\cos^3 \alpha - 3\cos\alpha. \tag{14}$$

$$\sin\alpha + \sin\beta = 2\sin\frac{\alpha+\beta}{2}\cos\frac{\alpha-\beta}{2}. \tag{15}$$

$$\sin\alpha - \sin\beta = 2\sin\frac{\alpha-\beta}{2}\cos\frac{\alpha+\beta}{2}. \tag{16}$$

$$\cos\alpha + \cos\beta = 2\cos\frac{\alpha+\beta}{2}\cos\frac{\alpha-\beta}{2}. \tag{17}$$

$$\cos\alpha - \cos\beta = -2\sin\frac{\alpha-\beta}{2}\sin\frac{\alpha+\beta}{2}. \tag{18}$$

$$\sin\alpha\cos\beta = \frac{1}{2}(\sin(\alpha-\beta)+\sin(\alpha+\beta)). \tag{19}$$

$$\sin\alpha\sin\beta = \frac{1}{2}(\cos(\alpha-\beta)-\cos(\alpha+\beta)). \tag{20}$$

$$\cos\alpha\cos\beta = \frac{1}{2}(\cos(\alpha-\beta)+\cos(\alpha+\beta)). \tag{21}$$

$$1 + \cos 2\alpha = 2\cos^2\alpha. \tag{22}$$

$$1 - \cos 2\alpha = 2\sin^2\alpha. \tag{23}$$

$$\cos^2\alpha = \frac{1+\cos 2\alpha}{2}. \tag{24}$$

$$\sin^2\alpha = \frac{1-\cos 2\alpha}{2}. \tag{25}$$

$$\left|\sin\frac{\alpha}{2}\right| = \sqrt{\frac{1-\cos\alpha}{2}}. \tag{26}$$

$$\left|\cos\frac{\alpha}{2}\right| = \sqrt{\frac{1+\cos\alpha}{2}}. \tag{27}$$

$$\left|\tan\frac{\alpha}{2}\right| = \sqrt{\frac{1-\cos\alpha}{1+\cos\alpha}}, \alpha \neq \pi + 2\pi n, n \in \mathbf{Z}. \tag{28}$$

$$\sin\alpha = \frac{2\tan\dfrac{\alpha}{2}}{1+\tan^2\dfrac{\alpha}{2}}, \alpha \neq \pi + 2\pi n, n \in \mathbf{Z}. \tag{29}$$

$$\cos\alpha = \frac{1-\tan^2\dfrac{\alpha}{2}}{1+\tan^2\dfrac{\alpha}{2}}, \alpha \neq \pi + \pi n, n \in \mathbf{Z}. \tag{30}$$

$$\tan\frac{\alpha}{2} = \frac{\sin\alpha}{1+\cos\alpha}, \alpha \neq 2\pi n, n \in \mathbf{Z}. \tag{31}$$

$$\tan\frac{\alpha}{2} = \frac{1-\cos\alpha}{\sin\alpha}, \alpha \neq \pi n, n \in \mathbf{Z}. \tag{32}$$

$$|\sin \alpha| = \frac{\tan \alpha}{\sqrt{1 + \tan^2 \alpha}}, \alpha \neq \frac{\pi}{2} + \pi n, n \in \mathbf{Z}. \tag{33}$$

$$|\sin \alpha| = \frac{1}{\sqrt{1 + \cot^2 \alpha}}, \alpha \neq \pi n, n \in \mathbf{Z}. \tag{34}$$

$$|\cos \alpha| = \frac{1}{\sqrt{1 + \tan^2 \alpha}}, \alpha \neq \frac{\pi}{2} + \pi n, n \in \mathbf{Z}. \tag{35}$$

$$|\cos \alpha| = \frac{|\cot \alpha|}{\sqrt{1 + \cot^2 \alpha}}, \alpha \neq \pi n, n \in \mathbf{Z}. \tag{36}$$

$$a \cos x + b \sin x = \sqrt{a^2 + b^2} \sin (x + \alpha), a^2 + b^2 \neq 0, \tag{37}$$

where

$$\sin \alpha = \frac{a}{\sqrt{a^2 + b^2}}, \cos \alpha = \frac{b}{\sqrt{a^2 + b^2}}.$$

## Identify Transformations of Trigonometric Expressions

Prove the following identities **(1-60)**.

1. $\dfrac{1 + \sin \alpha}{1 + \cos \alpha} \dfrac{1 + \sec \alpha}{1 + \operatorname{cosec} \alpha} = \tan \alpha.$

2. $\dfrac{\sin x + \cos x}{\cos^3 x} = \tan^3 x + \tan^2 x + \tan x + 1.$

3. $\dfrac{1}{2} (\cos t + \sqrt{3} \sin t) = \cos \left( \dfrac{\pi}{3} - t \right).$

4. $\dfrac{\sin^4 \alpha + 2 \sin \alpha \cos \alpha - \cos^4 \alpha}{\tan 2\alpha - 1} = \cos 2\alpha.$

5. $\dfrac{1 - 2\sin^2 \alpha}{1 + \sin 2\alpha} = \dfrac{1 - \tan \alpha}{1 + \tan \alpha}.$

6. $\dfrac{\cos^2 \beta (\tan^2 \alpha - \tan^2 \beta)}{\sec^2 \alpha} = \sin (\alpha + \beta) \sin (\alpha - \beta).$

7. $\tan \left( \dfrac{\pi}{4} + \alpha \right) = \dfrac{1 + \sin 2\alpha}{\cos 2\alpha}.$

8. $\sin^2 (\alpha + \beta) - \sin^2 \alpha - \sin^2 \beta = 2 \sin \alpha \sin \beta \cos (\alpha + \beta).$

9. $(1 + \tan \beta \tan 2\beta) \sin 2\beta = \tan 2\beta.$

10. $\sin \alpha - \sin 2\alpha + \sin 3\alpha = 4 \cos \dfrac{3\alpha}{2} \cos \alpha \sin \dfrac{\alpha}{2}.$

11. $\dfrac{\sin^2 4\alpha}{2 \cos \alpha + \cos 3\alpha + \cos 5\alpha} = 2 \sin \alpha \sin 2\alpha.$

12. $\sin \alpha + 2\sin 3\alpha + \sin 5\alpha + 4 \sin 3\alpha \cos^2 \alpha.$

13. $\sin 2\alpha + \sin 4\alpha - \sin 6\alpha = 4 \sin \alpha \sin 2\alpha \sin 3\alpha.$

**14.** $\dfrac{\sin\alpha + \sin 3\alpha + \sin 5\alpha}{\cos\alpha + \cos\alpha + \cos 5\alpha} = \tan 3\alpha.$

**15.** $\sin^2(45°+\alpha) - \sin^2(30°-\alpha) - \sin 15° \cos(15°+2\alpha) = \sin 2\alpha.$

**16.** $\dfrac{(\sin 2\alpha - \sin 6\alpha) + (\cos 2\alpha - \cos 6\alpha)}{\sin 4\alpha - \cos 4\alpha} = 2\sin 2\alpha.$

**17.** $\left( \dfrac{\tan^2\dfrac{\alpha-\pi}{4} - 1}{\tan^2\dfrac{\alpha-\pi}{4} + 1} + \cos\dfrac{\alpha}{2} \cdot \cot 4\alpha \right) \sec\dfrac{9}{2} = \operatorname{cosec} 4\alpha.$

**18.** $\cos^2\alpha - \sin^2 2\alpha = \cos^2\alpha \cos 2\alpha - 2\sin^2\alpha \cos^2\alpha.$

**19.** $\dfrac{\sin 2x - \sin 3x + \sin 4x}{\cos 2x - \cos 3x + \cos 4x} = \tan 3x.$

**20.** $1 - \cos(2x - \pi) - \cos(4x + \pi) + \cos(6x - 2\pi) = 4\cos x \times \cos 2x \cos 3x.$

**21.** $\dfrac{\tan 2x \tan x}{\tan 2x - \tan x} = \sin 2x.$

**22.** $\dfrac{\sin^4\alpha - \cos^4\alpha + \cos^2\alpha}{2(1 - \cos\alpha)} = \cos^2\dfrac{\alpha}{2}.$

**23.** $\dfrac{\sin(\pi-\alpha)}{\sin\alpha - \cos\alpha \tan\dfrac{\alpha}{2}} + \cos(\pi-\alpha) = 1.$

**24.** $\dfrac{\cos\alpha}{\cot^2\dfrac{\alpha}{2} - \tan^2\dfrac{\alpha}{2}} = \dfrac{1}{4}\sin^2\alpha.$

**25.** $\dfrac{1}{4\sin^2\alpha\cos^2\alpha} - \dfrac{(1 - \tan^2\alpha)^2}{4\tan^2\alpha} = 1.$

**26.** $\dfrac{\sqrt{2}}{2}(\cos\alpha + \sin\alpha) = \cos\left(\dfrac{\pi}{4} - \alpha\right).$

**27.** $\dfrac{1 - \sin 2\alpha}{1 + \sin 2\alpha} = \cot^2\left(\dfrac{\pi}{4} + \alpha\right).$

**28.** $\dfrac{\sin\alpha + \sin 5\alpha}{\cos\alpha + \cos 5\alpha} = \tan 3\alpha.$

**29.** $\dfrac{\tan\left(x - \dfrac{\pi}{2}\right)\cos\left(\dfrac{3}{2}\pi + x\right) - \sin^3\left(\dfrac{7}{2}\pi - x\right)}{\cos\left(x - \dfrac{\pi}{2}\right)\tan\left(\dfrac{3}{2}\pi + x\right)} = \sin^2 x.$

**30.** $\sin^6\alpha + \cos^6\alpha + 3\sin^2\alpha\cos^2\alpha = 1.$

**31.** $\dfrac{1 - 2\cos^2\alpha}{2\tan\left(\alpha - \dfrac{\pi}{4}\right)\sin^2\left(\dfrac{\pi}{4} + \alpha\right)} = 1.$

**32.** $\sin^2\left(\dfrac{\pi}{8} + \alpha\right) - \sin^2\left(\dfrac{\pi}{8} - \alpha\right) = \dfrac{\sin 2\alpha}{\sqrt{2}}.$

**33.** $\dfrac{\cot\alpha + \tan\alpha}{1 + \tan 2\alpha \tan\alpha} = 2\cot 2\alpha.$

**34.** $\dfrac{\cot\alpha - \tan\alpha}{2\sin\alpha + \cos(90° + 3\alpha) + \sin 5\alpha} = \operatorname{cosec}\alpha \operatorname{cosec} 4\alpha.$

**35.** $\dfrac{\tan 3\alpha}{\tan\alpha} = \dfrac{3 - \tan^2\alpha}{1 - 3\tan^2\alpha}.$

**36.** $\cos 4\alpha - \sin 4\alpha \cot 2\alpha = \cos 2\alpha - 2\cos^2\alpha.$

**37.** $1 + 2\cos 7a = \dfrac{\sin 10.5a}{\sin 3.5a}.$

**38.** $(\sin\alpha - \sin\beta)(\sin\alpha + \sin\beta) = \sin(\alpha - \beta)\sin(\alpha + \beta).$

**39.** $\cot\alpha - \tan\alpha - 2\tan 2\alpha - 4\tan 4\alpha = 8\cot 8\alpha.$

**40.** $\dfrac{\cos 4\alpha \tan 2\alpha - \sin 4\alpha}{\cos 4\alpha \cot 2\alpha + \sin 4\alpha} = -\tan^2 2\alpha.$

**41.** $\dfrac{1 - \cos 2\alpha + \sin 2\alpha}{1 + \cos 2\alpha + \sin 2\alpha} = \tan\alpha.$

**42.** $\dfrac{\sin\alpha + \cos\alpha}{\cos\alpha - \sin\alpha}\tan\left(\dfrac{\pi}{4} + \alpha\right) + 1 = \operatorname{cosec}^2\left(\dfrac{\pi}{4} - \alpha\right).$

**43.** $3 - 4\cos 2\alpha + \cos 4\alpha = 8\sin^4\alpha.$

**44.** $\dfrac{\sin^2 2\alpha + 4\sin^4\alpha - 4\sin^2\alpha\cos^2\alpha}{4 - \sin^2 2\alpha - 4\sin^2\alpha} = \tan^4\alpha.$

**45.** $\dfrac{\sin^2 2\alpha + 4\sin^2\alpha - 1}{1 - 8\sin^2\alpha - \cos 4\alpha} = \dfrac{1}{2}\cot^4\alpha.$

**46.** $\dfrac{\sqrt{1 + \cos\alpha} + \sqrt{1 - \cos\alpha}}{\sqrt{1 + \cos\alpha} - \sqrt{1 - \cos\alpha}} = \cot\left(\dfrac{\alpha}{2} + \dfrac{\pi}{4}\right),\ \pi < \alpha < 2\pi.$

**47.** $1 - \sin 4\alpha + \cot\left(\dfrac{3}{4}\pi - 2\alpha\right)\cos 4\alpha = 0.$

**48.** $\dfrac{\sin x + \cos x}{\sin x - \cos x} - \dfrac{\sec^2 x + 2}{\tan^2 x - 1} = \dfrac{2}{\tan x + 1}.$

**49.** $\dfrac{2\sin^2\left(\dfrac{\pi}{4} - \alpha\right)}{\cos 2\alpha} = \cot\left(\dfrac{\pi}{4} + \alpha\right).$

**50.** $\cos 2\alpha - \cos 3\alpha - \cos 4\alpha + \cos 5\alpha = -4\sin\left(\dfrac{\alpha}{2}\right)\sin\alpha\cos\left(\dfrac{7\alpha}{2}\right).$

**51.** $\dfrac{1 + \tan^4\alpha}{\tan^2\alpha + \cot^2\alpha} = \tan^2\alpha.$

**52.** $\dfrac{3 - 4\cos 2\alpha + \cos 4\alpha}{3 + 4\cos 2\alpha + \cos 4\alpha} = \tan^4\alpha.$

**53.** $\cos^6\alpha - \sin^6\alpha = \dfrac{(3 + \cos^2 2\alpha)\cos 2\alpha}{4}.$

**54.** $2(\sin^6\alpha + \cos^6\alpha) - 3(\sin^4\alpha + \cos^4\alpha) + 1 = 0.$

**55.** $\dfrac{\cot\alpha + \cot(270°+\alpha)}{\cot\alpha - \cot(270°+\alpha)} - 2\cos(135°+\alpha)\cos(315°-\alpha) = 2\cos 2\alpha.$

**56.** $\dfrac{1+\tan\alpha + \tan^2\alpha}{1+\cot\alpha + \cot^2\alpha} = \tan^2\alpha.$

**57.** $\dfrac{\sqrt{2}-\sin\alpha - \cos\alpha}{\sin\alpha - \cos\alpha} = \tan\left(\dfrac{\alpha}{2} - \dfrac{\pi}{8}\right).$

**58.** $\dfrac{1+\sin 2\alpha}{\sin\alpha + \cos\alpha} - \dfrac{1-\tan^2\dfrac{\alpha}{2}}{1+\tan^2\dfrac{\alpha}{2}} = \sin\alpha.$

**59.** $\dfrac{2\cos^2 2\alpha + \sqrt{3}\sin 4\alpha - 1}{2\sin^2 2\alpha + \sqrt{3}\sin 4\alpha - 1} = \dfrac{\sin(4\alpha + 30°)}{\sin(4\alpha - 30°)}.$

**60.** $1-\sin^4\alpha - \cos^4\alpha = \dfrac{1}{2}\sin^2 2\alpha.$

Simplify the following expressions **(61-79).**

**61.** $\dfrac{\tan(180°-\alpha)\cos(180°-\alpha)\tan(90°-\alpha)}{\sin(90°+\alpha)\cot(90°-\alpha)\tan(90°+\alpha)}.$

**62.** $\dfrac{2(\sin 2\alpha + 2\cos^2\alpha - 1)}{\cos\alpha - \sin\alpha - \cos 3\alpha + \sin 3\alpha}.$

**63.** $3\cos^2 x - 4\sin x \cos x - \sin^2 x - 1.$

**64.** $4\cos\left(2\alpha - \dfrac{3}{2}\pi\right) + \cos(2\alpha - \pi) + \sin\left(\dfrac{5}{2}\pi - 6\alpha\right).$

**65.** $\dfrac{1+\sin 2x}{(\sin x + \cos x)^2}.$

**66.** $\dfrac{\tan\alpha + \sin\alpha}{2\cos^2\dfrac{\alpha}{2}}.$

**67.** $\dfrac{1-\cos 4\alpha}{\sec^2 2\alpha - 1} + \dfrac{1+\cos 4\alpha}{\mathrm{cosec}^2 2\alpha - 1}.$

**68.** $\dfrac{\sin 3\alpha + \sin 5\alpha + \sin 7\alpha}{\cos 3\alpha + \cos 5\alpha + \cos 7\alpha}.$

**69.** $4\cos^4\alpha - 2\cos 2\alpha - \dfrac{1}{2}\cos 4\alpha.$

**70.** $\cos 0 + \cos\dfrac{\pi}{7} + \cos\dfrac{2\pi}{7} + \cos\dfrac{3\pi}{7} + \cos\dfrac{4\pi}{7} + \cos\dfrac{5\pi}{77} + \cos\dfrac{6\pi}{7}.$

**71.** $\dfrac{\cos\left(2\alpha - \dfrac{\pi}{2}\right) + \sin(3\pi - 4\alpha) - \cos\left(\dfrac{5}{2}\pi + 6\alpha\right)}{4\sin(5\pi - 3\alpha)\cos(\alpha - 2\pi)}.$

**72.** $\dfrac{\cos 2\alpha}{\cos^4\alpha - \sin^4\alpha} - \dfrac{\cos^4\alpha + \sin^4\alpha}{1-\dfrac{1}{2}\sin^2 2\alpha}.$

**73.** $\dfrac{1 + \sin 4\alpha - \cos 4\alpha}{1 + \cos 4\alpha + \sin 4\alpha}$.

**74.** $\dfrac{2\cos^2 \alpha - 1}{4\tan\left(\dfrac{\pi}{4} - \alpha\right)\sin^2\left(\dfrac{\pi}{4} + \alpha\right)}$.

**75.** $\dfrac{\cos 4\alpha + 1}{\cot \alpha - \tan \alpha}$.

**76.** $\dfrac{1 + \cos\alpha}{1 - \cos\alpha}\tan^2\dfrac{\alpha}{2} - \cos^2\alpha$.

**77.** $\tan\left(\dfrac{\alpha}{2} + \dfrac{\pi}{4}\right)\dfrac{1 - \sin\alpha}{\cos\alpha}$.

**78.** $\sin^2\alpha\left(1 + \dfrac{1}{\sin\alpha} + \cot\alpha\right)\left(1 - \dfrac{1}{\sin\alpha} + \cot\alpha\right)$.

**79.** $\dfrac{\cos 2\alpha}{\sin^2 2\alpha\,(\cot^2\alpha - \tan^2\alpha)}$.

Calculate without resorting to tables **(80-105)**.

**80.** $\cos 67°30'$ and $\cos 75°$.

**81.** $4\,(\cos 24° + \cos 48° - \cos 84° - \cos 12°)$.

**82.** $\dfrac{96\sin 80° \sin 65° \sin 35°}{\sin 20° + \sin 50° + \sin 110°}$.

**83.** $128\sin^2 20° \sin^2 40° \sin^2 60° \sin^2 80°$.

**84.** $\tan 20° \tan 40° \tan 80°$.

**85.** $\dfrac{\cot 15° + 1}{2\cot 15°}$.

**86.** $\cos 10° \cos 50° \cos 70°$.

**87.** $\sin 160° \cos 110° + \sin 250° \cos 340° + \tan 110° \tan 340°$.

**88.** $\tan 9° - \tan 63° + \tan 81° - \tan 27°$.

**89.** $96\sqrt{3}\sin\dfrac{\pi}{48}\cos\dfrac{\pi}{48}\cos\dfrac{\pi}{24}\cos\dfrac{\pi}{12}\cos\dfrac{\pi}{6}$.

**90.** $\sin\left(\dfrac{7\pi}{2} - 2\alpha\right)\cos\left(\dfrac{6\pi}{2} - \alpha\right)\sin\alpha$ for $\alpha = \dfrac{3\pi}{16}$.

**91.** $\dfrac{\sin 110° \sin 250° + \cos 540° \cos 290° \cos 430°}{\cos^2 1260°}$.

**92.** Verify the equality $\dfrac{1}{\cos 290°} + \dfrac{1}{\sqrt{3}\sin 250°} = \dfrac{4}{\sqrt{3}}$.

**93.** Verify $\cos\dfrac{\pi}{7}\cdot\cos\dfrac{4\pi}{7}\cdot\cos\dfrac{5\pi}{7} = \dfrac{1}{8}$.

**94.** Verify $\cos 20° + 2\sin^2 55° = 1 + \sqrt{2}\sin 65°$.

**95.** Calculate $(\sin 4\alpha + 2\sin 2\alpha)\cos\alpha$, if $\sin\alpha = \dfrac{1}{4}$.

**96.** Given : $20\sin^2\alpha + 21\cos\alpha - 24 = 0$, $\dfrac{7\pi}{4} < \alpha < 2\pi$. Find $\cot\dfrac{\alpha}{2}$.

97. $\dfrac{5}{6+7\sin 2\alpha}$ if $\tan\alpha = 0.2$.

98. $\sin 2\alpha$ if $\sin\dfrac{\alpha}{2} + \cos\dfrac{\alpha}{2} = -\dfrac{1}{2}$ and $\alpha$ belongs to the fourth quarter.

99. $\tan\left(\dfrac{\pi}{4} - 2\alpha\right)$ if $\tan\alpha = a$.

100. $\tan\left(\alpha - \dfrac{\pi}{4}\right)$ if $\cos\alpha = -\dfrac{9}{41}$, $\pi < \alpha < \dfrac{3\pi}{2}$.

101. $\cos\dfrac{\alpha}{2}$ if $\sin\alpha = -\dfrac{12}{13}$, $\pi < \alpha < \dfrac{3\pi}{2}$.

102. $\sin\dfrac{\alpha}{2}$ if $\sin\alpha = -\dfrac{5}{13}$ and $\pi < \alpha < \dfrac{3\pi}{2}$.

103. $\sin\dfrac{\alpha}{2}$ if $\sin\alpha = 0.8$ and $0 < \alpha < \dfrac{\pi}{2}$.

104. $\dfrac{\sin\alpha}{\sin^3\alpha + 3\cos^3\alpha}$ if $\tan\alpha = 2$.

105. $\sin^6\alpha + \cos^6\alpha$ if $\sin\alpha + \cos\alpha = m$.

Transform the following expressions into a product **(106-112)**.

106. $\sec\alpha - \cos\alpha + \sec 60° \cos 2\alpha \sin 3\alpha - \sin 5\alpha$.

107. $\sin\alpha + \sin 60° + \sin(\alpha + 60°)$.

108. $\sin(5\alpha + \beta) + \sin(3\alpha + \beta) + \sin 2\alpha$.

109. $\sin\dfrac{5}{2}\alpha \cos\dfrac{\alpha}{2} - \sin 3\alpha \cos\dfrac{\pi}{3} - \dfrac{1}{4}$.

110. $4\cos 11\alpha + (\sin 8\alpha - \sin 10\alpha - \sin 12\alpha + \sin 14\alpha)\operatorname{cosec}\alpha \times \operatorname{cosec} 2\alpha$.

111. $\tan^2\alpha - \tan^2\beta - \dfrac{1}{2}\sin(\alpha - \beta)\sec^2\alpha \sec^2\beta$.

112. $\dfrac{\tan\alpha + \tan\beta}{\cot\alpha + \cot\beta} + [\cos(\alpha - \beta)\sec(\alpha + \beta) + 1]^{-1}$.

## Inverse Trigonometric Functions

Arc sine of the number $x \in [-1, 1]$ (written arcsin x) is a number $y \in \left[-\dfrac{\pi}{2}, \dfrac{\pi}{2}\right]$, whose sine is x.

Arc cosine of the number $x \in [-1, 1]$ (written arccos x) is a number $y \in [0, \pi]$ whose cosine is x.

Arc tangent of the number $x \in R$ (written arctan x) is a number $y \in \left(-\dfrac{\pi}{2}, \dfrac{\pi}{2}\right)$ whose tangent is x.

**The fundamental identities.**

$\sin(\arcsin x) = x,$

$\cos(\arcsin x) = \sqrt{1-x^2},$

$\tan(\arcsin x) = \dfrac{x}{\sqrt{1-x^2}}, \quad -1 < x < 1,$

$\cos(\arccos x) = x,$

$\sin(\arccos x) = \sqrt{1-x^2},$

$\tan(\arccos x) = \dfrac{\sqrt{1-x^2}}{x}, \quad -1 < x < 0, \ 0 < x < 1.$

$\tan(\arctan x) = x,$

$\cos(\arctan x) = \dfrac{1}{\sqrt{1+x^2}},$

$\sin(\arctan x) = \dfrac{x}{\sqrt{1+x^2}},$

$\tan(\operatorname{arccot} x) = \dfrac{1}{x}, \quad x \neq 0,$

$\sin(\operatorname{arccot} x) = \dfrac{1}{\sqrt{1+x^2}},$

$\cos(\operatorname{arccot} x) = \dfrac{x}{\sqrt{1+x^2}},$

$\arcsin x = \arccos\sqrt{1-x^2}, \quad 0 \le x \le 1.$

$\arctan x = \operatorname{arccot}\dfrac{1}{x} = \arcsin\dfrac{x}{\sqrt{1+x^2}} = \arccos\dfrac{1}{\sqrt{1+x^2}}, \ x > 0.$

$\arcsin(-x) = -\arcsin x,$

$\arccos(-x) = \pi - \arccos x, \quad -1 \le x \le 1.$

$\arctan(-x) = -\arctan x.$

$\arcsin x + \arccos x = \dfrac{\pi}{2}, \quad -1 \le x \le 1.$

$\arctan x + \operatorname{arccot} x = \dfrac{\pi}{2}.$

Calculate **(113-126).**

**113.** $\arctan 1 + \arccos\left(-\dfrac{1}{2}\right) + \arcsin\left(-\dfrac{1}{2}\right).$

**114.** $\arcsin\left(-\dfrac{\sqrt{2}}{2}\right) + \arccos\left(-\dfrac{1}{2}\right) - \arctan(-\sqrt{3}) + \operatorname{arccot}\left(-\dfrac{1}{\sqrt{3}}\right).$

**115.** arccot [tan $(-37°)$].

**116.** arccos $\left( \cos\left( -\dfrac{\sqrt{3}}{2} \right) \right)$.

**117.** arctan $\sqrt{2}$ + arctan $\dfrac{1}{\sqrt{2}}$.

**118.** sin $\left( 2\arcsin \dfrac{3}{5} \right)$.

**119.** cos $\left( 2\arcsin \dfrac{2}{5} \right)$.

**120.** sin $(2\arctan 3)$.

**121.** cos $(2\arctan 2)$.

**122.** sin $\left( \dfrac{1}{2} \arccos \dfrac{1}{9} \right)$.

**123.** cos $\left( \dfrac{1}{2} \arccos \dfrac{1}{8} \right)$.

**124.** sin $\left( 2\arctan \dfrac{1}{3} \right) + \cos (\arctan 2\sqrt{2})$.

**125.** sin $\left( \arcsin \dfrac{3}{5} - \arccos \dfrac{3}{5} \right)$.

**126.** tan $\left( \arcsin \dfrac{4}{5} + \dfrac{3\pi}{2} \right)$.

Prove **(127-132).**

**127.** sin $(\arcsin x) = x, -1 \le x \le 1$.

**128.** cos $(\arcsin x) = \sqrt{1 - x^2}, -1 \le x \le 1$.

**129.** cos $(\arctan x) = \dfrac{1}{\sqrt{1 + x^2}}$.

**130.** tan $(\arccos x) = \dfrac{\sqrt{1 - x^2}}{x}, -1 \le x < 0, \ 0 < x \le 1$.

**131.** arcsin $x$ + arccos $x = \dfrac{\pi}{2}, -1 \le x \le 1$.

**132.** arctan $x$ + arccot $x = \dfrac{\pi}{2}$.

Simplify the following expressions **(133-141).**

**133.** sin $\left( \arccos \dfrac{3}{5} \right)$.

**134.** cos $\left( \arcsin \dfrac{4}{5} \right)$.

**135.** sin $(\arctan 2)$.

**136.** cos (arctan 2).

**137.** $\arcsin\left(\sin\dfrac{8\pi}{7}\right)$.

**138.** $\arctan\left(\tan\dfrac{10\pi}{7}\right)$.

**139.** $\arcsin\left(\sin\dfrac{6\pi}{7}\right)$.

**140.** $\arctan\left(\tan\dfrac{6\pi}{7}\right)$.

**141.** $\arccos\left(\cos\dfrac{6\pi}{7}\right)$.

**142.** Express $\arcsin\dfrac{5}{13}$ in terms of the values of all inverse trigonometric functions.

Solve the following equations **(143-145).**

**143.** $6\arcsin\ (x^2 - 6x + 8.5) = \pi$.

**144.** $4\arcsin x + \arccos x = \pi$.

**145.** $5\arctan x + 3\operatorname{arccot} x = 2\pi$.

**146.** Construct the graph $y = \arcsin(\cos x)$.

**147.** Solve the inequality $\arcsin(\sin 5) > x^2 - 4x$.

## 1.11. Solution of Trigonometric Equations, Inequalities and Systems of Equations

### Solution of Elementary Trigonometric Equations

**Example 1.** $\sin x = m,\ |m| \leq 1$.
$x = (-1)^n \arcsin m + \pi n,\ n \in \mathbf{Z}$.

**Example 2.** $\cos x = m,\ |m| \leq 1$.
$x = \pm\arccos m + 2\pi n,\ n \in \mathbf{Z}$.

**Example 3.** $\tan x = m,\ m \in \mathbf{R}$.
$x = \arctan m + \pi n,\ n \in \mathbf{Z}$.

**Example 4.** $\cot x = m,\ m \in \mathbf{R}$,
$x = \operatorname{arccot} m + \pi n,\ n \in \mathbf{Z}$.

**Remark 1.** In the simple examples given above, the solutions are written in a simple and unique form. In more complicated examples the *notation* for the set of solutions is not unique but the identity of different notations can always be proved by means of identity transformations. Different notations are usually due to the difference in the methods of solving the given problem. For instance, in problem 47 we have got a set of solutions.

$$-\frac{\pi}{4} + (-1)^n \arcsin \frac{1+\sqrt{3}}{2\sqrt{2}} + \pi n, \ n \in \mathbf{Z}.$$

When verifying the truth of the answer, we have used another method for solving the problem and have got the set of solutions

$$-\frac{\pi}{4} + (-1)^n \frac{5\pi}{12} + \pi n, \ n \in \mathbf{Z}.$$

This is the same set of solutions but in another notation. Indeed,

$$\arcsin \frac{1+\sqrt{3}}{2\sqrt{2}} = \arcsin \left( \frac{1}{2} \cdot \frac{1}{\sqrt{2}} + \frac{\sqrt{3}}{2} \cdot \frac{1}{\sqrt{2}} \right)$$

$$= \arcsin \left( \sin \frac{\pi}{6} \cos \frac{\pi}{4} + \cos \frac{\pi}{6} \sin \frac{\pi}{4} \right)$$

$$= \arcsin \left( \sin \left( \frac{\pi}{6} + \frac{\pi}{4} \right) \right) = \arcsin \left( \sin \frac{5\pi}{12} \right) = \frac{5\pi}{12}.$$

The last equality is based on the fact that $-\frac{\pi}{2} \leq \frac{5}{12} \leq \frac{\pi}{2}$ ; if the inequality were not satisfied, it would be necessary to use recursion formulas.

You should always remember this when solving problems. If you get a notation for the set of solutions which differs from that given in "Answers and Instructions", you must prove their identity. The operation of proving the identity will itself serve as a good exercise.

**Remark 2.** When seeking solutions, you often write the set of solutions of trigonometric equations by means of several formulas. You can sometimes unite them and obtain a simpler notation. In problem 187, for instance, we first wrote the set of solutions in the form

$$\pi n, \ \frac{\pi}{6} + \pi k, \ \frac{\pi}{3} + \pi m \quad (n, \ k, \ m \in \mathbf{Z}).$$

This set of solutions can also be written as

$$\pi n, \ (-1)^k \frac{\pi}{6} + \frac{\pi k}{2} \quad (n, \ k \in \mathbf{Z}).$$

Here, too, the difference in the notations for the set of solutions can be explained by the difference in the methods of solving the given equations.

**Remark 3.** If in the examples presented in remarks 1 and 2 the different notations are identical and the final notation for the set of solution is inessential, from the point of view of the correctness of the solution, it sometimes happens that one and the same set of solutions is given several times. Then it is necessary to exclude those repetitions from the final notation. For instance, when solving problem 76, we obtained two sets of solutions

$$\frac{\pi}{2} + \pi n, \quad \frac{\pi}{10} + \frac{\pi k}{5} \quad (n, \ k \in \mathbf{Z}).$$

In this case, all the solutions given by the first set form a part of the second set. Indeed, if we put $k = 5n + 2 \,(n \in \mathbf{Z})$, then the first set of solutions can be separated from the second set. Then, the final notation for the set of solutions looks like

$$\frac{\pi}{10} + \frac{\pi k}{5} \quad (k \in \mathbf{Z}).$$

i.e. the superfluous must be discarded.

**Remark 4.** In the case when the sets of solutions obtained overlap only partly, the common part in the final notation must be presented only once, without repetitions. Thus in problem 81 (solved by a certain method) we obtained two sets of solutions

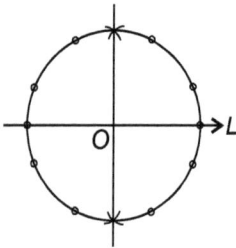

Fig. 1.1

$$\frac{\pi n}{5}, \frac{\pi k}{2} \quad (n, \ k \in \mathbf{Z}). \tag{1}$$

Neither of these sets of solutions is a part of the other set, but if we put $n = 5m$, but $k = 2m$, $(m \in \mathbf{Z})$, we get $\pi m$ in both cases. Consequently, this set of solutions must be excluded from one of the presented sets. For instance, the four values of $k$ must be excluded from the second set and only $k = 2l + 1$ $(l \in \mathbf{Z})$ must be left. Then all solutions of the problem will be written in the form

$$\frac{\pi n}{5}, \frac{\pi (2l+1)}{2} \quad (n, \ l \in \mathbf{Z}), \tag{2}$$

every solution being written only once. An alternative method immediately produces this notation. It is considered to be erroneous to give the final set of solutions in form (1).

**Remark 5.** The following technique is used to find the overlapping part of sets of solutions and to unite them. We draw a circle and mark solutions belonging to one set by dots and those belonging to the other set by crosses. In Fig. 1.1 dots denote the solutions belonging to the set $\dfrac{\pi n}{5}$ and the crosses those belonging to the set $\dfrac{\pi k}{2}$. The angles are reckoned from the $L$-axis counterclockwise. It can be seen from the figure that the values 0 and $\pi$ (and all the values that differ from them by a period) belong both to the first and to the second set. Thus, besides the solutions written in the form $\dfrac{\pi n}{\pi}$, two more dots,

$\dfrac{\pi}{2}$ and $\dfrac{3\pi}{2}$ (and those differing from them by a period), remain on the circle.

These solutions can be written as $\dfrac{\pi}{2} + \pi l$ $(l \in \mathbf{Z})$.

## Solution of Equations by Factoring

**Example 5.** $2\cos x \cos 2x = \cos x$.

*Solution.* The given equation is equivalent to the equation $\cos x$ $(2\cos 2x - 1) = 0$. This equation is equivalent to the collection of equations

$$\left[\begin{array}{l} \cos x = 0, \\ \cos 2x = \dfrac{1}{2}, \end{array}\right. \Rightarrow \left[\begin{array}{l} x = \dfrac{\pi}{2} + \pi n, \\ 2x = \pm \dfrac{\pi}{3} + 2\pi k, \text{ i. e. } x = \pm\dfrac{\pi}{6} + \pi k, \end{array}\right. \quad \begin{array}{l} n \in \mathbf{Z}, \\ \\ k \in \mathbf{Z}. \end{array}$$

*Answer.* $\dfrac{\pi}{2} + \pi n, \pm\dfrac{\pi}{6} + \pi k$ $(n, \, k \in \mathbf{Z})$.

## Solution of Equations Reducible to Quadratic Equations

**Example 6.** $3\cos^2 x - 10\cos x + 3 = 0$.

*Solution.* Assume $\cos x = y$. The given equation assumes the form $3y^2 - 10y + 3 = 0$. Solving it, we find that $y_1 = \dfrac{1}{3}$, $y_2 = 3$. The value $y_2 = 3$ does not satisfy the condition since $|\cos x| \le 1$. Consequently, $\cos x = \dfrac{1}{3}$;

$x = \pm \arccos \dfrac{1}{3} + 2\pi n, \, n \in \mathbf{Z}$.

## Solution of Homogeneous Equations and Equations Reducible to Them

Equations of the form

$$a_0 \sin^n x + a_1 \sin^{n-1} x \cos x + a_2 \sin^{n-2} x \cos^2 x + \dots$$
$$\dots + a_{n-1} \sin x \cos^{n-1} x + a_n \cos^n x = 0,$$

where $a_0, a_1, \dots, a_n$ are real numbers, are said to be homogeneous with respect to $\sin x$ and $\cos x$.

**Example 7.** $\cos 3x + \sin 3x = 0$.

*Solution.* $\sin 3x = -\cos 3x$. Since the values of $x$ for which $\cos 3x$ is equal to zero cannot serve as roots for the given equation, we divide both sides of the initial equation by $\cos 3x$ and obtain an equation which is equivalent to the initial equation:

$$\frac{\sin 3x}{\cos 3x} = -\frac{\cos 3x}{\cos 3x}, \text{or } \tan 3x = -1.$$

Hence
$$3x = -\frac{\pi}{4} + \pi n; \ x = -\frac{\pi}{12} + \frac{\pi}{3}n, \ n \in \mathbf{Z}.$$

*Answer.* $-\frac{\pi}{12} + \frac{\pi}{3}n \ (n \in \mathbf{Z}).$

**Example 8.** $6\sin^2 x - \sin x \cos x - \cos^2 x = 3.$

*Solution.* $6\sin^2 x - \sin x \cos x - \cos^2 x - 3(\sin^2 x + \cos^2 x) = 0.$ Removing the brackets and collecting like terms, we get

$$3\sin^2 x - \sin x \cos x - 4\cos^2 x = 0.$$

Since the values $x = \frac{\pi}{2} + \pi n$ are not roots of the equation and $\cos x \neq 0$, we divide both sides of the equation by $\cos^2 x$:

$$3\tan^2 x - \tan x - 4 = 0,$$

whence

$$\tan x = -1, \ x = -\frac{\pi}{4} + \pi n, \ n \in \mathbf{Z}$$

and

$$\tan x = \frac{4}{3}, \ x = \arctan \frac{4}{3} + \pi k, \ k \in \mathbf{Z}.$$

*Answer.* $-\frac{\pi}{4} + \pi n, \ \arctan \frac{4}{3} + \pi k \ (n, \ k \in \mathbf{Z}).$

## Solving Equations by Introducing an Auxiliary Argument

**Example 9.** $\sin x + \cos x = \sqrt{2}.$

*Solution.* $\sqrt{2}\left(\frac{1}{\sqrt{2}}\sin x + \frac{1}{\sqrt{2}}\cos x\right) = \sqrt{2}$

$$\sin x \cos \frac{\pi}{4} + \cos x \sin \frac{\pi}{4} = 1, \ \sin\left(x + \frac{\pi}{4}\right) = 1,$$

$$x + \frac{\pi}{4} = \frac{\pi}{2} + 2\pi n \ (n \in \mathbf{Z}), \ x = \frac{\pi}{2} - \frac{\pi}{4} + 2\pi n \quad (n \in \mathbf{Z}).$$

*Answer.* $\frac{x}{4} + 2\pi n \quad (n \in \mathbf{Z}).$

**Example 10.** $\frac{\sqrt{3}}{2}\cos x + \frac{1}{2}\sin x = 1.$

*Solution.* $\cos \frac{\pi}{6}\cos x + \sin \frac{\pi}{6}\sin x = 1,$

$$\cos\left(x - \frac{\pi}{6}\right) = 1, \ x - \frac{\pi}{6} = 2\pi n \ (n \in \mathbf{Z}), \ x = \frac{\pi}{6} + 2\pi n \ (n \in \mathbf{Z}).$$

*Answer.* $\frac{\pi}{6} + 2\pi n \quad (n \in \mathbf{Z}).$

## Solving Equations by Transforming a Sum of Trigonometric Functions into a Product

**Example 11.** $\cos 3x + \sin 2x - \sin 4x = 0$.

*Solution.*     $\cos 3x + (\sin 2x - \sin 4x) = 0$.

Transforming the expression in brackets by formula (16), Sec. 1.10, we obtain

$$\cos 3x + (-2\sin x \cos 3x) = 0,$$

$$\cos 3x (1 - 2\sin x) = 0.$$

The last equation is equivalent to the collection of equations

$$\cos 3x = 0, \quad \sin x = \frac{1}{2};$$

consequently,

$$x = \frac{\pi}{6} + \frac{\pi}{3}n, \quad x = (-1)^k \frac{\pi}{6} + \pi k \qquad (n, \ k \in \mathbf{Z}).$$

The set of solutions $x = (-1)^k \dfrac{\pi}{6} + \pi k \ (k \in \mathbf{Z})$ belongs entirely to the set of solutions $x = \dfrac{\pi}{6} + \dfrac{\pi n}{3} \ (n \in \mathbf{Z})$. Therefore, this set alone remains as a set of solutions.

*Answer.* $\dfrac{\pi}{6} + \dfrac{\pi}{3}n \ (n \in \mathbf{Z})$.

## Solving Equations by Transforming a Product of Trigonometric Functions into a Sum

**Example 12.** $\sin 5x \cos 3x = \sin 6x \cos 2x$.

*Solution.*     We apply formula (19), Sec. 1.10, to both sides of the equation :

$$\frac{1}{2}(\sin 8x + \sin 2x) = \frac{1}{2}(\sin 8x + \sin 4x),$$

$$\sin 2x - \sin 4x = 0.$$

Using formula (16), Sec. 1.10, we obtain

$-2\sin x \cos 3x = 0$

$$\Rightarrow \left[ \begin{array}{l} \sin x = 0, \\ \cos 3x = 0, \end{array} \right. \Rightarrow \left[ \begin{array}{ll} x = \pi n, & n \in \mathbf{Z}, \\ 3x = \dfrac{\pi}{2} + \pi k, \quad x = \dfrac{\pi}{6} + \dfrac{\pi}{3}k, & k \in \mathbf{Z}. \end{array} \right.$$

*Answer.* $\dfrac{\pi}{6} + \dfrac{\pi}{3}k \ (n, \ k \in \mathbf{Z})$.

## Solving Equations with the Use of Formulas for Lowering a Degree

(formulas (24) and (25), Sec. 1.10)

**Example 13.** $\sin^2 x + \sin^2 2x = 1$.

*Solution.*
$$\frac{1 - \cos 2x}{2} + \frac{1 - \cos 4x}{2} = 1$$
$$\Rightarrow \quad \cos 2x + \cos 4x = 0$$
$$\Rightarrow \quad 2 \cos 3x \cos x = 0.$$

The last equations is equivalent to the collection of two equations

(a) $\cos 3x = 0$, $3x = \dfrac{\pi}{2} + \pi n$, $x = \dfrac{\pi}{6} + \dfrac{\pi}{3} n$, $n \in \mathbf{Z}$.

(b) $\cos x = 0$, $x = \dfrac{\pi}{2} + \pi k$, $k \in \mathbf{Z}$.

The set of solutions of equation (b) is a subset of the set of solutions of (a) and therefore, in the answer we write only roots of equation (a).

*Answer.* $\dfrac{\pi}{6} + \dfrac{\pi n}{3}$ $(n \in \mathbf{Z})$.

## Solving Equations with the Use of Formulas for (22) and (23), Sec. 1.10

**Example 14.** $\cos x - 2 \sin^2 \dfrac{x}{2} = 0$

*Solution.* $\cos x - (1 - \cos x) = 0 \Rightarrow 2 \cos x - 1 = 0 \Rightarrow \cos x = \dfrac{1}{2}$

$$\Rightarrow \quad x = \pm \frac{\pi}{3} + 2\pi n \ (n \in \mathbf{Z}).$$

*Answer.* $\pm \dfrac{\pi}{3} + 2\pi n$ $(n \in \mathbf{Z})$.

## Solving Equations with the Use of Formulas for Double and Triple Arguments

(Formulas (10) - (14), Sec. 1.10 )

**Example 15.** $\sin x = \sqrt{2} \cos x$.

*Solution.* Using formula (10), Sec. 1.10, we obtain $2 \sin x \cos x = \sqrt{2} \cos x$ on the left-hand side of the equation. It is impossible to divide both sides of the equation by $\cos x$ since it would lead to the loss of solutions which are roots of the equation $\cos x = 0$. We transfer $\sqrt{2} \cos x$ to the left-hand side and get

$$2 \sin x \cos x - \sqrt{2} \cos x = 0 \Rightarrow \sqrt{2} \cos x (\sqrt{2} \sin x - 1) = 0$$

$$\Rightarrow \left[ \begin{array}{l} \cos x = 0, \\ \sqrt{2} \sin x = 1, \end{array} \right. \Rightarrow \left[ \begin{array}{ll} x = \dfrac{\pi}{2} + \pi n, & n \in \mathbf{Z}, \\ x = (-1)^k \dfrac{\pi}{4} + \pi k, & k \in \mathbf{Z}. \end{array} \right.$$

*Answer.* $\dfrac{\pi}{2} + \pi n$, $(-1)^k \dfrac{\pi}{4} + \pi k$  $(n, k \in \mathbf{Z})$.

**Example 16.** $2 \sin \dfrac{x}{2} \cos^2 x - 2 \sin \dfrac{x}{2} \sin^2 x = \cos^2 x - \sin^2 x$.

*Solution.*  On the left-hand side of the equation we put the factor $2 \sin \dfrac{x}{2}$ before the parentheses :

$$2 \sin \dfrac{x}{2} (\cos^2 x - \sin^2 x) = \cos^2 x - \sin^2 x,$$

Replacing the expression $\cos^2 x - \sin^2 x$ by $\cos 2x$ according to formula (11), Sec. 1.10, we get

$$2 \sin \dfrac{x}{2} \cos 2x = \cos 2x,$$

or $2 \sin \dfrac{x}{2} \cos 2x - \cos 2x = 0 \Rightarrow \cos 2x \left(2 \sin \dfrac{x}{2} - 1\right) = 0$

$$\Rightarrow \begin{bmatrix} \cos 2x = 0, \\ \sin \dfrac{x}{2} = \dfrac{1}{2}, \end{bmatrix} \Rightarrow \begin{bmatrix} x = \dfrac{\pi}{4} + \dfrac{\pi}{2} n, & n \in \mathbf{Z}, \\ x = (-1)^k \dfrac{\pi}{3} + 2\pi k, & n \in \mathbf{Z}. \end{bmatrix}$$

*Answer.* $\dfrac{\pi}{4} + \dfrac{\pi}{2} n$, $(-1)^k \dfrac{\pi}{3} + 2\pi k$  $(n, \ k \in \mathbf{Z})$.

## Solving Equations by a Change of Variable

(a) Equations of the form $P (\sin x \pm \cos x, \sin x \cos x) = 0$, where $P (y, z)$ is a polynomial, can be solved by the change

$$\cos x \pm \sin x = t \Rightarrow 1 \pm 2 \sin x \cos x = t^2.$$

Let us consider an example.

**Example 17.** $\sin x + \cos x = 1 + \sin x \cos x$.

*Solution.*  We introduce the designation $\sin x + \cos x = t$. Then $(\sin x + \cos x)^2 = t^2$, $1 + 2 \sin x \cos x = t^2$, $\sin x \cos x = \dfrac{t^2 - 1}{2}$. In the new designations the initial equation looks like

$$t = 1 + \dfrac{t^2 - 1}{2} \text{ or } t^2 - 2t + 1 = 0, (t - 1)^2 = 0, \ t = 1,$$

i.e.     $\sin x + \cos x = 1$, $\sqrt{2} \left(\dfrac{1}{\sqrt{2}} \sin x + \dfrac{1}{\sqrt{2}} \cos x\right) = 1$,

$$\cos \dfrac{\pi}{4} \cos x + \sin \dfrac{\pi}{4} \sin x = \dfrac{1}{\sqrt{2}},$$

$$\cos \left(x - \dfrac{\pi}{4}\right) = \dfrac{\sqrt{2}}{2},$$

$$x - \frac{\pi}{4} = \pm \frac{\pi}{4} + 2\pi n, \quad n \in \mathbf{Z},$$

$$x = \frac{\pi}{4} \pm \frac{\pi}{4} + 2\pi n, \quad n \in \mathbf{Z}.$$

*Answer.* $\frac{\pi}{2} + 2\pi n, \ 2\pi n \ (n \in \mathbf{Z})$.

(b) Equations of the form $a \sin x + b \cos x + d = 0$, where $a$, $b$, and $d$ are real numbers, and $a, \ b \ne 0$, can be solved by the change

$$\cos x = \frac{1 - \tan^2 \frac{x}{2}}{1 + \tan^2 \frac{x}{2}}, \quad \sin x \ \frac{2 \tan \frac{x}{2}}{1 + \tan^2 \frac{x}{2}}, \ x \ne \pi + 2\pi n \ (n \in \mathbf{Z}).$$

**Example 18.** $3 \cos x + 4 \sin x = 5.$

*Solution.* $3 \dfrac{1 - \tan^2 \frac{x}{2}}{1 + \tan^2 \frac{x}{2}} + 4 \dfrac{2 \tan \frac{x}{2}}{1 + \tan^2 \frac{x}{2}} = 5,$

$$3 - 3 \tan^2 \frac{x}{2} + 8 \tan \frac{x}{2} = 5 + 5 \tan^2 \frac{x}{2},$$

$$4 \tan^2 \frac{x}{2} - 4 \tan \frac{x}{2} + 1 = 0, \ \left( 2 \tan \frac{x}{2} - 1 \right)^2 = 0,$$

$$\tan \frac{x}{2} = \frac{1}{2}, \ x = 2 \arctan \frac{1}{2} + 2\pi n, \quad n \in \mathbf{Z}.$$

*Answer.* $2 \arctan \dfrac{1}{2} + 2\pi n \ (n \in \mathbf{Z})$.

(c) Many equations can be solved by introducing a new variable.

**Example 19.** $\sin^4 2x + \cos^4 2x = \sin 2x \cos 2x.$

*Solution.* $(\sin^2 2x + \cos^2 2x)^2 - 2 \sin^2 2x \cos^2 2x = \sin 2x \cos 2x,$

$2 \sin^2 2x \cos^2 2x + \sin 2x \cos 2x - 1 = 0.$

We introduce the designation $\sin 2x \cos 2x = y$. Then the last equation assumes the form $2y^2 + y - 1 = 0$ or

$$2(y + 1)\left( y - \frac{1}{2} \right) = 0.$$

We pass to the variable $x$ and obtain

(1) $\sin 2x \cos 2x = -1, \ 2 \sin 2x \cos 2x = -2, \ \sin 4x = -2, \ x \in \phi.$

(2) $\sin 2x \cos 2x = \dfrac{1}{2}, \ \sin 4x = 1, \ 4x = \dfrac{\pi}{2} + 2\pi n, n \in \mathbf{Z},$

$$x = \frac{\pi}{8} + \frac{\pi}{2}n, \ n \in \mathbf{Z}.$$

*Answer.* $\dfrac{\pi}{8} + \dfrac{\pi}{2}n \ (n \in \mathbf{Z})$.

## Solution of Trigonometric Equations of the Form $f(x) = \sqrt{\varphi(x)}$

**Example 20.** $\sqrt{1 - \cos x} = \sin x, \quad x \in [\pi, \ 3\pi].$

Solution.
$$\begin{cases} 1 - \cos x \geq 0, \\ \sin x \geq 0. \end{cases}$$

Under the condition that both sides of the equation are nonnegative, we square them :

$$1 - \cos x = \sin^2 x, \qquad 1 - \cos x = 1 - \cos^2 x,$$

$$\cos^2 x - \cos x = 0, \qquad \cos x \,(\cos x - 1) = 0.$$

(1) $\cos x = 0, \quad x = \dfrac{\pi}{2} + \pi n, \quad n \in \mathbf{Z}.$

(2) $\cos x = 1, \quad x = 2\pi k, \quad k \in \mathbf{Z}.$ But since $\sin x \geq 0$ and $x \in [\pi, \ 3\pi]$, we leave $x = 2\pi, \ \dfrac{5\pi}{2}.$

Answer. $2\pi, \ \dfrac{5\pi}{2}.$

## Solving Equations with the Use of the Boundedness of the Functions $\sin x$ and $\cos x$

**Example 21.** $\left( \cos \dfrac{x}{4} - 2 \sin x \right) \sin x + \left( 1 + \sin \dfrac{x}{4} - 2 \cos x \right) \times \cos x = 0.$

Solution.   $\cos \dfrac{x}{4} \sin x - 2 \sin^2 x + \cos x + \sin \dfrac{x}{4} \cos x - 2 \cos^2 x = 0.$

$$\sin \left( x + \dfrac{x}{4} \right) + \cos x - 2(\sin^2 x + \cos^2 x) = 0, \ \sin \dfrac{5x}{4} + \cos x = 2.$$

Since the functions $\sin \dfrac{5x}{4}$ and $\cos x$ have the greatest value equal to 1, their sum is equal to 2 if $\sin \dfrac{5x}{4} = 1$ and $\cos x = 1$ simultaneously, i.e.

$$\begin{cases} \sin \dfrac{5x}{4} = 1, \\ \cos x = 1, \end{cases} \Rightarrow \begin{cases} \dfrac{5x}{4} = \dfrac{\pi}{2} + 2\pi n, \\ x = 2\pi k \quad (n, \ k \in \mathbf{Z}); \end{cases}$$

$$2\pi k = \dfrac{2\pi}{5} + \dfrac{8\pi}{5} n, \qquad k = \dfrac{1 + 4n}{5}.$$

Since $k \in \mathbf{Z}$, it follows that $n = 1 + 5m \ (m \in \mathbf{Z})$ and then $x = 2\pi + 8\pi m$, $m \in \mathbf{Z}.$

Answer. $2\pi + 8\pi m, \ m \in \mathbf{Z}.$

## Trigonometric Systems

**Example 22.** $\begin{cases} \sin x \cos y = \dfrac{1}{4}, \\ 3 \tan x = \tan y. \end{cases}$

*Solution.*    We transform the second equation and get

$$3 \sin x \cos y - \sin y \cos x = 0.$$

Substituting now the value of the product $\sin x \cos y$ from the first equation into the equation obtained, we get a system

$$\begin{cases} \cos x \sin y = \dfrac{3}{4}, \\ \sin x \cos y = \dfrac{1}{4}. \end{cases} \tag{1}$$

Adding together the equations of system (1) and then subtracting the first equation from the second, we get a system which is equivalent to system :

$$\begin{cases} \sin (x + y) = 1, \\ \sin (x - y) = \dfrac{1}{2}. \end{cases} \tag{2}$$

Whence we have

$$\begin{cases} x + y = \dfrac{\pi}{2} + 2\pi k, \\ x - y = -\dfrac{\pi}{6} + 2\pi l. \end{cases} \tag{3}$$

and

$$\begin{cases} x + y = \dfrac{\pi}{2} + 2\pi k, \\ x - y = -\dfrac{5\pi}{6} + 2\pi l. \end{cases} \tag{4}$$

From system (3) we find

$$x = \frac{\pi}{6} + \pi (k + l),$$

$$y = \frac{\pi}{3} + \pi (k - l).$$

From system (4) we find

$$x = -\frac{\pi}{6} + \pi (k + l),$$

$$y = \frac{2\pi}{3} + \pi (k - l).$$

Pay attention to the fact that the integers by which $2\pi$ is multiplied in (3) must be represented by different letters (say, $k$ and $l$), since these sets are not interconnected. If the same letter is used in these sets, solutions will be lost.

Solve the following equations **(1-245)**.

**1.** $\sin x = \dfrac{1}{2}$.

**2.** $\sin x = -\dfrac{1}{3}$.

**3.** $\sin x = 0$.

**4.** $\sin x = 1$.

**5.** $\cos x = \dfrac{1}{2}$.

**6.** $\cos x = -1$.

**7.** $\cos x = 0$.

**8.** $\tan x = \sqrt{3}$.

**9.** $\cot x = -1$.

**10.** $\sin x = \dfrac{\sqrt{2}}{2}$.

**11.** $\sin x = -\dfrac{\sqrt{3}}{2}$.

**12.** $\sin x = -\dfrac{1}{2}$.

**13.** $\cos x = 1$.

**14.** $\cos 2x = 1$.

**15.** $\cos x = \sqrt{3}$.

**16.** $\tan x = -1$.

**17.** $\tan (x-1) = 7$.

**18.** $\tan (2x+3) = \sqrt{3}$.

**19.** $\cot\left(\dfrac{\pi}{4} - \dfrac{x}{3}\right) = \dfrac{\sqrt{3}}{3}$.

**20.** $\sin (3x-2) = -1$.

**21.** $\sqrt{2}\cos^2 7x - \cos 7x = 0$.

**22.** $2\sin x + \tan x = 0$.

**23.** $(2\sin x - \cos x)(1 + \cos x) = \sin^2 x$.

**24.** $4\cos^3 x - 4\cos^2 x - \cos (\pi + x) - 1 = 0$.

**25.** $1 + \sin x \cos 2x = \sin x + \cos 2x$.

**26.** $\tan^3 x - 1 + \dfrac{1}{\cos^2 x} - 3\cot\left(\dfrac{\pi}{2} - x\right) = 3$.

**27.** $\tan 2x \sin x + \sqrt{3}\,(\sin x - \sqrt{3}\tan 2x) = 3\sqrt{3}$.

**28.** $\sin^4 x = 1 - \cos^4 x$.

**29.** $\sqrt{3}\sin x - \tan x + \tan x \sin x - \sqrt{3} = 0$.

**30.** $\cos 2x + 3\sin x = 2$.

**31.** $\cos x + \sec x = 2$.

**32.** $1 + \cos x + \cos 2x = 0$.

**33.** $6\cos^2 x + 5\sin x - 7 = 0$.

**34.** $\cos x + 2\cos 2x = 1$.

**35.** $2\cos^2 x + 4\cos x = 3\sin^2 x$.

**36.** $4\sin^4 x + 12\cos^2 x = 7$.

**37.** $5\tan^4 x - \sec^4 x = 29$.

**38.** $\cos 4x + 2\cos^2 x = 0$.

**39.** $\sin^4 x + \cos^4 x = \cos 4x$.

**40.** $\sin x + \cos x = 0$.

**41.** $\cos^2 x - 3 \sin x \cos x = \sin \dfrac{3\pi}{2}$.

**42.** $3 \sin^2 x - 2 \sin x \cos x - \cos^2 x = 0$.

**43.** $\cos^2 x + 2\sqrt{2} \cos x \sin x + 1 = 0$.

**44.** $\sin^2 x \,(1 + \tan x) = 3 \sin x \,(\cos x - \sin x) + 3$.

**45.** $\sin 3x + \sin x = \sin 2x$.

**46.** $1 - \sin 2x = \cos x - \sin x$.

**47.** $\sin x + \cos x = \dfrac{1 + \sqrt{3}}{2}$.

**48.** $4 \sin 2x - 3 \sin \left( 2x - \dfrac{\pi}{2} \right) = 5$.

**49.** $(\sin 2x + \sqrt{3} \cos 2x)^2 - 5 = \cos \left( \dfrac{\pi}{6} - 2x \right)$.

**50.** $\sin (\pi - 6x) + \sqrt{3} \sin \left( \dfrac{\pi}{2} + 6x \right) = \sqrt{3}$.

**51.** $\cos \left( \dfrac{\pi}{2} - 3x \right) - \sin 2x = 0$.

**52.** $\cos 2x - \cos 8x + \cos 6x = 1$.

**53.** $\sin x + \sin 2x + \sin 3x = 0$.

**54.** $\sin x + \sin 2x + \sin 3x + \sin 4x = 0$.

**55.** $\tan 3x - \tan x = 0$.

**56.** $\sqrt{3} \sin 2x + \cos 5x - \cos 9x = 0$.

**57.** $\cos \left( \dfrac{\pi}{2} + 5x \right) + \sin x = 2 \cos 3x$.

**58.** $\sin x + \sin 2x + \sin 3x = \cos x + \cos 2x + \cos 3x$.

**59.** $\cos 3x + \sin 5x = 0$.

**60.** $\cos 3x + \sin (9x + 2) = 0$.

**61.** $\cos 3x = \dfrac{\sqrt{3}}{2} \cos x - \dfrac{1}{2} \sin x$.

**62.** $\sin x + \cos x = \sqrt{2} \sin 5x$.

**63*.** $\tan x + \dfrac{\tan \dfrac{\pi}{4} + \tan x}{1 - \tan \dfrac{\pi}{4} \tan x} = 2$

**64.** $\cos 2x \cos x = \sin 7x \ \sin 6x + 5 \cos \dfrac{\pi}{2}$.

**65.** $\cos 3 + \sin x \sin 2x = 0$.

**66.** $1 + 2 \cos 3x \cos x - \cos 2x = 0$.

**67.** $2 \cos x \sin 3x = \sin 4x + 1$.

**68.** $\sin x \sin 7x = \sin 3x \sin 5x$.

**69.** $\cos x \cos 3x = \cos 5x \cos 7x$.

**70.** $\sin x \sin 3x = \dfrac{1}{2}$.

**71.** $\sin^2 x + \sin^2 2x + \sin^2 3x + \sin^2 4x = 2$.

**72*.** $\cos 3x \tan 5x = \sin 7 x$.

**73.** $4 \sin x \sin \left( x + \dfrac{\pi}{3} \right) \sin \left( x + \dfrac{2\pi}{3} \right) + \cos 3x = 1$.

**74.** $\sin 2x \sin 6x = \cos x \cos 3x$.

**75.** $\sin 5x + \sin x + 2 \sin^2 x = 1$.

**76.** $\cos^2 2x + \cos^2 3x = 1$.

**77.** $\sin^4 \dfrac{x}{3} + \cos^4 \dfrac{x}{3} = \dfrac{5}{8}$.

**78.** $\sin^4 \dfrac{x}{2} - \cos^4 \dfrac{x}{2} = \dfrac{1}{4}$.

**79.** $\sin^4 x + \cos^4 x = \sin x \cos x$.

**80.** $\tan x + \sin (\pi + x) = 2 \sin^2 \dfrac{x}{2}$.

**81.** $\sin^2 x + \sin^2 2x - \sin^2 3x - \sin^2 4x = 0$.

**82.** $\sin^2 x + \sin^2 2x + \sin^2 3x = \dfrac{3}{2}$.

**83.** $\sin x \sin 2x + \cos^2 x = \sin 4x \sin 5x + \cos^2 4x$.

**84.** $\sin^2 x + \sin^2 2x = \cos^2 3x + \cos^2 4x$.

**85.** $\sin^2 (2 + 3x) + \cos^2 \left( \dfrac{\pi}{4} + 2x \right) = \cos^2 (2 - 5x) + \sin^2 \left( \dfrac{\pi}{4} - 6x \right)$.

**86.** $\sin^4 x + \cos^4 x = \dfrac{3 - \cos 6x}{4}$.

**87.** $2 \cos 4x - 2 \cos 2x = 4 \cos^2 x - 1$.

**88*.** $2 \cos^2 x - 1 = \sin 3x$.  **89.** $\tan \dfrac{x}{2} = 1 - \cos x$.

**90.** $3 \sin \left( \dfrac{\pi}{2} - x \right) - 4 \sin (\pi + x) \sin \left( \dfrac{5\pi}{2} + x \right) + 8 \cos^2 \dfrac{x}{2} = 4$.

**91.** $3 + 5 \sin 2x = \cos 4x$.

**92.** $(1 + \cos 4x) \sin 2x = \cos^2 2x$.

**93.** $\sin x - 2 \cos 2x = 1$.

**94.** $\left( 1 + \sin \left( \dfrac{\pi}{2} - 4x \right) \right) \sin 4x = \cos^2 (2x - \pi)$.

**95.** $\cos 2x - 3 \cos x = 4 \cos^2 \dfrac{x}{2}$.

**96.** $4 \cos x - 2 \cos 2x - \cos 4x = 1$.

**97*.** $\sin 3x + \sin^3 x = \dfrac{3\sqrt{3}}{4} \sin 2x$.

**98.** $\sin^3 x - \cos^3 x = 1 + \sin x \cos x$.

**99.** $\sin^4 x + \cos^4 x = \sin 2x - 0.5$.

**100.** $\cos^3 x \sin x - \sin^3 x \cos x = \dfrac{\sqrt{2}}{8}$.

**101.** $\sin 2x = \cos 2x - \sin^2 x + 1$.

**102.** $(\cos 2x - 1) \cot^2 x = -3 \sin x$.

**103.** $8 \cos x \cos 2x \cos 4x = \dfrac{\sin 6x}{\sin x}$.

**104.** $\sin \dfrac{x}{2} \cos 2x + \sin^2 x \cos \dfrac{x}{2} = \cos^2 x \cos \dfrac{x}{2}$.

**105\*.** $\cos 3x - \cos 2x = \sin 3x$.

**106.** $(\sin 7x + \cos 7x)^2 = 2 \sin^2 11x + \sin 30x$.

**107.** $\sin 3x - 4 \sin x \cos 2x = 0$.

**108\*.** $\sin^2 x \tan x + \cos^2 x \cot x - \sin 2x = 1 + \tan x + \tan^2 x$.

**109.** $\sec^2 x - \tan^2 x + \cot \left( \dfrac{\pi}{2} + x \right) = \cos 2x \sec^2 x$.

**110.** $\sin^4 x + \cos^4 x - 2 \sin 2x + \dfrac{3}{4} \sin^2 2x = 0$.

**111.** $(\cos 6x - 1) \cot 3x = \sin 3x$.

**112.** $\sin x + \sin \left( \dfrac{3}{2} \pi + x \right) = 1 - 0.5 \sin 2x$.

**113.** $(1 - \sin 2x)(\cos x - \sin x) = 1 - 2 \sin^2 x$.

**114\*.** $2 \sin 3x - \dfrac{1}{\sin x} = 2 \cos 3x + \dfrac{1}{\cos x}$.

**115.** $\sec x + \operatorname{cosec} x = 2\sqrt{2}$.

**116.** $\cos x + \sin x = \dfrac{\cos 2x}{1 - \sin 2x}$.

**117.** $4 \sin \left( 5x + \dfrac{\pi}{4} \right) + \sin 10x = \dfrac{3}{2}$.

**118.** $3 (\cos x - \sin x) = 1 + \cos 2x - \sin 2x$.

**119.** $\sin x - \cos x = 1$.

**120.** $3 \sin x = 2 (1 - \cos x)$.

**121\*.** $\cos^2 x - 2 \cos x = 4 \sin x - \sin 2x$.

**122.** $\sin 2x - 2\sqrt{2} \sin \left( \dfrac{3\pi}{2} - 2x \right) = 3$.

**123\*.** $\tan x + \tan 2x + \tan 3x = 0$.

**124.** $\cos 9x - 2 \cos 6x = 2$.

**125.** $\left( 4 \sqrt{\cos \dfrac{x}{2}} - 5 - \dfrac{\sqrt{2}}{2} \right)^2 + \sqrt{2} \left( 4 \sqrt{\cos \dfrac{x}{2}} - 5 - \dfrac{\sqrt{2}}{2} \right) - \dfrac{\cos x}{2} = 0$.

**126\*.** $2 \sin \left( 3x + \dfrac{\pi}{4} \right) = \sqrt{1 + 8 \sin 2x \cos^2 2x}$.

**127.** $\cot 2x + \cot x = 8\cos^2 x$.

**128.** $\dfrac{3}{2} - \sin 2x = \sqrt{9 + 10\sin 2x}$.

**129.** $\sqrt{13 - 18\tan x} = 6\tan x - 3$.

**130.** $4\sin 3x + 3 = \sqrt{2}\sin 3x + 2$.

**131.** $\sqrt{6 - \sin x - 7\cos^2 x} + \sin x = 0$.

**132\*.** $\tan x + \dfrac{\cos x}{\sqrt{1 + \sin 2x}} = 2$.

**133.** $\sin^4 x + \sin^4\left(\dfrac{x}{2} + \dfrac{\pi}{8}\right) + \cos^4 x = \dfrac{1}{2}\sin^2 2x$.

**134.** $\cos\left(\pi\sqrt{x - 4}\right)\cos\left(\pi\sqrt{x}\right) = 1$.

**135.** $\dfrac{1 - \cos 2x}{\sin 2x} = 0$.

**136.** $\cos^2 x + 5\cos x = 2\sin^2 x$.

**137.** $\sin 2x + \sin(\pi - 8x) = \sqrt{2}\cos 3x$.

**138.** $\dfrac{1 + \sin x + \cos x + \sin 2x + \cos 2x}{\tan 2x} = 0$.

**139.** $4\cos^2 6x + 16\cos^2 3x = 13$.

**140.** $\tan 5x + 2\sin 10x = 5\sin 5x$.

**141.** $\tan 3x - \tan x = \sec x \sec 3x \sin x$.

**142.** $\sin x + \sin 3x = 4\cos^3 x$.

**143.** $2\sin^2 2x + 6\sin^2 x = 5$.

**144.** $\cos^6 x - \sin^6 x = \dfrac{13}{8}\cos^2 2x$.

**145.** $4\sin^4 x + \sin^2 2x = 1$.

**146.** $1 + \sin 2x = (\cos 3x + \sin 3x)^2$.

**147.** $\sqrt{3} - \tan x = \tan\left(\dfrac{\pi}{4} - x\right)$.

**148.** $\cot^2 x - \tan^2 x = 8\cot^2 2x$.

**149.** $\cos x + \cos 4x + \cos 7x = 0$.

**150.** $\sin 5x + \cos 3x - \sin x = 0$.

**151.** $\cos^4\dfrac{x}{5} + \sin^2\dfrac{x}{5} = 1$.

**152.** $\sin^2 x = \sin 3x + \cos x\,(\cos x - 1)$.

**153.** $(3 - \cot^2 x)\sin 2x = 2(1 + \cos 2x)$.

**154.** $\dfrac{1 + \tan x}{1 - \tan x} = (\sin x + \cos x)^2$.

**155.** $\sin x + \sin 2x = \tan x$.

**156.** $\sin x - \sin 2x + \sin 5x + \sin 8x = 0$.

**157.** $8\cos^4 x - 8\cos^2 x - \cos x + 1 = 0$.

**158.** $\tan 3x + \cos 6x = 1$.

**159.** $\tan x \tan 3x = 1$.

**160.** $\cos^2 \dfrac{x}{2} + \sin^2 3x + \cos^2\left(\dfrac{\pi}{2} + 2x\right) = 3\cos\dfrac{\pi}{3}.$

**161.** $\sin 7x + \sin 9x = 2\left(\cos^2\left(\dfrac{\pi}{4} - x\right) - \cos\left(\dfrac{\pi}{4} + 2x\right)\right).$

**162.** $\sin 9x + \sqrt{3}\cos 7x = \sin 7x + \sqrt{3}\cos 9x.$

**163.** $\sin x + \sin 2x = \cos x + 2\cos^2 x.$

**164.** $\sin\left(2x + \dfrac{5\pi}{2}\right) - 3\cos\left(\dfrac{7\pi}{2} - x\right) = 1 + 2\sin x.$

**165\*.** $\left(\cos\dfrac{\pi}{4} - \sin\dfrac{\pi}{6}\right)\left(\dfrac{1}{\cos x} + \tan x\right) = \sin\dfrac{\pi}{4}\cos x.$

**166.** $\sin^6 x + \cos^6 x = \dfrac{7}{16}.$

**167.** $\dfrac{2}{\sqrt{3}}(\tan x - \cot x) = \tan^2 x + \cot^2 x - 2.$

**168.** $\cos 4x + 2\cos^2 x = 1.$

**169.** $\sin 3x - \cos 3x = \sqrt{\dfrac{3}{2}}.$

**170.** $\dfrac{\sin x}{1 + \cos x} = 2 - \cot x.$

**171.** $\cot x - 2\sin 2x = 1.$

**172.** $2\tan x - \cos 2x = 2.$

**173.** $\tan(3x - 1)\cot(x + 2) = 1.$

**174.** $|\cot x| = \cot x + \dfrac{1}{\sin x}.$

**175.** $2(1 - \cos 2x) = \sqrt{3}\tan x.$

**176.** $3\cos^2 x - \sin^2 x - \sin 2x = 0.$

**177.** $3\sin^2 2x + 7\cos 2x - 3 = 0.$

**178.** $2\cot^2 x\cos^2 x + 4\cos^2 x - \cot^2 x - 2 = 0.$

**179.** $4\sin x\sin\left(x + \dfrac{\pi}{3}\right)\sin\left(x + \dfrac{2\pi}{3}\right) + \cos 3x = 1.$

**180.** $\sin x + \sqrt{3}\sin\left(\dfrac{7}{2}\pi - x\right) + \tan x = \sqrt{3}.$

**181.** $\dfrac{5\sin x - 5\tan x}{\sin x + \tan x} + 4(1 - \cos x) = 0.$

**182.** $8\sin x = \dfrac{\sqrt{3}}{\cos x} + \dfrac{1}{\sin x}.$

**183.** $\cos x = \sin x - 1.$

**184.** $\cos^2 3x + \sin^2\left(\dfrac{\pi}{2} + 5x\right) = \cos^2 7x + \cos^2 9x + \cos\dfrac{3\pi}{2}.$

**185.** $\sin^2(1.5x) + \sin^2\left(\dfrac{\pi}{4} - 2.5x\right) = \sin^2(5.5x) + \sin^2\left(\dfrac{\pi}{4} - 6.5x\right).$

**186.** $\tan x + \cot x - \cos 4x = 3.$

**187\*.**  $\cos x - \sqrt{3} \sin x = \cos 3x.$

**188.**  $\dfrac{1}{\tan x} - 1 = \dfrac{\cos 2x}{1 + \tan x}.$

**189.**  $4 \sin x + \cos x = 4.$

**190.**  $\cos 5x \tan 6 \,|\, x \,| + \sin 5x = 0.$

**191.**  $\tan 4x \cos 7x = \sin 7 \,|\, x \,|.$

**192.**  $2 \cos 2x + 2 \cos x \sin^2 x = \cos x.$

**193.**  $\dfrac{1}{\cos x} + \dfrac{1}{\cos \left( x - \dfrac{3\pi}{2} \right)} = 4 \cos \left( \dfrac{7\pi}{4} - x \right).$

**194.**  $11 \cot x - 5 \tan x = \dfrac{16}{\sin x}.$

**195.**  $\cos^2 2x + \cos^2 4x - \sin^2 6x - \sin^2 8x = 0.$

**196\*.**  $\sin x + \sin^2 x + \cos^3 x = 0.$

**197.**  $\cos 3x - \sin 5x - \cos 7x = \sin 4x - \cos 2x.$

**198.**  $5 \sin x + \dfrac{3}{1 + \tan^2 x} = \cos 2x - 3 \sin^2 x.$

**199.**  $\cot x + \dfrac{\sin x}{1 + \cos x} = 2.$

**200.**  $\tan^2 2x = \dfrac{1 - \cos x}{1 + \cos x}.$

**201.**  $2 \sin 2x \sin 4x - \cos 2x = \sin 3x.$

**202.**  $3 \cot x - 3 \tan x + 4 \sin 2x = 0.$

**203.**  $\tan x + \cot x = 2.$

**204\*.**  $\dfrac{\sin^3 \dfrac{x}{2} - \cos^3 \dfrac{x}{2}}{2 + \sin x} = \dfrac{1}{3} \cos x.$

**205.**  $\sin 2x + 2 \cot x = 3.$

**206.**  $\sin x \cos x \sin 2x = \dfrac{1}{8}.$

**207.**  $\sin 6x + 2 = 2 \cos 4x.$

**208\*.**  $\tan^2 x \tan^2 3x \tan 4x = \tan^2 x - \tan^2 3x + \tan 4x.$

**209.**  $2 \cos 13x + 3 \cos 3x + 3 \cos 5x - 8 \cos x \cos^3 4x = 0.$

**210.**  $\tan 3x - \tan x = 2 (\sin 4x - \sin 2x).$

**211.**  $(\sin 13°)^{\cot 3x + \cot x} = \sin^2 (2\pi - x) - \cos (\pi - x) \sin \left( \dfrac{\pi}{2} + x \right).$

**212.**  $2 \sin 3x \sin x + (3\sqrt{2} - 1) \cos 2x = 3.$

**213.**  $\cot x - 2 \sin 2x = 1.$

**214.**  $\tan x \sin x - \cos x = 0.5 \sec x.$

**215.**  $\sin^2 x \tan x + \cos^2 x \cot x + 2 \sin x \cos x = \dfrac{4\sqrt{3}}{3}.$

**216.**  $\tan 2x \sin 2x - 3\sqrt{3} \cot 2x \cos 2x = 0.$

**217.** $\sin \dfrac{3x}{2} \cos \dfrac{x}{2} = \dfrac{\sin 2x}{2}.$

**218.** $\cot 2x - \tan 2x = \dfrac{2}{3} \tan 4x.$

**219.** $\sin x + \cos x \cot \dfrac{x}{2} = -\sqrt{3}.$

**220.** $\cos x \cos 2x \cos 4x = 1.$

**221\*.** $9 \cos^{12} x + \cos^2 2x + 1 + 2\cos 2x = 6\cos^6 x \cos 2x + 6\cos^6 x.$

**222.** $\sin 2x (0.1 - \cos x) = \sin 2x + 0.2\sin^3 x.$

**223\*.** $|\cos x| = \cos (x + a).$

**224\*.** $\sin \dfrac{2x+1}{x} + \sin \dfrac{2x+1}{3x} - 3\cos^2 \dfrac{2x+1}{3x} = 0.$

**225\*.** $a \sin \dfrac{x}{2} - \left( \sin x + \sin \dfrac{3x}{2} \right) = 0.$

**226.** $\cos (2\sin x + (1 + \sqrt{3})\cos x) = \sin ((1 - \sqrt{3})\cos x).$

**227\*.** $\sqrt{\sqrt{3}\cos x + \sin x - 2} + \sqrt{\cot 3x + \sin^2 x - \dfrac{1}{4}} = \sin \dfrac{3x}{2} + \dfrac{\sqrt{2}}{2}.$

**228\*.** $\sin^4 x + \cos^4 x + \sin 2x + a = 0.$

**229.** $\sqrt{\cos^2 2x + \left| \sin \left( 2x - \dfrac{3}{2}\pi \right) \right| + \dfrac{1}{4}} = \cos \left( \dfrac{20}{12}\pi \right).$

**230.** $\left| \cot \left( 2x - \dfrac{\pi}{2} \right) \right| = \dfrac{1}{\cos^2 2x} - 1.$

**231.** $\dfrac{1}{\cos x} + 1 = \sin (\pi - x) - \cos x \tan \dfrac{\pi + x}{2}.$

**232.** $\sin^3 x - \cos^3 x = 0.$

**233.** $\cos \dfrac{3x}{2} \cos \dfrac{x}{2} - 1 = \dfrac{1 + \sqrt{3}}{2} \cos x.$

**234.** $\dfrac{3}{\sqrt{4\cos 2x + 1}} = \sqrt{2\cos 2x + 2}.$

**235.** $\sin \dfrac{\pi x}{2\sqrt{3}} = x^2 - 2\sqrt{3}x + 4.$

**236.** $\sin 2x + \tan x = 2.$

**237.** $\sin^2 \left( \dfrac{\pi}{8} + x \right) = \sin x + \sin^2 \left( \dfrac{\pi}{8} - x \right).$

**238.** $6 \sin^2 x + \sin x \cos x - \cos^2 x = 2.$

**239.** $\dfrac{\cot 2x}{\cot x} + \dfrac{\cot x}{\cot 2x} + 2 = 0.$

**240.** $\sin x - 4\cos x + \tan x = 4.$

**241.** $8 \cos^6 x = 3\cos 4x + \cos 2x + 4.$

**242.** $(1 - \tan x)(1 + \sin 2x) = 1 + \tan x.$

**243.** $\sin 3x + 3\sin 4x + \sin 5x = 0.$

**244.** $2\tan^2 x + 3 = \dfrac{3}{\cos x}.$

**245.** $81^{\sin^2 x} + 81^{\cos^2 x} = 30.$

**246.** Find all values of $p$ for which the equation
$\sqrt{p}\cos x - 2\sin x = \sqrt{2} + \sqrt{12 - p}$ possesses a solution.

**247.** Find all values of $a$ for which the equation $\cos 2x + a\sin x = 2a - 7$ possesses a solution.

**248.** Find the roots of the equation $\sin\left(2x + \dfrac{\pi}{18}\right)\cos\left(2x - \dfrac{\pi}{9}\right) = -\dfrac{1}{4}$ in the interval $\left(0,\ \dfrac{\pi}{2}\right)$.

**249.** Find (in degrees) the solution $x$ of the equation
$1 + \cos 10x \cos 6x = 2\cos^2 8x + \sin^2 8x$, such that $20° < x < 80°$.

**250.** $\sin x \cos\dfrac{\pi}{8} + \cos x \sin\dfrac{\pi}{8} = \dfrac{1}{2},\ x \in \left[-\dfrac{3\pi}{2},\ \pi\right].$

**251.** Solve the equation $4\cos x\,(2 - 3\sin^2 x) + (\cos 2x + 1) = 0.$ Find the least distance between its positive roots.

**252.** $\cos(\pi + x) + \sqrt{3}\sin x = \sin\left(3x - \dfrac{\pi}{2}\right),\ x \in \left(-\dfrac{4\pi}{3},\ \dfrac{\pi}{2}\right].$

**253.** $2\sin^4 2x - \sin^2 2x \sin 4x = 2\sin^2 2x - \sin 4x,\ x \in [0,\ \pi].$

**254.** Find which of the numbers $\pi n - \arctan 3$, where $n \in \mathbf{Z}$, are solutions of the equation $12\tan 2x + \dfrac{\sqrt{10}}{\cos x} + 1 = 0.$

**255.** Find all the roots of the equation $\cos 4x + \dfrac{10\tan x}{1 + \tan^2 x} = 3$ which lie in the interval $\left[-\dfrac{3\pi}{4},\ \dfrac{\pi}{2}\right].$

**256.** Find all the roots of the equation $(1 - \cos 2x)\sin 2x = \sqrt{3}\sin^2 x$ which lie in the interval $\left[-\pi,\ \dfrac{\pi}{3}\right].$

**257.** Find all the roots of the equation
$\sin x \tan 2x + \sqrt{3}(\sin x - \sqrt{3}\tan 2x)3 = \sqrt{3}$ satisfying the inequality $2 + \log_{1/2} x \le 0.$

**258.** Solve the equation $\sin(\pi x^2) - \sin(\pi(x^2 + 2x)) = 0$ and find the seventh term of the increasing sequence of its positive roots.

**259.** Solve the equation $\cos 4x + 6 = 7\cos 2x$ and find the sum of its roots in the interval $[0,\ 314].$

**260.** Find all the solutions of the equation $3\sin^3 x - 3\cos^2 x + 7\sin x - \cos 2x + 1 = 0$ which are also solutions of the equation $\cos^2 x + 3\cos x \sin 2x - 8\sin x = 0.$

**261.** Solve the equation $\tan 2x = 3\tan x$ if $x \in (-2\pi, \ 2\pi)$.

**262.** Find all the solutions of the equations

$\sin\left(4x + \dfrac{\pi}{4}\right) + \cos\left(4x + \dfrac{5\pi}{4}\right) = \sqrt{2}$ which satisfy the inequality

$\dfrac{\cos 2x}{\cos 2 - \sin 2} > 2^{-\sin 4x}$.

## Trigonometric Inequalities and Systems of Inequalities

Solve the following inequalities **(263-278)**.

**263.** $\sin x > 0$.

**264.** $\sin x > \dfrac{1}{2}$.

**265.** $\log_2\left(\sin\dfrac{x}{2}\right) < -1$.

**266.** $\sin x \le \dfrac{1}{2}$.

**267.** $\sin x \le -1$.

**268.** $\cos x < -\dfrac{\sqrt{2}}{2}$.

**269.** $\tan x > 0$.

**270.** $\tan x < -\sqrt{3}$.

**271.** $\sin x < \cos x$.

**272.** $\sin 3x < \sin x$.

**273.** $\sin x + \sqrt{3}\cos x > 0$.

**274.** $\tan^2 x - (1 + \sqrt{3})\tan x + \sqrt{3} < 0$.

**275.** $\sin x\left(\cos x + \dfrac{1}{2}\right) \le 0$.

**276.** $2\cos^2 x + 5\cos x + 2 \ge 0$.

**277.** $\sqrt{5 - 2\sin x} \ge 6\sin x - 1$.

**278.** $2^{\frac{1}{\cos^2 x}}\sqrt{y^2 - y + \dfrac{1}{2}} \le 1$.

Solve the following systems **(279-312)**.

**279.** $\begin{cases} \sin^2 x + \sin^2 2x = \sin^2 3x, \\ \cos x < -\dfrac{1}{2}. \end{cases}$

**280.** $\begin{cases} \cos^3 x - \sin^3 x = \cos 2x, \\ 0 \le x \le \dfrac{3\pi}{2}. \end{cases}$

281. $\begin{cases} \sin^2 x + \sin^2 2x = \sin^2 3x, \\ -\dfrac{\pi}{2} \le x \le \pi. \end{cases}$

282. $\begin{cases} \sin\left(x - \dfrac{\pi}{4}\right) - \cos\left(x + \dfrac{3\pi}{4}\right) = 1, \\ \dfrac{2\cos 7x}{\cos 3 + \sin 3} > 2^{\cos 2x}. \end{cases}$

283. $\begin{cases} \sin x \sin y = \dfrac{1}{4}, \\ \cos x \cos y = \dfrac{3}{4}. \end{cases}$

284. $\begin{cases} \sin x \sin y = \dfrac{1}{4}, \\ \cot x \cot y = 3. \end{cases}$

285. $\begin{cases} \sin x + \cos y = 0, \\ \sin^2 x + \cos^2 y = \dfrac{1}{2}, \\ 0 < x < \pi, \quad 0 < y < \pi. \end{cases}$

286. $\begin{cases} y - x = \dfrac{1}{4}, \\ \cos(\pi x)\cos(\pi y) = \dfrac{\sqrt{2}}{2}. \end{cases}$

287. $\begin{cases} \cos(x - y) = \dfrac{1}{2}, \\ \cos(x + y) = -\dfrac{1}{2}. \end{cases}$

288. $\begin{cases} x + y = \dfrac{3\pi}{4}, \\ \tan x - \tan y = 2. \end{cases}$

289. $\begin{cases} \sin x + \cos y = 1, \\ x + y = \dfrac{\pi}{3}. \end{cases}$

290. $\begin{cases} x + y = \dfrac{2\pi}{3}, \\ \dfrac{\sin x}{\sin y} = 2. \end{cases}$

**291.** $\begin{cases} \tan x + \tan y = 2, \\ \cos x \cos y = \dfrac{1}{2}. \end{cases}$

**292.** $\begin{cases} x - y = \dfrac{13\pi}{2}, \\ 3\cos^2 x - 12\cos y = -4. \end{cases}$

**293.** $\begin{cases} \sin^2 x + \cos^2 y = \dfrac{11}{16}, \\ \sin \dfrac{x+y}{2} \cos \dfrac{x-y}{2} = \dfrac{5}{8}. \end{cases}$

**294.** $\begin{cases} \dfrac{1}{\sqrt{\sin\left(\dfrac{\pi}{4} - x\right)}}(3\cos 2x - 6\cot y + 2) = 0, \\ 18\sin^2 x - 2\tan y - 3 = 0. \end{cases}$

**295.** $\begin{cases} 4\sin y - 6\sqrt{2}\cos x = 5 + 4\cos^2 y, \\ \cos 2x = 0. \end{cases}$

**296.** $\begin{cases} 6\cos x + 4\cos y = 5, \\ 3\sin x + 2\sin y = 0. \end{cases}$

**297\*.** $\begin{cases} 3\tan \dfrac{y}{2} + 6\sin x = 2\sin(y - x), \\ \tan \dfrac{y}{2} - 2\sin x = 6\sin(y + x). \end{cases}$

**298.** $\begin{cases} \sqrt{1 + \sin x \sin y} = \cos x, \\ 2\sin x \cot y + 1 = 0. \end{cases}$

**299\*.** $\begin{cases} \cos 2y + \dfrac{1}{2} = \left(\cos y - \dfrac{1}{2}\right)(1 + 2\sin 2x), \\ \sin y\,(\tan^3 x + \cot^3 x) = 3\cot y. \end{cases}$

**300.** $\begin{cases} \tan x \tan z = 3, \\ \tan y \tan z = 6, \\ x + y + z = \pi. \end{cases}$

**301.** $\begin{cases} \sin x + \sin y = \sin(x + y), \\ |x| + |y| = 1. \end{cases}$

**302.** $\begin{cases} 4\tan 3x = 3\tan 2y, \\ 2\sin x \cos(x - y) = \sin y. \end{cases}$

**303.** $\begin{cases} \cos x + \cos y = a, \\ \quad x + y = \dfrac{\pi}{4}. \end{cases}$

**304.** $\begin{cases} \tan x + \sin y = 2a, \\ \tan x \sin y = a^2 - 1. \end{cases}$

**305.** $\begin{cases} \cos \dfrac{x+y}{2} \cos \dfrac{x-y}{2} = \dfrac{1}{2}, \\ \cos x \cos y = \dfrac{1}{4}. \end{cases}$

**306.** $\begin{cases} \cot x + \sin 2y = \sin 2x, \\ 2 \sin y \sin (x+y) = \cos x. \end{cases}$

**307.** $\begin{cases} \sin (x-y) = 3 \sin x \cos y - 1, \\ \sin (x+y) = -2 \cos x \sin y. \end{cases}$

**308.** $\begin{cases} \tan x + \cot x = 2 \sin \left( y - \dfrac{3}{4} \pi \right), \\ \tan y + \cot y = 2 \sin \left( x + \dfrac{\pi}{4} \right). \end{cases}$

**309.** $\begin{cases} \cot \sqrt[4]{x} = 1, \\ \cos \sqrt[4]{x} = -\dfrac{1}{\sqrt{2}}. \end{cases}$

**310.** $\begin{cases} \cos y \, (\cos x - \cos y) = 2 \cos \dfrac{x+y}{2} \sin y \sin \dfrac{y-x}{2}, \\ \qquad\qquad 2y - x = \dfrac{\pi}{2}. \end{cases}$

**311.** $\begin{cases} \cos x + \cos 2x + \cos 3x = 3, \\ \qquad \cos^3 \dfrac{x}{2} = \cos^4 2x. \end{cases}$

**312.** $\begin{cases} \sin^2 (-2x) - (3 - \sqrt{2}) \tan 5y = \dfrac{3\sqrt{2}-1}{2}, \\ \tan^2 (5y) + (3 - \sqrt{2}) \sin (-2x) = \dfrac{3\sqrt{2}-1}{2}. \end{cases}$

**313.** Find all solutions of the system of equations

$$\begin{cases} \sin (2x+3y) + \cos (2x+3y) = 1, \\ \cos \left( x + \dfrac{\pi+18}{12} y \right) + \sqrt{3} \sin \left( x + \dfrac{\pi+18}{12} y \right) = 2. \end{cases}$$

which satisfy the condition $|y| \le 1$.

**314.** Find all solutions of the system of equations

$$\begin{cases} \cot^2 (x-y)-(1+\sqrt{3})\cot (x-y)+\sqrt{3}=0, \\ \cos y = \dfrac{\sqrt{3}}{2}, \end{cases}$$

which satisfy the conditions $0 < x < \pi$, $0 \le y \le 2\pi$.

## 1.12. Progressions

**The arithmetic progression.** An *arithmetic progression* is a number sequence each term of which, beginning with the second, is equal to the sum of the preceding term and one and the same number $d$. The number $d$ is called the *common difference* or simply *difference* of the arithmetic progression.

The sequence $\{a_n\}$ is an arithmetic progression if and only if the equality

$$a_n = \frac{1}{2}(a_{n-1} + a_{n+1})$$

holds true for any $n > 1$. For the arithmetic progression $\{a_n\}$ we have

$$a_n = a_1 + (n-1)d, \quad S_n = \frac{a_1 + a_n}{2}n = \frac{2a_1 + d(n-1)}{2}n,$$

where $d$ is the difference of the progression and $S_n$ is the *sum* of its first $n$ terms.

To define an arithmetic progression, it is sufficient to specify its first term and the common difference.

**The geometric progression.** A *geometric progression* is a number sequence the first term of which is different from zero, and each term, beginning with the second, is equal to the preceding term multiplied by one and the same nonzero number $q$. The number $q$ is the *common ratio* of the geometric progression.

The sequence $\{b_n\}$ is a geometric progression if and only if the equality

$$b_n^2 = b_{n-1}b_{n+1}$$

holds true for any $n > 1$. For the geometric progression $\{b_n\}$ we have

$$b_n = b_1 q^{n-1}, \quad S_n = \frac{b_n q - b_1}{q-1} = \frac{b_1(q^n - 1)}{q-1},$$

where $q \ne 1$ is the common ratio of the progression and $S_n$ is the *sum* of its first $n$ terms. For $q = 1$ we have

$$S_n = b_1 n.$$

To define a geometric progression, it is sufficient to specify its first term and the common ratio.

The sum of the infinite geometric progression $\{b_n\}$ for $|q| < 1$ is defined by the equation

$$S = \lim_{n\to\infty} S_n, \text{ and in this case } S = \frac{b_1}{1-q}.$$

1. Find the first term $a_1$ and the common difference $d$ of the arithmetic progression in which
$$\begin{cases} a_2 + a_5 - a_3 = 10, \\ a_2 + a_9 = 17. \end{cases}$$

2. The sum of the first $n$ terms of the sequence $\{a_n\}$ is defined by the formula $S_n = 2n^2 + 3n$. Prove that it is an arithmetic progression.

3. The sum $S_n$ of the terms of an arithmetic progression is defined by the formula $S_n = 4n^2 - 3n$ for any $n$. Write the first three terms of the progression.

4. Find the sum of 20 terms of an arithmetic progression if its first term is 2 and the seventh term is 20.

5. Find the first term and the difference of an arithmetic progression if the sum of its first five even terms is equal to 15 and the sum of the first three terms is equal to (–3).

6. The sum of the first and the fifth term of an arithmetic progression is 26 and the product of the second by the fourth term is 160. Find the sum of the first six terms of the progression.

7. The sum of the third and the ninth term of an arithmetic progression is 8. Find the sum of the first 11 terms of the progression.

8. The sum of the squares of the fifth and the eleventh term of an arithmetic progression is 3 and the product of the second by the fourteenth term is equal to $k$. Find the product of the first by the fifteenth term of the progression.

9. The sum of the second and the fifth term of an arithmetic progression is 8 and that of the third and the seventh term is 14. Find the progression.

10. How many terms of an arithmetic progression must be taken for their sum to be equal to 91, if its third term is 9 and the difference between the seventh and the second term is 20?

11. Put five terms between the numbers 1 and 1.3 so that together with the given terms they will form an arithmetic progression.

12. Find four numbers between the numbers 4 and 40 such that an arithmetic progression results.

13. Find the sum of all three-digit natural numbers which leave a remainder 2 when they are divided by 3.

14. Solve the equation $5^2 \cdot 5^4 \cdot 5^6 \dots 5^{2x} = (0.04)^{-28}$.

15. Find an increasing arithmetic progression in which the sum of the first three terms is 27 and the sum of their squares is 275.

16. The sum of four numbers which form an arithmetic progression is 1 and the sum of the squares of those numbers is 0.3. Find the numbers.

17. An arithmetic progression consists of 12 terms whose sum is 354. The ratio of the sum of the even terms to the sum of the odd terms is 32 : 27. Find the common difference of the progression.

18. The product of the third by the sixth term of an arithmetic progression is 406. The division of the ninth term of the progression by the fourth term gives a quotient 2 and a remainder 6. Find the first term and the difference of the progression.

19. The first term of an arithmetic progression $a_1$, $a_2$, $a_3$, .... is equal to unity. At what value of the difference of the progression is $a_1a_3 + a_2a_3$ at a minimum?

20. In an arithmetic progression, $a_7 = 9$. At what value of its difference is the product $a_1a_2a_7$ the least?

21. Two arithmetic progressions contain the same number of terms. The ratio of the last term of the first progression to the first term of the second is equal to the ratio of the last term of the second progression to the first term of the first progression and is equal to 4. The ratio of the sum of the first progression to that of the second is 2. Find the ratio of the differences of the progressions.

22. All terms of an arithmetic progression are natural numbers. The sum of its nine successive terms, beginning with the first, is larger than 200 and smaller than 220. Find the progression if its second term is 12.

23. Each of the two triplets of numbers log $a$, log $b$, log $c$ and log $a$ – log $2b$, log $2b$ – log $3c$, log $3c$ – log $a$ is an arithmetic progression. Can the numbers $a$, $b$, $c$ be the lengths of the sides of a triangle? If they can, then what triangle is it? Find the angles of the triangle provided that it exists.

24. The fourth term of an arithmetic progression is 4. At what value of the difference of the progression is the sum of the pairwise products of the first three terms of the progression the least?

25. The sixth term of an arithmetic progression is 3 and the difference exceeds 0.5. At what value of the difference of the progression is the product of the first, the fourth, and the fifth term of the progression the greatest?

26. Certain numbers appear in both arithmetic progressions 17, 21, ...... and 16, 21,....... . Find the sum of the first hundred numbers appearing in both progressions.

27. Three numbers form an arithmetic progression. The sum of the numbers is equal to 3 and the sum of their cubes is equal to 4. Find the numbers.

28. Prove that the numbers $\dfrac{1}{\log_3 2}$, $\dfrac{1}{\log_6 2}$, and $\dfrac{1}{\log_{12} 2}$ form an arithmetic progression.

29. At what values of the parameter $a$ are there values of $x$ such that the numbers $5^{1+x} + 5^{1-x}, \dfrac{a}{2}, 25^x + 25^{-x}$ form an arithmetic progression?

30. Prove that the sequence, whose general term is $a_n = 2 \cdot 3^n$, is a geometric progression and find the sum of the first eight terms.

31. The fourth term of a geometric progression exceeds the second term by 24 and the sum of the second and the third term is 6. Find the progression.

32. The difference between the fourth and the first term of a geometric progression is 52 and the sum of the first three terms is 26. Calculate the sum of the first six terms of the progression.

33. The sum of the first four terms of a geometric progression is 30 and the sum of the next four terms is 480. Find the sum of the first twelve terms.

34. The sum of the first two terms of a geometric progression is 15. The first term exceeds the common ratio of the progression by $\dfrac{25}{3}$. Find the fourth term of the progression.

35. Find three numbers which form a geometric progression if their sum is 35 and the sum of their squares is 525.

36. The sequence $\{b_r\}$ is a geometric progression with $\dfrac{b_4}{b_6} = \dfrac{1}{4}$ and $b_2 + b_5 = 216$. Find $b_1$.

37. The sum of the first three terms of an increasing geometric progression is 13 and their product is 27. Calculate the sum of the first five terms of the progression.

38. Find the first term and the common ratio of a geometric progression if the sum of its first three terms is 10.5 and the difference between the first and the fourth term is equal to 31.5.

39. The numbers $a, b, c,$ and $d$ form a geometric progression. Find $(a - c)^2 + (b - c)^2 + (b - d)^2 - (a - d)^2$.

40. The sum of the first and the third term of a geometric progression is 20 and the sum of its first three terms is 26. Find the progression.

41. The difference between the first and the fifth term of a geometric progression, whose all terms are positive numbers, is 15 and the sum of the first and the third term of the progression is 20. Calculate the sum of the first five terms of the progression.

42. Find four numbers forming a geometric progression in which the sum of the extreme terms is 112 and the sum of the middle terms is 48.

43. The first term of the geometric progression $b_1, b_2, b_3, \ldots$ is unity. For what value of the common ratio of the progression is $4b_2 + 5b_3$ at a minimum?

**44.** The sum of three numbers which form a geometric progression is 13 and the sum of their squares is 91. Find the numbers.

**45.** The numbers $5x - y$, $2x + y$ and $x + 2y$ form an arithmetic progression, and the numbers $(y + 1)^2$, $xy + 1$, $(x - 1)^2$ form a geometric progression. Find $x$ and $y$.

**46.** The sum of the first and the last term of an increasing geometric progression is 66, the product of the second and the last but one term is 128, and the sum of all its terms is 126. How many terms are there in the progression?

**47.** The sum of the first three terms of a geometric progression is 31 and the sum of the first and the third term is 26. Find the seventh term of the progression.

**48.** The number of the terms of a geometric progression is even. The sum of all terms of the progression is thrice as large as the sum of its odd terms. Find the common ratio of the progression.

**49.** The first, the third, and the fifth term of a geometric progression are equal to the first, the fourth, and the sixteenth term of a certain arithmetic progression respectively. Calculate the fourth term of the arithmetic progression if its first term is 5.

**50.** Three numbers whose sum is equal to 28 form a geometric progression. If we add 3 to the first number and 1 to the second, and subtract 5 from the third number, the resulting numbers will form an arithmetic progression. Find the numbers.

**51.** Three numbers form an arithmetic progression. If we add 8 to the first number, we get a geometric progression with the sum of the terms equal to 26. Find the numbers.

**52.** The second, the first, and the third term of an arithmetic progression, whose common difference is nonzero, form a geometric progression in that order. Find its common ratio.

**53.** The first and the third term of an arithmetic progression are equal, respectively, to the first and the third term of a geometric progression and the second term of the arithmetic progression exceeds the second term of the geometric progression by 0.25. Calculate the sum of the first five terms of the arithmetic progression if its first term is 2.

**54.** Find four numbers the first three of which form a geometric progression and the last three form an arithmetic progression. The sum of the extreme terms is 14 and the sum of the middle terms is 12.

**55.** Find four numbers the first three of which form a geometric progression and the last three an arithmetic progression if the sum of the extreme numbers is 21 and the sum of the middle terms is 18.

**56.** Three integers whose sum is equal to 60 are three successive terms of an arithmetic progression. If we add 2.2, 4, and 7 to these numbers, respectively, the new numbers will be three successive terms of a geometric progression. Find the least of the original numbers.

**57.** The sum of the first three terms of a geometric progression is 6, and the sum of its first three odd terms is 10.5. Find the common ratio and the first term of the progression.

**58.** The difference between the third and the second term of a geometric progression is 12. If we add 10 to the first term and 8 to the second and leave the third term unchanged, the new three numbers will form an arithmetic progression. Find the sum of the first five terms of the geometric progression.

**59.** Three numbers whose sum is 26 form a geometric progression. If we add 1, 6, and 3 to these numbers respectively, an arithmetic progression will result. Find the numbers.

**60.** Between the number 3 and the unknown number there is one more number such that the three numbers form an arithmetic progression. If we diminish the middle term by 6, we get a geometric progression. Find the unknown number.

**61.** Three nonzero real numbers form an arithmetic progression, and the squares of those numbers, taken in the same order, form a geometric progression. Find all possible common ratios of the geometric progression.

**62.** Three distinct numbers $x$, $y$, $z$, form a geometric progression in that order, and the numbers $x + y$, $y + z$, $z + x$ form an arithmetic progression in that order. Find the common ratio of the geometric progression.

**63.** The common difference of an arithmetic progression is different from zero. The numbers, which are equal to the products of the first term of the progression by the second, of the second term by the third, and of the third term by the first form a geometric progression in the indicated order. Find its common ratio.

**64.** Three numbers form a geometric progression. If we subtract 4 from the third number, an arithmetic progression will result. Now if we subtract unity from the second and the third term of the arithmetic progression obtained, we again get a geometric progression. Find the numbers.

**65.** Find a three-digit number if its digits form a geometric progression and the digits of the number which is smaller by 400 form an arithmetic progression.

**66.** A computer solved several problems in succession. The time it took the computer to solve each successive problem was the same number of times smaller than the time it took it to solve the preceding problem. How many problems were suggested to the computer if it spent 63.5 min to solve all the problems except for the first, 127 min to solve all the problems except for the last one, and 31.5 min to solve all the problems except for the first two?

**67.** Given the first two terms of an infinitely decreasing geometric progression : $\sqrt{3}$, $2/(\sqrt{3}+1)$. Find the common ratio and the sum of the progression.

**68.** The sum of the second and the eighth term of an infinitely decreasing geometric progression is equal to 325/328, and the sum of the second and the sixth term, being diminished by 65/32, is equal to the fourth term of the progression. Find the sum of the squares of the terms of the progression.

**69.** The difference between the first and the fifth term of an infinitely decreasing geometric progression is 1.92, and the sum of the first and the third term is 2.4. Find the ratio of the square of the sum of the progression to the sum of the squares of its terms.

**70.** Find the first three terms of an infinitely decreasing geometric progression whose sum is 1.6 and the second term is (–0.5).

**71.** The sum of an infinitely decreasing geometric progression is equal to 4 and the sum of the cubes of its terms is equal to 64/7. Find the sixth term of the progression.

**72.** The sum of an infinitely decreasing geometric progression is equal to 4 and the sum of the cubes of its terms is equal to 192. Find the first term and the common ratio of the progression.

**73.** For what irrational value of $x$ can three numbers, $0.(27)$, $x$, and $0.(72)$, form a progression (arithmetic or geometric)? Find $x$ and the sum of four terms of the progression.

**74.** The sum $S$ of an infinitely decreasing geometric progression exceeds by 2 the sum of the first three terms of the progression. The sum of the first six terms is equal to 3. Find $S$.

**75.** Find the sum of an infinitely decreasing geometric progression if the sum of all its even terms is thrice as small than the sum of all its odd terms, and the sum of the first five terms of the progression is equal to 484.

**76.** Calculate the sum of the terms of the geometric progression $a_1 + a_1 q + a_1 q^2 + \ldots + a_1 q^{n-1} + \ldots$, where $a_1$ is the greatest value of the function $y = (6x^2 - x^3 - 16)/8$ on the interval $[1, 5]$ and the common ratio of the progression $q = \lim_{x \to 0} (1 - \cos x)/x^2$.

**77.** The first term of a certain infinitely decreasing geometric progression is 1 and its sum is $S$. Find the sum of the geometric progression which is formed by the squares of the terms of the initial progression.

**78.** The ratio of the sum of the cubes of the terms of an infinitely decreasing geometric progression to the sum of the squares of its terms is 12 : 13. The sum of the first two terms of the progression is equal to 4/3. Find the progression.

**79.** Find the sum of an infinitely decreasing geometric progression whose second term, which is the doubled product of the first term by the fourth, and the third term form, in that order, an arithmetic progression with the common difference 1/3.

80. Find three numbers which form a geometric progression if their product is 64 and the arithmetic mean is 14/3.

81. Three numbers, $a$, $b$, and 12, form, in that order, a geometric progression, and the numbers $a$, $b$, and 9 form an arithmetic progression. Find $a$ and $b$.

82. Find the value of $x$ for which $\log_2 (5 \cdot 2^x + 1)$, $\log_4 (2^{1-x} + 1)$, and 1 form an arithmetic progression.

83. Represent the periodic decimal fraction 7.2(3) as an ordinary fraction.

84. Calculate $(0.2)^{\log_{\sqrt{5}}\left(\frac{1}{4} + \frac{1}{8} + \frac{1}{16} + \dots\right)}$.

85. Find the sum of an infinitely decreasing geometric progression if the sum of its first three terms is 3 and the sum of its first three odd terms is $5\frac{1}{4}$.

86. The sum of an infinite geometric progression is 243 and the sum of its first five terms is 275. Find the progression.

87. The sum of an infinitely decreasing geometric progression is 1.5 and the sum of the squares of its terms is 1/8. Find the progression.

88. The sum of the terms of an infinitely decreasing geometric progression is 3 and the sum of the cubes of all its terms is 108/13. Write the progression.

## 1.13. Solution of Problems on Derivation of Equations

1. A train was delayed by a semaphore for 16 minutes and made up for the delay on a section of 80 km travelling with the speed 10 km per hour higher than that which accorded the schedule. Find the speed of the train which accorded the schedule.

2. A skier had to cover a distance of 30 km. Having started 3 min later than the fixed time, he glided with the speed which exceeded the assumed speed by 1 km/h and reached the finish at the time he would reach it if he began to race strictly at the appointed time and ran with the assumed speed. Find the speed of the skier.

3. A cyclist covered a distance of 96 km two hours faster than he assumed. Every hour he travelled 1 km more than he intended to cover in 1 h 15 min. What was his speed?

4. A tourist intended to cover $a$ km in a definite time. Having travelled $b$ km, he rested for 15 min and then increased his speed in order to be at the appointed place on time. Find the initial speed of the tourist.

5. Two people started simultaneously towards each other from points $A$ and $B$ which are 50 km apart. They met 5 hours later. After their meeting, the first person, who travelled from $A$ to $B$, decreased his speed by 1 km/h and the other person, who travelled from $B$ to $A$,

increased his speed by 1 km/h. The first person is known to arrive at $B$ 2 hours earlier than the second person arrived at $A$. Find the initial speed of the first person.

6. Find the speed and the length of the train being given that it travelled with a constant speed past a stationary observer for 7 seconds and needed 25 seconds to pass a platform 378 m long with the same speed.

7. A car travels from point $A$ to point $B$ with a constant speed. If the driver increased the speed of the car by 6 km/h, it would take him 4 hours less to cover that distance. And travelling with the speed 6 km/h lower than the initial speed, it would take him 6 hours more. Find the distance between $A$ and $B$.

8. A train left point $A$ at noon sharp. Two hours later another train started from point $A$ in the same direction. It overtook the first train at 8 p.m. Find the average speeds of the trains if the sum of their average speeds is 70 km/h.

9. A pedestrian and a cyclist start simultaneously towards each other from towns $A$ and $B$ which are 40 km apart and meet two hours after the start. Then they resumed their trips and the cyclist arrives at $A$ 7 h 30 min earlier than the pedestrian arrives at $B$. Find the speeds of the pedestrian and the cyclist.

10. After their meeting, one ship went south and the other went west. Two hours after their meeting, they were 60 km apart. Find the speed of each ship if the speed of one of them is known to be 6 km/h higher than that of the other.

11. A fisher had to sail 35 km to the meeting place and the other fisher had to sail $31\frac{3}{7}$ per cent less. To arrive at the meeting place at the same time as the second fisher, the first fisher started half and hour earlier and sailed with the speed exceeding by 2 km/h the speed of the second fisher. Find the speed of each fisher and the time it took each of them to cover the distance.

12. Two cyclists started simultaneously towards each other from points $A$ and $B$ which are 28 km apart. An hour later they met and kept pedalling with the same speed without stopping. The first cyclist arrived at $B$ 35 minutes earlier than the second arrived at $A$. Find the speed of each cyclist.

13. Three cars leave $A$ for $B$ in equal time intervals. They reach $B$ simultaneously and then leave for point $C$ which is 120 km from $B$. The first car arrives there an hour after the second car, and the third car, having reached $C$, immediately reverses the direction and 40 km from $C$ meets the first car. Find the speed of the first car.

14. Two ships left a sea port simultaneously in two mutually perpendicular directions. Half an hour later the shortest distance between them was 15 km and another 15 min later one ship was 4.5 km farther from the pier than the other. Find the speed of each ship.

15. Two ships take a straight route to the same port and travel with constant speeds. At the initial moment the positions of the ships and the port form an equilateral triangle. When the second ship covered a distance of 80 km the triangle became right-angled. When the first ship arrived at the port, the second ship was at a distance of 120 km from the port. Find the distance between the ships at the initial moment.

16. A car and a cyclist travel to point $A$ along two straight roads with constant speeds. At the initial moment the positions of the car, the cyclist, and point $A$ form a right triangle. After the car travelled 25 km, the triangle became equilateral. Find the distance between the car and the cyclist at the initial moment if at the time the car arrived at point $A$ it remained 12 km for the cyclist to travel.

17. Two cars started simultaneously from two towns towards each other. The first car covered 0.08 of the distance between the towns in 3 hours the second car covered 7/120 of that distance in 2.5 h. Find the speed (km/h) of the second car if the first car travelled 800 km to the meeting point.

18. A motor-cyclist left point $A$ for point $B$. Two hours later a car left point $A$ for $B$ and arrived at $B$ at the same time as the motor-cyclist. If the car and the motor-cyclist left simultaneously $A$ and $B$ towards each other, they would meet 1 h 20 min after the start. How much time did it take the motor-cyclist to travel from $A$ to $B$?

19. A cycle track is a right triangle with a difference of 2 km between the legs. Its hypotenuse passes along a side road and the two legs pass along a highway. One of the participants of the cycle race took the side road and raced with the speed of 30 km/h and then he covered the two intervals along the highway during the same time with the speed of 42 km/h. Find the length of the race track.

20. Two dots move along a circle 1.2 m long with constant speeds. When they move in different directions, they meet every 15 seconds. When they move in the same direction, one dot overtakes the other every 60 seconds. Find the speed of the dots.

21. Three cars started simultaneously from point $A$ to point $B$ along the same highway. The second car travelled with the speed 30 km/h higher than that of the first car and arrived at $B$ 3 hours earlier than the first car. The third car arrived at $B$ 2 hors earlier than the first car, travelling half the time with the speed of the first car and the other half with the speed of the second car. Find the distance between $A$ and $B$.

22. Three swimmers had to swim from $A$ to $B$ and back again. Five seconds after the start of the first swimmer the second swimmer took a start and another five seconds later the third swimmer started. The three swimmers passed a certain point $C$, which is somewhere between $A$ and $B$, simultaneously (none of them reached point $B$ yet). Having reached $B$

and reversed the direction, the third swimmer met the second one 9 km short of *B* and met the first swimmer 15 km short of *B*. Find the speed of the third swimmer if the distance *AB* is equal to 55 m.

23. A tourist leaves point *A* on a bicycle. Having travelled for 1.5 h at 16 km/h, he makes a stop for 1.5 h and then pedals on with the same speed. Four hours after the start of the first tourist, a second tourist leaves point *A* on a motor-cycle and rides with the speed of 56 km/h in the same direction. What distance will they cover before the second tourist overtakes the first?

24. Two cars start from the same point simultaneously and in the same direction. The first car travels at 40 km/h, and the speed of the second car is 125 per cent of the speed of the first car. Thirty minutes later a third car starts from the same point and in the same direction. It overtakes the second car 1.5 h later than it overtook the first car. What is the speed of the third car?

25. A passenger train left down *A* for town *B*. At the same time a goods train left *B* for *A*. The speed of each train is constant throughout the whole trip. Two hours after the trains met, they were 280 km apart. The passenger train arrived at the place of destination 9 hours after their meeting and the passenger train, 16 hours after the meeting. How long did it take each train to make the whole trip?

26. A tourist sailed along a river for 90 km and walked for 10 km. He spent 4 h less on walking than on sailing. If the tourist walked as long as he sailed and sailed as long as he walked, the distances would be equal. How long did he walk and how long did he sail?

27. Two cars move towards each other and meet 6 days after they started. If the first car travelled for 1.8 days and the second for 1.6 days, they would cover 520 km together. If the first car covered 2/3 of the way travelled by the second car, and the second car covered 1/3 of the way travelled by the first car, then the first car would need 2 days less than the second. How many kilometres a day did each car travel?

28. Two tourists walked towards each other, one from point *A* and the other from point *B*. The first tourist left point *A* 6 h later than the second left point *B*, and it turned out on their meeting that he had travelled 12 km less than the second tourist. After their meeting they kept walking with the same speed, and the first tourist arrived at *B* 8 hours later and the second arrived at *A* 9 hours later. Find the speeds of the tourists.

29. Two cyclists left simultaneously point *A* for point *B*. The first cyclist stopped 42 min later when he was 1 km short of *B* and the other one stopped 52 min later when he was 2 km short of *B*. If the first cyclist travelled as many kilometres as the second and the second as many kilometres as the first, the first one would need 17 min less than the second. Find the distance between points *A* and *B*?

30. Two objects move along a circle in the same direction and come alongside each other every 56 min. If they moved with the same speeds in opposite directions, they would meet every 8 minutes. It is known that when the objects moved along the circle in opposite directions, the distance between the $m$ decreased from 40 to 26 metres every 24 seconds. What is the speed of each object?

31. A car started from point $K$ to point $M$. At the same time a bus started from point $M$ towards the car. When the car covered 0.4 of the way from $K$ to $M$, the bus was 4 km away from the car. Now when the bus covered half the way, the car was at the distance of 10 km from it. Find the ratio of the time it took the car to cover the distance from $K$ to $M$ to the time it took the bus to cover the same distance. The speeds of the car and the bus are constant.

32. A tank of 2400 m$^3$ capacity is full of fuel. The discharging capacity of the pump is 10 m$^3$/ min higher than its filling capacity. As a result, the pump needs 8 min less to discharge the fuel than to fill up the tank. Find the filling capacity of the pump.

33. Three masons working together need $a$ hours to make a stone wall. The first of them, working alone, can make the wall twice as fast as the third mason and can complete the job an hour sooner than the second. How much time would it take each mason to make the wall if he worked alone?

34. Two workers, working together, can do the assigned job in 12 days. If the first worker does half the work and then the second worker does the other half, the job will be completed in 25 days. How many days would it take each worker to do the job if he worked alone?

35. If two pipes function simultaneously, the reservoir will be filled in 12 hours. Through one pipe the reservoir is filled up 10 hours faster than through the other. How many hours does it take the second pipe to fill the reservoir?

36. Two cranes worked for 15 h to unload the barge. One crane began operating 7 h later than the other. It is known that the first crane alone can unload the barge 5 h faster than the second crane. How many hours does it take each crane alone to unload the barge?

37. Two workers manufactured a batch of identical parts. One worker worked for 2 hours and the other for 5 hours and they did half the job. When they worked together for another 3 hours, they had to do 1/20 of the whole quota to complete the job. How much time does it take each worker to do the whole job?

38. Two students spent 7 h to prepare for a school quiz, reckoning from the moment when the first student began working. The second student began an hour and a half later than the first. If the job was assigned to each student separately, then it would take the first student 3 h more than the second to complete the job. How much time would it take each student to complete the job if they worked separately?

**39.** Two typists, working together, typed 65 pages, the first one typing an hour longer than the second. However, the second typist can do two pages more per hour than the first and, therefore, she did 5 pages more than the first. How many pages an hour can each typist do?

**40.** Two workers must work together to do a certain job. If the efficiency of the first worker were twice as high and that of the second 50% higher, then the time it took the workers to complete the job would be less by 14/85 than the time it took the first worker to do the job. The efficiency of which worker is higher and how many times?

**41.** Two workers do the same job. First, one worker worked for 1/3 of the time it takes the other to do the whole job, then the second worker worked for 1/3 of the time it would take the first worker to do the whole job. It turned out that they did 13/18 of the whole job. Calculate the time it would take each worker to do the job if he worked separately, provided that, working together, they can do it in $3\frac{3}{5}$ h.

**42.** There is a competition between three teams of woodcutters. The first and the third team felled twice as many trees as the second, and the second and the third team felled thrice as many trees as the first. Which team was the winner in the competition?

**43.** One worker can do the job 4 hours faster than the other. First, they worked for 2 hours together and then the remaining part of the work was done by the first worker alone in an hour. What time it would take the second worker to do the job if he worked alone?

**44.** Two workers, working together, completed the job in 5 days. If one worker worked twice as fast and the other worked half as fast, they would have to work for 4 days. How much time would it take the first worker to do the job if he worked alone?

**45.** Five people do a certain job. The first, the second, and the third person, working together, complete the job in 7.5 h, the first, the third, and the fifth in 5 h, the first, the third, and the fourth in 6 h, the fourth, the second, and the fifth in 4 h. How much time will it take to do the job if all the five people work together?

**46.** Two pipes, an inlet pipe and an outlet pipe, lead to a reservoir. It takes the inlet pipe 2 hours more to fill the reservoir than the outlet pipe to empty it. When the reservoir was one-third full, both valves were opened and 8 hours later the reservoir was empty. How many hours will it take the inlet pipe to fill the reservoir and the outlet pipe to empty it if they function separately?

**47.** Two towns are 220 km apart. Two cars start from these towns towards each other. They can meet halfway if the first car leaves 2 hours earlier than the second. If they start simultaneously, they will meet in 4 hours. Find the speeds of the cars.

**48.** A team of workers had to manufacture 360 parts. Manufacturing daily 4 parts more than it was planned, the team completed the job 1 day ahead of time. For how many days did the team work on the job?

**49.** Two grain thrashers managed to thrash the gathered wheat in 4 days. If one of them thrashed half the wheat and then the other thrashed the other half, they would complete the job in 9 days. How many days would it take each thrasher to do the job if they worked separately?

**50.** A motor boat whose speed in still water is 10 km/h went 91 km downstream and then returned. Calculate the speed of the river flow if the trip took 20 hours.

**51.** The piers $A$ and $B$ are 300 km apart. Two launches leave $A$ for $B$. The time difference between their starts is 5 h. The launches arrive at $B$ simultaneously. Find the time each launch spent on the trip if the speed of one of them is 10 km/h higher than that of the other.

**52.** A boat sails downstream from point $A$ to point $B$, which is 10 km away from $A$, and then returns to $A$. If the actual speed of the boat (in still water) is 3 km/h, the trip from $A$ to $B$ takes 2 h 30 min less than that from $B$ to $A$. What must the actual speed of the boat be for the trip from $A$ to $B$ to take 2 hours?

**53.** A launch provides for a regular transportation of passengers from point $A$ to point $B$ and back again, both points being located on the river bank. If the actual speed of the boat (in still water) increased 2 times, then the trip from $A$ to $B$ and back again would take a fifth of the time which the launch usually spends on the way from $A$ to $B$ and back again. How many times is the actual speed of the launch higher than the speed of the river flow?

**54.** The piers located on a river bank are 21 km apart. Leaving one of the piers for the other, a motor boat returns to the first pier in 4 h, spending 30 min of that time on taking the passengers at the second pier. Find the speed of the boat in still water if the speed of the river flow is 2.5 km/h.

**55.** A motor boat went downstream for 28 km and immediately returned. It took the boat 7 hours to make the trip. If the speed of the river flow were twice as high, the trip downstream and back would take 11 h 12 min. Find the speed of the boat in still water and the speed of the river flow.

**56.** A certain number of lorries were required to transport 60 t of cargo. Since each lorry could lake 0.5 t of cargo less, another 4 lorries were needed. How many lorries were initially planned to be used ?

**57.** A team of fishers planned to catch 1800 centners of fish during a certain number of days. Because of a storm, for a third of the planned period the team caught daily 20 centners of fish less than was planned. The rest of the days, however, the team overfulfilled the daily plan by 20 centners and the planned job was completed one day ahead of time. How many centners of fish did the team initially plan to catch?

58. One collective farm got an average harvest of buckwheat of 21 centners from a hectare, and another collective farm, which had 12 hectares of land less given to buckwheat, got 25 centners from a hectare. As a result, the second farm harvested 300 centners of buckwheat more than the first. How many centners of buckwheat did each farm harvest?

59. According to a plan, a team of woodcutters had to cut 216 m³ of wood in several days. The first three days, the team fulfilled the daily assignment and then it cut 8 m³ of wood over and above the plan every day. Therefore, a day before the planned date they cut 232 m³ of wood. How many cubic metres of wood a day did the team have to cut according to the plan?

60. Two kg of copper and one kg of lead are needed to manufacture an electric motor of type $A$, and 3 kg of copper and 2 kg of lead are needed to manufacture an electric motor of type $B$. How many electric motors of each type were manufactured if it is known that 130 kg of copper and 80 kg of lead were used?

61. There were 500 seats in a hall placed in similar rows. After the reconstruction of the hall the total number of seats became 1/10 less. The number of rows was reduced by 5 but each row contained 5 seats more than before. How many rows and how many seats in a row were there initially in the hall?

62. One fashion house has to make 810 dresses and another one 900 dresses during the same period of time. In the first house the order was ready 3 days ahead of time and in the second house 6 days ahead of time. How many dresses did each fashion house make a day if the second house made 21 dresses more a day than the first?

63. Several teams take part in a competition, each of which must play one game with all the other teams. How many teams took part in the competition if they played 45 games all in all?

64. A shop sold 64 kettles of two different capacities. The smaller kettle cost a rouble less than the larger one. The shop made 100 roubles from the sale of large kettles and 36 roubles from the sale of small ones. How many kettles of each capacity did the shop sold and what was the price of each kettle?

65. An enterprise got a bonus and decided to share it in equal parts between the exemplary workers. It turned out, however, that there were 3 more exemplary workers than it had been assumed. In that case, each of them would have got 4 roubles less. The administration had found the possibility to increase the total sum of the bonus by 90 roubles and as a result each exemplary worker got 25 roubles. How many people got a bonus?

66. The sum of the squares of the digits constituting a positive two-digit number is 13. If we subtract 9 from that number, we shall get a number written by the same digits in the reverse order. Find the number.

67. If we add the square of the digit in the tens place of a positive two digit number to the product of the digits of that number, we shall get 52, and if we add the square of the digit in the units place to the same product of the digits, we shall get 117. Find the two-digit number.

68. Find two numbers such that their sum, their product, and the difference of their squares are equal.

69. The sum of the digits of a three-digit number is 17, and the sum of the squares of its digits is 109. If we subtract 495 from that number, we shall get a number consisting of the same digits written in the reverse order. Find the number.

70. The geometric mean of two numbers is greater than the smaller of them by 12, and the arithmetic mean of the same numbers is less than the larger of them by 24. Find the numbers.

71. Find the pairs of natural numbers whose greatest common divisor is 5 and the least common multiple is 105.

72. The denominator of an irreducible fraction is greater than the numerator by 2. If we reduce the numerator of the reciprocal fraction by 3 and subtract the given fraction from the resulting one, we get 1/15. Find the given fraction.

73. A two-digit number exceeds by 19 the sum of the squares of its digits and by 44 the double product of its digits. Find the number.

74. The product of a natural number by the number written by the same digits in the reverse order is 2430. Find the number.

75. Find two natural numbers whose difference is 66 and the least common multiple is 360.

76. Find the pairs of natural numbers whose least common multiple is 78 and the greatest common divisor is 13.

77. Find two natural numbers whose sum is 85 and than least common multiple is 102.

78. Find the pairs of natural numbers the difference of whose squares is 55.

79. The sum of the digits of a three-digit number is 11. If we subtract 594 from the number consisting of the same digits written in the reverse order, we shall get a required number. Find that three-digit number if the sum of all pairwise products of the digits constituting that number is 31.

80. The sum of the squares of the digits constituting a certain positive three-digit number is 74. The hundreds digit of the number is equal to the doubled sum of the digits in the tens and units places. Find the

number if it is known that the difference between that number and the number written by the same digits in the reverse order is 495.

81. The sum of the squares of the digits constituting a two-digit positive number is 2.5 times as large as the sum of its digits and is larger by unity than the trebled product of its digits. Find the number.

82. The units digit of a two-digit number is greater than its tens digit by 2, and the product of that number by the sum of its digits is 144. Find the number.

83. An alloy consists of two metals taken in the ratio 1 : 2, and another alloy consists of the same metals taken in the ratio 2 : 3. How many parts of the two alloys must be taken to obtain a new alloy consisting of the same metals which are in the ratio 17/27?

84. The mass of two pieces of brass is 60 kg. The first piece contains 10 kg of pure copper and the second piece contains 8 kg of pure copper. What is the percentage of copper in the first piece of brass if the second piece contains 15 per cent more copper than the first?

85. The alloy of copper and tin with a mass of 8 kg contains $p$ per cent of copper. What piece of the copper-tin alloy containing 40 per cent of tin must be alloyed with the first piece in order to obtain a new alloy with the minimum percentage of copper if the mass of the second picece is 2 kg?

86. There are two volumes of water with the masses differing by 2 kg. The same quantity of heat, equal to 96 kcal, was imparted to each mass, and the larger mass of water was found to be 4 degrees cooler than the smaller mass. Find the mass of water in each of the two volumes.

87. Two ships sail in a fog towards each other with the same speed. When they are 4 km apart, the captains back the engines for 4 min with the acceleration of 0.1 m/s², and then the ships continue sailing with the speeds attained. At what values of the initial speed $v_0$ will the ships avoid collision?

88. There are two alloys of copper and zinc. In the first alloy there is twice as much copper as zinc and in the second alloy there is 5 times less copper than zinc. How many times more must we take of the second alloy than the first in order to obtain a new alloy in which there would be twice as much zinc as copper?

89. Calculate the weight and the percentage of silver in the silver-copper alloy being given that the latter's alloy with 3 kg of pure silver contains 90 per cent of silver and with 2 kg of 90% silver alloy contains 84 per cent of silver.

90. Two solutions, the first of which contains 0.8 kg and the second 0.6 kg of dry sulphuric acid, were poured together and 10 kg of a new sulphuric acid solution were obtained. Find the weight of the first and of the second solution in the mixture if the first solution is known to contains 10 per cent more of dry sulphuric acid than the second.

91. From a full tank containing 729 litres of an acid we pour off $a$ litres and add water to fill up the tank. After stirring the solution thoroughly, we pour off $a$ litres of the solution and again add water to fill up the tank. After the procedure was repeated 6 times, the solution in the tank contained 64 litres of the acid. Find $a$.

92. In two alloys, the ratios of copper to zinc are 5 : 2 and 3 : 4 (by weight). How many kg of the first alloy and of the second alloy should be alloyed together to obtain 28 kg of a new alloy with equal contents of copper and zinc?

93. There are two barrels of petrol of different prices, with the volumes of 220 litres and 180 litres. Equal amounts of petrol were poured off simultaneously from the two barrels, and the patrol poured off from the first barrel was poured into the second barrel and the petrol poured off from the second barrel was poured into the first barrel. Then the price of petrol in both barrels became the same. How much petrol was poured from one barrel into the other?

94. In two alloys copper and zinc are related as 4 : 1 and 1 : 3. After alloying together 10 kg of the first alloy, 16 kg of the second, and several kg of pure copper, an alloy was obtained in which the ratio of copper to zinc was 3 : 2. Find the weight of the new alloy.

95. Two litres of glycerine were poured off from a vessel filled up with pure glycerine to the brim and 2 litres of water were added. After the solution was stirred, 2 litres of the mixture were poured off and again 2 litres of water were added. Finally, after the solutions was stirred again, 2 litres of the mixture were poured off and 2 litres of water were added. As a result of these operations the volume of water in the vessel became larger by 3 litres than the volume of glycerine remaining in it. How many litres of glycerine and water were there in the vessel after the operations performed?

96. There are two alloys of zinc, copper, and tin. The first alloy is known to contain 40 per cent of tin and the second alloy 26 per cent of copper. The percentage of zinc is the same in both alloys. Having alloyed 150 kg of the first alloy and 250 kg of the second, we get a new alloy which contains 30 per cent of zinc. How many kg of tin are there in the new alloy?

97. There are two alloys with different percentages of lead. The first weighs 6 kg and the second 12 kg. A piece of the same weight was cut off from each of them and was alloyed with the remaining part of the other alloy. As a result, the percentage of lead became the same in both alloys. What was the weight of each cut- off piece?

98. After two successive rises, the salary became equal to 15/8 parts of the initial salary. By how many per cent was the salary raised first time if the second rise was twice as high (in per cent) as the first?

**99.** The ratio of the salary of a worker for October to that for November was $1\frac{1}{2} : 1\frac{1}{3}$, and the ratio of the salary for November to that for December was $2 : 2\frac{2}{3}$. The worker got 40 roubles more for December than for October and received a bonus constituting 40 per cent of the salary for three months for overfulfilling a quarterly plan. Find the bones. (Assume that the number of workdays is the same in every month.)

**100.** After two successive rises of the salary, equal in per cent, the sum of 100 roubles turned into 125 roubles and 44 kopecks. Find the rises of the salary in per cent.

**101.** The output of an enterprise increased by 4 per cent during a year and by 8 per cent during the next year. Find the average yearly increase in the output during that period.

**102.** The price of a certain article was raised by 25 per cent. By how many per cent must the price to result?

**103.** An object moved for several seconds and covered 3 m in the first second and 4 m more in each successive second than in its predecessor. If the object covered 1 m in the first second and 8 m more in each successive second than in its predecessor, then the difference between the path it would cover during the same time and the actual path would be more than 6 m but less than 30 m. Find the time for which the object moved (in seconds).

**104.** A railway embankment 100 m long must be constructed whose cross section would be an isosceles trapezoid with the lower base equal to 5 m, with the upper base not smaller than 2 m, and with the slope equal to 45°. How high should the embankment be for the volume of excavations to be not smaller than 400 $m^3$ but not larger than 500 $m^3$?

**105.** The boat sails down the river for 10 km and then up the river for 6 km. The speed of the river flow is 1 km/h. What should the range of the actual speed of the boat be for the trip to take from 3 to 4h?

**106.** The width of the river is $a$ m, the speed of its flow is $w$ m/s, the speed of a swimmer in still water is $v$ m/s ($v < w$), and the speed of his travelling over dry land is $u$ m/s. What is the least possible time that the swimmer should spend to reach a point located on the opposite bank just opposite the place from which the began crossing the river? Assume that in the water the swimmer does not change the direction he has chosen at the very beginning and that the vector of the speed of the river flow is parallel to the banks of the river which are assumed to be straight.

**107.** Points $A$, $B$, $C$ are located so that $|AB| = 285$ km, $\angle ABC = 60°$. A car leaves $A$ for $B$ at 90 km/h and at the same time a train starts from $B$ to $C$ at 60 km/h. In what time will the distance between the car and the train be the least?

**108.** Three points A, B, and C are located at the vertices of an equilateral triangle with sides equal to 168 km. A car starts from A to B at 60 km/h and at the same time a car starts from B to C at 30 km/h. In what time after their departure will the distance between the cars be the least?

**109.** Piers A, B, and C are located on a river bank, the speed of whose flow is 5 km/h, in the direction of its flow, pier B being located midway between A and C. A raft and a launch leave pier B at the same time, the raft travelling down the river to pier C and the launch travelling to A. The speed of the launch in still water is v km/h. Having reached A, the launch reverses its direction and starts to C. Find all the values of v for which the launch arrives at C later than the raft.

**110.** Points A and B are 120 km apart. A motor-cyclist starts from A to B along a straight road AB with the speed of 30 km/h. At the same time, a cyclist starts from B along a road, which is perpendicular to AB, with the speed of 10 km/h. When will the distance between them be the least?

**111.** Point B is 60 km distant from a straight railway. The distance, along the railway, from point A to point C, which is the nearest to B, is 285 km. At what distance from point C should a station be built in order to spend the least time to cover the distance between A and B, if the speed of travelling by train is 52 km/h and that of travelling along the road is 20 km/h?

**112\*.** Point N is located on the bank of a river 1 km wide, and the speed of the river flow is 1 km/h. A point M is not less than 3 km lower down the river, on the other bank. A fisherman starts from point M and walks along the bank towards N at 4 km/h. At the same time, a boatman starts from point N, crosses the river, waits for the fisherman and takes him to point N. The boatman rows across the river and back again along a straight line, having chosen the direction allowing him to spend the least possible time, 9/8 h, on the trip to N and the return trip. The speed of the boat in still water is 4 km/h. Find the distance between points M and N (down the river).

**113.** A self-propelled barge must transport a rush cargo from a river pier A to pier B located 24 km from point A up the river and return to A as soon as possible to take a new cargo. The speed of the river flow is 6 km/h. What must the least actual speed of the barge be for the trip from A to B and back again to take not more than three hours (not counting the time of loading and unloading)?

**114.** Represent the number 1.25 as a product of three positive factors so that the product of the first factor by the square of the second be equal to 5 and the sum of the three factors be the least.

**115.** Find a number x such that the sum of that number and its square is the least.

**116.** For what minuend is the difference the greatest if the subtrahend is equal to double the square of the minuend?

**117.** Decompose the number 20 into two terms such that their product is the greatest.

**118.** A tourist travels from point $A$, which is on a highway, to point $B$, which is 8 km from the highway. The distance between $A$ and $B$ along a straight line is 17 km. At what point would the tourist turn from the highway to reach point $B$ in the shortest possible time, if his speed along the highway is 5 km/h and over the land without roads, 3 km/h.

**119.** An object begins moving at time moment $t = 0$ and, 4 s after the beginning of motion, attains the acceleration of 3 m/s². Find the speed of the object 6 s after the beginning of motion and the path traversed by the object during that time if it is known that the speed of the body varies according to the law $v(t) = (t^2 + b \cdot t + 6)$m/s and the object moves along a straight line.

**120\*.** Two trains start simultaneously from points $A$ and $B$ towards each other. Each of them first moves with uniform acceleration (the initial speeds of the trains are equal to zero, the accelerations are different), and then, having attained a certain speed (at different time moments), uniformly. The ratio of the speeds of the uniform motion of the trains is 4 : 3. At the time of their meeting, the speeds of the trains were equal and they arrived at points $A$ and $B$ simultaneously. Find the ratio of the accelerations of the trains.

**121.** The distance between points $A$ and $B$ is 120 km. A motor-cyclist, travelling without stops, covers that distance in 8 h if he travels from $A$ to an intermediate point $C$ with the speed $v_0$ km/h and then continues on his way to $B$ with an acceleration of $a$ km/h². He will need the same time to cover the whole distance if he travels from $A$ to $C$ at $v_0$ km/h and from $C$ to $B$ at $v_1$ km/h, or from $A$ to $C$ at $v_1$ km/h and from $C$ to $B$ at $v_0$ km/h. Find $v_0$ if the parameter $a$ is equal to $2v_0$ in magnitude and $v_0 \neq v_1$.

## 1.14. Complex Number

*Complex numbers* are pairs $(x, y)$ of real numbers $x$ and $y$ for which the concept of an equality and the operations of addition and multiplication are defined as follows.

*Two complex numbers $(x_1, y_1)$ and $(x_2, y_2)$ are equal if $x_1 = x_2$ and $y_1 = y_2$. We write*

$$(x_1, y_1) = (x_2, y_2) \qquad (1)$$

*The sum of two complex numbers $(x_1, y_1)$ and $(x_2, y_2)$ is a complex number $(x_1 + x_2, y_1 + y_2)$. We write*

$$(x_1, y_1) + (x_2, y_2) = (x_1 + x_2, y_1 + y_2). \qquad (2)$$

*The product of two complex numbers* $(x_1, y_1)$ *and* $(x_2, y_2)$ *is a complex number* $(x_1 x_2 - y_1 y_2, x_1 y_2 + x_2 y_1)$. *We write*

$$(x_1, y_1) \cdot (x_2, y_2) = (x_1 x_2 - y_1 y_2, x_1 y_2 + x_2 y_1). \tag{3}$$

It follows from formulas (2) and (3) that $(x_1, 0) + (x_2, 0) = (x_1 + x_2, 0)$ and $(x_1, 0)(x_2, 0) = (x_1 x_2, 0)$ and, consequently, the operations of addition and multiplication of complex numbers of the form $(x, 0)$ coincide with the addition and multiplication of real numbers $x$, and, therefore, complex numbers of the form $(x, 0)$ are identified with real numbers $x : (x, 0) = x$.

A complex number $(0, 1)$ is called an imaginary unity and is denoted by the letter $i$, i.e. $i = (0, 1)$. Applying formula (3), we have

$$t^2 = i \cdot i = (0, 1)(0, 1) = (-1, 0) = -1.$$

If follows from formulas (2) and (3) that

$$(0, y) = (0, 1) \ (y, 0) = iy,$$
$$(x, y) = (x, 0) + (0, y) = x + iy.$$

The notation of a complex number as $x + iy$ is said to be its *algebraic form*. The complex number $iy = (0, y)$ is called an *imaginary number*. A complex number of the form $x + iy$ is usually denoted by one letter , say , $z : z = x + iy$, and in this case $x$ is said to be a *real* part and $y$, an *imaginary* part of the complex number $z$. We write $x = \text{Re } z$, $y = \text{Im } z$.

It is easy to verify that the operations of addition and multiplication we have introduced obey the familiar laws of algebra.

$(z_1 + z_2) + z_3 = z_1 + (z_2 + z_3)$, associative law of addition,

$z_1 + z_2 = z_2 + z_1$, commutative law of addition,

$(z_1 z_2) z_3 = z_1 (z_2 z_3)$, associative law of multiplication,

$z_1 z_2 = z_2 z_1$, commutative law of multiplication,

$(z_1 + z_2) z_3 = z_1 z_3 + z_2 z_3$, distributive law.

These properties make it possible to operate with complex numbers as with ordinary binomials, with due regard for the equality $t^2 = -1$.

In the set of complex numbers, the numbers $0 = (0, 0)$ and $1 = (1, 0)$ have the same properties as in the set of real numbers, namely :

$$z + 0 = z, \ z \cdot 0 = 0, \ z \cdot 1 = z.$$

The *difference* between any two complex numbers $z_1 = x_1 + iy_1, z_2 = x_2 + iy_2$ exists and is uniquely defined, and in that case $z_1 - z_2 = x_1 - x_2 + i (y_1 - y_2)$.

The number $x - iy$ is said to be a *complex conjugate* of the number $z = x + iy$ and is designated as $\bar{z}$.

For any complex number $z_1$ and any complex number $z_2 \neq 0$ the *quotient* $z_1 / z_2$ exists and is uniquely defined and in that case

$$\frac{z_1}{z_2} = \frac{z_1 \bar{z}_2}{z_2 \bar{z}_2} = \frac{x_1 x_2 + y_1 y_2}{x_2^2 + y_2^2} + i\,\frac{x_2 y_1 - x_1 y_2}{x_2^2 + y_2^2}.$$

The nonnegative number $r = |z| = \sqrt{z \bar{z}} = \sqrt{x^2 + y^2}$ is called a *modulus*, or an *absolute value*, of the complex number $z$. The absolute value of a complex number possesses the properties :

$$|z| = 0 \Leftrightarrow z = 0,$$

$$|z_1 z_2 \ldots z_n| = |z_1|\,|z_2| \ldots z_n|,$$

$$|z_1 + z_2| \leq |z_1| + |z_2|,$$

$$|z_1 - z_2| \geq |z_1| - |z_2|.$$

The magnitude of the angle $\varphi \in (-\pi,\ \pi)$, which satisfies the system of equations

$$\begin{cases} \cos\varphi = x / \sqrt{x^2 + y^2}, \\ \sin\varphi = y / \sqrt{x^2 + y^2}, \quad \sqrt{x^2 + y^2} \neq 0, \end{cases}$$

is the *principal argument* of the number $z$, and $\varphi + 2\pi n\ (n \in \mathbf{Z})$ is the argument of the number $z$.

The notation $z = r\,(\cos\varphi + i\,\sin\varphi) = r\,(\cos(\varphi + 2\pi n) + i\,\sin(\varphi + 2\pi n))$ is the *trigonometric notation* of a *complex number* $(r \neq 0)$.

If
$$z_1 = r_1\,(\cos\varphi_1 + i\,\sin\varphi_1),$$
$$z_2 = r_2\,(\cos\varphi_2 + i\,\sin\varphi_2),$$
then
$$z_1 z_2 = r_1 r_2 \cos(\varphi_1 + \varphi_2 + 2\pi n) + i\,\sin(\varphi_1 + \varphi_2 + 2\pi n)),$$
$$z_1 / z_2 = r_1 / r_2 [\cos(\varphi_1 - \varphi_2 + 2\pi n) + i\,\sin(\varphi_1 - \varphi_2 + 2\pi n)].$$

**De Moivre's theorem.** If $z = r\,(\cos\varphi + i\,\sin\varphi),\ r \neq 0,\ \ n \in \mathbf{N}, n \geq 2$, then
$$[r\,(\cos\varphi + i\,\sin\varphi)]^n = r^n\,(\cos n\varphi + i\,\sin n\varphi),$$
in particular, $(\cos\varphi + i\,\sin\varphi)^n = \cos n\varphi + i\,\sin n\varphi$.

**A root of a complex number.** If $n \in \mathbf{N}, n \geq 2,\ z = r\,(\cos\varphi + i\,\sin\varphi),\ r \neq 0$, then

$$\sqrt[n]{z} = \sqrt[n]{r}\left(\cos\frac{\varphi + 2\pi k}{n} + i\,\sin\frac{\varphi + 2\pi k}{n}\right),$$

$$k = 0,\ 1,\ \ldots, n-1.$$

In particular $\sqrt[n]{0} = 0$.

**The solution of the quadratic equation $ax^2 + bx + c = 0,\ \ a, b, c \in R.$** If $D = b^2 - 4ac \geq 0$, then the roots are real :

$$x_{1,2} = \frac{-b \pm \sqrt{D}}{2a}.$$

If $D < 0$, then the roots are complex numbers, and

$$x_1 = \frac{-b - i\sqrt{-D}}{2a}, \quad x_2 = \frac{-b + i\sqrt{-D}}{2a}.$$

The following assertion holds true (*Gauss' theorem*) ; *any polynomial of degree n has exactly n root (with due regard for their multiplicity).*

Represent the following complex numbers in trigonometric form **(1-16)**.

**1.** $-1$.      **2.** $-i$.      **3.** $-\sqrt{3} + i$.

**4.** $-\sqrt{3} - i$. **5.** $-3 - 4i$.    **6.** $3 - 4i$.    **7.** 5.

**8.** $\dfrac{3}{2} - i\dfrac{\sqrt{3}}{2}$. **9.** $\sin 32° + i \cos 32°$.

**10.** $\cos 12° - i \sin 12°$.    **11.** $-\sin 110° + i \cos 110°$.

**12.** $\sin \alpha - i \cos \alpha$.      **13.** $\sin \alpha + i \cos \alpha$.

**14.** $-\sin \alpha - i \cos \alpha$.      **15.** $1 + i \tan \alpha$.

**16.** $1 + i \cot \alpha$.

Perform the indicated operations **(17-23)**.

**17.** $(\cos 12° + i \sin 12°)$.

**18.** $(-\sqrt{2} + \sqrt{2}i)^4$.

**19.** $(-\sqrt{3} + i)^3$.        **20.** $(-1 + \sqrt{3}i)^3$.

**21.** $(1 + i)^{20}$.           **22.** $(1 + \cos\varphi + i \sin\varphi)^n$.

**23.** $(1 - \cos\varphi + i \sin\varphi)^n$.

**24.** Prove that if $x + \dfrac{1}{x} = 2\cos\alpha$, then $x^n + \dfrac{1}{x^n} = 2\cos n\alpha$.

**25.** Applying De Moivre's theorem, express $\cos 3\alpha$ and $\cos 4\alpha$ in terms of $\cos\alpha$, and $\sin 3\alpha$ in terms of $\sin\alpha$.

Calculate **(26-30)**.

**26.** $\sqrt{i}$.     **27.** $\sqrt[3]{-i}$.     **28.** $\sqrt[3]{-1}$.

**29.** $\sqrt[4]{-\dfrac{1}{2} - i\dfrac{\sqrt{3}}{2}}$.

**30.** $\sqrt[n]{1}$.

Solve the following equations **(31-32)**.

**31.** $2(1 + i)x^2 - 4(2 - i)x - 5 - 3i = 0$.      **32.** $\left(\dfrac{x+i}{x-i}\right)^n = 1$.

**33.** Prove that the polynomial $x^{3n} + x^{3m+1} + x^{3k+2}$ is exactly divisible by $x^2 + x + 1$ if $m$, $n$, $k$ are any nonnegative integers.

Indicate the points of the complex plane $z$ which satisfy the following equations (34-39).

**34.** $\text{Re } z^2 = 0$.    **35.** $\text{Im } z^2 = 0$.    **36.** $|z|+z = 0$.

**37.** $|z|^2 + z = 0$.    **38.** $z^2 + |\bar{z}| = 0$.

**39.** $z^2 + |z| = 0$.

**40.** Prove that

$$|z_1 + z_2|^2 + |z_1 - z_2|^2 = 2(|z_1|^2 + |z_2|^2)$$

for any two complex numbers $z_1$ and $z_2$.

**41.** Prove that $z + \bar{z}$, $\overline{z}z$ are real numbers.

**42.** Prove that $\overline{z_1 \pm z_2} = \bar{z}_1 \pm \bar{z}_2$,

$$\overline{z_1 + z_2 + \ldots + z_n} = \bar{z}_1 + \bar{z}_2 + \ldots + \bar{z}_n;\ \overline{z_1 z_2} = \bar{z}_1 \bar{z}_2,$$

$$\overline{z_1 z_2 \ldots z_n} = \bar{z}_1 \bar{z}_2 \ldots \bar{z}_n;\ \overline{z^n} = \bar{z}^n;\ \overline{(z_1/z_2)} = \bar{z}_1/\bar{z}_2.$$

**43.** Assume $P_n(z) = a_n z^n + a_{n-1} z^{n-1} + \ldots + a_1 z + a_0$.

Prove that

$$\overline{P_n(z)} = \bar{a}_n \bar{z}^n + \overline{a_{n-1}} \bar{z}^{n-1} + \ldots + \bar{a}_1 \bar{z} + \bar{a}_0.$$

If $a_n, a_{n-1}, \ldots, a_1, a_0$ are real numbers, then prove that $\overline{P_n(z)} = P_n(\bar{z})$.

Find the points of the plane which satisfy the following equations **(44-47)**.

**44.** $|z + i| = |z + 2|$.    **45.** $|z - 2| = |z + 2i|$.

**46.** $\left| \dfrac{z-2}{z+3} \right| = 1$.    **47.** $\left| \dfrac{z+i}{z-3i} \right| = 1$.

Find the points of the plane which satisfy the following inequalities **(48-57)**.

**48.** $|z - 1| > |z - i|$.    **49.** $|z + 2i| < |z - 1|$.

**50.** $\left| \dfrac{z-1}{z+1} \right| < 1$.    **51.** $|z + i| < 1$.

**52.** $|2z - i| < 2$    **53.** $1 < |z + 1 - i| < 2$

**54.** $1 < |3iz - 1| < 3$.

**55.** $1/\sqrt{2} < |(1+i)z + i| < \sqrt{2}$.

**56.** $|z - x| < |z - i|$.

**57.** $|z + i| < |z - x| < |z - 1|$.

**58.** Find the least value assumed by the function $w = \left| z + \dfrac{1}{z} \right|$, where $z$ is a complex variable, $|z| \geq 2$.

# Part-2

## Fundamentals of Mathematical Analysis

## 2.1. Sequences and Their Limits. An Infinitely Decreasing Geometric Progression. Limits of Functions

1. The sum of the terms of an infinitely decreasing geometric progression is equal to the greatest value of the function $f(x) = x^3 + 3x - 9$ on the interval $[-2, \ 3]$; the difference between the first and the second term of the progression is $f'(0)$. Find the common ratio of the progression.

Find the following limits (2-59).

2. $\lim\limits_{n \to \infty} \dfrac{\sqrt{2n^2 + 1} - \sqrt{n^2 + 1}}{n + 1}$.

3. $\lim\limits_{n \to \infty} \left( \dfrac{1}{n^2} + \dfrac{2}{n^2} + \ldots + \dfrac{n-1}{n^2} \right)$.

4. $\lim\limits_{n \to \infty} \left( 1 - \dfrac{1}{7} + \dfrac{1}{49} - \ldots + \dfrac{(-1)^{n-1}}{7^{n-1}} \right)$.

5. $\lim\limits_{n \to \infty} \left( \dfrac{1}{n^2 + 1} + \dfrac{2}{n^2 + 1} + \ldots + \dfrac{n-1}{n^2 + 1} \right)$.

6. $\lim\limits_{n \to \infty} \dfrac{n^2 + n + 1}{1 + 3 + 5 + \ldots + (2n - 1)}$.

7. $\lim\limits_{n \to \infty} \dfrac{2n^2 + n - 1}{(n + 1) + (n + 2) + \ldots + 2n}$.

8. $\lim\limits_{n \to \infty} \dfrac{1 + 5 + 5^2 + \ldots + 5^{n-1}}{1 - (25)^n}$.

9. $\lim\limits_{n \to \infty} \dfrac{n^2}{2^n}$.  10. $\lim\limits_{n \to \infty} \dfrac{2^n}{n!}$.  11. $\lim\limits_{n \to \infty} \dfrac{n!}{n^n}$.

12. $\lim\limits_{x \to 1} \dfrac{x^2 - 4x + 3}{x^2 + 3x - 4}$.

13. $\lim\limits_{n \to \infty} \dfrac{\sqrt[3]{4 + 3n + n^6}}{1 + 3n + 2n^2}$.

14. $\lim\limits_{x \to -2} \dfrac{x^4 + 5x^3 + 6x^2}{x^2 - 3x - 10}$.

15. $\lim\limits_{x \to 0} \dfrac{(4 + x)^3 - 64}{x}$.

16. $\lim\limits_{x \to 0} \dfrac{(1 + x)(1 + 2x)(1 + 3x) - 1}{x}$.

17. $\lim\limits_{x \to 0} \dfrac{(1 + x)^5 - (1 + 5x)}{x^2 + x^5}$.

18. $\lim\limits_{x \to 1} \dfrac{x^2 + 2x - 3}{x^2 + x - 2}$.

19. $\lim\limits_{x \to 2} \dfrac{x^2 - 5x + 6}{x^2 - 12x + 20}$.

20. $\lim\limits_{x \to -2} \dfrac{x^2 - x - 6}{x^2 + 7x + 10}$.

21. $\lim\limits_{x \to -3} \dfrac{x^2 - 2x - 15}{2x^2 - x - 21}$.

22. $\lim\limits_{x \to 3} \dfrac{x^2 - 5x + 6}{x^2 - 8x + 15}$.

23. $\lim\limits_{x \to 2} \dfrac{x^3 - 3x - 2}{x^3 - 8}$.

24. $\lim\limits_{x \to 1} \dfrac{x^4 - 2x^2 + 1}{x^3 - 1}$.

25. $\lim\limits_{x \to 1} \dfrac{x^4 - 3x + 2}{x^5 - 4x + 3}$.

26. $\lim\limits_{x \to -1} \dfrac{x^3 - 2x - 1}{x^5 - 2x - 1}$.

27. $\lim\limits_{x \to 1} \dfrac{x^m - 1}{x^n - 1}$.

28. $\lim\limits_{x \to 1} \dfrac{x + x^2 + \ldots + x^n - n}{x - 1}$.

29. $\lim\limits_{x \to 1} \dfrac{\sqrt{x} - 1}{x^2 - 1}$.

30. $\lim\limits_{x \to 6} \dfrac{\sqrt{x + 3} - 3}{x - 6}$.

31. $\lim\limits_{x \to 0} \dfrac{\sqrt{1 + x} - 1}{x}$.

32. $\lim\limits_{x \to 5} \dfrac{\sqrt{x - 1} - 2}{x - 5}$.

**33.** $\lim\limits_{x\to 2} \dfrac{x - \sqrt{3x - 2}}{x^2 - 4}$.

**34.** $\lim\limits_{x\to 4} \dfrac{\sqrt{1 + 2x} - 3}{\sqrt{x} - 2}$.

**35.** $\lim\limits_{x\to -8} \dfrac{\sqrt{1 - x} - 3}{2 + \sqrt[3]{x}}$.

**36.** $\lim\limits_{x\to 1} \dfrac{\sqrt[3]{x} - 1}{\sqrt{x} - 1}$.

**37.** $\lim\limits_{x\to 9} \dfrac{\sqrt[3]{x - 1} - 2}{x - 9}$.

**38.** $\lim\limits_{x\to 16} \dfrac{\sqrt[4]{x} - 2}{\sqrt{x} - 4}$.

**39.** $\lim\limits_{x\to 0} \dfrac{\sqrt{1 + x} - \sqrt{1 - x}}{\sqrt[3]{1 + x} - \sqrt[3]{1 - x}}$.

**40.** $\lim\limits_{x\to 9} \left( \dfrac{3 - \sqrt{x}}{9 - x} + \dfrac{1}{3 - \sqrt{x}} - 6 \cdot \dfrac{x^2 + 162}{729 - x^3} \right)$.

**41.** $\lim\limits_{x\to 1} \left[ \left( \dfrac{4}{x^2 - x^{-1}} - \dfrac{1 - 3x + x^2}{1 - x^3} \right)^{-1} + 3\dfrac{x^4 - 1}{x^3 - x^{-1}} \right]$.

**42.** $\lim\limits_{x\to 2} \left[ \left( \dfrac{x^3 - 4x}{x^3 - 8} \right)^{-1} - \left( \dfrac{x + \sqrt{2x}}{x - 2} - \dfrac{\sqrt{2}}{\sqrt{x} - \sqrt{2}} \right)^{-1} \right]$.

**43.** $\lim\limits_{x\to 0} x \cot 3x$.

**44.** $\lim\limits_{x\to 0} \dfrac{\cos(x/2) - \sin(x/2)}{\cos x}$.

**45.** $\lim\limits_{x\to 0} \dfrac{\sin 5x}{x}$.

**46.** $\lim\limits_{x\to 0} \dfrac{1 - \cos x}{x^2}$.

**47.** $\lim\limits_{x\to 0} \dfrac{\sin 5x - \sin 3x}{\sin x}$.

**48.** $\lim\limits_{x\to a} \dfrac{\cos x - \cos a}{x - a}$.

**49.** $\lim\limits_{x\to \frac{\pi}{6}} \dfrac{2\sin^2 x + \sin x - 1}{2\sin^2 x - 3\sin x + 1}$.

**50.** $\lim\limits_{x\to 0} \dfrac{\sqrt{1 - \cos x^2}}{1 - \cos x}$.

**51.** $\lim\limits_{x\to 0} \dfrac{\sin x - \tan x}{\sin^3 x}$.

**52.** $\lim\limits_{x\to\infty} \dfrac{x^2 + 3x - 4}{1 - 5x^2}$.

**53.** $\lim\limits_{x\to\infty} \left( \dfrac{3x}{5x - 1} \dfrac{2x^2 + 1}{x^2 + 2x - 1} \right)$.

**54.** $\lim\limits_{x\to\infty} \dfrac{2x^2 + 7x - 2}{6x^3 - 4x + 3}$.

**55.** $\lim\limits_{x\to\infty} \dfrac{(x - 1)(x - 2)(x - 3)(x - 4)(x - 5)}{(5x - 1)^5}$.

**56.** $\lim\limits_{x\to\infty} \dfrac{\sqrt{x^2 + 1} - x}{x + 1}$.

**57.** $\lim\limits_{x\to +\infty} (x - \sqrt{x^2 + 5x})$.

**58.** $\lim\limits_{x\to -4} \dfrac{\sqrt{5 + x} - 1}{x^2 + 4x}$.

**59.** $\lim\limits_{x\to 0} \dfrac{\int_0^x t\, dt}{x \tan (x + \pi)}$.

**60.** Using the definition, show that the function
$$f(x) = \begin{cases} x \sin (1/x) & \text{if } x \neq 0, \\ 0 & \text{if } x = 0 \end{cases}$$
is continuous at the point $x = 0$.

**61.** Using the definition, show that the function
$$f(x) = \begin{cases} \dfrac{\sin x}{|x|} & \text{if } x \neq 0, \\ 1 & \text{if } x = 0 \end{cases}$$
is discontinuous at the point $x = 0$.

## 2.2. The Derivative. Investigating the Behaviour of Functions with the Aid of the Derivative

**1.** Given a function $f(x) = \sin^2 2x$. Find
$$\frac{1}{2} \frac{f'(x)}{\cos 2x}.$$

**2.** Calculate $f'(x) + f(x) + 2$ if $f(x) = x \sin 2x$.

**3.** Simplify the expression for $f(x)$ and then find $f'(x)$ if
$$f(x) = \left( \frac{\sqrt{x - 2}}{\sqrt{x + 2} + \sqrt{x - 2}} + \frac{x - 2}{\sqrt{x^2 - 4} - x + 2} \right)^{-2} \left( \frac{x - 1}{2(\sqrt{x} + 1)} + 1 \right) \times \frac{2}{\sqrt{x} + 1}.$$

**4.** Simplify the expression for $f(x)$ and then find $f'(x)$ if

$$f(x) = \left( \frac{2\sqrt{x} + 3\sqrt[4]{x}}{\sqrt{16x} + 12\sqrt[4]{x} + 9} - \frac{\sqrt[4]{x} - 3}{2\sqrt[4]{x} + 3} \right)(2 \cdot 3^{\log_{81} x} + 3).$$

Prove the following identities **(5-8)**.

**5.** $f'(1) + f'(-1) = -4f(0)$ if $f(x) = x^5 + x^3 - 2x - 3.$

**6.** $f'(x) - 2xf(x) + \dfrac{1}{3}f(0) - f'(0) = 1$ if $f(x) = 3e^{x^2}.$

**7.** $f'(x) + f(x) + f\left(\dfrac{1}{x}\right) - \dfrac{1}{x} = 0$ if $f(x) = \ln x.$

**8.** $2f'\left(x + \dfrac{\pi}{3}\right)f'\left(x - \dfrac{\pi}{6}\right) = f'(0) - f\left(2x + \dfrac{\pi}{6}\right)$, where $f(x) = \cos x.$

**9.** Find the set of values of $x$ which satisfy the condition $[\varphi(x)]' + \varphi(x) = 0$ if $\varphi(x) = \cos x.$

Given two functions $f(x)$ and $g(x)$. For what $x$ does the equation $f'(x) = g(x)$ hold true? **(10-13)**.

**10.** $f(x) = \sin^4 3x$ and $g(x) = \sin 6x.$

**11.** $f(x) = \sin^3 2x$ and $g(x) = 4\cos 2x - 5\sin 4x.$

**12.** $f(x) = 2x^2 \cos^2(x/2)$ and $g(x) = x - x^2 \sin x.$

**13.** $f(x) = 4x \cos^2(x/2)$ and $g(x) = 8\cos(x/2) - 3 - 2x\sin x.$

For what $x$ is the derivative of the function $f(x)$ equal to zero? **(14-16)**.

**14.** $f(x) = 1 - \sin(\pi + x) + 2\cos((3\pi + x)/2).$

**15.** $f(x) = \sin 3x - \sqrt{3}\cos 3x + 3(\cos x - \sqrt{3}\sin x).$

**16.** $f(x) = 20\cos 3x + 12\cos 5x - 15\cos 4x.$

Solve the following inequalities **(17-21)**.

**17.** $f'(x) > g'(x)$ if $f(x) = x^3 + x - \sqrt{2}, g(x) = 3x^2 + x + \sqrt{2}.$

**18.** $f'(x) > g'(x)$ if $f(x) = 2x^3 - x^2 + \sqrt{3},\ g(x) = x^3 + \dfrac{x^2}{2} - \sqrt{3}.$

**19.** $f'(x) \le g'(x)$ if $f(x) = 2/x,\ g(x) = x - x^3.$

**20.** $f'(x) > g'(x)$ if $f(x) = x + \ln(x - 5),\ g(x) = \ln(x - 1).$

**21.** $f'(x) > g'(x)$ if $x = 5^{2x+1}/2, g(x) = 5^x + 4x\ln 5.$

**22.** Find all $f(x)$ which satisfy the equation
$$f^2(x) + 4f'(x) \times f(x) + [f'(x)]^2 = 0.$$

Find the derivatives of the following functions **(23-37)**.

**23.** $y = (x+1)(x+2)^2.$

**24.** $y = \sin(\cos^2(\tan^3 x)).$

**25.** $y = \dfrac{1}{x} + \dfrac{2}{x^2} + \dfrac{3}{x^3}.$

**26.** $y = \dfrac{2x}{1 - x^2}.$

**27.** $y = e^x(x^2 - 2x + 2).$

**28.** $y = \log^3(x^2).$

**29.** $y = \dfrac{1}{x} + \dfrac{1}{\sqrt{x}} + \dfrac{1}{\sqrt[3]{x}}.$

**30.** $y = \dfrac{1}{4}\ln\dfrac{x^2 - 1}{x^2 + 1}.$

**31.** $y = \dfrac{x}{\sqrt{a^2 - x^2}}.$

**32.** $y = \ln\tan(x/2).$

**33.** $y = \dfrac{\sin x - x\cos x}{\cos x + x\sin x}.$

**34.** $y = \ln\sqrt{\dfrac{1 - \sin y}{1 + \sin x}}.$

**35.** $y = x\sqrt{1 + x^2}.$

**36.** $y = \sin(\sin(\sin x)).$

**37.** $y = (2 - x^2)\cos x + 2x\sin x.$

**38.** Prove that $e^x - x > 1$ if $x > 0.$

Find the intervals of increase and decrease of the following functions **(39-43).**

**39.** $y = -x(x-2)^2.$

**40.** $y = 2x^3 + 3x^2 - 2.$

**41.** $y = \dfrac{2}{3}x^3 - x^2 - 4x + 5.$

**42.** $f(x) = 3x^4 - 8x^3 + 6x^2 + 1.$

**43.** $f(x) = (2^x - 1)(2^x - 2)^2.$

Find the points of maximum and minimum and the intervals of monotonicity of the following **(44-52).**

**44.** $y = 2x^3 - 6x^2 - 18x + 7.$

**45.** $y = x^4 + 4x^3 - 8x^2 + 3.$

**46.** $y = x^4 - 4x^3 - 8x^2 + 3.$

**47.** $y = 4x^4 - 2x^2 + 3.$

**48.** $f(x) = x^3/(x^2 + 3).$

**49.** $f(x) = \dfrac{(x-2)(8-x)}{x^2}.$

**50.** $f(x) = (x-1)e^{3x}.$

**51.** $f(x) = x \cdot e^{-3x}.$

**52.** $y = x - \ln x.$

Find the critical points of the following functions and test them for their maxima and minima **(53-63).**

**53.** $y = 2x^2 + x + 2.$

**54.** $f(x) = 3x^4 - 4x^3.$

**55.** $y = x^4 - 10x^2 + 9.$

**56.** $y = (x-3)^2(x-2)^2.$

**57.** $y = \dfrac{1}{5}x^5 - 4x^2.$

**58.** $y = x^5 - x^2 + 8.$

**59.** $f(x) = \dfrac{x}{3} + \dfrac{3}{x}.$

**60.** $f(x) = \dfrac{x^2 - 3x + 2}{x^2 + 3x + 2}$.

**61.** $f(x) = \dfrac{x^2}{17} - \ln(x^2 - 8)$.

**62.** $y = x \cdot e^x$.

**63.** $y = x \cdot e^{-x^2}$.

**64.** Find all values of $x$ for which the function $y = \sin x - \cos^2 x - 1$ assumes the least value. What is that value?

**65.** For what real values of $a$ and $b$ are all the extrema of the function $f(x) = \dfrac{5a^2}{3}x^3 + 2ax^2 - 9x + b$ positive and the maximum is at the point $x_0 = -5/9$?

**66.** For what real values of $a$ and $b$ are all the extrema of the function $f(x) = a^2 x^3 - 0.5ax^2 - 2x - b$ positive and the minimum is at the point $x_0 = 1/3$?

**67.** For what real values of $a$ and $b$ are all the extrema of the function $f(x) = a^2 x^3 + ax^2 - x + b$ negative and the maximum is at the point $x_0 = -1$?

**68.** Find the constant $p$ for which the function $f(x) = px^2 + \dfrac{2p^2 - 81}{2}x - 12$ has a maximum at the point $x = 9/4$.

**69\*.** Depending on $p$, indicate the values of $a$ for which the equation $x^3 + 2px^2 + p = a$ has three distinct real roots.

**70.** Find the greatest value of the function $y = 10 + 4x \ln 9 - 3^{x-1} - 3^{3-x}$.

**71.** Find the greatest value of the function $y = 7 + 2x \ln 25 - 5^{x-1} - 5^{2-x}$.

**72.** Find the least value of the function $y = 3^x + 2 \cdot 3^{3-x} - x \ln 27 - 9$.

**73.** For what value of $x$ does the expression
$$2^{x^2} - 1 + \dfrac{2}{2^{x^2} + 2}$$
assume the least value?

Find the greatest and the least values of the following functions **(74-93)**.

**74.** $y = \dfrac{4}{3}x^3 - 4x$ on the interval $[0,\ 2]$.

**75.** $f(x) = -x^3 + 3x^2 + 5$ on the interval $[0,\ 3]$.

**76.** $f(x) = 2x^3 - 9x^2 + 12x$ on the interval $[0,\ 3]$.

**77.** $f(x) = 3x^3 - 9x^2 + 2$ on the interval $[-1,\ 4]$.

**78.** $f(x) = 2x^3 + 3x^2 - 12x + 30$ on the interval $[-3,\ 3]$.

**79.** $f(x) = 2x^3 + 3x^2 - 12x + 1$ on the interval $[-1,\ 5]$.

**80.** $y = x^3 - 4x^2 + 4x + 3$ on the interval $[-1,\ 3]$.

**81.** $y = x^3 - 3x^2 + 3x + 2$ on the interval $[-2,\ 2]$.

**82.** $y = x^2 (2x - 3) - 12(3x - 2)$ on the interval $[-3, 6]$.

**83.** $y = |x^3 - 3x^2 + 5|$ on the interval $[0, 3]$.

**84.** $f(x) = x^4 - 8x^2 - 9$ on the interval $[0, 3]$.

**85.** $y = x^4 - 8x^2 - 9$ on the interval $[-1, 1]$.

**86.** $y = x^5 - x^3 + x + 2$ on the interval $[-1, 1]$.

**87.** $f(x) = \sqrt[3]{x^2} / (2x - 1)$ on the interval $[3/4, 2]$.

**88.** $f(x) = (2^x + 2^{-x}) / \ln 2$ on the interval $[-1, 2]$.

**89.** $f(x) = 2 \cdot 3^{3x} - 4 \cdot 3^{2x} + 2 \cdot 3^x$ on the interval $[-1, 1]$.

**90.** $f(x) = 2 \cdot 2^{3x} - 9 \cdot 2^{2x} + 12 \cdot 2^x$ on the interval $[-1, 1]$.

**91.** $f(x) = \sin x + \cos 2x$ on the interval $[0, \pi]$.

**92.** $f(x) = \cos 3x - 15 \cos x + 8$ on the interval $[\pi/3, 3\pi/2]$.

**93.** $f(x) = (5 + \sin x) \cos x + 3x$ on the interval $[0, \pi/2]$.

**94\*.** Find the greatest value of the function $f(x) = \dfrac{1}{2bx^2 - x^4 - 3b^2}$ on the interval $[-2, 1]$ depending on the parameter $b$.

**95\*.** Find the greatest value of the function $f(x) = x^4 - 6bx^2 + b^2$ on the interval $[-2, 1]$ depending on the parameter $b$.

**96.** Find the extrema of the function $f(x) = 2x \sin 2x + \cos 2x - \sqrt{3}$ on the interval $[-\pi/2, 3\pi/8]$.

**97.** Prove that the inequality

$$\min_{[-\pi, \pi]} f(x) > -7/18$$

holds for the function $f(x) = (\cos x)^2 (\sin x)$.

**98.** Prove that the inequality

$$\max_{[-\pi, \pi]} f(x) < 0.77$$

holds for the function $f(x) = (\sin x)(\sin 2x)$.

Set up an equation of a tangent to the graph of the following function **(99-108)**.

**99.** $y = x^2 - 2x$ at the points of its intersection with the abscissa axis.

**100.** $y = -x^2 - 1$ at the point $x = 2$.

**101.** $y = 2x^2 + 1$ at the point $(1, 5)$.

**102.** $y = 4x - x^2$ at the points of its intersection with the $Ox$ axis.

**103.** $y = x^2 - 2x + 5$ at the point of its intersection with the $Oy$ axis.

**104.** $y = \dfrac{1}{2}(e^{x/2} + e^{-x/2})$ at the point with abscissa $x = 2\ln 2$.

**105.** $y = 2^{-x} - 2^{-2x}$ at the point with abscissa $x = 2$.

**106.** $y = 3^x + 3^{-2x}$ at the point with abscissa $x = 1$.

**107.** $y = (2x - 1) e^{2(1-x)}$ at the point of its maximum.

**108.** $y = -x^2 - 2$ which is parallel to the straight line $y = 4x + 1$.

**109.** The tangent to the curve $y = 4 - x^2$ makes an angle of $75°$ with the $Ox$ axis. Find the coordinates of the point of tangency.

**110.** Find the value of the coefficient $k$ for which the curve $y = x^2 + kx + 4$ touches the $Ox$ axis.

**111.** Find the point on the curve $y = x^2 - x + 1$ at which the tangent is parallel to the straight line $y = 3x - 1$.

**112.** Find the point on the curve $y = 4x^2 - 6x + 3$ at which the tangent is parallel to the straight line $y = 2x$.

**113.** Two points with abscissas $x_1 = 1$ and $x_2 = 3$ are taken on the parabola $y = x^2$. A secant is drawn through those points. Find the point of the parabola at which the tangent drawn to it is parallel to the secant. Write the equation of the secant and the tangent.

**114\*.** Find the points at which the tangents to the curves $y = f(x) = x^3 - x - 1$ and $y = \varphi(x) = 3x^2 - 4x + 1$ are parallel. Write the equation of the tangents.

**115.** Find the angle between the tangents to the graph of the function $f(x) = x^3 - 4x^2 + 3x + 1$ drawn at the points with abscissas 0 and 1.

**116.** Find the angle between two tangents drawn to the parabola $y = x^2$ from the point $(0, -2)$.

**117.** Find the angle between the tangents drawn to the parabola $y = -3x^2$ from the point $(0, 2)$.

**118.** Find the equations of the common tangents to the parabolas $y = x^2$ and $y = -x^2 + 3x - 2$.

**119.** Find the equation of the tangent to the parabola $y = x^2 - 7x + 3$ if that tangent is parallel to the straight line $5x + y - 3 = 0$.

**120.** At what points does the tangent to the graph of the function $y = \dfrac{x + 2}{x - 2}$ make an angle of $135°$ with the $Ox$ axis?

**121.** Find the points at which the tangent to the curve $y = x - 0.5 \sin 2x - 0.5 \cos 2x + 16 \cos x$ is parallel to the abscissa axis.

**122.** Find the angle between the tangents to the graph of the function $y = x^3 - x$ at points with abscissas $x_1 = -1$ and $x_2 = 1$.

**123.** A tangent is drawn at the point $M(1, 8)$ to the curve $y = \sqrt{(5 - x^{2/3})^3}$. Find the length of its segment included between the coordinate axes.

**124.** Find the greatest volume $V$ of the cone with generatrix $a$.

**125.** A sector with a central angle $\alpha$ is cut off from a circle. A cone is made of the remaining part of the circle. At what value of $\alpha$ is the capacity of the cone the greatest?

**126.** The lower base of an isosceles trapezoid is $l$ and the base angle is $\alpha$. The diagonal of the trapezoid is perpendicular to one of its non parallel sides.

At what value of $\alpha$ is the area of the trapezoid the greatest? Find the greatest area.

**127.** The base of the pyramid is an isosceles triangle with the base angle $\alpha$. Each of the base dihedral angles is equal to $\varphi$. The radius of the sphere inscribed in the pyramid is $R$. Find the volume of the pyramid. At what value of $\alpha$ is the volume of the pyramid the least?

**128.** A cone is circumscribed about a sphere of radius $R$. The vertex angle in the axial section of the cone is $2\alpha$. Find the area of the axial section of the cone. At what value of $\alpha$ is the area of the cone the least?

**129.** The regular quadrilateral pyramid of volume $V$ is circumscribed about a hemisphere so that the centre of the base of the pyramid lies at the centre of the sphere. The angle between the lateral face of the pyramid and the plane of the base is $\alpha$. Find the volume of the hemisphere. At what value of $\alpha$ is the volume of the hemisphere the greatest?

**130.** Find the equations of the common tangents to the parabolas $y = x^2 - 5x + 6$ and $y = x^2 + x + 1$.

**131.** Prove that the curve $y = x^4 + 3x^2 + 2x$ does not meet the straight line $y = 2x - 1$ and find the distance between their nearest points.

**132.** The open body of a lorry is shaped as a rectangular parallelepiped with the surface area $2S$. What must be the length and the width of the body for its volume to be the greatest and the ratio of the length to the width to be equal to $5 : 2$?

**133.** The printed text takes 432 cm² of a page. The width of a margin at the top and bottom of the page is 2 cm and that of the lateral margins is 1.5 cm. What must be the width and the height of the page for the quantity of the paper used to be the least?

**134.** Represent the number 48 as the sum of two positive terms such that the sum of the cube of one of them and the square of the other is the least.

**135.** Given an open reservoir, with a square bottom, 32 m² in volume. What should be its dimensions for the material needed to face its walls and bottom to be the least?

**136.** At what angle $\alpha$ should a straight line be drawn through the point $(x_0, y_0)$ for the length of the segment intercepted between the coordinate axes to be the least?

**137.** The base of the pyramid is an isosceles triangle whose area is $S$ and the acute vertex angle is $\alpha$. Find the volume of the pyramid if the angle between each lateral edge and the altitude of the pyramid is $\beta$. For what value of $\alpha$ is the volume the least?

**138.** In a regular hexagonal pyramid a lateral edge is 1 cm long. For what length of the side of the base is the volume of the pyramid the greatest?

**139.** A chute has to be made from three boards of the same width. For what angle of inclination of the lateral sides will the cross-sectional area of the chute be the greatest?

**140.** Find a cylinder which would have the greatest volume for the given area $S$ of its total surface.

**141.** The volume of a right triangular prism is $V$. The base of the prism is an equilateral triangle. What must be the length of the side of the base for the total surface area of the prism to be the least?

**142.** A lateral edge of a regular rectangular pyramid is $a$ long. The lateral edge makes an angle $\alpha$ with the plane of the base. For what $\alpha$ is the volume of the pyramid the greatest?

**143.** Among all regular triangular prisms with volume $V$ find the prism with the least sum of the lengths of its edges. What is the length of a side of the base of that prism?

**144.** Among all rectangular parallelepipeds with square base inscribed in a given sphere find that which has the greatest lateral area.

**145.** The bases of a regular prism are squares. One of the bases of the prism belongs to the large circumference of a sphere of radius $R$, and the vertices of the other base lie on the surface of that sphere. What must be the length of the altitude of the prism for the sum of the lengths of all its edges to be the greatest?

**146.** A tin must be shaped as a cylinder with 1 dm³ capacity. What must be the radius of its bases for the area of the tin sheet. Used to make the tin to be at a minimum?

**147.** Find the altitude of the cylinder with the greatest lateral area which can be inscribed in a sphere of radius $R$.

**148.** The lateral area of a cone is $S$. For what radius of the base does the sphere inscribed in that cone have the greatest volume?

**149.** Given a sphere of radius $r$. Inscribe in it a cone which has the greatest lateral area. Find that area.

**150.** In a regular triangular prism the distance from the centre of one base to one of the vertices of the other base is $l$. For what length of the altitude of the prism is its volume the greatest? Find that greatest value of the volume.

**151.** Find the altitude of the cone of the least volume circumscribed about a hemisphere of radius $R = \sqrt{3}$ so that the centre of the base of the cone coincides with the centre of the sphere.

**152.** A right circular cylinder of altitude $H$ is inscribed in a sphere of radius $R$. Find the volume of the cylinder. For what $H$ is the volume the greatest? Find the greatest volume.

**153.** A cylinder with the greatest lateral area is inscribed in a sphere. Find the ratio of the radius of the sphere to that of the base of the cylinder.

**154.** A cylinder of the greatest volume is inscribed in a sphere. How many times is the volume of the sphere greater than that of the cylinder?

**155.** A cylinder with the greatest lateral area is inscribed in a sphere of radius $R$. Find the volume of the cylinder.

**156.** A cone of the greatest volume is inscribed in a sphere. Find the ratio of the radius of the sphere to the altitude of the cone.

**157.** A cone of the least volume is circumscribed about a cylinder the radius of whose base is $a$. The planes of the bases of the cylinder and the cone coincide. Find the radius of the base of the cone.

**158.** A sphere of unit radius is inscribed in a cone whose generatrix makes an angle of $2\varphi$ with the plane of the base. Find the lateral area of the cone. For what values of $\varphi$ is that area the least?

**159.** A regular rectangular pyramid is inscribed in a sphere of radius $R$ so that all the vertices of the pyramid belong to the sphere. What must be the altitude of the pyramid for its volume to be the greatest? Find the greatest value of the volume.

**160.** A regular hexagonal prism is inscribed in a cone of altitude $H$ with the base radius $R$ so that one base of the prism lies in the plane of the base of the cone and the vertices of the other base belong to the lateral surface of the cone. What must be the altitude of the prism for its volume to be the greatest? Find the greatest value of the volume of the prism.

**161.** A cylinder with the greatest lateral surface is inscribed in a cone. Find the ratio of the altitude of the cone to that of the cylinder.

**162.** A cylinder of the greatest volume is inscribed in a cone. Find the ratio of the radius of the base of the cone to the radius of the base of the cylinder.

**163.** A cylinder is inscribed in a cone with altitude $H$ and the base radius $R$ so that its one base lies in the plane of the base of the cone and the circumference of the other base belongs to the lateral surface of the cone. What must be the altitude of the cylinder for its volume to be the greatest? Find the greatest value of the volume.

**164.** All the edges of a triangular prism $ABCA_1B_1C_1$ are of the same length and all the plane vertex angles $A$ are congruent. $K$ and $L$ are the midpoints of the edges $[AA_1]$ and $[AB]$. Where on the edge $[AC]$ must a point $M$ be chosen for the area of the triangle $KLM$ to be the least?

**165.** A straight line is drawn through a point $N$ (2, 4) ; its segment forms a right triangle with the segments $(x > 0, y > 0)$ of the coordinate axes. What must be the length of the larger leg for the area of the triangle to be the least?

**166.** A boat is at a point $Q$ of the lake, which is 6 km distant from the nearest point $A$ of the bank. The boatman must get to the point $B$ which is 11 km away from $A$ on the bank. The speed of the boat is 3 km/h and the speed with which the boatman can walk along the bank is 5 km/h. The boatman has calculated that if he first rows to the point $C$ which is between $A$ and $B$ and then walks to the point $B$, then it will take him the least time to get from $Q$ to $B$. Find the distance between the points $A$ and $C$, assuming that the boat travels along a straight line and that the bank of the lake is a straight line.

167. A car travels from point $A$ to point $C$. Its speed from point $A$ to point $B$, which is between $A$ and $C$, is 48 km/h. At point B the driver decreases the speed of the car by $a$ km/h ($0 < a < 48$) and covers with that speed 1/3 of the way from $B$ to $C$. The remaining part of the way from $B$ to $C$ he drives with the speed which exceeds the initial speed (48 km/h) by $2a$ km/h. For what value of $a$ will it take the car the least time to cover the way from $B$ to $C$?

## 2.3. Graphs of Functions

1. Investigate the behaviour of the function $y = 3x^2 - 6x + 5$ and construct its graph. Find the greatest and the least value of the function on the interval $[0, 2]$.

2. Find the greatest and the least value of the function $f(x) = x^3 - 3x^2 + 2$ on the interval $[-1, 3]$ and construct its graph on that interval.

3. Investigate the behaviour of the function $y = x^3 - 3x + 2$ and construct its graph.

4. Investigate the behaviour of the function $y = x^3 - 4x^2 - 3x + 12$ and construct its graph.

5. Given a function $y = x^3 - 9x + 1$. Investigate its behaviour with the aid of the derivative and construct its graph.

6. Find the intervals of monotonicity and the extrema of the function $f(x) = x - \dfrac{12}{5}x^2 - \dfrac{1}{3}x^3$ and construct its graph.

7. Find the least and the greatest value of the function $f(x) = \dfrac{1}{15}x^3 + \dfrac{9}{20}x^2 - \dfrac{1}{2}x$ on the interval $[-2, 1]$ and construct its graph on the interval.

8. Find the intervals of monotonicity and the extrema of the function $f(x) = (x^3 - 5x^2 - 8x)/3$ and construct its graph.

9. Test the function $y = (x^3 - 4x)/4$ for monotonicity and extrema and construct its graph. Calculate the area of the figure bounded by the graph of the function and the straight line $y = 2(2 + x)$.

10. Construct the graph of the function $y = x(x^2 + 3x + 2)$. Write the equation of the tangent to the graph at the point with abscissa $x_0 = 0$. Find the coordinates of the points of intersection of the tangent and the graph of the function.

11. Construct the graph of the function $y = (x^2 + x)(x - 2)$. Write the equation of the tangent to the graph at the point with abscissa $x_0 = 0$. Find the coordinates of the points of intersection of the tangent and the graph of the function.

12. Investigate the behaviour of the function $y = (x^4 - 2x^2)/4$ with the aid of the derivative and construct its graph.

13. Given a function $y = (6x^2 - x^4)/9$. Investigate its behaviour and make a rough drawing of the graph.

14. Given a function $y = x^4 - 2x^2 + 5$. Investigate its behaviour with the aid of the derivative and construct its graph.

15. Given a function $y = x^4 - 10x^2 + 9$. Investigate its behaviour and construct a rough drawing of the graph.

16. Given a function $y = (x - 1)^2(x - 2)^3$. Investigate its variation and construct its graph.

17. Given a function $y = x^2/(x - 2)$. Investigate its behaviour and make a rough drawing of the graph.

18. Given a function $y = (-x^2 + 3x - 1)/x$. Investigate its behaviour and make a rough drawing of the graph.

19. Investigate the behaviour of the function $y = (x^3 - 4)/(x - 1)^3$ and construct its graph. How many roots does the equation $(x^3 - 4)/(x - 1)^3 = c$ possess?

20. Investigate the behaviour of the function $y = (x^3 + 4)/(x + 1)^3$ and construct its graph. How many solutions does the equation $(x^3 + 4)/(x + 1)^3 = c$ possess?

21. Investigate the behaviour of the function $y = 8(x^3 + x)/(2x - 1)^3$ and construct its graph. How many roots does the equation $8(x^3 + x)/(2x - 1)^3 = c$ possess?

22. Investigate the behaviour of the function $y = 8(x^3 + x)/(2x + 1)^3$ and construct its graph. How many roots does the equation $8(x^3 + x)/(2x + 1)^3 = c$ possess?

23. Investigate the behaviour of the function $y = (x^4 - 8)/(x + 1)^4$ and construct its graph. How many roots does the equation $(x^4 - 8)/(x + 1)^4 = c$ possess?

Construct the graphs of the following functions and carry out a complete investigation (24-28).

24. $y = \sin^4 x \cos^4 x$.

25. $y = (1.3)^{-2x} \sin^2 x$.

26. $y = \arcsin (2x/(1 + x^2))$.

27. $y = |x^2 - 4x + 3| + 2x$.

28. $y = \dfrac{x^3}{6} - x^2$.

29. Construct the graph of the function $y = f(x)$, where
$$f(x) = \lim_{n \to \infty} \frac{x^n - x^{-n}}{x^n + x^{-n}}, \quad x > 0.$$

## 2.4. The Antiderivative. The Integral. The Area of a Curvilinear Trapezoid

Find $F(x)$ from the given $F'(x)$ **(1-7)**.

**1.** $F'(x) = 4x + 1$ and $F(-1) = 2$.

**2.** $F'(x) = 3x^2 - 4x$ and $F(0) = 1$.

**3.** $F'(x) = 7x^2 - 2x + 3$, whose graph passes through the point $M(1, 5)$.

**4.** $F'(x) = 1 + x + \cos 2x$ and $F(0) = 1$.

**5.** $F'(x) = \sin 2x + 3x^2$ and $F(0) = 2$.

**6.** $F'(x) = 1 / \sin^2 x$, whose graph passes through the point $A\left(\dfrac{\pi}{6}, 0\right)$.

**7.** $F'(x) = 2\sin 5x + 3\cos\left(\dfrac{x}{2}\right)$ which is zero for $x = \dfrac{\pi}{3}$.

Calculate the following integrals **(8-10)**.

**8.** $\displaystyle\int_0^\pi \dfrac{dx}{\cos^2(x/5)}$.

**9.** $\displaystyle\int_{\pi/8}^{\pi/4} \cot^2 2x\, dx$.

**10.** $\displaystyle\int_0^3 \left[ 3^{1-x} + \left(\dfrac{1}{3}\right)^{2x-1} \right] dx$.

**11.** Find the numbers $A$ and $B$ such that the function $f(x) = A \cdot 2^x + B$ satisfies the conditions $f'(1) = 2$, $\int_0^3 f(x)\, dx = 7$.

**12.** Find all values of $a$ which satisfy the equation $\int_0^2 (t - \log_2 a)\, dt = 2\log_2\left(\dfrac{2}{a}\right)$.

**13.** Find all numbers $b > 1$ for which $\int_1^b (b - 4x)\, dx \geq 6 - 5b$.

**14.** Find all numbers $\alpha$ for which $\int_1^2 [\alpha^2 + (4 - 4\alpha)x + 4x^3]\, dx \leq 12$.

**15.** Find $\alpha > 0$ for which the inequality $\int_{-\alpha}^{\alpha} e^x dx > 3/2$ holds.

**16.** For what $a < 0$ does the inequality $\int_a^0 (3^{-2x} - 2 \cdot 3^{-x})dx \geq 0$ hold true?

Find the area of the closed figure bounded by the following curves **(17-97).**

**17.** $y = 2x - x^2, \ y = 0.$
**18.** $y = 3x + 18 - x^2, \ y = 0.$
**19.** $y = 4x - x^2$ and the $Ox$ axis.
**20.** $y = 1 + x^2, \ y = 2.$
**21.** $y = 0, \ y = 2x^2 + 1, \ x = -1, \ x = 1.$ Make a drawing.
**22.** $y = x^2, \ y = x + 2.$
**23.** $y = x^2 - x, \ y = 3x.$
**24.** $y = x^2 - 2x + 3, \ y = 3x - 1.$
**25.** $y = \frac{1}{3}x^2 - 2x + 4, \ y = 10 - x.$ Make a drawing.
**26.** $y = x, \ y = 2x - x^2.$
**27.** $y = 7x - 2x^2, \ x + y = 7/2.$
**28.** $y = x^2, \ y = 0, \ x = 3.$
**29.** $y = \frac{x^2}{2} - x + 2, \ y = x, \ x = 0.$ Make a drawing.
**30.** $y = 2(1 - x), \ y = 1 - x^2, \ x = 0.$ Make a drawing.
**31.** $y = x^2, \ y = 2 - x, \ y = 0.$
**32.** $y = 0, \ y = 4(x - 2), \ y = (x - 1)^2.$
**33.** $x = 0, y = -2x + 4,$ the $Ox$ axis, and $y = x^2 + 1.$
**34.** $y = 0, \ x = 0, \ y = -x + 1, y = 2 - x^2.$
**35.** $y = x^2, \ y = 1 + \frac{1}{4}x^2.$
**36.** $y = x^2, \ y = 2x - x^2.$
**37.** $y = x^2 - 2x + 2, \ y = 2 + 4x - x^2.$
**38.** $y = x^2 + 2, \ y = 1 - x^2, \ x = 0, \ x = 1.$
**39.** $y = x^3 - 3x^2 - 9x + 1, \ x = 0, \ y = 6 \ (x < 0).$
**40.** $y = (6x^2 - x^4)/9, \ y = 1.$
**41.** $y = x^4 - 10x^2 + 9,$ the $Oy$ axis, and the $Ox$ axis.
**42.** $x^2 = 4y, \ x + y = 3.$
**43.** $y^2 = 8x, \ 2x - 3y + 8 = 0.$
**44.** $y = \cos x, \ y = 0, \ x = \frac{3\pi}{4}, \ x = -\frac{\pi}{4}.$
**45.** $xy = 2, \ x + 2y - 5 = 0.$
**46.** $y = -x^2, \ y = 2e^x, \ x = 0, \ x = 1.$
**47.** $y = x^2, \ y = 3x + 4.$
**48.** $y = \frac{4}{x^2}, \ x = 1, \ y = x - 1.$
**49.** $y = x^2 - 2x + 3, \ y = 4 - 2x.$
**50.** $y = e^x, \ y = 0, \ x = 2.$
**51.** $y = x, \ y = -x,$ and the tangent to the curve $y = \sqrt{x^2 - 5}$ at the point $M \ (3, \ 2).$

**52.** $y = x^3$, $y = \sqrt{x}$.    **53.** $y = \dfrac{1}{x}$, $x = 1$, $x = 2$, $y = 0$.

**54.** $y = \dfrac{1}{x^2}$, $y = 0$, $x = \dfrac{1}{2}$, $x = 2.5$.

**55.** $y = \sin((\pi x)/2)$, $y = x^2$.

**56.** $y = x^4 - 2x^2 + 5$, $y = 1$, $x = 0$, $x = 1$.

**57.** $y = \dfrac{5}{x}$, $y = 6 - x$.   **58.** $xy = 3$, $x + y = 4$.

**59.** $y = 5 - x$, $y = 6/x$.  **60.** $y = \dfrac{5}{x}$, $y = 6 - x$, $x = 6$.

**61.** $y = \dfrac{1}{x}$, $y = x$, $x = 2$. Make a drawing.

**62.** $y = \dfrac{9}{x}$, $y = x$, $x = 9$, $y = 0$.

**63.** $y = x^2$, $y = 0$, $x = 4$, $y = x^2/(x-2)$.

**64.** $y = \dfrac{-x^2 + 3x - 1}{x}$, $x = 1$, $x = 2$, and the $Ox$ axis.

**65.** $y = 2\sqrt{x}$, $6 - y = 0$, $x = 0$.   **66.** $y = 0$, $y = -x + 2$, $y = \sqrt{x}$.
**67.** $y = x^2$, $y = 2\sqrt{2x}$.      **68.** $y = x^2$, $y = \sqrt[3]{x}$.
**69.** $y = \sqrt{x}$, $y = \sqrt{4 - 3x}$, $y = 0$.
**70.** $y = \sin x$, $y = 0$, with $0 \le x \le \pi$.
**71.** $y = \sin 6x$, $x = 0$, $x = \pi$, and the abscissa axis.
**72.** $y = \sin 2x$, $y = 1$, $x = 0$ ($x \ge 0$). Make a drawing.
**73.** $y = \cos x$, $y = 1 + \dfrac{2}{\pi} x$, $x = \dfrac{\pi}{2}$.

**74.** $y = \dfrac{1}{\cos^2 x}$, $y = 0$, $x = 0$, $x = \dfrac{\pi}{4}$.

**75.** $y = 4\sin^2 x \, (1 + \cos^2 x)$ on the interval $[-\pi, \pi]$ and the abscissa axis.

**76.** $y = 2\cos^2 x \, (1 + \sin^2 x)$ on the interval $[0, 2\pi]$ and the abscissa axis.

**77.** $y = 8\sin^4 x + 4\cos 2x$ on the interval $[0, \pi]$ and the abscissa axis.

**78.** $y = |4 - x^2|/4$ and $y = 7 - |x|$.

**79.** $y = 2 - |2 - x|$ and $y = 3/|x|$.

**80.** $y = 4 - \dfrac{6}{|1 + x|}$ and $y = |-x + 2|$.

**81.** $y = 3 - |3 - x|$ and $y = 6/|x + 1|$.

**82.** $x = -1$, $x = 2$,
$$f(x) = \begin{cases} -x^2 + 2, & x \le 1, \\ 2x - 1, & x > 1, \end{cases}$$

and the abscissa axis.

**83.** $y = 0.5x^2 - 2x + 2$ and the tangents to it : $y = -x + 1.5$ at the point $(1, 1/2)$, and $y = 2x - 6$ at the point $(4, 2)$.

**84.** $y = 2x^2 - 8x$, the tangent to that parabola at its vertex, and the axis of ordinates.

85. $y^2 = -x - 16$ and the tangents to that parabola drawn from the origin.

86. $y = x^2 + 10$ and the tangent to that parabola drawn from the point $(0, 1)$.

87. $y = x^2 - 2x + 2$, the tangent to that curve drawn at the point of its intersection with the axis of ordinates, and the straight line $x = 1$.

88. $y = 2x^2$, the $Ox$ axis, and the tangent to the curve $y = 2x^2$ at the point $A(x_0, y_0)$, where $x_0 = 2$.

89. $y = x^2 + 3$, the coordinate axes, and the tangent to the curve $y = x^2 + 3$ at the point $A(x_0, y_0)$, where $x_0 = 2$.

90. $x = -1$, $y = 0$, $y = x^2 + x + 1$, and the tangent to the curve $y = x^2 + x + 1$ at the point $A(x_0, y_0)$, where $x_0 = 1$.

91. $y = 1/x$, $x = 1$, and the tangent to the curve $y = 1/x$ drawn through the point $x = 2$.

92. $y = 0$, $x = 6$, $y = 4/x$, and the tangent to the curve $y = 4/x$ drawn through the point $A(x_0, y_0)$, where $x_0 = 2$.

93. At a certain point of the graph of the function $y = \sqrt{x}$ the tangent makes an angle of $45°$ with the abscissa axis. Calculate the area of the figure bounded by that tangent and the straight lines $y = 0$ and $x = \dfrac{1}{4}$.

94. $y = x^2 - x + 2$ and the tangent to the curve $y = \ln x + 3$ at the point with abscissa $x = 1$.

95. $y = e^{3x}$, $x = 3$, and the straight line which is a tangent to the curve $y = e^{3x}$ at the point with abscissa $x = 0$.

96. $y = \sin x \left(0 \le x \le \dfrac{\pi}{2}\right)$, $y = 0$, $y = 1$, and a straight line which is a tangent to the curve $y = \log x$ at the point with abscissa $x = 1$.

97. $y = \cos x \left(0 \le x \le \dfrac{\pi}{2}\right)$, $y = 0$, $x = 0$, and a straight line which is a tangent to the curve $y = \cos x$ at the point $x = \dfrac{\pi}{4}$.

Calculate the areas of the figures which are defined on the coordinate plane by the following inequalities **(98-103)**.

98. $|y| + \dfrac{1}{2} \le \sqrt{1 - |x|}$.

99. $|y| + \dfrac{1}{2} \le e^{-|x|}$.

100. $|x^2 + y^2 - 2| \le 2(x + y)$.

101. $|y| + 2|x| \le x^2 + 1$.

102. $x^2 + y^2 \le 2(|x| + |y|)$.

103. $4 \le x^2 + y^2 \le 2(|x| + |y|)$.

104. Through the point of intersection of the curve $y = 2x^2 + 4x - 3$ with the axis of ordinates a tangent is drawn to that curve. Calculate the area of the figure bounded by that tangent and the straight lines $y = 0$ and $x = 0$.

**105.** Let us designate as $S(k)$ the area included between the parabola $y = x^2 + 2x - 3$ and the straight line $y = kx + 1$. Find $S(-1)$ and calculate the least value of $S(k)$.

**106.** A point $M(1/2, 1)$ is given on a plane. A straight line which passes through that point forms a triangle with the positive coordinate semi-axes. What minimum value can the area of that triangle assume?

**107.** A figure is bounded by the curves $y = x^2 + 1$, $y = 0$, $x = 0$, and $x = 1$. At what point $(x_0, y_0)$ of the graph of the function $y = x^2 + 1$ must a tangent to the graph be drawn for it to cut off a trapezoid of the greatest area from the figure?

**108.** A figure is bounded by the curves $y = 1/x$, $y = 0$, $x = 1$, and $x = 2$. At what point $(x_0, y_0)$ of the graph of the function $y = 1/x$ must a tangent be drawn to it so that it cuts off a trapezoid of the greatest area from the figure?

**109.** Find the value of the parameter $p$ $(p < 0)$ for which the area of a figure bounded by the parabola $y = (1 + p^2)^2 x^2 + p$ and the straight line $y = 0$ attains its greatest value.

**110.** A figure is bounded by the curves $y = (x + 3)^2$, $y = 0$, and $x = 0$. At what angles to the $Ox$ axis must straight lines be drawn through the point $(0, 9)$ for them to partition the figure into three parts of the same size?

**111.** A figure is bounded by the curves $y = |\sin(\pi x / 2)|$, $y = 0$, and $x = -1$. At what angles to the $Ox$ axis must straight lines be drawn through the point $(0, 0)$ for them to partition the figure into three parts of the same size?

**112.** An equilateral triangle with a side of 10 cm rotates about the external axis which is parallel to the side of the triangle and is at the distance equal to half the altitude of the triangle from it. Find the volume of the body of revolution.

**113.** The sum of the legs of a right triangle is 9 cm. When the triangle rotates about one of the legs, a cone results which has the maximum volume. Find the lateral area of the cone.

# Part-3

## Geometry and Vector Algebra

This part of the book includes problems on the elements of vector algebra studied at school and geometry problems. The former are given to check the knowledge of the concepts such as vectors, vector coordinates in a given system of coordinates, a scalar product of vectors. Problems of this kind usually require calculations. It should be pointed out, however, that concepts and methods of vector algebra can often be efficiently used in solving purely geometrical problems.

Let us dwell in more detail on geometry problems. Not infrequently do they constitute the most laborious part of an examination paper since a geometry problem must always be considered from many aspects. One must first construct the necessary elements of the figure (straight lines, sections, and so on), make a drawing, prove the relations between the given and constructed elements of the figure (perpendicularity, parallelism of straight lines and planes, and so on), and, proceeding from those relations, calculate the required quantities. The most essential part of the solution of a geometry problem is a strict substantiation of the relations between the elements of the figure based on the theorems of geometry. In this way the knowledge of the theoretical material is checked as well as the ability of a student to use it in problem solving. That is why problems requiring proof and a number of construction problems are separated into special sections. Although such problems, in a "pure" state, are seldom encountered in examination papers, they constitute an inalienable part of the solution of any geometry problem. In addition, the solution of these problems will undoubtedly deepen the knowledge of the theory and will help prepare for the oral exams in mathematics.

## 3.1. Vector Algebra

Calculate the scalar product of the following vectors **(1-4)**.

1. $\mathbf{a} = \{2, 4, 1\}$ and $\mathbf{b} = \{3, 5, 7\}$.

2. $\mathbf{a} = \{-2, 3, 11\}$ and $\mathbf{b} = \{5, 7, -4\}$.

3. $\mathbf{a} = 2\mathbf{i} + 3\mathbf{j} - 4\mathbf{k}$ and $\mathbf{b} = \mathbf{i} - 2\mathbf{j} + \mathbf{k}$.

4. $(2\alpha + 3\beta) \cdot (4\alpha - 6\beta)$, where $\alpha, \beta$ are mutually perpendicular unit vectors

5. Resolve the vector $\mathbf{d} = \{1, 1, 1\}$ into components with respect to three noncoplanar vectors $\mathbf{a} = \{1, 1, -2\}$, $\mathbf{b} = \{1, -1, 0\}$, and $\mathbf{c} = \{0, 2, 3\}$.

6. For what value of $\alpha$ are the vectors $\mathbf{a} = \{2, 3, -4\}$ and $\mathbf{b} = \{\alpha, -6, 8\}$ parallel?

7. For what value of $\alpha$ are the vectors $\mathbf{a} = \{1, \alpha, -2\}$ and $\mathbf{b} = \{\alpha, 3, -4\}$ mutually perpendicular?

8. Being given that $|\mathbf{a}| = 2$, $|\mathbf{b}| = 5$, and $(\widehat{\mathbf{a}, \mathbf{b}}) = 2\pi / 3$, find the value of $\alpha$ for which the vectors $\mathbf{p} = \alpha\mathbf{a} + 17\mathbf{b}$ and $\mathbf{q} = 3\mathbf{a} - \mathbf{b}$ are perpendicular.

9. For what value of $\alpha$ are the vectors $l = \{6, \alpha, -8\}$ and $\mathbf{m} = \{-3, -1, 4\}$ parallel?

10. The vectors $\mathbf{a}$ and $\mathbf{b}$ make an angle of $120°$, $|\mathbf{a}| = 3$, and $|\mathbf{b}| = 5$. Find $|\mathbf{a} - \mathbf{b}|$.

11. Find the angle between the vectors $\mathbf{a} = \{-1, 2, -2\}$ and $\mathbf{b} = \{6, 3, -6\}$.

12. Find the cosine of the angle between the vectors $\mathbf{a} - \mathbf{b}$ and $\mathbf{a} + \mathbf{b}$ if $\mathbf{a} = \{1, 2, 1\}$ and $\mathbf{b} = \{2, -1, 0\}$.

13. Given three forces $\mathbf{M} = \{3, -4, 2\}$, $\mathbf{W} = \{2, 3, -5\}$, and $\mathbf{P} = \{-3, -2, 4\}$ applied to the same point. Calculate the work performed by the resultant of these three forces when the point of application of the resultant moves along a straight line and is displaced from the position $M_1(5, 3, -7)$ to the position $M_2(4, -1, -4)$.

14. For what values of $x$ are the vectors $\mathbf{a} = \{x, 3, 4\}$ and $\mathbf{b} = \{5, 6, 3\}$ perpendicular?

15. Given three vectors $\mathbf{a}$, $\mathbf{b}$, and $\mathbf{c}$. Prove that the vector $(\mathbf{b} \cdot \mathbf{c})\mathbf{a} - (\mathbf{a} \cdot \mathbf{c})\mathbf{b}$ is perpendicular to the vector $\mathbf{c}$.

16. Given three vectors $\mathbf{a} = \{2, 3, -5\}$, $\mathbf{b} = \{3, 0, 1\}$, and $\mathbf{c} = \{4, -3, 2\}$. Find the coordinates and the length of the vector $\mathbf{d} = 3\mathbf{a} + \mathbf{b} - \mathbf{c}$.

17. Find (in degrees) the angle between the vectors $\mathbf{a} = 2\mathbf{i} + 5\mathbf{j} - \mathbf{k}$ and $\mathbf{b} = \mathbf{i} - \mathbf{j} - 3\mathbf{k}$.

18. Find the angle between the vectors $2\mathbf{a}$ and $\mathbf{b} / 2$, if $\mathbf{a} = \{-4, 2, 4\}$ and $\mathbf{b} = \{\sqrt{2}, \sqrt{2}, 0\}$.

19. The vector $\mathbf{a}$ and $\mathbf{b}$ make an angle $\varphi = 2\pi / 3$. Being given that $|\mathbf{a}| = 3$ and $|\mathbf{b}| = 4$, calculate $(3\mathbf{a} - 2\mathbf{b}) \cdot (\mathbf{a} + 2\mathbf{b})$.

20. Given four points $A$ $(-2, -3, 8)$, $B$ $(2, 1, 7)$, $C(1, 4, 5)$, and $D$ $(-7, -4, 7)$. Are the vectors $\vec{AB}$ and $\vec{CD}$ collinear?

21. Find the vector **a** which is collinear with the vector **b** = {3, 6, 6} and satisfies the conditions **a** · **b** = 27.

22. Find the vector **a** which is collinear with the vector **b** = {2, -1, 0} if **a** · **b** = 10.

23. Find the coordinates of the vector **x** which is collinear with the vector **a** = $2\mathbf{i} + \mathbf{j} - \mathbf{k}$ and satisfies the condition **x** · **a** = 3.

24. Find the vector **a** which is collinear with the vector **b** = {1, -3, 1} and satisfies the condition **a** · **b** = 22.

25. Find the vector **b** = {x, y, z} which is collinear with the vector **a** = $\{2\sqrt{2}, -1, 4\}$ if $|\mathbf{b}| = 10$.

26. Find the vector **c** being given that it is perpendicular to the vectors **a** = {2, 3, -1}, **b** = {1, -2, 3} and satisfies the condition **c** · $(2\mathbf{i} - \mathbf{j} + \mathbf{k}) = -6$.

27. The vector **b**, which is collinear with the vector **a** = {8, -10, 13}, makes an acute angle with the $Oz$ axis. Being given that $|\mathbf{b}| = \sqrt{37}$, find its coordinates. In the answer write the sum of the coordinates of the vector **b** accurate to within 0.01.

28. Find the vector **b** being given that it satisfies the following conditions : the scalar product **b** · **c** = 3, where **c** = {1, -1, 2}, the vector **b** is perpendicular to **a** and **d**, where **a** = {-2, -1, 1} and **d** = {3, 5, -2}.

29. Find the cosine of the angle between the vectors **p** and **q** which satisfy the system of equations
$$\begin{cases} 2\mathbf{p} + \mathbf{q} = \mathbf{a} \\ \mathbf{p} + 2\mathbf{q} = \mathbf{b} \end{cases}$$
if it is known that in a rectangular system of coordinates the vectors **a** and **b** have the forms **a** = {1, 1} and **b** = {1, -1}.

30. Given three vectors **a** = {3, -1}, **b** = {1, -2} and **c** = {-1, 7}. Determine the resolution of the vector **p** = **a** + **b** + **c** into components with respect to the vectors **a** and **b**.

31. Given two vectors **a** = {1, -1, 3} and **b** = {3, -5, 6}. Calculate Proj $_{(a+b)}$ $(2\mathbf{a} - \mathbf{b})$.

32. The vector **x**, which is perpendicular to the vectors **a** = $3\mathbf{i} + 2\mathbf{j} + 2\mathbf{k}$ and **b** = $18\mathbf{i} - 22\mathbf{j} - 5\mathbf{k}$, makes an obtuse angle with the $Oy$ axis. Find its coordinates if the length of the vector **x** is 14.

33. The vector **x** satisfies the following conditions : (a) **x** is collinear with the vector **a** = $6\mathbf{i} - 8\mathbf{j} - 7.5\mathbf{k}$; (b) **x** makes an acute angle with the $Oz$ axis; (c) $|\mathbf{x}| = 50$. Find the coordinates of the vector **x**.

**34.** Given two vectors $\mathbf{a} = \{-1, 1, 1\}$ and $\mathbf{b} = \{2, 0, 1\}$. Find the vector $\mathbf{x}$ if it is known that it is coplanar with the plane of the vectors $\mathbf{a}$ and $\mathbf{b}$, is perpendicular to the vector $\mathbf{b}$, and $\mathbf{a} \cdot \mathbf{x} = 7$.

**35.** Given two vectors in space ; $\mathbf{a} = \{1, 1, 2\}$ and $\mathbf{b} = \{-1, 3, 1\}$. Find the unit vector which lies in the plane of the vectors $\mathbf{a}$ and $\mathbf{b}$ and makes an angle $\alpha = \pi / 4$ with the vector $\mathbf{a}$.

**36.** Given three nonzero vectors $\mathbf{a}$, $\mathbf{b}$, and $\mathbf{c}$, each two of which are noncollinear. Find their sum if $\mathbf{a} + \mathbf{b}$ is collinear with $\mathbf{c}$, and $\mathbf{b} + \mathbf{c}$ is collinear with $\mathbf{a}$.

**37.** Given the vertices $A(3, 2, -3)$, $B(5, 1, -1)$, and $C(1, -2, 1)$ of a triangle. Find its interior vertex angle $A$.

**38.** Prove that the points $A(-2, -3)$, $B(-3, 1)$, $C(7, 7)$, and $D(3, 0)$ are the vertices of a trapezoid. Find the length of the median of the trapezoid.

**39.** A triangle is specified by the coordinates of its vertices $A(3, 2, -3)$, $B(5, 1, -1)$, and $C(1, -2, 1)$. Calculate the exterior vertex angle $A$ of the triangle and the coordinates of the vector $\mathbf{a}$ which is of the same direction as the vector $\vec{AB}$ and has the same length as the vector $\vec{AC}$.

**40.** Given three successive vertices of a parallelogram : $A(-3, -2, 0)$, $B(3, -3, 1)$, and $C(5, 0, 2)$. Find its fourth vertex $D$ and the angle between the vectors $\vec{AC}$ and $\vec{BD}$

**41.** A triangle is defined by the coordinates of its vertices $A(2, 1, 2)$, $B(1, 0, 0)$, and $C(1 + \sqrt{3}, \sqrt{3}, -\sqrt{6})$. Calculate the angles of the triangle and the length of the median $m$ drawn to the side $BC$.

**42.** In a triangle $ABC$ the point $A(-1, 2, 3)$ is the vertex of the right angle. Find the coordinates of the vertices $B$ and $C$ if it is known that $B$ and $C$ lie on the straight line $(MN)$, where $M(-1, 3, 2)$, $N(1, 1, 3)$, and $\angle ABC = 30°$.

**43.** In a trapezoid $ABCD$ the vector $\vec{BC} = \lambda \vec{AD}$. Prove that the vector $\mathbf{p} = \vec{AC} + \vec{BD}$ is collinear with $\vec{AD}$ (and, hence, with $\vec{BC}$) and find the coefficient $\alpha$ in the notation $\mathbf{p} = \alpha \vec{AD}$.

**44.** Given a triangle $ABC$. The lengths of the vectors $\vec{CB}$, $\vec{CA}$, and $\vec{AB}$ are $a$, $b$, and $c$ respectively. Find the scalar product of the vectors $\vec{CA}$ and $\vec{CD}$, where $[CD]$ is the median of the triangle $ABC$.

**45.** Prove that the sum of the vectors which connect the centre of a regular triangle with its vertices is zero.

**46.** A straight line is drawn through the vertex $C$ of the square $ABCD$ parallel to the diagonal $[BD]$ which cuts the straight line $(AD)$ at a point $E$; $Q$ is the point at which the diagonals of the squares meet. Express the sum of the vectors $\vec{AB}$ and $\vec{CE}$ in terms of the vectors $\vec{DC}$ and $\vec{CQ}$.

**47.** What angle do the unit vectors **a** and **d** make if the vectors **c** = **a** + 2**b** and **d** = 5**a** − 4**b** are known to be mutually perpendicular?

**48.** Given a parallelogram *ABCD*. The lengths of the vectors $\vec{AB}$, $\vec{AD}$, and $\vec{BD}$ are $a$, $b$, and $c$ respectively. Find the scalar product of the vectors $\vec{AC}$ and $\vec{AD}$.

**49.** Given a parallelogram *ABCD*. The lengths of the vectors $\vec{AB}$, $\vec{AD}$, and $\vec{AC}$ are $a$, $b$, and $c$ respectively. Find the scalar product of the vectors $\vec{DB}$ and $\vec{AB}$.

**50.** Being given vectors **p** and **q** on which a parallelogram is constructed, use them to express the vector which coincides with the altitude of the parallelogram, which is perpendicular to the side **p**.

**51.** Given a trapezoid *ABCD*. The length of the base [*AD*] is thrice as large as that of the base [*BC*]. The lengths of the vectors $\vec{AB}$, $\vec{BC}$, and $\vec{AC}$ are $a$, $b$, and $c$ respectively. Find the scalar product of the vectors $\vec{BA}$ and $\vec{AD}$.

**52.** Medians [*AD*], [*BE*], and [*CF*] are drawn in a triangle. Calculate $\vec{BC} \cdot \vec{AD}$ + $\vec{CA} \cdot \vec{BE}$ + $\vec{AB} \cdot \vec{CF}$.

**53.** Given a parallelogram *ABCD* ([*AD*] || [*BC*], [*AB*] || [*CD*]). A point *K* is chosen on the side [*AD*], and a point *L* on the side [*AC*] such that $|\vec{AK}| = |\vec{AD}|/5, |\vec{AL}| = |\vec{AC}|/6$. Prove that the vectors $\vec{KL}$ and $\vec{BL}$ are collinear and find the proportionality factor $\lambda$ in the notation $\vec{KL} = \lambda \vec{BL}$

**54.** Assume that *ABCD* is a parallelogram, and [*AD*] || [*BC*], *K* is the midpoint of [*BC*], *L* is the midpoint of [*DC*]. Introduce the designations $\vec{AK} = $ **a**, $\vec{AL} = $ **b** and express the vectors $\vec{BD}$ and $\vec{AC}$ in terms of **a** and **b**.

**55.** Given a regular hexagon *ABCDEF*, *M* is the midpoint of [*DE*], *N* is the midpoint of [*AM*], *P* is the midpoint of [*BC*]. Resolve the vector $\vec{NP}$ into components with respect to the vectors $\vec{AB}$ and $\vec{AF}$.

**56.** In a triangle *ABC* the lengths of two sides are known : |*AB*| = 6, |*AB*| = 8, and the angle *A* = 90°; [*AM*] and [*BN*] are the bisectors of the angles *A* and *B*. Find the cosine of the angle between the vectors $\vec{AM}$ and $\vec{BN}$.

**57.** Squares with centres $A_1$, $B_1$, and $C_1$ are constructed on the sides $[BC]$, $[CA]$, and $[AB]$ of a triangle $ABC$, respectively, in its exterior. Find $\vec{AA_1} + \vec{BB_1} + \vec{CC_1}$.

**58.** $M$ is the point of intersection of the medians of a triangle $ABC$. Prove that $\vec{MA} + \vec{MB} + \vec{MC} = \mathbf{0}$.

**59.** The length of the hypotenuse $AB$ of a right triangle $ABC$ is $c$. Find the sum $\vec{AB} \cdot \vec{AC} + \vec{BC} \cdot \vec{BA} + \vec{CA} \cdot \vec{CB}$.

**60.** A point $M$ is the centre of a circle inscribed in a triangle $ABC$. Find the vector $\alpha \vec{MA} + \beta \vec{MB} + \gamma \vec{MC}$ if $\alpha = |\vec{BC}|$, $\beta = |\vec{CA}|$, and $\gamma = |\vec{AB}|$.

**61.** A point $O$ is the centre of a circle circumscribed about a triangle $ABC$. Prove that $\vec{OA} \sin(2A) + \vec{OB} \sin(2B) + \vec{OC} \sin(2C) = \mathbf{0}$.

**62.** Given a rectangular system of coordinates with the origin at point $O$. A straight line cuts the abscissa axis at a point $C$ and the axis of ordinates at a point $A$ so that $\angle OCA = \alpha$ and $\alpha > \dfrac{\pi}{4}$. A point $D$ is the midpoint of $[AC]$, and a point $A_1$ is symmetric with respect to the point $A$ about the straight line $(OD)$. Find the lengths of the sides of the triangle $ACA_1$ if the coordinates of the point $C$ $(1, 0)$ are known.

**63.** Given a rectangular system of coordinates with origin at a point $O$. A straight line cuts the abscissa axis at a point $C$ and the axis of ordinates at a point $A$ so that $\angle OCA = \alpha$ and $\alpha > \dfrac{\pi}{4}$. A point $D$ is the midpoint of $[AC]$, and a point $A_1$ is symmetric with respect to the point $A$ about the straight line $(OD)$. Find the ratio of the areas of the tiangles $ODC$ and $OCA_1$.

**64.** Prove that the points $A(5, 0)$, $B(0, 2)$, and $C(2, 7)$ are the vertices of a right triangle. Find its area and indicate all the displacements of the plane which convert it into a triangle with vertices $(-5, 0)$, $(0, -2)$, and $(-2, -7)$.

**65.** Prove that the points $A(3, 0)$, $B(0, 1)$, $C(2, 7)$, and $D(5, 6)$ are the vertices of a rectangle $ABCD$. Calculate its area and indicate all the displacements of the plane upon which it passes into itself.

**66.** For what $x$ and $y$ are the points with coordinates $A(2, 0)$, $B(0, 2)$, $C(0, 7)$, and $D(x, y)$ successive vertices of an isosceles trapezoid $ABCD$? For each of these trapezoids find the area and indicate all the displacements of the plane which convert it into a trapezoid with vertices $(2, 0)$, $(7, 0)$, $(0, -7)$, and $(0, -2)$.

**67.** Three forces applied to a vertex of a cube are equal to 1, 2, and 3 in magnitude and are directed along the diagonals of the faces of the cube, which meet at that vertex. Find the magnitude of the resultant of these three forces.

**68.** Given a cube $ABCD A_1 B_1 C_1 D_1$ with lower base $ABCD$, upper base $A_1 B_1 C_1 D_1$, and lateral edges $AA_1$, $BB_1$, $CC_1$, and $DD_1$. Prove that on the straight line $(MM_1)$, where $M$ and $M_1$ are the centres of the faces $ABCD$ and $A_1 B_1 C_1 D_1$, respectively, there is a point $O$ such that $\vec{OA} + \vec{OB} + \vec{OC} + \vec{OD} = -\vec{OM_1}$. Show that $\vec{MO} = \lambda \vec{OM}$ and find $\lambda$.

**69.** Find the sum of the scalar products of vectors whose origins are at the centre of a face of a cube and the terminal points are at the vertices (there are 8 vectors of this kind). The edge of the cube is $b$ long.

**70.** In a regular triangular pyramid $ABCS$ the edge angle is a right angle. Points $D$ and $E$ are the midpoints of the edges $AC$ and $SB$ respectively. Find the angle between the vectors $\vec{SD}$ and $\vec{EC}$.

**71.** A point $A(x_1, y_1)$ with abscissa $x_1 = 1$ and a point $B(x_2, y_2)$ with ordinate $y_2 = 1$ are given in the rectangular Cartesian system of coordinates $Oxy$ on the curve $y = x^2 - 4x + 5$. Find the scalar product of the vectors $\vec{OA}$ and $\vec{OB}$.

**72.** A point $A(x_1, y_1)$ with abscissa $x_1 = 1$ and a point $B(x_2, y_2)$ with ordinate $y_2 = 11$ are given in the rectangular Cartesian system of coordinates $Oxy$ on the part of the curve $y = x^2 - 2x + 3$ which lies in the first quadrant. Find the scalar product of $\vec{OA}$ and $\vec{OB}$.

**73.** A point $A(x_1, y_1)$ with abscissa $x_1 = 1$ and a point $B$, which is the point of intersection of the curves $y = 2x^2 - 3x + 5$ and $y = 2x^2 - 2x + 3$, are given in the rectangular Cartesian system of coordinates $Oxy$ on the curve $y = 2x^2 - 3x + 5$. Find the scalar product of the vectors $\vec{OA}$ and $\vec{OB}$.

**74.** Two points $A$ and $B$ are given in the rectangular Cartesian system of coordinates $Oxy$ on the curve $y = x^2$, the points being such that $\vec{OA} \cdot i = 1$ and $\vec{OB} \cdot i = -2$, where $i$ is a unit vector of the $Ox$ axis. Find the length of the vector $2\vec{OA} - 3\vec{OB}$.

**75.** Two points $A$ and $B$ are given in the rectangular Cartesian system of coordinates $Oxy$ on the curve $y = 2^{x+2}$, the points being such that $\vec{OA} \cdot i = -1$ and $\vec{OB} \cdot i = 2$, where $i$ is a unit vector of the $Ox$ axis. Find the length of the vector $-4\vec{OA} + \vec{OB}$.

**76.** Two points $A$ and $B$ are given in the rectangular Cartesian system of coordinates $Oxy$ on the curve $y = 6 / x$, the points being such that $\vec{OA} \cdot i = -2$ and $\vec{OB} \cdot i = 3$, where $i$ is a unit vector of the $Ox$ axis. Find the length of the vector $2\vec{OA} + 3\vec{OB}$.

77. In the rectangular Cartesian system of coordinates $Oxy$ a tangent is drawn to the curve $y = x^2 + x + 10$ at a point $A$ $(x_0, y_0)$, where $x_0 = 1$. The tangent cuts the $Ox$ axis at a point $B$. Find the scalar product of the vectors $\vec{OA}$ and $\vec{AB}$.

78. In the rectangular Cartesian system of coordinates $Oxy$ a tangent is drawn to the curve $y = 2\sqrt{x}$ at a point $A$ $(x_0, y_0)$, where $x_0 = 1$. The tangent cuts the $Ox$ axis at a point $B$. Find the scalar product of the vectors $\vec{AB}$ and $\vec{OA}$.

79. In the rectangular Cartesian system of coordinates $Oxy$ a tangent is drawn to the curve $y = 8 / x^2$ at a point $A$ $(x_0, y_0)$, where $x_0 = 2$. The tangent cuts the $Ox$ axis at a point $B$. Find the scalar product of the vectors $\vec{AB}$ and $\vec{OB}$.

80. Write the equation of the image of the parabola $y = x^2 - 2x + 1$ under a translation $\mathbf{p} = \mathbf{i} + 3\mathbf{j}$.

## 3.2. Plane Geometry. Problems on Proof

1. Prove that $\angle C$ of a triangle $ABC$ is a right angle if and only if the lengths of the sides of the triangle are related as $| AB |^2 = | AC |^2 + | BC |^2$.

2. The midpoints of the sides of two convex quadrangles coincide. Prove that the areas of the quadrangles are equal.

3. Prove that a circle can be inscribed in a convex quadrangle if and only if the sums of the lengths of the opposite sides of the quadrangle are equal.

4. A circle is circumscribed about a trapezoid. Prove that this is possible if and only if the trapezoid is isosceles.

5. An isosceles trapezoid $ABCD$ is circumscribed about a circle. $E$ and $K$ are the points of tangency of the circle and the nonparallel sides $[AB]$ and $[CD]$. Prove that the segment $[EK]$ is parallel to the bases of the trapezoid.

6. Tangents are drawn through the terminal points of the arc of a circle equal to $120°$, and a circle is inscribed in the figure bounded by the tangents and the given arc. Prove that the length of the circle is equal to the length of the initial arc.

7. Prove that in a right triangle the sum of the lengths of the legs is equal to the sum of the diameters of the inscribed and the circumscribed circle.

8. In an isosceles triangle with base $a$ and a lateral side $b$ the vertex angle is equal to $20°$. Prove that $| a |^3 + b^3 = 3| a | | b |^2$.

9. The lengths of the sides $a$, $b$, and $c$ in a triangle are related as $| a |^2 + | b |^2 = 5| c |^2$. Prove that the medians drawn to the sides $a$ and $b$ are mutually perpendicular.

10. Two circles are externally tangent to each other at a point $C$. A common external tangent $(AB)$ is drawn to them, where $A$ and $B$ are points of tangency. Prove that $\angle ACB = \pi/2$.

11. Prove that the radius $r$ of the circle inscribed in a polygon is equal to $2\,S/P$, where $S$ and $P$ are the area and the perimeter of the polygon respectively.

12. A point $O$ is taken in the interior of a parallelogram $ABCD$. Prove that the sum of the areas of the triangles $OAB$ and $OCD$ is constant at any choice of the point $O$.

13. $ABCD$ is a square. A point $M$ is taken on the side $[CD]$, $K$ is the point of intersection of the side $[BC]$ and the bisector of the angle $BAM$. Prove that $|MA| = |BK| + |DM|$.

14. In a right triangle $ABC$, $\angle B$ is a right angle and $[BD]$ is an altitude drawn to the hypotenuse $[AC]$. Prove that $|BD|$ is equal to the sum of the radii of the circles inscribed in the triangle $ABC$, triangle $ADB$, and triangle $CDB$.

15. Each side of a convex quadrangle is cut by a certain circle at two points, the lengths of the segments of the sides lying in the interior of the circle being equal. Prove that a circle can be inscribed in the given quadrangle.

16. One of the angles of a triangle is 30°. Prove that the length of the side which is opposite to that angle is equal to the radius of the circumscribed circle.

17. Prove that the midpoints of the sides of a convex quadrangle are the vertices of a parallelogram.

18. Prove that the sum of the lengths of the medians of a triangle is larger than its half-perimeter but smaller than its perimeter.

19. The centres of the inscribed and the circumscribed circle in a triangle coincide. Prove that the triangle is equilateral.

20. Prove that if the straight line which connects the midpoints of the bases of a trapezoid is perpendicular to the bases, then the trapezoid is isosceles.

21. Prove that the straight line, which connects the midpoints of the bases of a trapezoid, and the extensions of the nonparallel sides of the trapezoid meet at one point.

22. Prove that the line of centres of two intersecting circles bisects their common chord.

23. Prove that the common external tangents of two circles intersect along the line of centres or are parallel to it; the common internal tangents intersect along the line of centres.

24. Prove that in an isosceles triangle the sum of the distances from any point of the base to the lateral sides is constant.

25*. A straight line is drawn through the centre of a regular triangle. Prove that the sum of the squares of the distances from the vertices of the triangle to that straight line does not depend on the choice of the line.

26. Prove that the straight lines that successively connect the centres of the squares which are constructed on the sides of the parallelogram and are externally tangent to it form a square.

27. Given an isosceles triangle *ABC*. *R* and *r* are the radii of the circumscribed and the inscribed circle. Prove that the distance between the centres of the circles is equal to $\sqrt{R\,(R-2r)}$.

28. Prove that the distance from any point of the circle circumscribed about a regular triangle to one of its vertices is equal to the sum of the distances from that point to the other two vertices.

29. Prove that if the sides of one trapezoid are equal, respectively, to the sides of the other trapezoid, then the trapezoids are equal.

30. Prove that in the trapezoid, whose diagonals are the bisectors of the angles at one of the bases, the lengths of the three sides are equal.

31. Assume that *a* and *b* (*a* > *b*) are the lengths of the bases of a trapezoid. Prove that the line segment which connects the midpoints of the diagonals of the trapezoid is parallel to the bases of the trapezoid, and its length is (*a* – *b*)/2.

32. A straight line is drawn through the point *O* of intersection of the diagonals of a trapezoid parallel to its bases. Prove that the point *O* bisects the segment intercepted on the straight line by the nonparallel sides of the trapezoid.

33. Prove that the bisector of the interior angle of a triangle divides the opposite side into segments which are proportional to the adjacent sides. Prove that the converse statement is also true, i.e. if a straight line which passes through the vertex divides the opposite side into segments which are proportional to the adjacent sides, then the straight line is a bisector.

34. Assume that *m*, β, and *h* are a median, a bisector and an altitude of the triangle respectively, which are drawn to the same side of the triangle. Prove that the point of intersection of the bisector and that side lies between the points of intersection of the median and the altitude with that side (or its extension). Prove that these points coincide if and only if the triangle is isosceles.

35. Assume that *a*, *b*, and *c* are the lengths of the sides of the triangle *ABC*, which lie opposite the angles *A*, *B*, and *C* respectively, $p = (a+b+c)/2$ is a half-perimeter, *S* is the area, *R* and *r* are the radii of the circumscribed and the inscribed circle. Prove that the following relations hold true :

(1) $\dfrac{a}{\sin A} = \dfrac{b}{\sin B} = \dfrac{c}{\sin C} = 2R$, the *sine rule*;

(2) $a^2 = b^2 + c^2 - 2bc\cos A$, the *cosine rule*;

(3) $S = \dfrac{1}{2}bc\sin A$;

(4) $S = \sqrt{p\,(p-a)\,(p-b)\,(p-c)}$, *Hero's formula*;

(5) $S = pr$ ;

(6) $R = \dfrac{abc}{4S}$ ;

(7) if $m_a$, $\beta_a$, and $h_a$ are the lengths of a median, a bisector, and an altitude drawn from $\angle A$, then

$$m_a = \frac{1}{2}\sqrt{2c^2 + 2b^2 - a^2} \, ;$$

$$\beta_a = \frac{\sin B}{\sin (A/2)} \frac{ac}{b+c} \, ;$$

$$h_a = \frac{2}{a}\sqrt{p\,(p-a)\,(p-b)\,(p-c)}.$$

36. Two straight lines are drawn through a point $A$ which lies in the exterior of a circle, and one of them touches the circumference of the circle at a point $B$ and the other cuts that circumference at points $C$ and $D$. Prove that $|AD|\,|AC| = |AB|^2$ (the theorem on a tangent and a secant).

37. An altitude and a median drawn from the same vertex of a triangle divide the angle at that vertex into three equal parts. Prove that the angles of that triangle are equal to $30°$, $60°$, and $90°$.

38. The sum of the angles at the larger base of a trapezoid is $90°$. Prove that the length of the segment connecting the midpoints of the bases is equal to half the difference of the lengths of the bases.

39. Prove that the interaction of the bisectors of the interior angles of a parallelogram, which is not a rhombus, results in a rectangle the lengths of whose diagonals are equal to the difference between the adjacent sides of the parallelogram.

40. Prove that if at least one of the diagonals of a trapezoid is bisected by the point of intersection with the other diagonal, then the trapezoid is a parallelogram.

41. Prove that if the segment connecting the midpoints of two opposite sides of a quadrangle is equal to half the sum of the other two sides, then the quadrangle is a trapezoid.

42. Prove that if the lengths of two medians of a triangle are equal, then the triangle is isosceles.

43. Prove that if the lengths of two altitudes of a triangle are equal, then the triangle is isosceles.

44*. Prove that a bounded figure cannot have two centres of symmetry.

45. Prove that if a convex quadrangle has an axis of symmetry, then either a circle can be circumscribed about it or a circle can be inscribed in it.

46. Prove that if a quadrangle has a centre of symmetry, then it is a parallelogram.

**47.** Prove that a bisector of a triangle bisects the angle between the altitude and the radius of the circumscribed circle drawn to the same vertex.

**48.** Two mutually perpendicular chords $[AB]$ and $[CD]$ are drawn in a circle with diameter $d$. Prove that $|AD|^2 + |CB|^2 = d^2$.

**49.** An equilateral triangle is constructed on a diameter of a circle of radius 1 as a side. Prove that the area of the part of the triangle lying in the exterior of the circle is equal to $(3\sqrt{3} - \pi)/6$.

**50.** Given three equal squares $ABCD$, $DCEF$, and $FEPQ$ adjoining each other. Prove that $\angle CAD + \angle EAF + \angle PAQ = 90°$.

**51.** Prove that the bisectors of an interior angle and of an adjacent exterior angle of a triangle are mutually perpendicular. Prove that the converse statement is true : the straight line which passes through a vertex of the triangle at right angles to the bisector of the interior angle at that vertex is a bisector of the exterior angle.

**52.** Given an isosceles triangle. A perpendicular is drawn from an arbitrary point $P$ of the base. Prove that the sum of the lengths of the segments drawn from the point $P$ to the points of intersection of the perpendicular with the lateral sides or their extensions does not depend on the choice of the point $P$ on the base.

**53.** A square is inscribed in a rhombus which is not a square. Prove that its sides are parallel to the diagonals of the rhombus.

**54.** Prove that the inequality $R \geq 2r$ ($R$ and $r$ are the radii of the circumscribed and the inscribed circle) holds true in any triangle, and the equality $R = 2r$ holds only for a regular triangle.

**55.** Prove that the inequality $\cos A + \cos B + \cos C \leq 3/2$ holds true for every $\triangle ABC$.

**56.** Assume that a, $b$, c, and $d$ are the lengths of the successive sides of a quadrangle, and $S$ is its area. Prove that

(1) $(a + c)(b + d) \geq 4S$;       (2) $ac + bd \geq 2S$;

(3) $ab + cd \geq 2S$;       (4) $ad + bc \geq 2S$.

**57.** A chord $(AB)$ is drawn through a point $M$ taken in the interior of a circle (points $A$ and $B$ lie on the circle). Prove that the quantity $|AB| \cdot |MB|$ is constant for all such chords.

The following problems on seeking the loci of points (i.e. the sets of all points which satisfy some required property) are closely connected with problems on proof. When solving problems of this kind, it is often convenient to use the method of coordinates. (See Answers and Hints.)

**58.** Two mutually perpendicular straight lines are given on a plane. Find the set of all points of the plane the sum of whose distances from the given straight lines is equal to the sum of the quantities which are the reciprocals of the given distances.

59. Two points are given on a plane the distance between which is 1. Find the set of points whose distances from the two given points are related as $m : n$.

60. Two mutually perpendicular straight lines are given on a plane. Find the set of all points of the plane (1) the product of whose distances from the given straight lines is equal to the absolute value of the difference between the distances; (2) the absolute value of the difference of whose distances from the given straight lines is equal to the absolute value of the difference between the quantities which are the reciprocals of the given distances.

61. A straight line and a point not lying on it are given on a plane. Find the set of points which are equidistant from the given straight line and the given point.

62. Find the set of points the difference of the squares of whose distances from two given points $A$ and $B$ is constant.

63. Find the set of points the sum of whose distances from two intersecting straight lines is equal to a constant quantity.

64. Find the set of points the difference of whose distances from two intersecting straight lines is equal to a constant quantity.

65. Given two nonparallel segments $[AB]$ and $[CD]$. Find the set of points $M$ for which the areas of $\Delta\,AMB$ and $\Delta\,CMD$ are equal.

66. Find the set of midpoints of the chords which pass through a given point in the interior of a given circle.

67. The terminal points of a segment of length $a$ slide along the sides of a given right angle. Find the set of points traversed by the midpoint of the given segment.

68. Find the locus of points of the plane such that the tangents drawn from those points to a given circle make an angle $\alpha$.

## 3.3.  Plane Geometry. Construction Problems

Unless otherwise specified, all the constructions in this section are carried out only with the aid of a pair of compasses and a one sided rule without divisions. In other words, we can construct a circle whose radius is equal to the length of a given segment, with centre at a given point (the compasses), and draw a straight line through any two given points (a rule). More intricate constructions are based on various relations between the elements of the figures which we have to construct and which follow from the axioms and theorems of plane geometry and on successive performance of the two indicated elementary constructions. Let us carry out simple constructions indicating the theorems from the course of plane geometry on which they are based.

**Example 1.**   Given a straight line $l$ and a point $M$. Draw a straight line $l'$ which passes through the point $M$ at right angles to the line $l$ .

**Construction.** (a) Let us first consider the case when $M \in l$ (Fig. 3.1).

(1) We construct a circle with centre at the point $M \in l$ of any radius $r > 0$. Assume that points $M_1$ and $M_2$ are the points of intersection of the circle and the line $l$ Then the point $M$ is the midpoint of the segment $[M_1 M_2]$.

(2) We construct circles of the same radius $R > r$ with centres at the points $M_1$ and $M_2$. Assume that $M'$ is one of the two points of intersection of these circles. By construction, the point $M'$ is equidistant

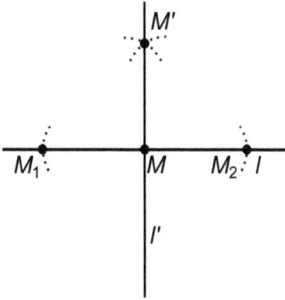

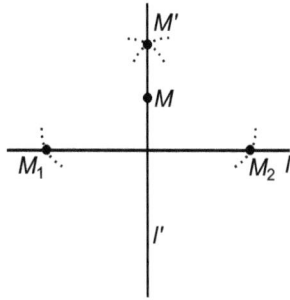

Fig. 3.1              Fig. 3.2

from the points $M_1$ and $M_2$ and, consequently, it lies on the perpendicular drawn through the midpoint of the segment $[M_1 M_2]$ i.e. through the point $M$.

(3) Then we draw the required straight line $l'$ through the points $M$ and $M'$. We have completed the construction.

Thus, in this case, we have used the theorem on a perpendicular drawn to the midpoint of a segment and, in addition, what is essential, implicitly used the theorem on the uniqueness of a perpendicular to a given straight line, the perpendicular being drawn through a given point.

(b) Assume now that $M \notin l$ (Fig. 3.2).

(1) We construct a circle, with centre at a point $M$, of a sufficiently large radius $R$ (the radius of the circle is such that it cuts the straight line $l$ at two points). Assume that $M_1$ and $M_2$ are the points of intersection of the line $l$ and the circle. By construction, the points $M_1$ and $M_2$ are equidistant from the point $M$ and, consequently, the point $M$ lies on the midperpendicular drawn to the segment $[M_1 M_2]$.

(2) We construct circles of the same radius $R_1 > R$ with centres at the points $M_1$ and $M_2$. Assume that $M'$ is one of the points of intersection of the circles. Then, by construction, $M'$ is also equidistant from $M_1$ and $M_2$ and, consequently, it also lies on the midperpendicular drawn to the segment $[M_1 M_2]$.

(3) Then we draw the required straight line $l'$ through the points $M$ and $M'$.

We have again used here the theorem on a midperpendicular and the uniqueness of that perpendicular. In addition, in both cases we implicitly based our arguments on the axioms of plane geometry which guarantee the possibility of carrying out our two elementary constructions (say, the axiom stating that a straight line can be drawn through two given points and that the straight line thus drawn is unique). In what follows we shall not repeat this remark.

**Example 2.** Given a straight line $l$ and a point $M \notin l$. Draw a straight line $l'$ through the point $M$ parallel to $l$.

**Construction** (Fig. 3.3).

(1) We draw a perpendicular $l_1$ from the point $M$ to the line $l$. Assume that $M_1$ is the point of intersection of the lines $l$ and $l$.

(2) We draw a perpendicular $l_2$ from some point $M_2 \in l$, which does not coincide with $M_1$, to the line $l$.

(3) Then we draw a circle with centre at the point $M_2$ whose radius is equal to $|MM_1|$. Assume that $M'$ is the point of intersection of the circle and the line $l_2$ which lies in the same (relative to $l$) half-plane as the point $M$.

Then, by construction, the points $M$ and $M'$ are equidistant from the line $l$ and lie in the same half-plane. Consequently (theorem!), a straight line which passes through the points $M$ and $M'$ is parallel to the line $l$.

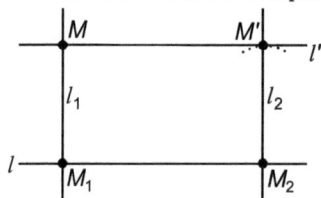

(4) Now we draw the required straight line $l'$ through the points $M$ and $M'$.

When performing the constructions, besides using the theorems employed in the preceding problem, we proceeded, in the third item, from the theorem which can be considered as a "criterion of a parallelogram". Indeed, the segments $[MM_1]$ and $[M'M_2]$ are parallel (since they are perpendicular to the straight line $l$) and equal in length. Consequently, the quadrangle $MM_1M_2M'$ is a parallelogram. Hence it follows that $l' \parallel l$.

Fig. 3.3

**Example 3.** Bisect the given segment $[AB]$.

**Construction** (Fig. 3.4). This construction has, in fact, been carried out in Example 1.

(1) We draw circles whose radius is equal to $|AB|$, with centres at points $A$ and $B$. Assume that $M_1$ and $M_2$ are the points of intersection of the circles. They are equidistant from the terminal points of the segment $[AB]$ and, consequently, they both lie on the midperpendicular drawn to the segment $[AB]$.

(2) Then we draw a straight line $l_1$ through two points $M_1$ and $M_2$. Their point of intersection $C$ is the required point.

**Example 4.** Given : a straight line $l$, a point $O \in l$, and an angle $A$. Lay off an angle, which is congruent to the angle $A$, from the line $l$ so that its vertex coincides with the point $O$ and one of the sides is directed along the line $L$ .

**Construction** (Fig. 3.5).

(1) From the vertex A of the given angle $A$ we draw a circle of radius $R > 0$. Assume that $M_1$ and $M_2$ are the points of intersection of the circle and the sides of the angle.

(2) We construct a circle of the same radius $R$ with centre at the point $O$. Assume that $K_1$ is one of the (two) points of intersection of the circle and

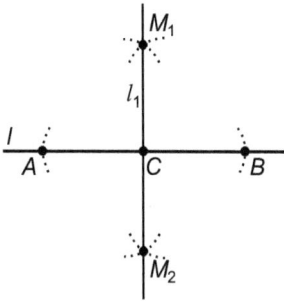   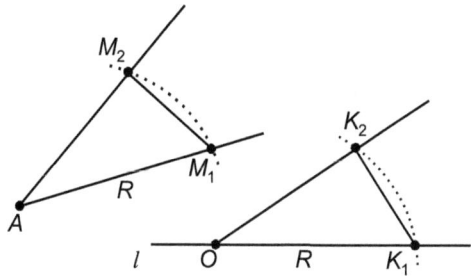

Fig. 3.4                      Fig. 3.5

the line $l$. We draw a circle with centre at $K_1$ whose radius is equal to $|M_1 M_2|$. Assume that $K_2$ is the point of intersection of that circle and the circle of radius $R$ with centre at $O$ (again one of the two).

(3) Then we connect the points $O$ and $K_2$ and have

$$|OK_1| = |OK_2| = |AM_1| = |AM_2|, \ |K_1K_2| = |M_1M_2|$$

by construction. Consequently, $\triangle M_1AM_2$ is congruent to $\triangle K_1OK_2$. Therefore, the angle we have constructed is congruent to the angle given in the hypothesis.

Note that we have constructed one of the four angles satisfying the hypothesis. Besides the angle we have constructed, the hypothesis is satisfied by the angle which is symmetric to the given angle about the straight line $l$, as well as the angles which are centrosymmetric with respect to those angles about the point $O$. In addition, we must separately consider the case when the given angle is equal to two straight lines. In that case the straight line $l$ itself with the point $O$ on it gives the required construction.

**Example 5.** Given a segment $[AB]$ and $\angle K$. Construct a set of all points $C$ such that $\angle ACB$ is congruent to the given $\angle K$.

**Construction** (Fig. 3.6).

(1) Assume that $l$ is a straight line passing through the points $A$ and $B$. We lay off an angle $BAM$, which is congruent to the given angle $K$, from the line $l$ at the point $A$ and assume that $l_1$ is a straight line passing through the points $A$ and $M$.

(2) We erect a midperpendicular $l_3$ to the segment $[AB]$ ($D$ is the midpoint of the segment $[AB]$) and a perpendicular $l_2$ to the line $l_1$ at the point $A$.

Assume that $O$ is the point of intersection of the perpendiculars (if $\angle K$ is a right angle, then the point $O$ coincides with $D$).

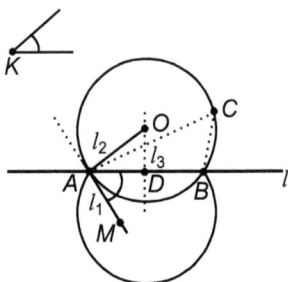

Fig. 3.6

(3) We draw a circle with centre at a point $O$ and the radius of length $|OA|$; next we consider the arc $A$ $B$, which lies in the same half-plane (relative to the line $l$) as the point $O$, with pricked-out points $A$ and $B$ and an arc which is symmetric with respect to it about the line $L$. The union of the two arcs of the circles yields the required set of points. Taking an arbitrary point $C$, which belongs to the constructed set, we find that $\angle ACB$ is measured by half the arc $\hat{AB}$ by which it is subtended (an inscribed angle) and $\angle BAM$ is also measured by the same half-arc (as an angle between the tangent $l_1$ to the circle and the chord $[AB]$ drawn from the point of tangency). Consequently, $\angle ACB$ is congruent to the given $\angle K$. Now if we take a point C, which does not belong to the constructed set, then the hypothesis will not be satisfied (verify this fact yourself).

As can be seen, in the construction carried out we have used many theorems of plane geometry. In addition to the theorems employed in the problems considered earlier, we have used here the theorem on an angle inscribed in a circle, a theorem on a tangent and a chord, a theorem on a tangent and a radius drawn to the point of tangency (namely, we have used the fact that the straight line $l_1$ is a tangent to the circle with centre at a point $O$ of radius $|OA|$). Hence we can see that the solution of construction problems provides for an active mastering of the theoretical material.

Let us consider one more simple construction problem based on the theorem on the similarity of triangles.

**Example 6.** Given segments $[AB]$ and $[CD]$. Construct a segment $[MN]$ whose length is equal to the geometric mean of the lengths of the given segments, i.e.

$$|MN| = \sqrt{|AB| \cdot |CD|}.$$

**Construction** (Fig. 3.7). The given relation between the lengths of the segments can be written in the following equivalent form :

$$\frac{|AB|}{|MN|} = \frac{|MN|}{|CD|},$$

and, therefore, this problem is sometimes called the problem on the construction of the mean proportional of two given segments.

(1) On some straight line $l$ we lay off a segment $[AB]$ and a segment

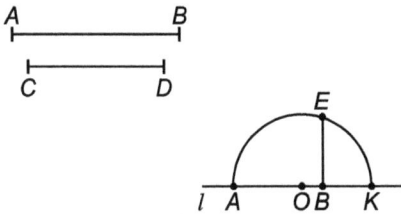

Fig. 3.7

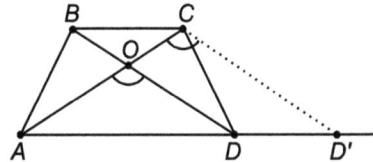

Fig. 3.8

$[BK]$ which is congruent to $[CD]$. Assume that $O$ is the midpoint of the segment $[AK]$. We construct a circle with centre at $O$ of radius

$$|OK| = |AK|/2 = (|AB| + |CD|)/2.$$

(2) From the point $B$ we erect a perpendicular to the straight line $l$. Assume that $E$ is one of the points of intersection of the perpendicular with the constructed circle. Then $[BE]$ is the required segment. This follows from the similarity of $\triangle ABE$ and $\triangle EBK$ (present the arguments).

Let us consider now some more complicated construction problems which can be solved by certain standard techniques.

**Example 7.** Construct a trapezoid from the given diagonals, the angle between the diagonals, and one of the nonparallel sides.

**Construction** (Fig. 3.8). Let us analyse the solution of this problem. Assume that $ABCD$ is the required trapezoid in which we are given the diagonals $[AC]$ and $[BD]$, the angle $AOD$ between the diagonals, and one of the non parallel sides $[CD]$. We extend the side $[AD]$ and lay off a segment $[DD']$ which is congruent to the segment $[BC]$. Then in the triangle $ACD'$ we know two sides $[AC]$ and $[CD']$ (since $[CD']$ is congruent to $[BD]$ ) and the angle between them. Having constructed the triangle and knowing the segment $[CD]$, we can find the position of the point $D$ on $[AD']$. Then we draw through the point $C$ a straight line parallel to the line $(AD')$. Having the segment $[BD]$ and the point $D \in [AD']$, we find the position of the point $B'$. We have thus constructed the four vertices of the required trapezoid. In analysing the problem, we have followed all the main stages of the construction. It is now simple enough to carry out the construction in detail. Note that in solving the problem we have used the technique known as a parallel displacement, or a translation, of an element of a figure, namely, instead of [BD] we first considered [CD′] resulting from [BD] upon a parallel displacement.

**Example 8.** Given an angle A and a point M lying in its interior. On the side of the angle A find points K and N such that the perimeter of the triangle KMN is the least.

**Construction** (Fig. 3.9). Assume that the points $M_1$ and $M_2$ are symmetric with respect to the given point M about the sides of the given angle. For any choice of the points N′and K′ on the sides of the angle the perimeter of the triangle

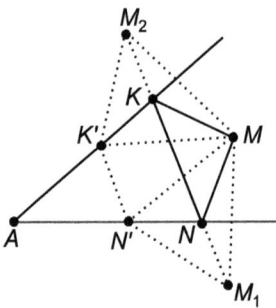

Fig. 3.9

Fig. 3.10

K′ MN′ is equal to the length of the polygonal line $M_1N′ K′ M_2$. This length is the least if the points K′and N′ lie on the straight line $(M_1 M_2)$. Thus we must take the points of intersection of the line $(M_1 M_2)$ with the sides of the angle as the points K and N. The corresponding constructions are simple enough. Note that, when solving this problem, we have used the technique of reflection of an element (a point in this case) with respect to a straight line.

**Example 9.** Given an angle A and a point M lying in its interior. Construct a circle which passes through the given point M and touches the sides of the given angle.

**Construction** (Fig. 3.10). Note that the centres of all circles which touch the sides of the given angle lie on the bisector of that angle. Assume that l is the bisector of ∠ A. We take a point O ∈ l and construct a circle with centre at that point, which touches the sides of ∠ A. Assume that M′and M ″ are the points of intersection of that circle and the straight line (AM). We draw straight lines, parallel to the segments [OM ″] and [OM′], through the point M and assume that $K_1$ and $K_2$ are the points of intersection of those straight lines and the bisector of ∠ A. The points $K_1$ and $K_2$ are exactly the centres of the required

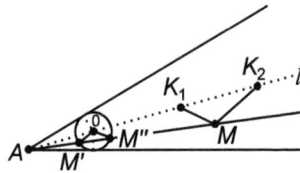

circles (carry out the complete proof and make the necessary constructions). When solving this problem, we have used the technique of a homothetic transformation with centre at a given point.

In a triangle $ABC$ the letters $a$, $b$, and $c$ denote the sides lying opposite the angles $A$, $B$, and $C$ respectively, $m_a$, $m_b$, and $m_c$ are medians, $h_a$, $h_b$, and $h_c$ are altitudes, $\beta_a$, $\beta_b$, and $\beta_c$ are the bisectors drawn to the respective sides, $P$ is a line segment whose length is equal to the perimeter of the given triangle $ABC$, $R$ and $r$ are the radii of the circumscribed and the inscribed circle.

Construct $\triangle ABC$ proceeding from the following elements* **(1- 40).**

1. $a$, $\angle A$, $\angle B$.  **2.** $a$, $B$, $\angle A$.  **3.** $a$, $\angle B$, $h_a$.

4. $a$, $\angle A$, $h_a$.  **5.** $a$, $\angle A$, $h_b$.  **6.** $a$, $\angle B$, $h_b$.

7. $a$, $h_a$, $h_b$.  **8.** $a$, $h_b$, $h_c$.  **9.** $a$, $b$, $m_a$.

10. $a$, $b$, $m_c$.  **11.** $a$, $\angle A$, $m_a$.  **12.** $a$, $\angle A$, $m_b$.

13. $a$, $\angle B$, $m_c$.  **14.** $a$, $\angle B$, $m_b$.  **15.** $a$, $m_a$, $m_b$.

16. $a$, $m_b$, $m_c$.  **17.** $m_a$, $m_b$, $m_c$.  **18.** $a$, $\angle B$, $\beta_c$.

19. $a$, $\angle B$, $\beta_b$.  **20.** $a$, $h_b$, $h_c$.

21. $a$, $\angle A$, and the point $M \in a$ through which $\beta_a$ passes.

22. $P$, $\angle A$, $\angle B$.  **23.** $P$, $\angle A$, $r$.  **24.** $a$, $h_b$, $m_a$.

25. $a$, $m_a$, $|b|/|c|$, where $|b|$ is the length of the line segment $b$ (i.e. it is assumed that some two line segments are given whose lengths are related as $|b| : |c|$).

26. $\angle A$, $r$, and the radius of one of the escribed circles.

27. $a$, $\angle A$, and a segment of length $||b|^2 - |c|^2|$.

28. $\angle A$, $h_a$, and the ratio of the lengths of the segments into which the altitude $h_a$ divides the base $a$.

29. $a$, $h_a$, and the angle which is equal to the difference between the angles $B$ and $C$.

30. $h_a$, $m_a$, and $\beta_a$.

31. The vertex $B$, the centre of a circumscribed angle, and the centre of gravity i.e, the point of intersection of the medians).

32. $h_b$, $h_c$, and $m_a$.  **33.** $a$, $b+c$, and $\angle A$.  **34.** $a$, $b+c$, and $\angle B$.

35. $a$, $b+c$, and $\angle B + \angle C$.  **36.** $a$, $b-c$, and $\angle A$.  **37.** $a$, $b-c$, and $\angle B$.

38. The centre of gravity and the midpoints of two medians.

39. The centre of gravity, the centre of a circumscribed circle, and one of the vertices.

40. The points of intersection of the altitudes with a circumscribed circle.

41. Construct a right triangle from the hypotenuse and the sum of the legs.

42. Construct a right triangle from the hypotenuse and the median of a leg.

* It is assumed that the data given are such that the necessary constructions are possible.

**43.** Construct an isosceles right triangle from its perimeter.

**44.** Construct a trapezoid, being given its bases and the non parallel sides.

**45.** Construct a parallelogram in which the midpoints of three sides lie at given points.

**46.** Construct a square whose three vertices lie on three given parallel straight lines.

**47.** Construct a quadrangle ABCD, being given all its sides and the angle between the extensions of the sides [AB] and [CD].

**48.** Inscribe a square in a given triangle.

**49.** Inscribe a rectangle in a given triangle when (a) the diagonal of the rectangle is given; (b) the diagonal of the rectangle is the least.

**50.** Inscribe a triangle in another triangle, the sides of the former being parallel to three given straight lines.

**51.** Inscribe a parallelogram in a given convex quadrangle. (Show that the problem has infinitely many solutions.)

**52.** Given an angle and a point M in its interior. Draw a straight line through the point M such that its segment lying between the sides of the angle is divided by that point in the ratio 2 : 3.

**53.** Given an angle and a point M lying in its exterior. Draw a straight line through the point M such that it cuts off a triangle of a given perimeter from the given angle.

**54.** Given a circle. Construct its centre.

**55.** Given two circles. Construct their common tangents.

**56\*.** Construct a circle of a given radius, (1) which touches two given straight lines; (2) which touches a given straight line and a given circle; (3) which touches two given circles.

**57.** Given two circles and their common external tangent. Find a point on the tangent such that the sum of the angles at which the circles can be seen from that point is equal to the given angle.

**58.** Given line segments of lengths $a$ and $b$. Construct line segments whose lengths are (a) $\sqrt{ab}$; (b) $\sqrt{a^2 - ab + b^2}$; (c) $\sqrt{a^2 - 4ab + 7b^2}$.

**59.** Given a circle and a point in its interior. Draw a chord of a specified length through the given point.

**60\*.** Given two circles. Draw a common secant such that its parts lying in the interior of the circles are equal to a given segment.

**61.** Inscribe a right triangle in a circle, being given the acute angle of the triangle and the point through which one of the legs passes.

**62.** Given two concentric circles. Draw a circle through a given point which touches the given concentric circles.

**63\*.** Given three concentric circles. Construct an equilateral triangle whose vertices belong to those circles.

**64.** On a given circle find a point M such that the distances from M to the sides of the given angle are related as 2 : 3.

**65\***. Draw a straight line through the point of intersection of two given circles so that the circles intercept equal chords on the given straight line.

**66\***. Given a triangle and a point belonging to one of its sides. Draw two straight lines through the given point such that they divide the triangle into three parts of the same size.

**67.** A circle of radius $R$ is inscribed in an angle of $\alpha$ radians. Between the vertex of the angle and the centre of the circle a tangent is drawn to the circle such that the triangle it cuts off from the angle has the greatest area. Carry out the construction and calculate the area of the cut-off triangle.

**68.** Given a triangle. Construct a square of the same size.

**69.** In $\triangle ABC$ draw a secant $(DE)$ such that
$| AD | = | DE | = | EC | (D \in [AB], E \in [BC])$.

**70\***. Given an acute triangle $ABC$. Inscribe in it a triangle whose sides are perpendicular to those of the given triangle. Prove that there are two triangles of that kind.

**71\***. Cut a trapezoid by a straight line, which is parallel to the bases, such that its segment lying in the interior of the trapezoid is divided by the diagonals into three equal parts.

**72.** Inscribe a square in a given segment so that one of its sides lies on the chord.

**73.** Construct a square such that its two vertices lie on a given straight line and the other two, on a given circle.

**74.** Given a straight line $(CD)$ and two points $A$ and $B$ not lying on it. On the straight line find a point $M$ such that (1) $\angle AMC = \angle BMD$;
(2) $2\angle AMC = \angle BMD$.

**75.** Circumscribe a square about an equilateral triangle so that the two figures should have a common vertex.

**76.** Given three rays drawn from the same point. Draw a straight line, passing through the given point, such that the lengths of the segments intercepted on it by the rays are related as 3 : 5.

Perform the following constructions with limited possibilities **(77-81)**.

**77.** Using only a pair of compasses, divide a given segment in half.

**78.** Given a triangle with the lengths of the sides equal to 3 cm, 4 cm, and 5 cm. Inscribe a circle in it using only a pair of compasses.

**79.** Using only a two-sided rule, bisect a given angle.

**80.** Given an angle of 19°. Using only a pair of compasses, construct an arc of 7°.

**81.** Using a pair of compasses and a rule, divide an angle of 54° into three equal parts.

## 3.4. Plane Geometry. Calculation Problems

In this section we present the fundamental "calculation" formulas and some statements often used in solving plane geometry problems (see also problems in Sec. 3.2).

I.  Assume that $a$, $b$, and $c$ are the lengths of the sides $\triangle ABC$ which lie opposite the angles $A$, $B$, and $C$ respectively, $p = (a+b+c)/2$ is half the perimeter, $S$ is the area, $R$ and $r$ are the radii of the circles circumscribed about and inscribed in the triangle, $h_a$, $m_a$, and $\beta_a$ are the lengths of the altitude, the median, and the bisector drawn to the side which is opposite to $\angle A$. The following statements hold true.

(1) The *sine rule* : $\dfrac{a}{\sin A} = \dfrac{b}{\sin B} = \dfrac{c}{\sin C} = 2R.$

(2) The *cosine rule* : $a^2 = b^2 + c^2 - 2bc \cos A$. In particular, if $A$ is a right angle, then we obtain the *Pythagorean theorem*.

(3) The formulas for calculating areas of triangles.

$$S = \frac{1}{2} a h_a; \ S = \frac{1}{2} ab \sin C;$$

$$S = \sqrt{p\,(p-a)\,(p-b)\,(p-c)} - Hero's\ formula;$$

$$S = pr; \ S = \frac{abc}{4R}.$$

(4) *Three medians of a triangle meet at one point which lies strictly in the interior of the triangle (the centre of gravity). The point of inter section divides the medians into segments whose lengths are related as* $2 : 1$, *reckoning from the corresponding vertex.*

$$m_a = \sqrt{\frac{a^2}{4} + c^2 - ac \cos B}; \quad m_a = \frac{1}{2}\sqrt{2c^2 + 2b^2 - a^2}.$$

(5) *Three bisectors of a triangle meet at one point which lies strictly in the interior of the triangle. The point of intersection of the bisectors is equidistant from the sides of the triangle (the centre of an inscribed circle).*

*When a bisector cuts a side of a triangle, it divides it into segments which are proportional to the adjacent sides of the triangle.*

$$\beta_a = \frac{\sin B}{\sin (A/2)} \cdot \frac{ac}{b+c}.$$

(6) *Three altitudes of a triangle meet at one point (the orthocentre).*

$$h_a = b \sin C; \ h_a = \frac{2S}{a} = \frac{2}{a}\sqrt{p\,(p-a)\,(p-b)\,(p-c)}.$$

(7) *Three perpendiculars drawn to the midpoints of the sides of a triangle meet at one point. That point is equidistant from the vertices of the*

triangle and is the centre of a circumscribed circle.

$$R = \frac{abc}{4S}; \quad R = \frac{a}{2\sin A} = \frac{b}{2\sin B} = \frac{c}{2\sin C}.$$

$$r = \frac{S}{P} = \left[\frac{1}{p}(p-a)(p-b)(p-c)\right]^{1/2}$$

**II.** Assume that $a$ and $b$ are the lengths of the adjacent sides of a parallelogram $ABCD$, $A$ is the magnitude of the angle between those sides, $h_a$ is the altitude dropped to the side of length $a$, $d_1$ and $d_2$ are the lengths of the diagonals, and $S$ is the area of the parallelogram. The following statements hold true.

(1) $h_a = b \sin A.$

(2) $S = ah_a = ab \sin A.$

(3) $d_1^2 = a^2 + b^2 - 2ab \cos A.$

$d_2^2 = a^2 + b^2 + 2ab \cos A.$

$d_1^2 + d_2^2 = 2(a^2 + b^2).$

(4) *The point of intersection of the diagonals of a parallelogram is the centre of its symmetry. Hence it follows, in particular, that the diagonals' are bisected by the point of intersection.*

(5) *A parallelogram can be inscribed in a circle if and only if it is a rectangle.*

(6) *A circle can be inscribed in a parallelogram if and only if it is a rhombus.*

**III.** Assume that $a$ and $b$ are the lengths of the bases of a trapezoid, $c$ and $d$ are the lengths of its nonparallel sides, $h$ is the altitude, and. $S$ is the area of the trapezoid. The following statements hold true.

(1) $S = \dfrac{1}{2}(a+b)h.$

(2) *A circle can be inscribed in a trapezoid if and only if $a + b = c + d$.*

(3) *A trapezoid can be inscribed in a circle if and only if it is isosceles.*

**IV.** Assume, finally, that $R$ is the length of the radius of a circle, $S$ is its area, and $l$ is the length of the circumference which is the boundary of the circle. Then $l = 2\pi R$ and $S = \pi R^2$.

1. The length of one of the legs of a right triangle exceeds the length of the other leg by 10 cm but is smaller than that of the hypotenuse by 10 cm. Find the length of the hypotenuse of the triangle.

2. In a triangle $ABC$, $[BD]$ is a median, $|BD| = |AB|\sqrt{3}/4$, and $\angle DBC = \pi/2$. Find the magnitude of $\angle ABD$.

3. The length of the base of a triangle is 4 cm smaller than the length of the altitude and the area of the triangle is 96 cm$^2$. Find the lengths of the base and the altitude of the triangle.

4. The lengths of the sides of a triangle are 11 cm, 13 cm, and 12 cm. Calculate the length of the median drawn to the larger side.

5. The length of the base of an isosceles triangle is $a$ and the vertex angle is $\alpha$. Find the length of the bisector drawn to a lateral side.

6. In an isosceles triangle the vertex angle is equal to $\alpha$ and the area is $S$. Find the length of the base of the triangle.

7. In a right triangle the lengths of the medians of the acute angles are $\sqrt{156}$ cm and $\sqrt{89}$ cm. Find the length of the hypotenuse of the triangle.

8. The legs of a right triangle are $a$ and $b$ long. Find the length of the bisector of the right angle of the triangle.

9. The bisector of the angle $N$ of a triangle $MNP$ divides the side $[MP]$ into segments whose lengths are 28 and 12. Find the perimeter of the triangle $MNP$ if $|MN| - |NP| = 18$.

10. The ratios of the lengths of the sides $[BC]$ and $[AC]$ of a triangle $ABC$ to the radius of a circumscribed circle are equal to 2 and 1.5 respectively. Find the ratio of the lengths of the bisectors of the interior angles $B$ and $C$.

11. The side $[AB]$ of a triangle $ABC$ is 2 cm long. A median $[BD]$, which is 1 cm long, is drawn from the vertex $B$ to the side $[AC]$. Find the area of the triangle $ABC$ if $\angle BDA = 30°$.

12. Find the angles of the triangle in which the altitude and the median drawn from the same vertex divide the angle at that vertex into three equal parts.

13. The points $M$ and $N$ in a rhombus $ABCD$ are the midpoints of the sides $[BC]$ and $[CD]$ respectively. Find $\angle MAN$ if $\angle BAD = 60°$.

14. The perimeter of a rhombus is equal to 48, and the sum of the lengths of the diagonals is equal to 26. Find the area of the rhombus.

15. Find the angle between the diagonals of a rectangle with perimeter $2p$ and area $\dfrac{3}{16} p^2$.

16. The point $M$ in a square $ABCD$ is the midpoint of $[BC]$ and $O$ is the point of intersection of $[DM]$ and $[AC]$. Find the angle $\angle MOC$.

17. The midline of a trapezoid is 10 cm and divides the area of the trapezoid in the ratio 3 : 5. Find the lengths of the bases of the trapezoid.

18. In an isosceles trapezoid $ABCD$ the leg $[AB]$ and the smaller base $[BC]$ are 2 cm long, and $[BD] \perp [AB]$. Calculate the area of the trapezoid.

19. In a parallelogram $ABCD$ the angle $BAD$ is equal to $\pi/3$, and the side $[AB]$ is 3 cm. The bisector of the angle $A$ cuts the side $[BC]$ at a point $E$. Find the area of the triangle $ABE$.

20. A parallelogram with the perimeter of 44 cm is divided by the diagonals into four triangles. The difference between the perimeters of two adjacent triangles is 6 cm. Find the lengths of the sides of the parallelogram.

21. Given parallelogram whose acute angle is 60°. Find the ratio of the lengths of the sides of the parallelogram if the squares of the lengths of the diagonals are related as 1 : 3.

22. The lengths of the bases of a trapezoid are 5 cm and 15 cm, and those of the diagonals are 12 cm and 16 cm. Find the area of the trapezoid.

23. The bases of a trapezoid are $a$ and $b$ long. Find the length of the segment of the straight line connecting the midpoints of its diagonals.

24. Find the area of an isosceles trapezoid, being given the length of its diagonal $l$ and the angle $\alpha$ between that diagonal and the larger base.

25. In an isosceles trapezoid $ABCD$ ($[AD] \parallel [BC]$) the distance from the vertex $A$ to the straight line $(CD)$ is equal to the length of a leg. Find the angles of the trapezoid if $|AD| : |BC| = 5 : 1$.

26*. The nonparallel sides of a trapezoid are extended to their intersection point and a straight line, which is parallel to the bases of the trapezoid, is drawn through that point. Find the length of the segment of that straight line which is bounded by the extensions of the diagonals if the bases of the trapezoid are $a$ and $b$ long.

27. In a rectangular trapezoid the ratio of the lengths of the bases is 4 and that of the lengths of the diagonals is 2. Find the acute angle of the trapezoid.

28. Given an isosceles trapezoid $ABCD$. It is known that $|AD| = 10$ cm, $|BC| = 2$ cm, and $|AB| = |CD| = 5$ cm. The bisector of the angle $BAD$ cuts the ray $[BC)$ at a point $K$. Find the length of the bisector of the angle $ABK$ in the triangle $ABK$.

29. In a right triangle $ABC$ the bisector $[AD]$ of the acute angle $A$ is divided by the centre $O$ of the inscribed circle in the ratio $|AO| : |OD|$
$= (\sqrt{3} + 1) : (\sqrt{3} - 1)$. Find the acute angles of the triangle.

30. In a triangle $ABC$ the side $[AB]$ is 3 m and the altitude $[CD]$ drawn to the side $[AB]$ is $\sqrt{3}$ m. The foot $D$ of the altitude $[CD]$ lies on the side $[AB]$, and the length of the segment $[AD]$ is equal to that of the side $[BC]$. Find the length of the side $[AC]$.

31. The length of the diagonal $[BD]$ of a trapezoid $ABCD$ is equal to $m$, and the length of the leg $[AD]$ is equal to $n$. Find the length of the base $[CD]$ if it is known that the lengths of the bases, the diagonal, and the lateral side drawn from the vertex $C$ are equal.

32. In a convex quadrangle $MNLQ$ the vertex angles $N$ and $L$ are right angles and the vertex angle $M$ is equal to arctan $(2/3)$. Find the length of the diagonal $[NQ]$ if it is known that the length of the side $[LQ]$ is half that of the side $[MN]$ and is longer than the side $[LN]$ by 21 m.

33. The area of a triangle $ABC$ is equal to $15\sqrt{3}$ m$^2$. The angle $BAC$ is $120°$. The angle $ABC$ is larger than the angle $ACB$. The distance from the vertex $A$ to the centre of the circle inscribed in the triangle $ABC$ is 2 m. Find the length of the median of the triangle $ABC$ drawn from the vertex $B$.

**34.** Given a square *ABCD* with a side of length unity. A point *K* belongs to the side [*CD*], and | *CK* |/| *KD* | = 1/2. Find the distance from the vertex *C* to the straight line (*AK*).

**35.** The angles at one of the bases of a trapezoid are 20° and 70°, and the length of the segment connecting the midpoints of the bases is equal to 2. Find the lengths of the bases of the trapezoid if the length of the midline of the trapezoid is 4.

**36.** One of the bases of a trapezoid serves as a diameter of a circle of radius *R* and the other base intercepts an arc of α radians on the circle (0 < α < π). Find the area of the trapezoid.

**37.** The base angle of an isosceles triangle is equal to φ. Find the ratio of the length to the radius of the circle inscribed in the given triangle to that of the radius of the circumscribed circle.

**38.** In an isosceles triangle the length of the base is 24 cm and that of a lateral side is 15 cm. Find the radii of the inscribed and the circumscribed circle.

**39.** In a circle of radius equal to 12 cm the length of the chord [*AB*] is 6 cm and that of the chord [ *BC* ] is 4 cm. Find the length of the chord connecting the ends of the arc *AC*.

**40.** Semicircles are constructed on the sides [*AB*] and [*AC*] of an angle *BAC*, equal to 2π / 3, as diameters. A circle with the maximum radius is inscribed in the common part of the two resulting semicircles. Find the radius of the circle if | *AB* | = 4 and | *AC* | = 2.

**41.** Given a triangle *ABC* in which | *AC* | = *b* and ∠ *ABC* = α. Find the radius of the circle which passes through the centre of the circle inscribed in the triangle *ABC* and the vertices *A* and *C*.

**42.** A chord of length *r*/2 is drawn in a circle of radius *r*. A tangent to the circle is drawn through one end of the chord, and a secant, which is parallel to the tangent, is drawn through the other end. Find the distance between the tangent and the secant.

**43.** Tangents are drawn through the ends of the arc of a circle equal to 120°, and a circle is inscribed in the figure bounded by the tangents and the given arc. Calculate the length of that circle if the radius of the initial circle is equal to *R*.

**44.** There is an external tangency at a point *C* between two circles of radii *R* and *r*. A common external tangent [ *AB* ], where *A* and *B* are the points of tangency, is drawn to them. Calculate the lengths of the sides of the triangle *ARC*.

**45.** There is an external tangency between two circles of radii *R* and *r* (*R* > *r*). Find the radii of the circles touching the two given circles and their common external tangent.

**46.** There is an external tangency at a point *A* between two circles of radii *R* and *r*. A point *B*, which is diametrically opposite to the point *A*, is taken on the circle of radius *r* and a tangent *l* is constructed at that point. Find the radius of the circle which touches the two given circles and the straight line *l*.

**47.** Points $O_1$ and $O_2$ are the centres of the circles $K_1$ and $K_2$ which are externally tangent. The radii of the circles are $r_1$ and $r_2$ respectively. A circle $K_3$ is constructed on the segment $[O_1 \ O_2]$ as a diameter. Calculate the radius of the circle which is externally tangent to the circles $K_1$ and $K_2$ and internally tangent to the circle $K_3$.

**48.** The angle $QRP$ in a triangle $PQR$ is equal to $\pi/3$. Find the distance between the points of tangency of a circle of radius 2 inscribed in the triangle and a circle of radius 3 which touches the straight lines $(PQ)$ and $(PR)$ with the side $[QR]$.

**49.** Given a triangle $ABC$ whose sides are $|\ AB\ | = 15$, $|\ BC\ | = 12$, and $|\ AC\ | = 18$. The centre $O$ of the inscribed circle divides the bisector of the angle $C$ into two parts $[CO]$ and $[OD]$. How many times is the length of $[CO]$ larger than that of $[OD]$?

**50.** A circle of radius $R$ is inscribed in an angle of $\alpha$ radians. A tangent, which is perpendicular to the bisector of the given angle, is drawn between the vertex of the angle and the centre of the circle. Find the area of the cut-off triangle.

**51.** Find the area of a right triangle if it is known that the radius of the circle inscribed in the triangle is $r$ and that of the circumscribed circle is $R$.

**52.** A circle touches the larger leg of a triangle, passes through the vertex of the opposite acute angle, and has a centre lying on the hypotenuse. Find its radius if the lengths of the legs of the triangle are 3 and 4.

**53.** Given a triangle with sides $a$, $b$, and $c$ long and a semicircle inscribed in it, with its diameter lying on the side $c$. Find the radius of the semicircle.

**54.** The base angle of an isosceles triangle is equal to $\pi/6$. The length of the altitude drawn to the base exceeds the radius of an inscribed circle by 2. Find the length of the base of the triangle.

**55.** A right-angled trapezoid whose smaller side is equal to $3r/2$ is circumscribed about a circle of radius $r$. Calculate the area of the trapezoid.

**56.** The area of an isosceles trapezoid circumscribed about a circle is $S$. Find the length of the midline of the trapezoid if the acute angle at its base is $\alpha$.

**57.** A trapezoid one of whose non parallel sides is congruent to the smaller base and the angular measure of the arc subtended by that base is $\alpha$ is inscribed in a circle of radius $R$. Find the area of the trapezoid.

**58.** A circle is inscribed in a regular triangle with a side of 10 cm. A regular triangle is again inscribed in that circle and a circle is again inscribed in that triangle and so on. Find the sum of the areas of all the circles resulting from the successive inscriptions.

**59.** The angles of a triangle are related as $2 : 3 : 7$. The length of the smallest side is $a$. Find the radius of the circle circumscribed about the triangle.

**60.** A circle of radius $r$ is inscribed in an isosceles triangle. The altitude drawn to the base is divided by the circle in the ratio 1 : 2. Find the area of the triangle.

**61.** In a right triangle $ABC$ the angle $A$ is a right angle, the angle $B$ is 30°, and the radius of the circumscribed circle is equal to $\sqrt{3}$. Find the distance from the vertex $C$ to the point of tangency of the inscribed circle and the leg $[AB]$.

**62.** A circle inscribed in a square $ABCD$ with side $a$ touches the side $[CD]$ at a point $E$. Find the length of the chord connecting the points at which the circle meets the straight line $(AE)$.

**63.** A circle is inscribed in a right triangle whose perimeter is equal to 36 cm. The point of tangency with the circle divides the hypotenuse in the ratio 2 : 3. Find the lengths of the sides of the triangle.

**64.** A circle is inscribed in a right triangle. The point of tangency with the circle divides one of the legs into segments 6 cm and 10 cm in length, reckoning from the vertex of the right angle. Find the area of the triangle.

**65.** A semicircle is circumscribed on the larger leg as a diameter. Find its length if the smaller leg is 30 cm and the chord which connects the vertex of the right angle with the point of intersection of the hypotenuse and the semicircle is equal to 24 cm.

**66.** In a parallelogram $ABCD$ the diagonal $[AC]$ is perpendicular to the side $[AB]$. A circle touches the side $[BC]$ of the parallelogram $ABCD$ at a point $P$ and touches the straight line, which passes through the vertices $A$ and $B$ of the parallelogram, at a point $A$. A perpendicular $[PQ]$ is drawn to the side $[AB]$ through the point $P$ (point $Q$ is the foot of the perpendicular). Find the angle $ABC$ if it is known that the area of the parallelogram $ABCD$ is equal to 1/2 and that of the pentagon $QPCDA$ is $S$.

**67.** A circle is circumscribed about a trapezoid whose altitude is $H$ in length. The bases of the trapezoid can be seen from the centre of the circle at the angles $\alpha$ and $\beta$. Find the radius of the circle and the area of the trapezoid.

**68.** A circle of radius 13 cm touches two adjacent sides of a square the length of whose side is 18 cm. Into what two segments does the circle divide each of the other two sides of the square?

**69.** Given a triangle $ABC$ with $\angle BAC = \alpha$, $\angle BCA = \beta$, and $|AC| = b$. A point $D$ taken on the side $[BC]$ is such that $|BD| = 3|DC|$. A circle drawn through the points $B$ and $D$ touches the side $[AC]$ or its extension beyond the point $A$. Find the radius of the circle.

**70.** Points $C$ and $D$ are taken on the side of an angle with vertex $A$ ($C$ is between $A$ and $D$). These points are such that $| AC | = 2 | CD |$. A circle is drawn through the points $C$ and $D$, which touches the other side of the angle at a point $B$. A point $E$ is taken between $A$ and $B$. It is known that $\angle DAE = \alpha$, $\angle DEA = \beta$ and $| AE | = k$. Find the radius of the circle.

**71.** The lengths of the nonparallel sides of a trapezoid are equal to 6 and 10. It is known that a circle can be inscribed in the trapezoid. The midline divides the trapezoid into two parts whose areas are related as 5 : 11. Find the lengths of the bases of the trapezoid.

**72.** The length of the midline of an isosceles trapezoid is equal to 5. It is known that a circle can be inscribed in the trapezoid. The midline divides the trapezoid into two parts whose areas are related as 7 : 13. Find the length of the altitude of the trapezoid.

**73.** A circle of radius 3 inscribed in a triangle $ABC$ touches the side $[BC]$ at a point $D$. A circle of radius 4 touches the extensions of the sides $[A\ B]$ and $[AC]$, and also touches the side $[BC]$ at a point $E$. Find $| ED |$ if the angle $BCA$ is equal to $2\pi/3$.

**74.** A triangle is defined by the coordinates of its vertices $A$ (3, –2, 1), $B$ (3, 1, 5), and $C$ (4, 0, 3). Calculate the length of the median $| BB_1 |$ and the magnitude of the angle $B$.

**75.** Find the area of an isosceles trapezoid in which the bases are 10 cm and 26 cm long and the diagonals are perpendicular to the nonparallel sides.

**76.** Calculate the area of a rectangular trapezoid with the bases 7 cm and 3 cm long and the acute angle equal to 60°.

**77.** Given a trapezoid $MNPQ$ with bases $[MQ]$ and $[NP]$. The straight line which is parallel to the bases cuts the leg $[MN]$ at a point $A$ and the side $[PQ]$ at a point $B$. $S_{ANPQ} : S_{MABQ} = 2 : 7$. Find $| AB |$ if $| NP | = 4$ and $| MQ | = 6$.

**78.** A median is drawn to a lateral side, which is 4 cm long, of an isosceles triangle. Find the length of the base of the triangle if the median is 3 cm long.

**79.** The legs of a right triangle are 12 cm and 5 cm long. Find the distance between the centres of the inscribed and the circumscribed circle.

**80.** A square inscribed in a right triangle with legs $a$ and $b$ has a common right angle with the triangle. Find the perimeter of the square.

**81.** The magnitudes of the angles $A$, $B$, and $C$ in a triangle triangle $ABC$ form an arithmetic progression. The smallest side is a quarter of the largest side. Find the tangent of the smallest angle.

**82.** The lengths of the sides of a triangle are proportional to the numbers 5, 12, and 13. The largest side of the triangle exceeds the smallest side by 1.6 m. Find the perimeter and the area of the triangle.

**83.** In an isosceles triangle the sine of the base angle is three times as large as the cosine of the vertex angle. Find the sine of the base angle.

**84.** Find the lengths of the sides of a right triangle if $R = 15$ cm and $r = 6$ cm, where $R$ and $r$ are the radii of the circumscribed and the inscribed circle respectively.

**85.** Given two sides $b$ and $c$ of a triangle and its area $S = 2bc/5$. Find the third side of the triangle.

**86.** Find the angle between the medians of the legs of an isosceles right triangle.

**87.** In a triangle $ABC$ a straight line is drawn from the point $E$ of the side $[BC]$ parallel to the altitude $[BD]$ and cuts the side $[AC]$ at a point $F$. A line segment $[EF]$ divides the triangle $ABC$ into two figures of the same area. Find the length of $|EF|$ if $|BD| = 6$ cm and $|AD| : |DC| = 2 : 7$.

**88.** A semicircle is inscribed in a right triangle so that its diameter lies on the hypotenuse and the centre divides the hypotenuse into segments 15 cm and 20 cm long. Find the length of the arc of the semicircle included between its points of tangency with the legs.

**89.** Given a triangle whose sides are 10 cm, 12 cm, and 18 cm long. A circle is drawn which touches the two smaller sides, whose centre is on the larger side. Find its radius.

**90.** Two tangents 12 cm long are drawn to a circle from the same point, and the distance between the points of tangency is 14.4 cm. Find the radius of the circle.

**91.** Two tangents drawn to a circle with its centre at $N$ from a point $A$ touch that circle at points $B$ and $M$. The chord $[BM]$ cuts the segment $[NA]$ at a point $K$. The length of the segment $|NK|$ is $1\dfrac{3}{4}$ times as small as the length of the segment $|KA|$; $|AB| = 4$ cm. Find the area of the triangle $BAK$.

**92.** An isosceles trapezoid is circumscribed about a circle, the midline of the trapezoid being 5 cm long. Find the perimeter and the length of the nonparallel sides of the trapezoid.

**93.** The lengths of the bases $|MF|$ and $|PQ|$ of a trapezoid $MPQF$ are 24 cm and 4 cm respectively. The altitude of the trapezoid is 5 cm in length. A point $N$ divides the side $[MP]$ into segments $[MN]$ and $[NP]$. The length of the segment $|MN|$ is three times as large as that of the segment $|NP|$. Find the area of the triangle $NQF$.

**94.** A circle of radius 2 cm long is inscribed in a rhombus which is divided by its diagonal into two equilateral triangles. Find the side of the rhombus.

**95.** Given a right triangle $ABC$. A circle is constructed on its altitude $[CK]$ as a diameter which cuts the legs $[CA]$ and $[CB]$ at points $M$

and $N$ respectively $| CM | = 12$ cm and $| CN | = 18$ cm. Find $| CA |$ and $| CB |$.

96. An isosceles triangle with an angle of 120° is circumscribed about a circle of radius $R$. Find the sides of the triangle.

97. In an isosceles trapezoid $ABCD$ the larger base $[AD]$ is 12 cm and $|AB| = 6$ cm. Find the distance between the point $O$ of intersection of the diagonals and the point $K$ of intersection of the extensions of the nonparallel sides if the altitude of the trapezoid is 1 cm.

98. Find the diagonal and the nonparallel sides of an isosceles trapezoid with bases 20 cm and 12 cm long if the centre of a circumscribed circle is known to lie on the larger base of the trapezoid.

99. There is an internal tangency between a circle of radius $r$ and two circles of radii $R_1$ and $R_2$, the centres of the three circles lying on the same straight line. Find the radius of the circle which touches the three given circles.

100. A point $P$ chosen on the leg $[AC]$ of an isosceles right triangle $ABC$ is such that the semicircle constructed on the segment $[PC]$ as a diameter touches the hypotenuse $[AB]$. In what ratio does the semicircle divide the segment $[PB]$?

101. For what value of the length of the altitude does a right angled trapezoid with an acute angle of 45° and perimeter equal to 4 cm have the greatest area?

102. The sum of the lengths of two sides of a triangle is equal to $a$ and the angle between them is equal to 30°. What must be the lengths of the sides of the triangle for its area to be the greatest?

103. The sum of the lengths of the diagonals of a parallelogram is 8 cm. Find the minimum sum of the squares of the lengths of all the sides of the parallelogram.

104. A rectangle is inscribed in an isosceles triangle whose base is 20 cm long and the altitude is 8 cm long, one of the sides of the rectangle lying on the base of the triangle. How long must be the altitude of the rectangle for the rectangle to have the greatest area?

105. A rectangle of the greatest area is inscribed in an isosceles triangle with a vertex angle of 120° and a base 8 cm long, two vertices of the rectangle lying on the base of the triangle. Find the area of the rectangle.

106. A rectangle of the greatest area is inscribed in a triangle whose base is $a$ long and the altitude is $h$ long (the base of the rectangle lies on the base of the triangle). Find the lengths of the sides of the rectangle.

107. Two sides of a parallelogram lie on the sides of a given triangle, and one of its vertices belongs to the third side. Under what conditions is the area of the parallelogram the greatest?

108. A leg of an isosceles trapezoid is congruent to its smaller base whose length is $a$. How long must be the larger base of the trapezoid for its area to be the greatest?

109. A point $A$ is given on a circle of radius $R$. At what distance from the point $A$ must the chord $[BC]$ be drawn, parallel to the tangent to the circle at the point $A$, for the area of the triangle $ABC$ to be the greatest?

110. The nonparallel sides of a trapezoid are perpendicular. What can be the maximum value of the area of the triangle, formed by the diagonals and the midline of the trapezoid, if the lengths of the bases of the trapezoid are known to be $a$ and $b$?

111. Among all the rectangles inscribed in a circle of radius $R$ find that which has the greatest area.

112. Among all the rectangles with the area equal to 9 dm$^2$ find that whose perimeter is the smallest.

113. Given a right triangle with a hypotenuse equal to 24 cm and an angle of 60°, and a rectangle inscribed in it, whose base lies on the hypotenuse. How long must be the sides of the rectangle for its area to be the greatest?

114. Find the lengths of the sides of a rectangle of the greatest area inscribed in a right-angled trapezoid with the lengths of the bases equal to 24 cm and 8 cm and the altitude equal to 12 cm.

115. Find the lengths of the sides of a rectangle of the greatest area inscribed in a right triangle with sides 18 cm, 24 cm and 30 cm long and having an angle in common with it.

116. A parallelogram of the greatest area is inscribed in an isosceles triangle with sides 15 cm, 15 cm, and 18 cm long so that they have a common base angle. Find the lengths of the sides of the parallelogram.

117. In what circle can a rectangle of the greatest area with perimeter equal to 56 cm be inscribed?

118. The hypotenuse of a right triangle is equal to $c$. What must be the legs of the triangle for its area to be the greatest?

## 3.5. Solid Geometry. Problems on Proof

1. How many planes are there which are equidistant from four given points not lying in the same plane?

2. Prove that if three straight lines in space possess the property that any two of them meet, then either they pass through a common point or they all lie in the same plane.

3. Points $A$, $B$, $C$, and $D$ are given in space, with $[AB] \perp [CD]$ and $[AC] \perp [BD]$. Prove that $[AD] \perp [BC]$.

4. Two segments $[AB]$ and $[CD]$, not lying in the same plane, are considered in space. Assume that $[MN]$ is a segment connecting their midpoints. Prove that $|AC| + |BD| > 2|MN|$.

5. An oblique line makes equal angles with three pairwise nonparallel straight lines lying in the same plane. Prove that the oblique line is perpendicular to the plane.

6. Segments [*AB*] and [*CD*] lie on two parallel planes. The terminal points of the segments are the vertices of a triangular pyramid. Prove that the volume of the pyramid does not vary when the segments are displaced in the planes parallel to themselves.

7. Prove that the straight line which cuts two faces of a dihedral angle makes equal angles with them if and only if the intersection points are equidistant from the edge.

8. Prove that every convex tetrahedral angle can be cut by a plane so that a parallelogram results in the section.

9. Can a trihedral angle be always cut by a plane so that a regular triangle results in the section?

10. Prove that if all the dihedral angles of a convex trihedral angle are acute, then all the plane angles are also acute.

11. How many planes of symmetry can a triangular pyramid have?

12. Prove that two triangular pyramids are equal or symmetric if their respective edges are equal.

13. Prove that the following four conditions are equivalent :
    (1) the lateral edges of a pyramid are equal;
    (2) the lateral edges make the same angle with the plane of the base of a pyramid;
    (3) the lateral edges make the same angle with the altitude of a pyramid;
    (4) a circle can be circumscribed about the base of a pyramid, and the altitude of the pyramid passes through the centre of the circle.

14. Prove that the following three conditions are equivalent :
    (1) the altitudes of the lateral faces of a pyramid are equal;
    (2) the altitude of a pyramid makes the same angle with the lateral faces;
    (3) the lateral faces of a pyramid make the same angle with the plane of the base (note that in that case the dihedral angles at the base of the pyramid may prove to be different).

15. Prove that the dihedral angles at the base of a pyramid are equal if and only if a circle can be inscribed in the base of the pyramid, and the altitude of the pyramid passes through the centre of the circle.

16. Prove that if all the dihedral angles of a certain triangular pyramid are equal, then all the edges of the pyramid are also equal.

17. Prove that if in a triangular pyramid the sum of the lengths of the opposite edges is the same for any pair of such edges, then the vertices of the pyramid are the centres of four spheres which are pairwise tangent.

18. Prove that in a triangular pyramid the altitude passes through the point of intersection of the altitudes of the triangle lying at the base if and only if the opposite edges of the pyramid are perpendicular.

218 | Part-3 • Geometry and Vector Algebra

**19.** All the edges of one pyramid are respectively smaller than those of another pyramid. Can we assert that the volume of the first pyramid is smaller than that of the second?

**20.** Given a regular triangular pyramid. A perpendicular is drawn from an arbitrary point $P$ of its base to the plane of the base. Prove that the sum of the lengths of the segments from the point $P$ to the points of intersection of the perpendicular and the planes of the faces does not depend on the choice of the point $P$ on the base.

**21.** What regular polygons can result when a cube is cut by a plane?

**22.** Prove that if all the diagonals of a parallelepiped are equal in length, then it is rectangular.

**23.** Prove that if all the faces of a parallelepiped are equal parallelograms, then they are rhombi.

**24.** Is there a polyhedron whose all faces are parallelograms and are pairwise parallel but which, however, is not a prism?

**25.** Prove that the volume of a regular truncated pyramid is
$V = \dfrac{1}{3} H (S_1 + S_2 + \sqrt{S_1 S_2})$, where $H$ is its altitude, and $S_1$ and $S_2$ are the areas of the bases.

**26.** Prove that all the tangents drawn to a sphere from the same point are equal in length.

**27.** Prove that a triangular prism can be inscribed in a sphere if and only if the prism is right-angled.

**28.** Prove that a sphere can be inscribed in every triangular pyramid and a sphere can be circumscribed about that pyramid. In that case (1) all the bisectors of the planes of the dihedral angles of the pyramid meet at one point, and that point is the centre of the inscribed sphere (2) all the planes drawn through the midpoints of the edges of the given pyramid are perpendicular to those edges, meet at one point, and that point is the centre of the circumscribed sphere.

**29.** Prove that if the opposite edges of a tetrahedron are pairwise equal, then the spheres inscribed in and circumscribed about the tetrahedron are concentric.

**30.** Assume that $ABCD$ is a triangular pyramid, $S_1, S_2, S_3$, and $S_4$ are the areas of its four faces, and $r$ is the radius of the sphere inscribed in the pyramid. Prove that the volume $V$ of the pyramid can be calculated by the formula $V = \dfrac{1}{3} r (S_1 + S_2 + S_3 + S_4)$.

**31.** A sphere is said to be escribed in a triangular pyramid if it touches one of the faces of the pyramid and the planes of all the other faces. Prove that every triangular pyramid has four escribed spheres.

**32\*.** Assume that $r$ is the radius of the inscribed sphere and $R_1, R_2, R_3$, and $R_4$ are the radii of the spheres escribed in the triangular pyramid.

Prove that

$$\frac{2}{r} = \frac{1}{R_1} + \frac{1}{R_2} + \frac{1}{R_3} + \frac{1}{R_4}.$$

33. Prove that to circumscribe a sphere about a pyramid, it is necessary and sufficient that a circle can be circumscribed about the base of the pyramid.

34. A sphere with centre at a point $O$ is inscribed in a trihedral angle with vertex $S$. Prove that the plane which passes through three points of tangency is perpendicular to the straight line $(OS)$.

35. Prove that in a rectangular parallelepiped $ABCD\,A'\,B'\,C'\,D'$ the square of the sectional area $A'\,BD$ is eight times smaller than the sum of the squares of the areas of the faces.

36. Prove that if all the edges of a tetrahedron touch the same sphere, then the sums of the lengths of all the pairs of the opposite edges are the same.

37. What is the greatest lateral area that a rectangular parallelepiped can possess if its diagonal is $a$ long? Prove that a cube has the greatest lateral area.

38. All the dihedral angles at the base of a pyramid are $\alpha$ in magnitude. Prove that the lateral area and the base area of the pyramid are related as $S_{\text{base}} = S_{\text{lat}}\cos\alpha$.

39. Two planes $\alpha$ and $\beta$ intersect in space. A point $A$ is given on the line of their intersection. Prove that of all the straight lines, which lie in the plane $\alpha$ and pass through the point $A$, the straight line which is perpendicular to the line of intersection of the planes $\alpha$ and $\beta$ makes the greatest angle with the plane $\beta$. What is that angle equal to?

40. Given a triangular pyramid $SABC$ and the plane angles at its vertex $S : \angle BSC = 90°$ and $\angle ASC = \angle ASB = 60°$. The vertices $A$ and $S$ and the midpoints of the edges $[SB]$, $[SC]$, $[AB]$, and $[AC]$ lie on the surface of a sphere. Prove that the edge $[SA]$ is a diameter of the sphere.

41. A sphere touches three sides of the base of a triangular pyramid at their midpoints and cuts the lateral edges at their midpoints. Prove that the pyramid is regular.

42. A sphere touches all the lateral faces of a triangular pyramid at the centres of the circles circumscribed about them. The edge angles of the pyramid are equal. Prove that the pyramid is regular.

43. Prove that if a sphere is inscribed in a prism (not necessarily a right prism), then (1) the altitude of the prism is equal to the diameter of the sphere ; (2) the points of tangency of the sphere and the lateral faces lie on the section of the prism by the plane which passes through the centre of the sphere at right angles to the lateral edges.

**44.** Prove that if a right circular cylinder can be inscribed in a prism, then it is a right prism, the length of its lateral edge is equal to that of the generatrix of the cylinder, and a circle can be inscribed in the base of the prism.

**45.** If a prism is inscribed in a right circular cylinder, then it is a right prism, its altitude is equal to the generatrix of the cylinder, and the base of the prism is an inscribed polygon. Prove these statements.

**46.** A sphere is inscribed in a truncated cone. Prove that the surface area of the sphere is smaller than the lateral area of the cone.

**47.** A tetragonal truncated pyramid is circumscribed about a sphere. Prove that the volumes of the pyramid and the sphere are related as their total surfaces.

## 3.6. Solid Geometry. Calculation Problems

Here are the formulas which will be of help in calculating the surface areas and the volumes of space figures.

$$S_1 = S \cos \alpha, \tag{1}$$

where $S$ is the area of a given plane figure, $S_1$ is the area of its orthogonal projection onto another plane, and $\alpha$ is the angle between the planes.

$$S = Pl, \tag{2}$$

where $S$ is the lateral area of a prism, $P$ is the length of the perimeter of the orthogonal section, and $l$ is the length of the generatrix of the prism.

$$\frac{S}{S_1} = \frac{a^2}{a_1^2} = \frac{h^2}{h_1^2}, \tag{3}$$

where $S$ and $S_1$ are the areas of the parallel sections of a pyramid, $a$ and $a_1$ are the lengths of the corresponding elements of the section, $h$ and $h_1$ are the distances of the sections from the vertex of the pyramid or the distances of some corresponding elements of the sections from the vertex of the pyramid.

$$S = \frac{Ph}{2}, \tag{4}$$

where $S$ is the lateral area of a regular pyramid, $P$ is the length of the perimeter of the base, and $h$ is the apothem of a lateral face.

$$S = 2\pi R h, \quad S_1 = 2\pi R (R + h), \tag{5}$$

where $S$ is the lateral area of a right cylinder, $S_1$ is the total surface area of the cylinder, $R$ is the length of the radius of the circle of the base of the cylinder, and $h$ is the length of the altitude of the cylinder.

$$S = \pi Rl, \quad S_1 = \pi R (R + l), \tag{6}$$

where $S$ is the lateral area of a right cone, $S_1$ is the total surface area of the cone, $R$ is the length of the radius of the circle of the base of the cone, and $l$ is the length of the generatrix of the cone.

$$S = \pi (R + r) l, \tag{7}$$

where $S$ is the lateral area of a right truncated cone, $R$ and $r$ are the lengths of the radii of the bases of the truncated cone, and $l$ is the length of the generatrix of the cone.

$$S = 2\pi ra = 2\pi Rh \quad \text{( Fig. 3.11), (8)}$$

where S is the area of the surface of revolution of the segment [AB] of length $a$ (Fig. 3.11) about the axis (ab) which does not cut the .segment, $r$ is the distance between the midpoint of the segment and the axis of revolution, $h$ is

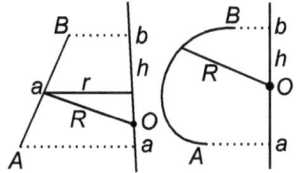

Fig. 3.11  Fig. 3.12

the length of the projection [ab] of the segment [AB] onto the axis of revolution, $R$ is the length of the radius of the circle whose centre is on the axis of revolution and which touches the segment [AB] at its midpoint, and $a$ is the length of the segment [AB].

$$S = 2\pi Rh \text{ (Fig. 3.12), } \tag{9}$$

where $S$ is the area of the surface of revolution of the arc $\hat{AB}$ of the circle with a radius of length $R$ about the axis (ab) on which the centre of the circle lies (the point A or B or both can lie on the axis of revolution), and $h$ is the length of the projection of the arc $\hat{AB}$ onto the axis of revolution.

$$S = 4\pi R^2, \tag{10}$$

where $S$ is the surface area of a sphere and $R$ is the length of the radius of the sphere.

$$V = abc \tag{11}$$

where $V$ is the volume of a rectangular parallelepiped and $a$, $b$, $c$ are the lengths of its sides.

$$V = Sh \tag{12}$$

where $V$ is the volume of a prism, $S$ is the area of the polygon which is the base of the prism, and $h$ is the length of the altitude of the prism.

$$V = Sl, \tag{13}$$

where $V$ is the volume of an oblique prism, $S$ is the area of the section perpendicular to the generatrix of the prism, and $l$ is the length of a lateral edge of the prism.

$$V = \frac{1}{3} Sh, \qquad (14)$$

where $V$ is the volume of a pyramid, $S$ is the area of the polygon which is the base of the pyramid, and $h$ is the length of the altitude of the pyramid.

$$V = \frac{1}{3} (S + s + \sqrt{Ss})h, \qquad (15)$$

where $V$ is the volume of a truncated pyramid, $S$ and $s$ are the areas of its bases, and $h$ is the length of the altitude of the truncated pyramid.

$$V = \pi R^2 h, \qquad (16)$$

where $V$ is the volume of a circular cylinder, $R$ is the length of the radius of the base of the cylinder, and $h$ is the length of the altitude of the cylinder.

$$V = \frac{1}{3} \pi R^2 h, \qquad (17)$$

where $V$ is the volume of a circular cone, $R$ is the length of the radius of the base of the cone, and $h$ is the length of the altitude of the cone.

$$V = \frac{1}{3} \pi (R^2 + r^2 + Rr)h, \qquad (18)$$

where $V$ is the volume of a circular truncated cone, $R$ and $r$ are the lengths of the radii of the bases of the cone, and $h$ is the length of the altitude of the cone.

$$V = \frac{2}{3} \pi R^2 h \text{ (Fig. 3.13),} \qquad (19)$$

where $V$ is the volume of a spherical sector and centre on the axis of revolution $(ab)$, $R$ is the length of the radius of the spherical sector, and $h$ is the length of the altitude of the spherical zone which serves as the base of the spherical sector (the point $A$ or $B$ or both can lie on the axis of revolution $(ab)$). Figure 3.13 shows only the section of the spherical sector by an axial half-plane.

$$V = \frac{4}{3} \pi R^3, \qquad (20)$$

where $V$ is the volume of a sphere and $R$ is the length of the radius of the sphere.

$$V = \frac{1}{6} \pi l^2 h \text{ (Fig. 3.14),} \qquad (21)$$

where $V$ is the volume of a ring with centre on the axis of revolution $(ab)$, $l$ is the length of the chord $[AB]$ of the ring, and $h$ is the length of the projection $[ab]$

of the chord [AB] onto the axis of revolution (the point A or B or both can lie on the axis of revolution (ab)). Figure 3.14 shows only the section of the ring by an axial half-plane.

$$V = \frac{1}{6}\pi h^3 + \frac{1}{2}\pi\,(r_1^2 + r_2^2)\,h \text{ (Fig. 3.15),}\tag{22}$$

where $V$ is the volume of a spherical layer with centre on the axis of revolution (ab), $r_1$ and $r_2$ are the distances from the points A and B to the axis of revolution (ab) ($r_1$ and $r_2$ are the lengths of the radii of the circles bounding the layer), and $h$ is the length of the altitude of the spherical zone which serves as

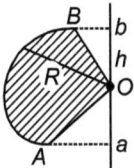
Fig. 3.13

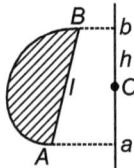
Fig. 3.14

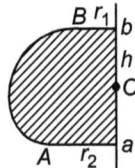
Fig. 3.15

the base of the spherical layer (the point A or B or both can lie on the axis of revolution). Figure 3.15 shows only the section of the spherical layer by an axial half-plane.

$$V = \frac{1}{6}h\,(S + 4S_1 + s),\tag{23}$$

where $V$ is (1) the volume of any polyhedron whose bases are various irregular polygons lying in parallel planes, and the lateral surface is formed by triangles or trapezoids whose vertices coincide with those of the polygons serving as the bases ; (2) the volume of a cylinder ; (3) the volume of a cone; (4) the volume of a spherical layer; $S$ and $s$ are the areas of the figures lying in parallel planes (bases), $S_1$ is the area of the section of a body by a plane which is parallel to the bases and is equidistant from them ; and $h$ is the length of the altitude (the distance between the bases).

1. Find the angle between two skew straight lines $L_1$ and $L_2$ if the distance between the points $A \in L_1$ and $B \in L_2$, which are equidistant from the feet $C \in L_1$ and $D \in L_2$ of the common perpendicular to those lines, is equal to $2l$, and $|DC| = |AC| = |BD| = l$.

2. Given two skew straight lines $L_1$ and $L_2$ which make an angle $\alpha$ with each other. The distance between the points $A \in L_1$ and $B \in L_2$, which are equidistant from the feet $C \in L_1$ and $D \in L_2$ of the common

perpendicular to those lines, is equal to $m$. Find the distance between the straight lines if $|AC| = |BD| = l$.

3. The length of the altitude of a right triangle $ABC$, drawn to the hypotenuse $[AB]$, is equal to 9.6. A perpendicular $[CM]$ is drawn from the vertex $C$ of the right angle to the plane of the triangle $ABC$ and $|CM| = 28$. Find the distance from the point $M$ to the hypotenuse $[AB]$.

4. Points $A$, $B$, and $C$, belonging to a circle, divide the length of the circle in the ratio $1 : 2 : 2$ ($\widehat{AC} = \widehat{BC}$). Find the distance from the centre $O$ of the circle to a plane $\gamma$ if it is known that the plane $\gamma$ is at the distance $d$ from the points $A$ and $B$ and at the distance $b$ from the point $C$.

5*. Given a rhombus $ABCD$ and a plane $\beta$. Find the distance from the vertex $D$ to the plane $\beta$ which passes through the vertex $A$ if the distances from the points $B$ and $C$ to the plane $\beta$ are equal to $b$ and $c$ respectively.

6. The planes drawn through each vertex of a unit cube are perpendicular to the same diagonal. Into what parts is the diagonal divided by the planes?

7. The centre of the upper base of a cube is connected with the midpoints of the sides of the lower base. Determine the lateral surface of the pyramid obtained if the edge of the cube is $a$ long.

8. The diagonals of the lateral faces of a rectangular parallelepiped make angles $\alpha$ and $\beta$ with the plane of the base. Find the angle between the diagonal of the parallelepiped and the plane of the base.

9. A plane cuts a rectangular parallelepiped with a square base along a rhombus with an acute angle $\alpha$. At what angle does the plane cut the lateral edges of the parallelepiped?

10. In a rectangular parallelepiped $ABCD$ $A_1 B_1 C_1 D_1$ the diagonals $[AC]$ and $[BD]$ of the base meet at a point $M$ and $\angle AMB = \alpha$. Determine the lateral area of the parallelepiped if $|B_1 M| = b$ and $\angle BMB_1 = \beta$.

11. The bases of a parallelepiped are squares with a side $b$ and all the lateral faces are rhombi. One of the vertices of the upper base is equidistant from all the vertices of the lower base. Find the volume of the parallelepiped.

12. In a parallelepiped $ABCD$ $A_1 B_1 C_1 D_1$ the face $ABCD$ is a square with a side of 5 cm; the edge $[AA_1]$ is also 5 cm long and makes angles of 60° with the edges $[AB]$ and $[AD]$. Find the length of the diagonal $[BD_1]$.

13. The diagonal of a rectangular parallelepiped is equal to $l$ and makes an angle $\alpha$ with the plane of the base. Find the lateral area of the parallelepiped if the area of its base is equal to $S$.

14. Find the lateral area and the volume of a right parallelepiped, being given that its altitude is $h$, its diagonals make angles $\alpha$ and $\beta$ with the base, and its base is a rhombus.

15. The base of a right parallelepiped is a rhombus. The areas of the diagonal sections are equal to $S_1$ and $S_2$. Find the lateral area of the parallelepiped.

16. The base of a right prism is a rhombus with an acute angle $\alpha$. Find the volume of the prism if its larger diagonal is $l$ in length and makes an angle $\beta$ with the plane of the base.

17. A plane drawn through a side of the lower base of a regular triangular prism and the opposite vertex of the upper base makes an angle $\alpha$ with the plane of the base. The area of the section of the prism by that plane is $S$. Find the volume of the cut-off pyramid.

18. The base of a right prism is an equilateral triangle. A plane which passes through one of the sides of the lower base and the opposite vertex of the upper base makes an angle $\varphi$ with the plane of the lower base. The area of the section is $Q$. Find the volume of the prism.

19. The base of a right prism is an isosceles trapezoid with an acute angle $\alpha$. The nonparallel sides of the trapezoid and its smaller base are equal in length. Find the volume of the prism if the diagonal of the prism is $a$ and makes an angle $\beta$ with the plane of the base.

20. The base of a right prism is an isosceles trapezoid whose diagonal is equal to $a$ and the angle between the diagonal and the larger base is $\alpha$. The diagonal of the prism makes an angle $\beta$ with the base. Find the volume of the prism.

21. Find the volume of a right prism whose base is a right triangle with an acute angle $\alpha$ if a lateral edge of the prism is $l$ in length and makes an angle $\beta$ with the diagonal of the larger lateral face.

22. The base of a right prism is a right triangle with a hypotenuse of length $c$ and an acute angle of $30°$. A plane drawn through the hypotenuse of the lower base and the vertex of the right angle of the upper base makes an angle of $45°$ with the plane of the base. Find the volume of the triangular pyramid cut off from the prism by the plane.

23. The base of a right prism is an equilateral triangle. The plane which passes through one of its sides at an angle $\alpha$ to the plane of the base cuts off a triangular pyramid of volume $V$ from the prism. Find the area of the section.

24. Each edge of an oblique triangular prism is equal to 2, one of the lateral edges makes angles of $60°$ with the adjacent sides of the base. Find the total surface area of the prism.

25. The base of a right prism is a right triangle with the perimeter $2p$ and an acute angle $\alpha$. Find the lateral area of the prism if it is known that a sphere can be inscribed in it.

26. The length of the diagonal of a regular tetragonal prism is equal to $l$ and the diagonal makes an angle equal to $\alpha$ with the plane of the base. Find the volume of the prism.

27. In a regular tetragonal prism a plane is drawn through the midpoints of two adjacent sides of the base, which cuts that base at an angle $\alpha$ and also cuts three lateral edges of the prism. Find the area of the resulting section and its acute angle if a side of the base of the prism is equal to $b$.

28. Find the volume of a regular tetragonal prism if its diagonal makes an angle $\alpha$ with a lateral face and the length of a side of the base is $a$.

29. The base of a prism is a regular triangle whose side is 4 cm in length. One of the lateral faces, which is perpendicular to the plane of the base, is a rhombus whose diagonal is 6 cm long. Find the volume of the prism.

30. The base of a right prism is a rhombus with an acute angle $\alpha$ ; the angle between the smaller diagonal of the rhombus and the smaller diagonal of the prism is $\beta$. Find the volume of the prism if the smaller diagonal of the rhombus is $d$ .

31. The base of a right prism is an isosceles triangle whose base is $a$ in length and whose base angle is $\alpha$. Find the volume of the prism if its lateral area is equal to the sum of the areas of the bases.

32. The largest diagonal of a regular hexagonal prism, which is $d$ in length, makes an angle $\alpha$ with a lateral edge. Find the volume of the prism.

33. The base of a prism is an isosceles right triangle with a leg $a$ cm in length. The lateral edge, which is opposite to the hypotenuse, makes angles $\alpha$ and $\beta$ with the legs. The length of a lateral edge is $b$ cm. Find the volume of the prism.

34. The total area of a regular triangular pyramid is $S$. Being given that the angle between a lateral face and the base of the pyramid is $\alpha$, find a side of the base.

35. In a regular triangular pyramid a side of the base is $a$ and the edge angle is $\alpha$. Find the volume of the pyramid.

36. Find the angle between the apothem of a lateral face of a regular triangular pyramid and the plane of its base, being given that the difference between that angle and the angle which a lateral edge of the pyramid makes with the plane of the base is $\alpha$.

37. The length of a side of the base of a regular triangular pyramid is 10 cm and the dihedral angle at the base is 30°. Find the volume of the pyramid.

38. A lateral edge of a regular triangular pyramid is $l$ in length and makes an angle $\alpha$ with the plane of the base. Find the volume of the pyramid.

39. Find the volume of a regular triangular pyramid whose edge angle is $\alpha$ and the shortest distance between a lateral edge and the opposite side of the base is $d$ .

40. Construct a section of a regular triangular pyramid by a plane passing through the centre of the base parallel to one of the lateral faces of the

pyramid. Find the area of the section if a lateral face of the pyramid makes an angle $\alpha$ with the base and a side of the base of the pyramid is equal to $a$.

41. The base of a pyramid is a right triangle with a leg of length $a$ and the opposite angle $\alpha$. Each lateral edge makes an angle $\beta$ with the plane of the base. Find the area of the lateral face which passes through the smaller leg of the base if $\tan \alpha = 3/2$, $\tan \beta = 2/3$, and $a = 8$ cm.

42. The base of a pyramid is a right triangle $ABC$ whose leg $|AC| = b$ and $\angle B = \beta$. The lateral faces of the pyramid, which pass through the legs $[AC]$ and $[BC]$, are perpendicular to the plane of the base, and the third lateral face makes an angle $\alpha$ with the base. Find the volume of the pyramid.

43. Given a pyramid $SACB$ with vertex $S$ whose base is a right triangle $ACB$ in which $[AB]$ is a hypotenuse $2\sqrt{3}$ cm in length. The lateral edge $[SA]$ is perpendicular to the plane of the base. The dihedral angle formed by the lateral faces $SAC$ and $SAB$ is $30°$. The altitude of the pyramid is $4$ cm in length. Find the lateral area.

44. Find the volume of a pyramid whose base is a right triangle with hypotenuse $c$ in length and an acute angle $\alpha$ if the lateral edges make an angle $\beta$ with the plane of the base.

45. The base of the pyramid is an isosceles right triangle whose hypotenuse is $5$ cm in length. Each lateral edge makes an angle $\alpha$ with the plane of the base. Find the total surface area of the pyramid.

46. The base of a triangular pyramid is an isosceles triangle whose area is $S$ and the vertex angle is $\alpha$. Find the volume of the pyramid if the angle between each lateral edge and the altitude of the pyramid is $\beta$.

47. The base of a pyramid is an isosceles triangle in which the angle between the equal sides is $\alpha$ and the side opposite to it is $a$ in length. Each lateral face of the pyramid makes an angle $\beta$ with the base. Find the total surface area of the pyramid.

48. Find the volume of a pyramid whose base is a triangle two angles of which are equal to $\alpha$ and $\beta$ and the radius of the circle circumscribed about the base is $R$. Each lateral edge makes an angle $\varphi$ with the plane of the base.

49. In a regular triangular pyramid, a side of the base is $a$ in length and the dihedral angle between the lateral faces is $\alpha$. Find the volume of the pyramid.

50. The angle between the altitude of a regular triangular pyramid and the apothem is $\alpha$, and the length of a lateral edge of the pyramid is $l$. Find the volume of the pyramid.

51. The edge angle of a regular triangular pyramid is $\alpha$ and the radius of the circle inscribed in a lateral face is $r$. Find the lateral area of the pyramid.

**52.** A side of the base of a regular triangular pyramid is $a$ and the dihedral angle at the base is $45°$. Find the volume of the pyramid.

**53.** Find the altitude of a regular tetrahedron whose volume is equal to $V$.

**54.** Find the volume of a regular tetrahedron whose altitude is equal to $H$.

**55.** In a regular triangular pyramid, whose volume is $V$, each lateral face makes an angle $\alpha$ with the plane of the base. Find the total surface area of the pyramid.

**56.** Find the volume and the lateral surface of a regular triangular pyramid, being given that the plane, which passes through a side of the base and the midpoint of the altitude of the pyramid, makes an angle $\varphi$ with the base, and a side of the base is $a$.

**57.** Find the volume of a regular triangular pyramid whose lateral edge makes an angle $\alpha$ with the plane of the base and is at the distance $d$ from the midpoint of the opposite side.

**58.** A regular triangular pyramid is cut by a plane which is perpendicular to the base and bisects two sides of the base. Find the area of the section of the pyramid by that plane if it is known that a side of the base is $a$ and the altitude of the pyramid is $H$.

**59.** In a regular triangular pyramid, a side of whose base is $a$ and a lateral edge is $2a$, a plane is drawn through the midpoint of a lateral edge at right angles to it. Find the area of the resulting section.

**60.** A plane is drawn through the vertex $C$ of the base of a regular triangular pyramid $SABC$ at right angles to the lateral edge $[SA]$. The plane makes an angle with the plane of the base whose cosine is equal to $2/3$. Find the cosine of the angle between the lateral faces of the pyramid.

**61.** A section is drawn through the vertex of a regular triangular pyramid and the midpoints of two sides of the base. Find the area of the section and the volume of the pyramid, being given the length $a$ of a side of the base and the angle $\alpha$ between the section and the base.

**62.** The altitude of a regular triangular prism is $H$. A plane is drawn through one of the edges of the lower base and the vertex of the upper base which is opposite to it. Find the area of the section if the angle of the resulting triangle at the given vertex of the prism is $\alpha$.

**63.** Find the volume of a regular triangular pyramid if the edge angle is $\alpha$ and the radius of the circle circumscribed about a lateral face is $R$.

**64.** A lateral edge of a regular triangular pyramid is $a$ and the edge angle is $2\alpha$. Find the surface area of the sphere circumscribed about the pyramid.

65. The lateral edges and two sides of the base of a triangular pyramid are equal to each other and to $l$. The angle between the equal sides of the triangle serving as the base is $\alpha$. Find the volume of the pyramid.

66. The base of a pyramid is an isosceles triangle whose lateral side is $a$ and the vertex angle is $\alpha$. Each lateral edge of the pyramid makes an angle $\beta$ with the plane of the base. Find the volume of the pyramid.

67. A lateral face of a regular triangular pyramid makes an angle $\alpha$ with the plane of the base, the sum of the lengths of the altitude of the pyramid and the radius of the circle inscribed in the base of the pyramid is equal to $a$. Find the volume of the pyramid.

68. The lateral edges of a triangular pyramid are mutually perpendicular and each of them is equal to $b$. Find the volume of the pyramid.

69. In a tetrahedron $ABCD$ find the angle between the straight lines $(AD)$ and $(BC)$ if $|AB| = |AC|$ and $\angle DAB = \angle DAC$.

70. The sides of the triangle which serves as the base of a pyramid are equal to 13 cm, 14 cm, and 15 cm. The dihedral angles at the base of the pyramid are 45° each. Find the lateral surface of the pyramid.

71. The lateral edges of a triangular pyramid are equal and the base is a right triangle whose altitude dropped from the vertex of the right angle is $h$. The dihedral angles formed by the faces of the pyramid, which intersect along the legs of the base, are equal to $\alpha$ and $\beta$. Find the volume of the pyramid.

72. The base of a pyramid is an isosceles triangle with the vertex angle $\varphi$. All the lateral edges of the pyramid are equal to $a$ in length. Find the volume of the pyramid if the length of the radius of the circle inscribed in the base is $r$.

73. The base of a pyramid is an isosceles triangle with the vertex angle $\alpha$ in magnitude. All the dihedral angles at the base of the pyramid are equal to $\beta$. Find the volume of the pyramid if the radius of the circle circumscribed about the base is $R$.

74. The base of a pyramid is an isosceles triangle whose equal sides are $b$ in length and form an angle $\alpha$. Each lateral edge of the pyramid makes an angle $\beta$ with the altitude of the pyramid. Find the volume of the pyramid.

75. In a triangular pyramid $SABC$ two equal lateral faces $ASB$ and $BSC$ are perpendicular to the plane of the base and the face $ASC$ makes an angle $\beta$ with the plane of the base. Find the radius of the sphere circumscribed about the pyramid if the radius of the circle circumscribed about the base is $r$ and $\angle ABC = \alpha$.

76. A side of the base of a regular triangular pyramid is $m$ and the dihedral angle between the lateral faces of the pyramid is $2\alpha$. A plane drawn through one of the sides of the base is perpendicular to a lateral edge.

Find the volume of the part of the pyramid which lies below the plane.

**77.** The lateral surface of a triangular pyramid is $S$ and each of the lateral edges is $l$. Find the edge angles of the pyramid, being given that they form an arithmetic progression with the difference $\pi/4$.

**78.** The faces of a triangular pyramid are equal isosceles triangles. The base and the angle opposite to it of each triangle are equal to $a$ and $\alpha$ respectively. Find the volume of the pyramid.

**79*.** The base of a pyramid is a regular triangle with side $a$. One of the lateral faces is perpendicular to the base and the areas of the other two are equal to $P$ and $Q$ respectively. In what ratio does the altitude of the pyramid divide a side of the base?

**80*.** The base of a pyramid is an isosceles right triangle. The lateral face, which rests on the hypotenuse, is perpendicular to the plane of the base. The areas of the other two faces are equal to $S$ and $T$ respectively. Find the length of the hypotenuse of the base if it is known that the altitude divides it in the ratio $1 : p$.

**81.** Find the radius of a sphere inscribed in a triangular pyramid if all the vertex angles of the pyramid are right angles and the lengths of the lateral edges are equal to $a$, $b$, and $c$.

**82*.** Find the volume of a triangular pyramid if the areas of its faces are equal to $S_0$, $S_1$, $S_2$, and $S_3$ and the dihedral angles at the face with the area $S_0$ are equal to each other.

**83*.** $SABC$ is a regular tetrahedron with an edge of length unity and $O$ is the centre of a sphere of radius $\sqrt{2}$, which touches the edges $(AS)$, $(AC)$, and $(AB)$ (or their extensions). Find the length of the segment $[OK]$, where $K$ is the midpoint of the edge $[SC]$.

**84*.** In a tetrahedron $ABCD$ the edges $[AC]$, $[BC]$, and $[DC]$ are mutually perpendicular. A point $M$ lies in the plane $ABC$ and is equidistant from the edges $[AB]$, $[BC]$, and $[CD]$. A point $N$ lies in the plane $BCD$ and is equidistant from the same edges. Find $|MN|$ if $|BC| = |CD| = \sqrt{3}$ and $|AC| = 3$.

**85.** In a regular tetragonal pyramid the altitude is 3 cm long and a lateral edge is 5 cm long. Find the volume of the pyramid.

**86.** Find the volume of a regular tetragonal pyramid if a side of its base is $a$ long and the base dihedral angle is $\alpha$.

**87.** The edge angle of a lateral face of a regular tetragonal pyramid is $\varphi$ in magnitude. Find the angle between a lateral edge and the plane of the base.

**88.** In a regular tetragonal pyramid the dihedral angle at a lateral edge is $\alpha$. Find the edge angle of the pyramid.

**89.** The base of a pyramid is a rectangle, its two lateral faces are perpendicular to the base, and the other two faces make angles $\alpha$ and $\beta$ with the base respectively. Find the volume of the pyramid if the largest lateral edge is $l$ in length.

90. The altitude of a regular tetragonal pyramid is $H$ and the base dihedral angle is φ. Find the radius of the sphere inscribed in the pyramid.

91. The angle between a lateral edge and the base of a regular tetragonal pyramid is 60° and the lateral edge is $a$ long. A plane is drawn through the midpoint of one of the lateral edges at right angles to it. Find the area of the section.

92. The altitude of a regular tetragonal pyramid is $h$ and makes an angle α with a lateral face. A plane is drawn through a side of the base of the pyramid at right angles to the opposite face. Find the volume of the pyramid cut off by that plane from the given pyramid.

93. A side of the base of a regular tetragonal pyramid is $a$ and the base dihedral angle is α. Find the area of the section of the pyramid by a plane which divides the base dihedral angle in half.

94. The length of each edge of a regular tetragonal pyramid $SABCD$ is unity. A point $M$ lies in the base $ABCD$ of the pyramid and is equidistant from the edges $[AS]$ and $[DS]$, and $|MS| = |MC|$. A point $N$ lies on the face $BSC$ and is also equidistant from the same edges, with $|NS| = [ NC]$. Find the area of the triangle $BMN$.

95. The base of a pyramid is a rhombus whose diagonals are 6 m and 8 m long. The altitude of the pyramid passes through the point of intersection of the diagonals of the rhombus and is 1 m long. Find the lateral area of the pyramid.

96. The base of a pyramid is a rhombus with a side $a$ and an acute angle α. Two lateral faces are perpendicular to the base and the other two make an angle β with it. Find the volume of the pyramid and its lateral area.

97. The base of a pyramid is an isosceles trapezoid whose nonparallel sides are equal to $a$ each and the acute angle is α. Each lateral face makes an angle β with the base of the pyramid. Find the volume of the pyramid.

98. The base of a pyramid is a rectangle with an angle α between the diagonals. Each lateral edge makes an angle β with the plane of the base. Find the volume of the pyramid if the radius of the sphere circumscribed about it is equal to $R$.

99. Find the volume of a regular tetragonal pyramid if the angle at the base of a lateral face is α and the radius of the circle inscribed in that face is $R$ .

100. The altitude of a regular truncated tetragonal pyramid is $H$, and a lateral edge and the diagonal of a lateral face of the pyramid make angles α and β with the plane of the base. Find the lateral area of the pyramid.

101. In a regular hexagonal pyramid the angle between a lateral edge and the adjacent edge of the base is α, and the sum of the radii of the circles

inscribed in the base and circumscribed about the base is equal to $m$. Find the lateral area of the pyramid.

102. The apothem of a regular hexagonal pyramid is $m$. The base dihedral angle is $\alpha$. Find the total surface area of the pyramid.

103. The altitude of a regular truncated tetragonal pyramid is 3 cm, its volume is 38 cm$^3$, and the areas of the bases are related as 9 : 4. Find the lateral area of the truncated pyramid.

104. The areas of the bases of a truncated pyramid are equal to $S_1$ and $S_2$. Find the area $S$ of the section of the pyramid by a plane which is parallel to the bases and is equidistant from them.

105. The total surface area of a regular tetragonal pyramid is $S$, and the plane angle of a lateral face at the vertex is $\alpha$. Find the altitude of the pyramid.

106. The difference between the length of the apothem and that of the altitude of a regular tetragonal pyramid is equal to $m$, and the angle between them is $\alpha$. Find the volume of the pyramid.

107. The altitude of a regular tetragonal pyramid is $h$ long. A section, which is drawn through the diagonal of the base of the pyramid and the midpoint of the opposite edge, makes an angle $\alpha$ with the diagonal plane drawn through the same diagonal of the base. Find the area of the section.

108. Find the edge angle of a regular tetragonal pyramid if that angle is equal to the angle between a lateral edge and the plane of the base of the pyramid.

109. The base of a pyramid is a rectangle whose area is 12 cm$^2$. Two lateral faces of the pyramid are perpendicular to the base and the other two make angles of 30° and 60° with the base. Find the total surface area of the pyramid.

110. The base of a tetragonal pyramid is a rhombus with a side $a$ long and an acute angle $\beta$. Each lateral face makes an angle $\gamma$ with the plane of the base. Find the total surface area of the pyramid.

111. The base of a tetragonal pyramid is a rhombus whose smaller diagonal is $d$ long and the acute angle is equal to $\alpha$. Each lateral face makes an angle $\beta$ with the plane of the base. Calculate the total surface area of the pyramid.

112. The base of a pyramid is a parallelogram with an obtuse angle $\alpha$. Two lateral faces are perpendicular to the plane of the base and the other two make angles $\beta$ and $\gamma$ with the plane of the base. Find the area of the smaller lateral face if the smaller lateral edge is 8 cm long, $\sin \beta = 1/3$, $\sin \gamma = 3/5$, and $\tan \alpha = -1/3$.

113. A side of the base of a regular tetragonal pyramid is $a$ long and a lateral face makes an angle $\alpha$ with the plane of the base. Find the radius of the circumscribed sphere.

114. A side of the base of a regular triangular pyramid is $a$ long and the edge angle of the pyramid is $\alpha$. Find the length of the radius of the sphere inscribed in the pyramid.

115. Find the radius of the sphere inscribed in a regular triangular pyramid whose altitude is $H$ in length and the angle between a lateral edge and the plane of the base is $\alpha$.

116. Given a cone and a regular triangular pyramid inscribed in it whose lateral edge makes an angle $\alpha$ with the plane of the base. Find the volume of the cone if a side of the base of the pyramid is $a$ in length.

117. A pyramid is inscribed in a cone whose generatrix is $l$ in length and makes an angle $\alpha$ with the plane of the base. The base of the pyramid is a rectangle with an acute angle $2\alpha$ between the diagonals. Find the distance from the foot of the altitude to the lateral face which passes through the smaller side of the base.

118. A sphere of radius $a$ long is inscribed in a truncated cone. The generatrix of the cone makes an angle $\alpha$ with the plane of the base. Find the volume of the truncated cone.

119. The volume of the cone' is $V$. A pyramid is inscribed in the cone whose base is an isosceles triangle with an angle $\alpha$ between the lateral sides. Find the volume of the pyramid.

120. The altitude of the cylinder inscribed in a cone is equal to the radius of the base of the cone. Find the angle between the axis of the cone and its generatrix, being given that the total surface area of the cylinder is related to the area of the base of the cone as 3 : 2.

121. Find the volume and the total surface area of a cone if in its base a chord of length $a$ subtends an arc $\alpha$, and the angle between the altitude and the generatrix of the cone is $\beta$.

122. Find the total surface area of a cylinder whose axial section is a square and the lateral surface is $S$.

123. A plane is drawn through the vertex of a cone at an angle $\alpha$ to the base of the cone. The plane cuts the base of the cone along a chord of length $a$ which subtends an arc $\beta$. Find the volume and the lateral surface of the cone.

124. A plane is drawn through two generatrices of a cone, which cuts an arc of 120° from the base. Find the area of the section if the radius of the base of the cone is $R$ in length and the plane of the section makes an angle $\alpha$ with the plane of the base.

125. A triangle $ABC$ rotates about the side $[BC]$ and $\angle A = 120°$. Find the area of the surface which results from the rotation of the polygonal line formed by the sides $[CA]$ and $[AB]$ if $|AB| = 2\sqrt{3}$ and $|BC| = 5$.

126. A cone is inscribed in a sphere. The area of the axial section of the cone is $S$. The angle between its altitude and generatrix is $\alpha$. Find the volume of the sphere.

127. A cone and a cylinder have a common base, and the vertex of the cone is at the centre of the other base of the cylinder. Find the angle between the axis of the cone and its generatrix if it is known that the ratio of the

total surface area of the cylinder to the total surface area of the cone is 7 : 4.

128. The generatrix of a cone makes an angle φ with the plane of the base of the cone. The area of the section of the cone by the plane which passes through the centre of the sphere inscribed in the cone parallel to the base is equal to Q. Find the volume of the cone.

129. A cylinder is inscribed in a right cone whose axial section is a right triangle (the lower base of the cylinder lies in the plane of the base of the cone). The ratio of the lateral area of the cone to that of the cylinder is equal to $4\sqrt{2}$. Find the angle between the plane of the base of the cone and the straight line passing through the centre of the upper base of the cylinder and an arbitrary point of the circle which is the base of the cone.

130. Find the lateral surface of the cone, being given the length of the radius $R$ of the sphere circumscribed about it and the angle $\alpha$ at which the generatrix of the cone can be seen from the centre of the sphere.

131. A cylinder is inscribed in a cone, the altitude of the cylinder being equal to the diameter of the base of the cone. The total surface area of the cylinder is equal to the area of the base of the cone. Find the angle between the generatrix of the cone and the plane of the base.

132. Given a sphere of radius $R$ and a right circular cone circumscribed about it in which the angle between the generatrix and the plane of the base is $\alpha$. Find the total surface area and the volume of the cone.

133. Given a sphere of radius $R$ and a cone inscribed in it. A cylinder with a square axial section is inscribed in the cone. Find the total surface area of the cylinder if the angle between the generatrix of the cone and the plane of the base is $\alpha$.

134. A sphere is inscribed in a cone. The radius of the circle of tangency of the surface of the sphere and the lateral surface of the cone is equal to $r$ in length. The straight line which connects the centre of the sphere and an arbitrary point of the circle serving as the base of the cone makes an angle $\alpha$ with the altitude of the cone. Find the volume of the cone.

135. The radius of the base of the cylinder is 26 cm and the length of the generatrix is 48 cm. At what distance from the axis of the cylinder must a section be drawn, parallel to the axis of the cylinder, for that section to be shaped as a square?

136. A plane is drawn in a cylinder, parallel to its axis, at the distance $a$ from it, which cuts off an arc of $\alpha$ radians from the circle of the base. The area of the section is $S$. Find the volume of the cylinder.

137. A plane drawn parallel to the axis of a cylinder divides the circle of the base in the ratio $m : n$. The area of the section is $S$. Find the lateral surface area of the cylinder.

**138.** The altitude of a cone, equal to $h$, is a diameter of a sphere which divides the lateral area of the cone in the ratio $m : n$ (reckoning from the vertex). Find the radius of the base of the cone.

**139*.** Two equal cones with a common vertex $A$ lie on different sides of the plane $P$ so that only one generatrix of each cone ([AB] for one cone and [AC] for the other) belongs to the plane $P$. It is known that $\angle BAC = \beta$ and the angle between the altitude and the generatrix of each cone is $\varphi$. Find the angle between the line of intersection of the planes of the bases of the cones and the plane $P$.

**140*.** Point $O$ is a common vertex of two congruent cones which lie on the same side of the plane $\alpha$ so that only one generatrix of each cone ([OA] for one cone and [OB] for the other) belongs to the plane $\alpha$. It is known that the angle between the altitudes of the cones is $\beta$, and the angle between the altitude and the generatrix of each cone is $\varphi$, with $2\varphi < \beta$. Find the angle between the generatrix $(OA)$ and the plane of the base of the other cone to which the point $B$ belongs.

**141.** The diagonals of the axial section of a truncated cone are divided by their point of intersection in the ratio $2 : 1$, reckoning from the larger base. The angle between the diagonals facing the bases of the cone is $\alpha$. The diagonal is $l$ long. Find the volume of the truncated cone.

**142.** A sphere of radius equal to $\sqrt[3]{2}$ cm is of the same volume as a right cone whose lateral surface is three times as large as the area of the base. Find the altitude of the cone.

**143.** In a sphere of radius $R$ three equal chords are drawn from a point of its surface at an angle $\alpha$ to each other. Find their length.

**144*.** Four spheres of radius $r$ lie so that each of them touches the other three. Find the radius of the sphere which touches each of the given ones.

**145*.** Four spheres lie on a plane : two spheres of radius $a$ and two similar spheres of an unknown radius $x$ lie so that each sphere touches the other three and the plane. Find the radius $x$.

**146.** A sphere whose centre is at the vertex of a cone touches its base. Find the vertex angle in the axial section of the cone if the sphere divides the cone into parts of equal volumes.

**147.** A sphere with centre at the vertex of a cone divides the cone into two parts of the same volume. Find the radius of the sphere if the radius of the base of the cone is $a$ and the vertex angle of its axial section is $\alpha$.

**148*.** A sphere of radius $R$ is inscribed in a cone whose axial section vertex angle is $\alpha$. Find the volume of the part of the cone which lies above the sphere.

**149.** The distance from the centre of a sphere inscribed in a cone to the vertex of the cone is $a$. The angle between the generatrix and the plane of the base of the cone is $\alpha$. Find the volume of the cone.

150. A sphere the area of whose larger circle is $Q$ is inscribed in a cone. Find the total surface area of the cone if the vertex angle of its axial section is $2\alpha$.

151*. A sphere is inscribed in a right circular cone with an angle $\alpha$ at the vertex of the axial section. A cone with the same axial section vertex angle is inscribed in the sphere. Find the angle $\alpha$ if the ratio of the volume of the first cone to that of the second is equal to 27.

152. The centre of a sphere coincides with the centre of the base of a cone, and its radius is equal to the radius of the base of the cone. The altitude of the cone is larger than the radius of the base of the cone. A plane is drawn through the circle along which the sphere cuts the lateral surface of the cone. What must be the axial section vertex angle of the cone for the plane to divide the cone into two parts equal in volume?

153. A sphere touches all the faces of a cube. Find the ratio of the surface areas and the ratio of the volumes of the figures.

154. The surface area of a sphere inscribed in a cone is $Q$. Find the total surface area of the cone if the largest angle between its generatrices is $\alpha$.

155. Find the ratio of the total surface area of a right circular cone inscribed in a sphere to the surface area of the sphere if it is known that the vertex angle of the axial section of the cone is $\alpha$ and $\alpha > \pi / 2$.

156. Three identical spheres of radius $r$ are inserted into a right circular cone with an angle of $60°$ at the vertex of the axial section. Each inserted sphere touches the other two spheres, the base, and the lateral surface of the cone. Find the radius of the base of the cone.

157. A sphere of radius $r$ is inscribed in a cone. Find the volume of the cone if it is known that the plane which touches the sphere and is perpendicular to one of the generatrices of the cone is at the distance $d$ from the vertex of the cone.

158. Find the ratio of the volume of a sphere to that of a right circular cylinder inscribed in the sphere it is known that the smaller angle between the diagonals of the axial section of the cylinder is $\alpha$, and the diameter of the base exceeds the altitude of the cylinder.

159. A cone is inserted into a cylinder so that the base of the cone coincides with the lower base of the cylinder, and the vertex of the cone coincides with the centre of the upper base of the cylinder. The generatrix of the cone makes an angle $\alpha$ with the plane of the base. Find the volume of the cylinder if the total surface area of the cone is $S$.

160. A cylinder is inscribed in a cone the radius of whose base is $r$ and the angle which the altitude makes with the generatrix is $\alpha$. The cylinder is inscribed so that its lateral surface is related to that of the cone as $m : n$. Find the volume of the cylinder.

**161.** A hemisphere is inscribed in a cone so that the larger circle of the hemisphere lies in the plane of the base of the cone, and the circular surface touches the surface of the cone. Find the total surface area of the hemisphere and its volume if the generatrix of the cone is $l$ and makes an angle $\beta$ with the plane of the base.

**162.** A triangular pyramid which has a right triangle with an acute angle $\alpha$ as its base is inscribed in a cone whose lateral area is $S$ and the angle the generatrix makes with the plane of the base is $\varphi$. Find the volume of the pyramid.

**163.** The angle between the plane of the base and a lateral face of a regular tetragonal pyramid is $\varphi$. The surface area of a sphere inscribed in the pyramid is $S$. Find the lateral area of the pyramid.

**164\*.** A sphere is inscribed in a triangular pyramid whose all edges are $a$ in length. A sphere is again inscribed in one of the trihedral angles of the pyramid so that it touches the first sphere. Find the volume of the second sphere.

**165.** The altitude of a regular tetragonal pyramid and the radius of the circumscribed sphere are equal to $h$ and $r$ respectively $(r \le h)$. Find the area of the base of the pyramid.

**166.** A sphere is inscribed in a regular tetragonal pyramid. The distance from the centre of the sphere to the vertex of the pyramid is $a$ and the angle a lateral face of the pyramid makes with the plane of the base is $\alpha$. Find the lateral area and the volume of the pyramid.

**167.** A side of the base of a regular tetragonal pyramid is $a$ and the base dihedral angle is $\alpha$. A sphere is inscribed in the pyramid and a tangent plane is drawn to the sphere, parallel to the base of the pyramid. Find the lateral area of the resulting truncated pyramid.

**168.** A pyramid is inscribed in a cone whose generatrix makes an angle $\alpha$ with the plane of the base. The base of the pyramid is a right triangle with legs $a$ and $b$. Find the volume of the pyramid.

**169.** A right cone is inscribed in a regular hexagonal pyramid and a right cone is circumscribed about the pyramid. The altitude of the pyramid is $H$ long and the radius of the base of the circumscribed cone is $R$. Find the difference between the volumes of the circumscribed and the inscribed cone.

**170.** The radius of the base of a cone is $r$ and the generatrix makes an angle $\varphi$ with the plane of the base. A pyramid whose base is a right triangle with an acute angle $2\varphi$ is circumscribed about the cone. Find the volume of the pyramid.

**171.** Find the lateral area of a cone inscribed in a regular triangular pyramid if the length of a lateral edge of the pyramid is $l$ and a lateral face of the pyramid makes an angle $\alpha$ with the plane of the base.

**172.** A cylinder with radius $r$ of the base is inscribed in a regular tetragonal pyramid. The altitude of the cylinder is half that of the pyramid. The edge angle of the pyramid is α. Find the volume of the pyramid.

**173.** A parallelepiped is inscribed in a cylinder. The diagonal of the parallelepiped makes an angle α with the plane of the base and an angle β with the larger lateral face. Find the volume of the cylinder if a side of the base of the larger lateral face of the parallelepiped is $a$.

**174.** The base of a right prism is a rhombus with an acute angle α and the smaller diagonal $d$. The plane which passes through that diagonal and a vertex of the second base of the prism makes an angle β with the plane of the base. Find the volume of the cylinder inscribed in the prism.

**175\*.** A triangle whose vertices are the midpoints of the edges of the base of a regular pyramid is the base of a regular prism. What part of the volume of the prism is not in the pyramid if the altitude of the pyramid is known to be three times smaller than that of the prism?

**176\*.** *SABC* is a regular unit tetrahedron. A sphere touches the edges [*AS*], [*AC*], and [*AB*] and passes through the midpoint of the edge [*BC*]. Find the radius of the sphere if its centre is known to lie in the interior of the tetrahedron.

**177\*.** A sphere touches the lateral edge [*AA'*] and the non parallel edges [*AB*] and [*A'D'*] of the bases of a unit cube *ABCDA' B' C' D'* and passes through a point *M* of the edge [*CC'*], with $|CM| = 1/3$. Find the radius of the sphere.

**178\*.** Three identical spheres of radius $a$ are inserted into a cylinder with altitude $3a$ so that each sphere touches the other two spheres and the lateral surface of the cylinder, with two spheres touching the lower base and the third sphere, the upper base. Find the radius of the base of the cylinder.

**179.** The generatrix of a cone is $l$ and makes an angle φ with the plane of the base. For what value of φ is the volume of the cone the greatest? What is the volume equal to?

**180.** A cone is inscribed in a hemisphere of radius $R$ so that the vertex of the cone is at the centre of the hemisphere. Find the radius of the base of the cone for which the volume of the cone is the greatest.

**181.** The base of a pyramid is a right triangle with area $Q$ and an acute angle α. The lateral face passing through a leg, which is opposite to the given angle, is perpendicular to the plane of the base, and the other two faces make angles equal to β with the base. Find the volume of the pyramid. For what value of α is the volume the greatest?

**182.** Find the ratio of the altitude to the radius of the base of a cylinder which has the least total surface area for a given volume.

**183.** A cylinder is inscribed in a hemisphere of radius $R$ so that the plane of the base of the cylinder coincides with the plane bounding the hemisphere. Find the altitude of the cylinder of the greatest volume.

**184.** A cylindrical tank must hold $V$ litres of water. What must be its dimensions for the surface without the lid to be the least?

**185.** A cone of volume $V$ is circumscribed about a sphere. The angle between the generatrix of the cone and the plane of the base is $\alpha$. Find the volume of the sphere. For what value of $\alpha$ is the volume the greatest?

**186.** Circumscribe a cone of the least volume about a cylinder (the radius of whose base is $r$ and the altitude is $h$) if the plane of the base of the cylinder and that of the base of the cone coincide. Find the volume of the cone.

**187.** A plane section with the least perimeter is drawn through the edge [*AB*] of a regular pyramid *SABC* with the vertex *S*. Find the area of the section if the altitude of the pyramid is $h$ and $|AB| = a$.

**188.** The base of a pyramid *SABC* is a triangle *ABC* in which $\angle ABC = 90°$, $\angle BAC = \varphi$ and $|AC| = b$. The lateral edges of the pyramid make the same angle with the plane of the base, and the angle between the face *SBC* and the plane of the base is $\alpha$. Find the volume of the pyramid. For what value of $\varphi$ is the volume of the pyramid the greatest?

**189.** A sphere of radius $r$ is inscribed in a right circular cone of radius $R$. A plane which cuts the sphere is drawn through the vertex of the cone. Find the area of the section of the cone by that plane if that area is known to have the greatest of all possible values.

**190.** A cone is inscribed in a regular tetragonal pyramid whose altitude is $H$ in length and whose lateral edge makes an angle $\beta$ with the plane of the base. A plane which cuts the surface of the cone is drawn through the apothem of a lateral face of the pyramid. Find the area of the section of the cone by that plane if it is known that the area has the greatest of all possible values.

**191.** Find the ratio of the surface area resulting from the rotation of a rhombus about the larger diagonal to the surface area resulting from the rotation of the rhombus about the smaller diagonal if it is known that the smaller angle between the sides of the rhombus is $\alpha$.

**192.** Find the ratio of the volume of the body resulting from the rotation of a rectangle about the larger side to that of the body resulting from the rotation of that rectangle about the smaller side if it is known that the smaller angle between the diagonals of the rectangle is $\alpha$.

**193.** A right triangle with a leg $a$ long and an adjacent acute angle $\alpha$ rotates about a straight line which passes through the vertex of the given angle and is perpendicular to the bisector of that angle. Find the volume of the body of revolution.

**194.** The length of the smaller side of a parallelogram is $a$, the acute angle of the parallelogram is $\alpha$, and the angle between the smaller diagonal and the larger side is $\beta$. Find the volume of the body resulting from the rotation of the parallelogram about its larger side.

**195.** Given a triangle $ABC$, with $|BC| = a$, $\angle ABC = \alpha$, and $\angle ACB = 90° + \alpha$. Find the volume of the body resulting from the rotation of the triangle about its altitude drawn from the vertex $A$.

**196.** The area of a rectangular trapezoid $ABCD$ is $S$, the length of the altitude $[AB]$ is $h$, and the acute angle $ADC$ is equal to $\alpha$. Find the volume of the body resulting from the rotation of the quadrangle $ABED$ about the straight line $(AB)$ if the point $E$ is the midpoint of the segment $[CD]$.

**197.** The base of a pyramid is a rhombus with an acute angle $\alpha$. All the lateral faces make the same angle $\beta$ with the plane of the base. Find the radius of the sphere inscribed in the pyramid if the volume of the pyramid is equal to $V$. For what values of $\beta$ is the radius of the sphere the greatest?

**198.** A cone is circumscribed about a hemisphere of radius $R$ so that the centre of the base of the cone lies at the centre of the sphere. The vertex angle of the axial section of the cone is $\alpha$. Find the volume of the cone. For what value of $\alpha$ is the volume the least?

# Part-4

## Oral Examination Problems and Questions

The oral examinations set by all the educational establishments of the USSR may only cover the mathematics outlined in the programme for entrance candidates. Each examination paper includes two or three theoretical questions, formulated as is usual at school and two or three problems. In what follows we present sample oral examinations set at some Soviet higher schools and more than four hundred problems of various kinds.

## 4.1. Sample Examination Papers

### I

1. Natural numbers. Prime and composite numbers. The divisor and the multiple. Common divisors. The least common multiple. The criteria for divisibility by 2, 3, 5, 10.
2. The criterion of parallelism of planes.

### II

1. The concept of an antiderivative. The theorem on the general form of all the antiderivatives of a given function. The integral.
2. The existence of a circle circumscribed about a triangle.

### III

1. The properties of the function $y = k / x$ and its graph.
2. The formula for the area of $a$ regular polygon (in terms of the radius of a circle circumscribed about it).

Additional questions

3. Find the domain of the function $y = \log (x^2 - 6x + 6)$.
4. If the given function is periodic, find the least period
   $T : f(x) = 10 \sin 3x$.
5. Can a composition of two rotations with different centres be a parallel displacement?

**IV**

1. Number functions. The graph of a number function. An increase and a decrease of a function; periodicity, evenness, oddness.
2. Vectors. Operations on vectors. Collinear vectors. Coplanar vectors.
3. The derivative of the sum of two functions.

**V**

1. Vectors. Operations on vectors. Collinear vectors. Coplanar vectors.
2. A composition including seven bodies must be formed from three pyramids and ten cubes. In how many ways can the composition be formed if at least one pyramid must enter into it?
3. Calculate the area of the figure on the interval $[0, \pi]$ bounded by the $Oy$ axis, the straight line $x = \pi$, and the curves $y = \sin x$ and $y = |\cos x|$.

**VI**

1. Express a side of a regular polygon in terms of the radius of the circle inscribed in it.
2. Solve the equation $\sin x + \sin 3x = 0$.
3. Vectors and operations on them.
   **Problem.** Calculate the length of the vector $4\mathbf{a} + 3\mathbf{b}$ if $\mathbf{a} = (1, 1, -1)$ and $\mathbf{b} = (2, 0, 1)$.

**VII**

1. The surface area and the volume of a cylinder.
2. Trigonometric functions of half the argument.
3. The derivative of the function $y = \log_a x$.

**VIII**

1. The cosine rule.
2. Factoring a quadratic trinomial into linear factors.

**IX**

1. The formula for an $n$th term and the sum of the first $n$ terms of an arithmetic progression.
2. The property of the midpoint of a diagonal of a parallelepiped.
3. The derivative of the function $y = \tan x$.

**X**

1. The logarithm of a product, of a power, and of a quotient.
2. The criterion of parallelism of planes.
3. Find the limit $\lim\limits_{x \to \infty} \dfrac{x^2 - x + 1}{2x^2 + x}$.

**XI**

1. The derivative of the product of two functions.
2. The sum of the angles of a triangle.
3. Solve the inequality $\dfrac{\ln (x-1)^2}{3x^2 - 4x + 75} > 0$.
4. Find the greatest and the least value of the function $y = 2x - \sqrt{x}$ on the interval $[0, 4]$.

## XII

1. The properties of an exponential function.
2. The formulas for the surface area and the volume of a cylinder.
3. Solve the equation $\sin x \sin 3x = \sin 2x \sin 4x$.
4. Find the least and the greatest value of the function $y = x^3 - 3x^2 + 3$ on the interval $[1, 3]$.

## XIII

1. Trigonometric functions of double the argument and of half the argument.
2. The existence of a circle circumscribed about a triangle.
3. Find the intervals of monotonicity of the function $y = x^2 - 3x$.
4. Solve the inequality $\log_2 x < \log_2 x^2$.

## XIV

1. The derivative of the function $y = \cos x$.
2. Solve the inequality $\log_{0.3} (x^2 - 1) \geq 0$.
3. Factoring a quadratic trinomial into linear factors.
4. The theorem on three perpendiculars.

## XV

1. The properties of the function $y = ax^2 + bx + c$ and its graph.
2. The criterion of parallelism of a straight line and a plane.
3. Solve the equation $5^{x+1} + 5^x = 150$.
4. Simplify the expression $\left(\dfrac{b}{a^2+ab} - \dfrac{2}{a+b} + \dfrac{a}{b^2+ab}\right) : \left(\dfrac{b}{a} - 2 + \dfrac{a}{b}\right)$.

## XVI

1. The sufficient condition for an extremum of a function. **(Prove.)**
2. The property of a median of a trapezoid. **(Prove.)**
3. Solve the inequality $\log_{0.3}(x+1) > -1$.
4. Using the definition, calculate the derivative of the function $y = \sqrt{x} + 1$.

## XVII

1. A logarithmic function, its properties, and its graph.
2. The pyramid. The formulas for the surface and the volume of a pyramid.
3. Solve the inequality $\log (3x^2 + 1) - \log (3x - 2) < 1$.
4. Simplify the expression $\dfrac{\dfrac{\cos^2 (\alpha - 270°)}{1} + \dfrac{\sin^2 (\alpha + 270°)}{1}}{\dfrac{1}{\sin^2 (\alpha + 90°)} - 1 \dfrac{1}{\cos^2 (\alpha - 90°)} - 1}$.

5A. Find the greatest and the least value of the function $y = \dfrac{x}{3} + \dfrac{2}{x}$ on the interval $[1, 6]$.

5B. Solve the equation $\dfrac{1}{\sin^2 x} - 3\cot^2 x + 1 = 0$.

## XVIII

1. The function. Its definition. The ways of defining a function.
2. The criterion of perpendicularity of two planes.
3. Find the derivative of the function $y = \cot^3 x$.

## XIX

1. The sum of an infinitely decreasing geometric progression.
2. The properties of points which are equidistant from the terminal points of a line segment.
3. $\sin\left(\alpha - \dfrac{\pi}{2}\right) = -\dfrac{2}{3}$, $\alpha \in [3\pi / 2,\ 2\pi]$. Find $\tan 2\alpha$.

## XX

1. The properties of the function $y = k / x$ and its graph.
2. Derive a formula for calculating the area of a sphere.
3. Calculate $\arctan 1 + \arccos (-1 / 2) + \arcsin (-1 / 2)$.

## XXI

1. The derivatives of the functions $y = \sin x$, $y = \cos x$, $y = \tan x$, $y = a^x$, and $y = \log_a x$.
2. The property of points which are equidistant from the terminal points of a line segment.
3. Find the limit $\lim\limits_{x \to \infty} \dfrac{2x^2 + 7x - 2}{6x^3 - 4x + 3}$.

## XXII

1. The derivative of the function $y = \sin x$.
2. The existence of a circle circumscribed about a triangle.
3. Solve the equation $\sin x + \cos 2x = 1$.

## XXIII

1. Solving equations of the form $\sin x = a$, $\cos x = a$, and $\tan x = a$.
2. The properties of an exponential function.
3. The centre of symmetry of a parallelogram.
4. Solve the equation $\cos x + \sin 2x = 0$.

## 4.2.  Problems Set at an Oral Examination

1. Prove that the fraction $(14n + 3)/(21n + 4)$ is in its lowest terms $(n \in \mathbf{Z})$.
2. Can the number $n^4 + 4$ $(n \in \mathbf{N})$ be prime? If it can, find the prime numbers.
3. Assume that $p$ and $q$ are two successive prime numbers. Can their sum be a prime number?
4. Show that every odd number can be represented as a difference of the squares of two integers.
5. Find the sum of all even numbers from 12 to 82.

Compare the following numbers **(6-7)**.

**6.** (a) sin 3 and cos 3; (b) $\sqrt[3]{3}$ and $\sqrt{2}$.

**7.** $A = 4 + \dfrac{5}{8} + \dfrac{6}{8^2} + \dfrac{3}{8^3} + \dfrac{7}{8^4}$ and $B = 4 + \dfrac{5}{8} + \dfrac{5}{8^2} + \dfrac{7}{8^3} + \dfrac{6}{8^4}$.

**8.** Prove that $\sqrt{2} + \sqrt{3}$ is an irrational number.

**9.** Prove that $3 < \pi < 4$.

**10.** Reduce the fraction $\dfrac{\sqrt{(x-1)^2}}{x-1}$.

Simplify the following expressions **(11-13)**.

**11.** $\dfrac{1 - a^{-1/2}}{1 + a^{1/2}} - \dfrac{a^{1/2} + a^{-1/2}}{a - 1}$.

**12.** $\sqrt{a}\sqrt[4]{a}\sqrt[8]{a} \ldots . \sqrt[512]{a}$.

**13.** $x + 2x^2 + 3x^3 + \ldots . + nx^n$.

Prove that the following equations hold for any natural $n$ **(14-16)**.

**14.** $\left(1 - \dfrac{1}{4}\right)\left(1 - \dfrac{1}{9}\right)\left(1 - \dfrac{1}{16}\right) \ldots \left(1 - \dfrac{1}{n^2}\right) = \dfrac{n+1}{2n}$.

**15.** $\dfrac{1}{1 \cdot 2} + \dfrac{1}{2 \cdot 3} + \dfrac{1}{3 \cdot 4} + \ldots \ldots + \dfrac{1}{n(n+1)} = \dfrac{n}{n+1}$.

**16.** $1^2 + 2^2 + 3^2 + \ldots . + n^2 = \dfrac{n(n+1)(2n+1)}{6}$.

**17.** How many different odd five-digit numbers with unrepeated digits can be formed from the digits 0, 1, 3, 4, 5?

**18.** How many telephone numbers are there which contain the combination 12 if the number consists of 5 digits?

**19.** How many five-digit numbers are there in whose notation each successive digit exceeds its predecessor?

**20.** How many diagonals are there in a convex $n$ –gon?

**21.** The straight lines $l_1$, $l_2$, and $l_3$ are parallel and lie in the same plane. A total of $m$ points are taken on the line $l_1$, $n$ points on $l_2$, and $k$ points on $l_3$. How many triangles are there whose vertices are at these points?

**22.** Prove that $(C_n^0)^2 + (C_n^1)^2 + \ldots + (C_n^n)^2 = C_{2n}^n$.

Construct the graphs of the following functions **(23-67)**.

**23.** $y = (|x+1| - |x-1|)/2$.

**24.** $y = |x-1| - x$.

**25.** $y = |x-1| + |x-2| + x$.

**26.** $y = |x^2 - x|$.

**27.** $y = x|x-3|$.

**28.** $y = (x+1)^3 - (x-1)^3$.

**29.** $y = \sqrt{(x^2+1)^2 - 4x^2}$.

**30.** $y = \sqrt{1 - |x|}$.

**31.** $y = x^2 - 2|x+1| - 1$.

**32.** $y = x|x| - 4x - 5$.

**33.** $y = (|1-x| + 2)(x+1)$.

**34.** $y = |2x^2 - 3x + |x-1||$.

**35.** $y = \dfrac{|x-1|}{x-1}(x^2 + 3)$.

**36.** $y = x(x-3)^2$.

**37.** $y = 1 / |x|$.

**38.** $y = (x+3) / (x-1)$.

**39.** $y = (2x+1) / (x-1)$.

**40.** $y = (2|x| - 1) / (x-3)$.

**41.** $y = 2^{|x|} + 1$.

**42.** $y = 2^{(|x| + x)/x}$.

**43.** $y = |\sin x| / \sin x$.

**44.** $y = -2^{-|x|}$.

**45.** $y = 2^{|1-x|}$.

**46.** $y = 3 \cdot 2^x - 2$.

**47.** $y = 2^{|\log_2 x|}$.

**48.** $y = \log_2 |x-1| - 1$.

**49.** $y = |\log_{1/4}(x/4)|$.

**50.** $y = \log_x \sqrt{x}$.

**51.** $y = \log|x| - \log x^2$.

**52.** $y = \log_2 (4x - x^2)$.

**53.** $y = \log_2 (2-x)$.

**54.** $y = \log_{1/2}(1-x)$.

**55.** $y = |\log_2 x| / \log_2 x$.

**56.** $y = \log_2 (x^2 - 2x + 1)$.

**57.** $y = \log_{1/2}\left|\dfrac{2x-1}{x+1}\right|$.

**58.** $y = \dfrac{2\tan(x/2)}{1 - \tan^2(x/2)}$.

**59.** $y = \sin|2x|$.

**60.** $y = \sin^2 x$.

**61.** $y = x + \sin x$.

**62.** $y = \cos\left(|x| + \dfrac{\pi}{2}\right) / \sin x$.

**63.** $y = \log_{1/2} \sin x$.

**64.** $y = \log \tan x + \log \cot x$.

**65.** $y = \log_{|\sin x|}(1/2)$.

**66.** $y = \begin{cases} 4^x - 1 & \text{if } x < 0; \\ \sqrt{4x - x^2} & \text{if } x \geq 0. \end{cases}$

**67.** $y = \begin{cases} 1 - \sqrt{1-x^2} & \text{if } x \le 1; \\ 1 + \log_{1/2} x & \text{if } x > 1. \end{cases}$

Represent the following sets of points on the plane $(x, \ y)$ **(68-73).**

**68.** $\{(x, y)|\ x^2 + y^2 - 2x - 2y + 1 = 0\}$.

**69.** $\{(x, y)|\ (x - |\ x\ |)^2 + (y - |\ y\ |)^2 \le 4\}$.

**70.** $\{(x, y)|\ |\ x + y\ | + |\ y - x\ | \le 4\}$.

**71.** $\{(x, y)|\ \log_{(|\ x\ | - 0.5)} (x^2 + y^2) \le \log_{(|\ x\ | - 0.5)} 4\}$.

**72.** $\{(x, y)|\ \cos(x + y) = \cos(x - y)\}$.

**73.** $\{(x, y)|\ \sin 2x = \sin 2y\}$.

## Linear and Quadratic Equations and Inequalities

**74.** For what values of $x$ does the function $y = |\ x - 1\ | + |\ x - 3\ |$ possess the least value? Find that value.

**75.** Solve the equation $x|\ x\ | + 2x + 1 = 0$

**76.** Find the least value assumed by $z$ if $z = x^2 + 2xy + 3y^2 + 2x + 6y + 4$

Solve the following inequalities **(77-80).**

**77.** $9x - 14 - x^2 > 0$.

**78.** $x^2 - 5|\ x\ | + 4 < 0$.

**79.** $|\ x^2 - 4x\ | > 1$.

**80.** $2x^2 - 5|\ x\ | + 3 \ge 0$.

**81.** Prove that the roots of the quadratic equations $ax^2 + bx + c = 0$ and $cx^2 + bx + a = 0$ are mutually inverse.

**82.** Prove that if the value of the quadratic trinomial $ax^2 - bx + c$ is an integer for $x_1 = 0$, $x_2 = 1$, and $x_3 = 2$, then the value of the given trinomial is an integer for any integral $x$.

**83.** Prove that if $ax^2 + bx + c \in \mathbf{Z}$ for all $x \in \mathbf{Z}$, then $a, b, c \in \mathbf{Z}$.

**84.** Find $a$ in the equation $ax^2 - 5x + 6 = 0$ if the ratio of its roots is $x_1 / x_2 = 2/3$.

**85.** The inequalities $y(-1) > -4$, $y(1) < 0$, and $y(3) > 5$ are known to hold for $y = ax^2 + bx + c$. Determine the sign of the coefficient $a$.

## Higher-Degree Equations. Bezout's Theorem. Rational Inequalities. Solving Inequalities by the Method of Intervals

Solve the following equations **(86-87).**

**86.** $(x^2 - 6x)^2 - 2(x - 3)^2 = 81$.

**87.** $6x^4 - 13x^3 + 12x^2 - 13x + 6 = 0$.

**88.** Find the remainder of the division of the polynomial $x^3 + 2x^2 - 2x + 5$ by $x^2 - 1$ without performing the division.

**89.** How many roots does the equation $x^4 = 5x + 2a$ possess depending on $a$?

**90.** How many roots does the equation $x^3 + ax + 2 = 0$ possess depending on $a$?

**91.** Prove that the polynomial $P(x) = ax^3 + bx^2 + cx + d$ cannot have all integral roots if $P(0)$ and $P(-1)$ are odd.

**92.** Solve the equation $f'(x) = g'(x)$ if $f(x) = 3 + 2x^2$ and $g(x) = 3x^2 - (5x - 3)(2 - x)$.

Solve the following inequalities **(93-99)**.

**93.** $x + \dfrac{1}{x} < 0.$

**94.** $\dfrac{2 - 5x}{x + 1} > 2.$

**95.** $\dfrac{x^2 + 4}{x^2 + 7x + 12} < 0.$

**96.** $\dfrac{x}{x + 6} < \dfrac{1}{x}.$

**97.** $\left|\dfrac{2x - 4}{x + 1}\right| > 2.$

**98.** $\dfrac{x^2 - 7x + 12}{x - 4} \geq 0.$

**99.** $\dfrac{(x - 1)^2 x}{x + 1} \leq 0.$

## Irrational Equations and Inequalities

Solve the following equations **(100-101)**.

**100.** $\sqrt{x} - \sqrt{x - 3} = 1.$

**101.** $\dfrac{x}{\sqrt{1 + x^2}} + \dfrac{x}{\sqrt{1 - x^2}} = 0.$

Solve the following inequalities **(102-106)**.

**102.** $\sqrt{4 - x^2} \geq 1/x.$

**103.** $\sqrt{1 - x} + \sqrt{2x + 3} < 5.$

**104.** $x + 1 < \sqrt{11 - x}.$

**105.** $\dfrac{1 - \sqrt{1 - 8x^2}}{2x} < 1.$

**106.** $\sqrt{5x^2 + a^2} \geq -3x.$

## Systems of Equations and Inequalities

Solve the following systems of equations **(107-112)**.

**107.** $\begin{cases} |\,x+y\,| = 1, \\ |\,x\,|+|\,y\,| = 1. \end{cases}$

**108.** $\begin{cases} x+y+z = 0, \\ 2xy - z^2 = 4. \end{cases}$

**109.** $\begin{cases} 2x+3y = 5, \\ x-y = 2, \\ x+4y = a. \end{cases}$

**110.** $\begin{cases} x+2y = 3, \\ ax-4y = -6, \\ x+y = 1. \end{cases}$

**111.** $\begin{cases} x+y = 9, \\ x^{1/3}+y^{1/3} = 3. \end{cases}$

**112.** $\begin{cases} \sqrt{\dfrac{2x-1}{y+2}} + \sqrt{\dfrac{y+2}{2x-1}} = 2, \\ x+y = 12. \end{cases}$

**113.** How many solutions does the system of equations $\begin{cases} |\,x\,|+|\,y\,| = 1, \\ x^2+y^2 = a^2 \end{cases}$ possess?

**114.** For what values of $a$ does the system of equations $\begin{cases} x-y = a\,(1+xy), \\ 2+x+y+xy = 0 \end{cases}$ possess only one solution?

## Logarithmic and Exponential Equations, Inequalities, and Systems

Simplify the following expressions **(115-120)**.

**115.** $\log_{1/4}(\log_2 3 \cdot \log_3 4)$.

**116.** (a) $5^{\log 5/\log 25}$; (b) $\log_2 \log 100$.

**117.** $\log_3 64 \cdot \log_2 (1/27)$.

**118.** $(\log_3 4 + \log_2 9)^2 - (\log_3 4 - \log_2 9)^2$.

**119.** $0.8 \cdot (1 + 9^{\log_3 8})^{\log_{65} 5}$.

**120.** $\log \tan 3° \cdot \log \tan 6° \cdot \log \tan 9° \ldots \ldots \log \tan 87°$.

Determine the signs of the following numbers **(121-122)**.

**121.** $\log_3 4 - \log_4 3$.

**122.** $\log_{1/3}^2 0.4 - \log_{1/4} 0.4$.

**123.** Calculate $\log_5 9.8$ if $\log 2 = a$ and $\log 7 = b$.

**124.** Calculate $\log_9 40$ if $\log 15 = a$ and $\log_{20} 50 = b$.

Find the domains of definition of the following functions **(125-131)**.

**125.** $y = \sqrt{2^x - 3^x}$.

**126.** $y = \dfrac{\log x}{\arcsin (x - 3)}$.

**127.** $f(x) = \sqrt{9 - x^2} + \log(x - 1) - \sqrt{x}$.

**128.** $y = \log_3 \log_{1/2} x$.

**129.** $f(x) = \log_{2x-5}(x^2 - 3x - 10)$.

**130.** $y = \log(\sqrt{x^2 - 5x - 24} - x - 2)$.

**131.** $y = \log_2 \sin x$.

Solve the following equations **(132-154)**.

**132.** $\dfrac{3^x + 3^{-x}}{3^x - 3^{-x}} = 2$.

**133.** $6 \cdot 9^{0.5x-2} + 2 \cdot 3^{x-6} = 56$.

**134.** $(0.25)^{2-x} = \dfrac{1}{2^{x+3}}$.

**135.** $\log(5 - x) + \log(3 - x) = 1$.

**136.** $\log_3 \log_4 \log_2 x = 0$.

**137.** $\log_2 |x - 1| = 1$.

**138.** $\log(3x - 2) - 2 = \dfrac{1}{2}\log(x + 2) - \log 50$.

**139.** $\log(3 + 2\log(1 + x)) = 0$.

**140.** $\log \sqrt{x - 3} + \log \sqrt{2x - 1} = \log \sqrt{3x + 3}$.

**141.** $3^{2x^2} - 2 \cdot 3^{x^2+x+6} + 3^{2(x+6)} = 0$.

**142.** $2^{\sin^2 x} + 4 \cdot 2^{\cos^2 x} = 6$.

**143.** $\dfrac{\log_5(2x + 3)}{1 - \log_5(2x + 3)} + \dfrac{3}{5 - \log_5(2x + 3)} = 0$.

**144.** $\log_2(4^x + 4) - \log_2(2^{x+1} - 3) = x$.

**145.** $\log_8 x + \log_8^2 x + \log_8^3 x + \ldots = 1/2$.

**146.** $\log_3 \cos x + \log_{1/3}(1 - \sin x) = 0$.

**147.** $\log_2 \sin x + \log_{1/2}(-\cos x) = 0$.

**148.** $2^{\tan\left(x - \frac{\pi}{4}\right)} - 2 \cdot 0.25^{\sin^2\left(x - \frac{\pi}{4}\right)/\cos 2x} + 1 = 0$.

**149.** $\log_2 \sin x - \log_2 \cos x - \log_2(1 - \tan x) - \log_2(1 + \tan x) = 1$.

**150.** $\log_{a^4} x - \log_{a^2} x + \log_a x = 0.75$.

**151.** $x^{\log x - 3} = 0.01$.

**152.** $x^{\log x - 3} = 10^{\log (10/3)^{-1}}$.

**153.** $(\sqrt[3]{x})^{\log_x 11-2} = 11$.

**154.** $5^{\log x} = 50 - x^{\log 5}$.

**155.** For what $a$ does the equation $3x \log x = 1 + a \log x$ possess
(a) one solution ; (b) two solutions?

**156.** How many roots does the equation $x^2 e^{2-|x|} = 4a$ possess depending on $a$?

Solve the following systems of equations **(157-160)**.

**157.** $\begin{cases} \log_2 x + \log_2 y = 1, \\ \qquad\quad x + y = 3. \end{cases}$

**158.** $\begin{cases} \quad \log_3 (xy) = 3, \\ \log_{1/3}(x / y) = 1. \end{cases}$

**159.** $\begin{cases} (3x + y)^{x-y} = 9, \\ \sqrt[x-y]{324} = 18x^2 + 12xy + 2y^2. \end{cases}$

**160.** $\begin{cases} \quad 3^{\frac{2}{3}\sqrt{x}-\sqrt{y}} = 81, \\ \log \sqrt{xy} = 1 + \log 3. \end{cases}$

Solve the following inequalities **(161-186)**.

**161.** $2^x + 3^x \geq 2$.

**162.** $2 \cdot 2^x < 3^{x-1}$.

**163.** $4^x - 5 \cdot 2^x - 1 > 0$.

**164.** $\log_{0.5} ((2-x)/3) < 0$.

**165.** $\log_{1/2} (x^2 - 2x + 4) > -2$.

**166.** $\log_{1/2} (x + 4) < 2$.

**167.** $\log_{1/3} (x - 1) - \log_{1/3} (2x - 3) < 0$.

**168.** $\log_{0.5} (3x - 1) > \log_{0.5} (3 - x)$.

**169.** $\log_2 | x | < 3$.

**170.** $| \log_3 x | < 2$.

**171.** $\dfrac{1}{\log_3 x} < 1$.

**172.** $\log_{1/2} \log_3 (1 - x) > -1$.

**173.** $(1 / 2)^{\log_2 (x^2 - 2x - 3)} > 1$.

**174.** $\log_{1/3} \log_2 (x^2 - 8) \geq -1$.

**175.** $\log_x 2 \cdot \log_{2x} 2 \cdot \log_2 4x > 1$.

**176.** $\log (6 / x) > \log (x + 5)$.

**177.** $\log_{x-3} (x - 4) < 2$.

**178.** $\log_{x-3} (x - 1) < 2$.

**179.** $\log_{x^2} (2 + x) < 1$.

**180.** $\log_a \dfrac{4}{x - 3} \geq 0, \quad 0 < a < 1$.

**181.** $\log_{x+0.2} 2 < \log_x 4$.

**182.** $\log_{2x+3} x^2 < 1$.

**183.** $\log_x \dfrac{x+3}{x-1} > 1$.

**184.** $\log_{1/2} (2^{-x} - 100 \sin x) < x$.

**185.** $\log_a (1 - x^2) \geq 1$.

**186.** $|x - 2|^{\log_4 (x+2) - \log_2 x} < 1$.

**187.** Solve the system of inequalities $\log_{1/2} \cos x < \log_{1/2} \tan x$, $0 \leq x \leq \pi$.

**188.** For what values of $a$ does the equation $x \ln |x| = a$ possess one root?

## Transformation of Trigonometric Expressions

Find which of the following given numbers is greater **(189-191).**

**189.** $\sin 1980°$ or $\cos 1980°$?

**190.** $\tan 1$ or $\arctan 1$?

**191.** $\sin 2$ or $\cos 3$?

Calculate **(192-214).**

**192.** $\arcsin (-1) + 2 \arctan (-\sqrt{3})$.

**193.** $\sin (\arcsin(3/5) - \arccos (3/5))$.

**194.** $\tan (\arccos (1/2) + \arcsin (\sqrt{3}/2))$.

**195.** $\arctan 1 + \arcsin (1/3) + \arccos (1/3)$.

**196.** $\arctan \sqrt{2} - \text{arccot} (1/\sqrt{2})$.

**197.** $\sin (2 \arctan 2)$.

**198.** $\sin ((1/2) \arccos (7/32))$.

**199.** $\sin ((1/2) \arctan (4/3))$.

**200.** $\arcsin (\sin 3)$.

**201.** $\arccos (\cos (8\pi/7))$.

**202.** $\arctan (\tan (8\pi/7))$.

**203.** $\sin \alpha$ if $\sin(\alpha/2) + \cos (\alpha/2) = 7/5$.

**204.** $\sin 2\alpha$ if $\sin \alpha + \cos \alpha = b$.

**205.** $\sin 2\alpha$ if $\cot \alpha = -7$.

**206.** $\dfrac{\cos^4 \alpha + \sin^3 \alpha \cos \alpha}{\sin^2 2\alpha}$ if $\tan \alpha = 3$.

**207.** $\dfrac{\sin \alpha}{\sin^3 \alpha + 3\cos^3 \alpha}$ if $\tan \alpha = 2$.

**208.** $\tan^2 \alpha + \cot^2 \alpha$ if $\tan \alpha + \cot \alpha = a$.

**209.** $\cot (\alpha/2)$ if $\sin \alpha = -3/5$.

**210.** $\tan (\alpha/2)$ if $\cos \alpha = -3/5$, $\pi/2 < \alpha < \pi$.

**211.** $\tan (\alpha/2)$ if $\sin 2\alpha = 0.6$, $0 < \alpha < \pi/4$.

**212.** $65 \cos (B - A)$ if $\sin A = -4/5$, $3\pi/2 < A < 2\pi$; $\cos B = 5/13$, $0 < B < \pi/2$.

**213.** $\sin \alpha$, $\cos \alpha$, and $\cot \alpha$ if $\tan \alpha = 12/5$, $\pi < \alpha < 3\pi/2$.

**214.** $\sin 2\alpha$ and $\sin(\alpha/2)$ if $\tan \alpha = 5/12$, $\pi < \alpha < 3\pi/2$.

**215.** Transform $1 + \dfrac{\sin\left(\dfrac{3\pi}{2} - \alpha\right)}{\sin\left(\dfrac{\pi}{2} + \alpha\right) - 1}$ into a product.

**216.** Prove that if $\alpha + \beta = \pi/4$, then $(1 + \tan \alpha)(1 + \tan \beta) = 2$.

**217.** Prove that if $\alpha$, $\beta$, and $\gamma$ are angles of a triangle, then
$\sin \alpha \times \sin \beta - \cos \gamma = \cos\alpha \cos\beta$.

**218.** Simplify the expression $\dfrac{\sin \alpha + \sin 2\alpha - \sin(\pi + 3\alpha)}{2\cos\alpha + 1}$.

Prove the following identities **(219-223)**.

**219.** $\sin^4 \alpha \cos^2 \alpha = \dfrac{1}{16} - \dfrac{1}{32}\cos 2\alpha - \dfrac{1}{16}\cos 4\alpha + \dfrac{1}{32}\cos 6\alpha$.

**220.** $2(\sin^6 \alpha + \cos^6 \alpha) - 3(\sin^4 \alpha + \cos^4 \alpha) + 1 = 0$.

**221.** $\tan^2 \alpha - \sin^2 \alpha = \tan^2 \alpha \sin^2 \alpha$.

**222.** $\dfrac{\sin 2\alpha}{1 + \cos 2\alpha} \cdot \dfrac{\cos \alpha}{1 + \cos\alpha} = \tan \dfrac{\alpha}{2}$.

**223.** $\dfrac{(1 + \cos x)(1 + \cos 2x)}{(1 + \sin x)(1 - \cos 2x)} = \dfrac{1 - \sin x}{1 - \cos x}$.

## Trigonometric Equations and Inequalities

Solve the following equations **(224-234)**.

**224.** $\sin 5x - \sin 3x = 0$.

**225.** $\sin^3 x + \sin^2 x = 1 + \sin x$.

**226.** $\sqrt{3} \sin x + \cos x = 1$.

**227.** $\sin \alpha = \sin 2\alpha$.

**228.** $\cos x - \cos 2x + \cos 3x = 0$.

**229.** $\tan^2 33x = \cos 2x - 1$.

**230.** $6\cos^2 x + 11 \sin x - 10 = 0$.

**231.** $\sin 3x + \cos 4x - 4 \sin 7x = \cos 10x + \sin 17x$.

**232.** $\sin^4 x + \cos^4 x = \cos^2 2x + 0.25$.

**233.** $\sin x \cos x \cos 2x = -1/2$.

**234.** $\tan |x| = |\tan x|$.

**235.** Does the equation $4 \sin 2x + \cos x = 5$ have a solution?

**236.** Solve the equation $f'(0) = f'(x)$ if $f(x) = 5\sin x + 3\cos x$.

**237.** For what $a$ does the equation $1 + \sin^2 ax = \cos x$ possess a unique solution?

**238.** How many roots does the equation $\log_{(5\pi/2)} x = \cos x$ possess?

Solve the following inequalities **(239-240)**.

**239.** $\sin\left(\dfrac{\pi}{4} - x\right) < \dfrac{1}{2}$.

**240.** $2\sin^2\left(x - \dfrac{\pi}{3}\right) - 5\sin\left(x - \dfrac{\pi}{3}\right) + 2 > 0$.

**241.** Find the maximum value of the function $f(x) = 3\sin x + 4\cos x$.

## Progressions

**242.** The numbers $a^2$, $b^2$, $c^2$ form an arithmetic progression. Prove that the numbers $\dfrac{1}{b+c}$, $\dfrac{1}{c+a}$, $\dfrac{1}{a+b}$ also form an arithmetic progression.

**243.** The sum of an infinitely decreasing geometric progression is 12 and the sum of the squares of its terms is 48. Find the sum of the first ten terms of the progression.

## Limits

**244.** Is the sequence $a_n = \dfrac{n^2 - n + 1}{n^2 + n + 1}$, $n \in \boldsymbol{N}$, monotonic?

Calculate the following limits **(245-266)**.

**245.** $\lim\limits_{n\to\infty} \dfrac{n^2}{1 - 9n^2}$, $n \in \boldsymbol{N}$.

**246.** $\lim\limits_{n\to\infty} \dfrac{10n + 1}{0.1n^2 + 0.001n}$, $n \in \boldsymbol{N}$.

**247.** $\lim\limits_{n\to\infty} \dfrac{3n^4 + 2n^2 + 3}{12n^4 - 7n^2 + n}$, $n \in \boldsymbol{N}$.

**248.** $\lim\limits_{n\to\infty} \dfrac{\dfrac{1}{2} + 1 + \dfrac{3}{2} + \ldots + \dfrac{n}{2}}{0.25n^2 + n + 3}$, $n \in \boldsymbol{N}$.

**249.** $\lim\limits_{n\to\infty} \left(\dfrac{1}{n^2} + \dfrac{2}{n^2} + \ldots + \dfrac{n-1}{n^2}\right)$, $n \in \boldsymbol{N}$.

**250.** $\lim\limits_{n\to\infty} \dfrac{2^n + 5^n}{3^n + 5^n}$, $n \in \boldsymbol{N}$.

**251.** $\lim\limits_{n\to\infty} \dfrac{5^{n+1} + 3^n - 2^{2n}}{5^n + 2^n + 3^{n+3}}$, $n \in \boldsymbol{N}$.

**252.** $\lim\limits_{x\to 3} \dfrac{x^2 - 5x + 6}{x - 3}$.

**253.** $\lim\limits_{x\to -3} \dfrac{x^2 - 9}{x^2 + x - 6}$.

**254.** $\lim\limits_{x\to 5} \dfrac{x^2-8x+15}{(x+3)(x-5)}$.

**255.** $\lim\limits_{x\to 1} \dfrac{x^2+5x-6}{x-x^2}$.

**256.** $\lim\limits_{x\to -2/3} \dfrac{6x^2-5x-6}{3x^2-x-2}$.

**257.** $\lim\limits_{x\to -3} \dfrac{x+3}{\sqrt{x+4}-1}$.

**258.** $\lim\limits_{x\to 0} \dfrac{\sqrt{6+x}-\sqrt{6-x}}{x}$.

**259.** $\lim\limits_{x\to 7} \dfrac{2-\sqrt{x-3}}{x^2-49}$.

**260.** $\lim\limits_{x\to -1} \dfrac{\sqrt{2x+3}-1}{\sqrt{5+x}-2}$.

**261.** $\lim\limits_{x\to 2} \dfrac{\sqrt{x^2-1}-\sqrt{x^2+x-3}}{x^2-4}$.

**262.** $\lim\limits_{x\to 64} \dfrac{\sqrt{x}-8}{\sqrt[3]{x}-4}$.

**263.** $\lim\limits_{x\to 0} \dfrac{\sin 3x}{2x}$.

**264.** $\lim\limits_{x\to 0} \dfrac{\cos 4x - \cos 6x}{\sin^2 5x}$.

**265.** $\lim\limits_{x\to 0} \dfrac{\sqrt[3]{1+\sin x}-\sqrt[3]{1-\sin x}}{x}$.

**266.** $\lim\limits_{x\to 2} \dfrac{2^x+2^{3-x}-6}{\sqrt{2^{-x}-2^{1-x}}}$.

**267.** Two rays are drawn through a point $A$ at the angle of $30°$. A point $B$ is taken on one of them at the distance $a$ from the point $A$. A perpendicular is drawn from the point $B$ to the other ray, another perpendicular is drawn from its foot to [$AB$], and so on. Find the length of the resulting infinite polygonal line.

## The Derivative. Investigation of a Function with the Aid of the Derivative

Find the derivatives of the following functions (268-271).

**268.** (a) $\cot^3 x$; (b) $\sin\sqrt{x}$.

**269.** $\log_2 (2x^2-3x+1)$.

**270.** $(2x^3 - 5) \tan x$.

**271.** (a) $\sqrt{\ln x}$; (b) $\sqrt{\sin 2x}$; (c) $(\sin 2x + 8)^3$.

Find the derivatives of the following functions at the indicated points (**272-275**).

**272.** $y = \ln (2 - \sqrt{2x + 1})$, $y'(0) = ?$

**273.** $y = (4x + 5)^2$, $y'(0) = ?$

**274.** $f(x) = \dfrac{\sqrt{x + 1}}{\sqrt{x + 1} + 1}$, $f'(0) = ?$

**275.** $f(x) = \sin 4x \cos 4x$, $f'(\pi / 3) = ?$

Find the intervals of increase and decrease of the following functions (**276-280**).

**276.** $f(x) = 2x + \dfrac{2}{x}$.

**277.** $f(x) = x + \ln (1 - 4x)$.

**278.** $f(x) = x^2 e^{-x}$.

**279.** $f(x) = x / \ln x$.

**280.** $y = 3 + 8x + 4x^4$.

Prove that the following functions are increasing (**281-283**).

**281.** $y = \dfrac{x^4}{4} - \dfrac{1}{x} + 2$ for $x > 0$.

**282.** $f(x) = \dfrac{2}{3}x^9 - x^6 + 2x^3 - 3x^2 + 6x - 1$ for $x \in \mathbf{R}$.

**283.** $y = 2x + \sin x$ for $x \in \mathbf{R}$.

Find the values of $a$ for which the following functions increase throughout the number axis (**284-286**).

**284.** $f(x) = \dfrac{a^2 - 1}{3}x^3 + (a - 1)x^2 + 2x + 1$.

**285.** $f(x) = 2e^x - ae^{-x} + (2a + 1)x - 3$.

**286.** $y = \sin x - a \sin 2x - \dfrac{1}{3}\sin 3x + 2ax$.

**287.** For what values of $a$ does the function $y = (a + 2)x^3 - 3ax^2 + 9ax - 1$ decrease monotonically throughout the number axis?

Find the critical points of the following functions (**288-289**).

**288.** $f(x) = (x^2 - 4)^{10}$.

**289.** $y = \dfrac{x^3}{3} + 2x^2 - 5x + 4$.

**290.** Find the extremum of the function $y = x^2 - \ln (1 + 2x)$.

**291.** For what values of $a$ do the points of extremum of the function $y = x^3 - 3ax^2 + 3(a^2 - 1)x + 1$ lie in the interval $(-2, 4)$?

Find the greatest and the least values of the following functions **(292-296)**.

**292.** $f(x) = \cos^2 x + \cos x + 3$.

**293.** $f(x) = 4x^2 + \dfrac{1}{x}$ on $[1/4, 1]$.

**294.** $f(x) = \sqrt{x} - 2\sqrt[4]{x}$ on $[0, 100]$.

**295.** $f(x) = e^{x^2 - 4x + 3}$ on $[-5, 5]$.

**296.** $y = 2x - \sqrt{x}$ on $[0, 4]$.

**297.** For what values of $a$ does the function $f(x) = x^3 + 3(a-7)x^2 + 3(a^2 - 9)x - 1$ have a positive point of maximum?

**298.** Assume that $x_1$ and $x_2$ are a point of maximum and a point of minimum of the function $f(x) = 2x^3 - 9ax^2 + 12a^2x + 1$ respectively. For what $a$ does the equality $x_1^2 = x_2$ hold true?

**299.** For what $a$ does the function $f(x) = (a/3)x^3 + (a+2)x^2 + (a-1)x + 2$ possess a negative point of minimum?

**300.** Find all the values of the parameter $a$ for which the points of minimum of the function $y = 1 + a^2x - x^3$ satisfy the inequality $\dfrac{x^2 + x + 2}{x^2 + 5x + 6} \le 0$.

**301.** At what point of the interval $(0, \pi/2)$ does the function $y = (\tan x + 1)^2 / \tan x$ assume the least value?

**302.** Find the number which, being added to its square, yields the least sum.

**303.** Find the positive number which, being added to its inverse, yields the least sum.

**304.** Among isosceles triangles with a given length $a$ of a lateral side find the triangle with the greatest area.

**305.** Find the vertex angle of an isosceles triangle of the greatest area inscribed in a circle of radius $R$.

## Tangents to Curves

**306.** Find the slope of the tangent drawn to the graph of the function $y = \tan x$ at the point with abscissa $x_0 = \pi/4$.

**307.** On the curve $y = 4x^2 - 6x + 3$ find the point at which the tangent is parallel to the straight line $y = 2x$.

**308.** At what angle does the sine curve $y = (1/\sqrt{3})\sin 3x$ cut the abscissa axis at the origin?

**309.** Construct a tangent to the graph of the function $y = \ln x^2$, parallel to the straight line $y = -x$.

Write the equation of a tangent to the graphs of the following curves at the indicated points **(310-315)**.

**310.** $y = \sin 2x$ at the point $x = \pi / 12$

**311.** $y = x^3$ at the point $x = 2$.

**312.** $y = x^2 e^{-x}$ at the point $x = 1$.

**313.** $y = \sin x + 1$ at the point $x = \pi / 2$

**314.** $y = x^2 - 4$ at the point of its intersection with the axis of ordinates.

**315.** $y = 2x^2 - 4x$ at the points of intersection of this graph with the abscissa axis.

**316.** Write the equation of the tangent to the curve $y = x^2 - 7x + 3$ which is parallel to the straight line $5x + y - 3 = 0$.

**317.** Prove that the tangent to the hyperbola $y = a^2 / x$ forms a triangle of constant area with the coordinate axes.

## Miscellaneous Problems

**318.** Differentiating the identity $\sin 2x = 2 \sin x \cos x$ term-by-term, prove the identity $\cos 2x = \cos^2 x - \sin^2 x$.

**319.** At what points does the derivative of the function $y = x^3$ coincide with the value of the function itself?

**320.** Does the straight line $x + 4y - 4 = 0$ touch the hyperbola $y = 1 / x$?

**321.** Onto what interval does the derivative of the function $y = \sqrt[4]{x^3}$ map the interval $[1 / 16, \ 81]$?

**322.** Are the following functions even or odd : (a) $f(x) = 2^x + 2^{-x}$;
(b) $f(x) = \sqrt{1 + x + x^2} - \sqrt{1 - x + x^2}$; (c) $f(x) = \log \dfrac{1+x}{1-x}$.

**323.** Find $f(g(x))$ and $g(f(x))$ if $f(x) = 2^x$, and $g(x) = x^2$.

**324.** Find the quadratic function $f(x)$ if $f(0) = 1$, $f(1) = 0$, and $f(3) = 5$.

**325.** Find the inverse of the function $y = x^2 - 1$, $x \in (-\infty, \ 0]$.

**326.** Through what point $A$ on the curve $y = -x^2 + 2x$ must the tangent to that curve pass for the trapezoid formed by the tangent and the straight lines $x = 0$, $y = 0$, and $x = 1$ to have the least area?

## The Anti derivative. The Integral

**327.** Find the antiderivative of the function $y = 2 / \sin^2 3x$ whose graph passes through the point $(\pi / 12, \ 1)$.

**328.** Find the antiderivative $F(x)$ of the function $f(x) = \cos 4x$ if $F(\pi / 24) = 1$.

**329.** For what values of $x$ does the antiderivative of the function $f(x) = \pi \sin \pi x + 2x - 4$ which has the value 3 for $x = 1$ vanish?

**330.** Find the antiderivative of the function $(\sin(x / 2) + \cos(x / 2))^2$.

Calculate the following integrals (331-338).

**331.** $\int_0^{\pi/2} \sin 2x \, dx.$

**332.** $\int_0^{\pi/4} 8 \cos 3x \, dx.$

**333.** $\int_1^2 \dfrac{dx}{x^3}.$

**334.** $\int_0^{2e} \dfrac{dx}{0.5x+1}.$

**335.** $\int_0^{\pi} \cos^2 x \, dx.$

**336.** $\int_0^{\pi/2} \sin x \cos x \, dx.$

**337.** $\int_{-\pi/2}^{\pi/2} |\sin x| \, dx.$

**338.** $\int_0^{\pi} \sin x \cos 3x \, dx.$

Find the area of the figure bounded by the following curves (339-349).

**339.** $y = x^2, \ y = 0, \ x = 5.$

**340.** $y = x^2 - 2x + 5, \ y = 0, \ x = 2, \ x = 4.$

**341.** $y = x^3, y = 27, x = 0.$

**342.** $y = x^3, y^2 = x.$

**343.** $y = \sin x, \ y = 2x / \pi.$

**344.** $y = \sin x, \ y = 2x / \pi, \ 0 \le x \le \pi / 2.$

**345.** $y = \dfrac{x^2}{4} - 1, \ y = 2 - x.$

**346.** $y = (x-1)^2, \ y = 0, \ x = 2, \ x = 3.$

**347.** $y = 2 + \sin x, \ y = 1 + \cos^2 x, \ x = 0, \ x = \pi.$

**348.** $y = -3x^2 - |x| + 2, \ y = 0.$

**349.** $|y| = 1 - x^2.$

**350.** Is it true that the area of the figure bounded by the curves $y = e^{x-1}, y = 0,$ $x = 2,$ and $x = 0$ is smaller than 2?

**351.** Calculate the area of the figure bounded by the graph of the function $y = x^2 + 1$ and the tangents drawn to that graph at the points with abscissas $x = 0,$ and $x = 2.$

352. Through the origin of coordinates draw a straight line which divides the curvilinear triangle with vertex at the origin, bounded by the curves $y = 2x - x^2$, $y = 0$, and $x = 1$, into two parts of the same size.

353. For what $k$ is the area of the figure bounded by the curves $y = x^2 - 3$ and $y = kx + 2$ the least? Calculate that area.

354. Calculate the volume of a space figure resulting from the rotation of a plane figure, bounded by the graph of the function $y = 1 + \cos^2 x$, about the straight line $y = 1$ on the interval $[-\pi / 2, \pi / 2]$ and that straight line.

355. Find all positive $a$ which satisfy the condition $\int_0^a (3x^2 + 4x - 5)\, dx$ $= a^3 - 2$.

356. Find all values of $a$ for which the inequality $\int_0^a x\, dx \leq a + 4$ is satisfied.

## Vector Algebra

357. Calculate the length of the vector $2\mathbf{a} + 3\mathbf{b}$ if $\mathbf{a} = (1, 1, -1)$ and $\mathbf{b} = (2, 0, 0)$.

358. For what value of $k$ is the length of the vector $\mathbf{a} = (-2, 2, 4k)$ half that of the vector $\mathbf{b} = (3, 3k, 0)$?

359. Calculate the length of the vector $\mathbf{a}$ if $\mathbf{b} = (3, -2, 1)$ and $\mathbf{a} \cdot \mathbf{b} = 7$, $\mathbf{a} \| \mathbf{b}$.

360. Find the angle between the vectors $\mathbf{a} = (3, 1, -2)$ and $\mathbf{b} = (-2, 3, 4)$.

361. Find the value of $m$ for which the vectors $\mathbf{a} = 2\mathbf{i} + m\mathbf{j} - 3\mathbf{k}$ and $\mathbf{a} = \mathbf{i} - 2\mathbf{j} + \mathbf{k}$ are perpendicular.

362. For what value of $a$ is the angle between the vectors $\mathbf{x} = \mathbf{i} + \mathbf{j} + \mathbf{k}$ and $\mathbf{y} = a\mathbf{i} + \mathbf{j} - \mathbf{k}$ equal to arccos $(1/2 \sqrt{3})$?

363. The vector $\mathbf{a}$ is perpendicular to the vectors $\mathbf{b} = (1, 2, 3)$ and $\mathbf{c} = (-2, 4, 1)$ and satisfies the conditions $\mathbf{a} (\mathbf{i} - 2\mathbf{j} + \mathbf{k}) = -6$. Find $\mathbf{a}$.

364. Find the length of the diagonals of the parallelogram constructed on the vectors $\mathbf{a} = 5\mathbf{p} + 2\mathbf{q}$ and $\mathbf{b} = \mathbf{p} - 3\mathbf{q}$ if it is known that $|\mathbf{p}| = 2\sqrt{2}$, $|\mathbf{q}| = 3$, and $(\widehat{\mathbf{p}, \mathbf{q}}) = \pi / 4$.

365. It is known that $|\mathbf{a}| = |\mathbf{b}| = |\mathbf{c}| = 1$ and $\mathbf{a} + \mathbf{b} + \mathbf{c} = 0$. Prove that $\mathbf{ab} + \mathbf{bc} + \mathbf{ca} = -3/2$.

366. Use vectors to prove that the midline of a triangle is parallel to the third side, and its length is equal to half the length of the third side.

**367.** Given a triangle $ABC$ and an arbitrary point $M$ on the side $AB$. The straight line drawn through the point $M$ is parallel to the median $CC_1$, cuts $(CA)$ at a point $P$ and $(CB)$ at a point $O$. Prove that $\overrightarrow{PM} + \overrightarrow{OM} + 2\overrightarrow{CC_1}$.

**368.** Prove that the relation $\overrightarrow{MP} = \dfrac{1}{2}(\overrightarrow{AD} + \overrightarrow{BC})$ holds true in any quadrangle $ABCD$, where $M$ and $P$ are the midpoints of the segments $AB$ and $CD$ respectively.

## Plane Geometry

**369.** Construct a right triangle from its hypotenuse and the altitude dropped onto the hypotenuse.

**370.** Construct circles externally tangent to each other with centres at the vertices of a given triangle.

**371.** Given an angle and a point $A$ in its exterior. Construct a straight line which passes through the point $A$ and cuts off a triangle with a given perimeter $p$ from the angle.

**372.** Given a circumference $C$ and a point $A$ which lies outside the circle bounded by the circumference $C$. Construct straight lines which pass through the point $A$ and are tangent to the circumference $C$.

**373.** Can a triangle be constructed from the segments which are equal to the medians of the triangle?

**374.** Two trapezoids with parallel respective sides are inscribed in a circle. Prove that the diagonals of the trapezoids are equal.

**375.** Prove that by connecting the midpoints of the sides of a convex quadrangle we get a parallelogram. When is that parallelogram a rhombus? A square?

**376.** Prove that if a circle can be inscribed in a polygon, then $r = S / p$, where $r$ is the radius of the circle, $S$ is the area, and $p$ is half the perimeter of the polygon.

**377.** Prove that any point of a convex quadrangle belongs to at least one of the circles whose diameters are the sides of the quadrangle.

**378.** Prove that the ratio of the sum of the squares of the lengths of the medians of a triangle to the sum of the squares of the lengths of its sides is $3/4$.

**379.** Prove that the area of a triangle is smaller than unity if the lengths of all the bisectors are smaller than unity.

**380.** Prove that the sum of the squares of the lengths of all sides and all diagonals of a regular polygon is equal to $n^2 r^2$, where $n$ is the number of the sides of the polygon and $r$ is the radius of the circumscribed circle.

381. The straight line, which is parallel to the base of a triangle, divides its area in half. In what ratio does it divide it lateral sides?

382. Find the angles of a rhombus whose diagonal is equal to a side.

383. Find the lengths of the sides of a triangle if they are expressed by integers, form an arithmetic progression, and the perimeter of the triangle is equal to 15.

384. Find the angles of a triangle in which the centres of the inscribed and the circumscribed circle are symmetric about one of the sides of the triangle.

385. Find the acute angle between the medians of an isosceles right triangle which are drawn from the vertices of its acute angles.

386. Calculate the area of an isosceles trapezoid if its altitude is $h$, and a lateral side can be seen from the centre of the circumscribed circle at an angle $\alpha$.

387. A square with side $a$ is revolved about the centre through 45°. Find the area of the common part of the "old" and the "new" square.

388. Given a square and a circle of the same area. What is larger, the length of the circle or the perimeter of the square?

389. Is there a triangle whose all altitudes are smaller than 1 cm and the area is larger than or equal to 10 cm$^2$?

390. Find the radius of the sector if its area is 144 cm$^2$ and the arc contains 4/9 radians.

391. The perimeter of a circular sector is $l$. Find the central angle of the sector for which the area of the sector is the greatest.

## Solid Geometry

392. The radius of a sphere increased by 50 percent. By how many per cent did the surface area of the sphere increase?

393. Find the volume of a parallelepiped with edges $a$, $b$, and $c$ making angles $\pi / 2$, $\alpha$, and $\alpha$ with each other.

394. The total surface area of a regular tetragonal pyramid is $S$ and the edge angle of a lateral face is $\alpha$. Find the altitude of the pyramid.

395. The volume of a regular tetragonal pyramid is $V$ and its altitude is $H$. Find the length of the apothem of the pyramid.

396. The edge angle of a regular tetragonal pyramid is $\alpha$. Find the dihedral angle between a lateral face and the base of the pyramid.

397. Find the angle between the nonintersecting edges of a regular triangular pyramid.

398. A cube with an edge $a$ is cut from the angles by planes so that a regular octagon remains from each face. Find the volume of the polygon obtained.

399. A cylinder with the altitude $h$ is inscribed in a cone with the altitude $H$ and the radius of the base $R$. Find the radius of the base of the cylinder.

**400.** A regular triangular pyramid with an edge angle $\alpha$ is inscribed in a sphere. Find the ratio of the volume of the sphere to that of the pyramid.

## Miscellaneous Problems

**401.** Assume $f(x) = \ln \dfrac{1-x}{1+x}$. For what $a$ and $b$ is the equality $f(a) + f(b) = f\left(\dfrac{a+b}{1+ab}\right)$ satisfied?

**402.** Calculate $x_1^4 + x_2^4$, where $x_1$ and $x_2$ are the roots of the equation $3x^2 - 5x - 1 = 0$.

Solve the following equations **(403-409)**.

**403.** $3^x + 1 - |3^x - 1| = 2\log_5 |6 - x|$.

**404.** $|x-1|^{\log^2 x - \log x^2} = |x-1|^3$.

**405.** $2\pi \cos x = |x| - |x - \pi|$.

**406.** $\log(2^x + x - 41) = x(1 - \log 5)$.

**407.** $\cos x + \cos 2x + \cos 3x + \cos 4x + \cos 5x = 0$.

**408.** $\sqrt{5 - \log_2 x} = 3 - \log_2 x$.

**409.** $\cos 3x + \sin\left(2x - \dfrac{7\pi}{6}\right) = -2$.

**410.** How many roots of the equation $\cos^2 x + \dfrac{\sqrt{3}+1}{2}\sin x - \dfrac{\sqrt{3}}{4} - 1 = 0$ lie in the interval $[-\pi,\ \pi]$?

How many roots do the following equations possess ? **(411-413)**.

**411.** $x^4 + x^3 = 10$.

**412.** $3^{|x|} |2 - |x|| = 1$.

**413.** $x^2 - 2x - \log_2 |1 - x| = 3$.

**414.** For what values of $a$ does the equation $x^3 + ax + 2 = 0$ possess three roots?

**415.** For what values of $a$ does the equation $x^2 e^x = a$ possess three roots?

**416.** For what values of $a$ does the equation $|\ln x| - ax = 0$ possess three roots?

**417.** Find the solution of the equation $\cos^2 x = 1$ for which $x^2 \leq 20$.

**418.** Find $\sin 2\alpha$ if $\cos\alpha$ satisfies the equation $25\cos^2\alpha + 5\cos\alpha - 12 = 0$ and $\pi/2 < \alpha < \pi$.

Solve the following inequalities **(419-423)**.

**419.** $\sqrt{25 - x^2} \leq 12/x$.

**420.** $\log_{(1+x^2)/2|x|}(5 - x^2) \geq 0$.

**421.** $0.11^{\log_3((4x-1)/(3x+2))} > 1$.

**422.** $3\sqrt{6 + x - x^2} > 4x - 2.$

**423.** $\dfrac{x+1}{x+5} + \dfrac{x}{x-1} < \dfrac{2x^2 + 5x}{x^2 + 4x - 5}.$

Find the following limits **(424-425)**.

**424.** $\displaystyle\lim_{x\to 0} \dfrac{x^2}{1 - \cos x}.$

**425.** $\displaystyle\lim_{x\to -1} \dfrac{\cos 2 - \cos 2x}{x^2 - |x|}.$

Find the intervals of monotonicity and the points of extremum of the following functions **(426-427)**.

**426.** $y = xe^{-3x}.$

**427.** $y = x / \ln x.$

**428.** Prove that the function $y = 3x^4 - 4x^3 + 6x^2 + ax + b$ has only one point of extremum for any $a$ and $b$.

**429.** For what values of $a$ does the function $y = (\sqrt{3}\cot x + a^2)^3 \times \tan x$ attain its least value on the interval $(0,\, \pi / 2)$ at the point $x = \pi / 3$?

**430.** For what values of $a$ does the minimum value of the function $y = x^2 - 4ax - a^4$ assume the greatest value?

**431.** Using the geometric meaning of the integral, calculate $\displaystyle\int_0^2 \sqrt{2x - x^2}\,dx.$

**432.** Calculate the integral $\displaystyle\int_0^3 (3x - x^2)\,dx$ and give its geometrical interpretation.

**433.** A particle moves according to the law $s(t) = -2t^2 + 8t + 7$ until its velocity vanishes. Find the path traversed.

**434.** Find the greatest volume of a triangular pyramid $MABC$ whose base is an isosceles right triangle $ABC$ ($|AB| = |BC|$) if $[MB] \perp (ABC)$ and $|MA| = \sqrt{3}.$

**435.** Calculate the volume of the figure resulting from the rotation, about the abscissa axis, of a plane figure bounded by the curve $y = \cos x$ and the abscissa axis, $x \in [-\pi / 2,\, \pi / 2]$.

**436.** If the point $(x,\, y,\, z)$ lies on the plane $x + y + z = 3$, then $x^2 + y^2 + z^2 \geq 3$. Prove this fact.

**437.** The pyramid is defined by the coordinates of the vertices $S\,(0,\, 0,\, 2)$, $A\,(0,\, 0,\, 0)$, $B\,(1,\, 0,\, 0)$, and $C\,(0,\, 1,\, 0)$. Find the coordinates of the point $M$ which lies on the $Oz$ axis and the coordinates of the point $N$ which lies in the plane $(SBC)$ if it is known that $\overrightarrow{MN} = (1/3,\, 1/3,\, 0)$.

**438.** Given points $A\,(2,\, 1,\, -1)$, $B\,(3,\, 2,\, -1)$, and $C\,(3,\, 1,\, 0)$. Find the angle between the vectors $\overrightarrow{AB}$ and $\overrightarrow{AC}$.

# Hints and Answers

# Part-1

## Sec. 1.1

**1.** 24. *Solution.* Assume that $x$ is the number of tens in the required number. Then the required number can be written in the form $10x + y$. Under these assumptions the problem reduces to the solution of the system of equations

$$\begin{cases} x + y = 6, \\ 10x + y + 18 = 10y + x, \end{cases}$$

solving which we get $x = 2$, $y = 4$.

**2.** 45. *Solution.* We write the required number in the form $10x + y$, where $x$ is the tens digit and $y$ is the unit digit in the decimal notation of the number. We write the hypothesis as a system of equations

$$\begin{cases} (10x + y)(x + y) = 405, \\ (10y + x)(x + y) = 486, \end{cases}$$

whence we get $(10x + y)/(10y + x) = 5/6$, or $x = 4y/5$. Substituting this value of $x$ into the first equation of the system, we get $y^2 = 25$, $y = 5$ ($y = -5$ does not satisfy the hypothesis). Then we find that $x = 4$.

**3.** 8.5, 10, 11.5.   **4.** 3, 12.   **5.** 49, 1.   **6.** 6, 54.

**7.** 24.   **8.** 63.

**9.** 24. *Solution.* We write the required number in the form $10x + y$, where $x$ is the tens digit and $y$ is the units digit in the decimal notation of the number. By the hypothesis

$$\begin{cases} 10x + y = 3xy, \\ 10x + y + 18 = 10y + x. \end{cases}$$

Solving the system of equations and neglecting the extraneous solution ($x = -1/3$, $y = 5/3$), we get 24.

**10.** 32.   **11.** 27.

**12.** 64. *Hint.* Assume that $x$ is the tens digit and $y$ is the units digit in the decimal notation of the required number. By the hypothesis,

$$\begin{cases} 10x + y = x^2 + y^2 + 12, \\ 10x + y = 2xy + 16. \end{cases}$$

Subtracting the second equation from the first equation of the system, we get $(x - y)^2 = 4$, whence $x - y = 2$, $x - y = -2$. Solving each of these equations simultaneously, say, with the second equation of the system and neglecting the extraneous solutions, we get $x = 6$, $y = 4$.

**13.** 13, 31. *Solution.* We write the required number in the form $10x + y$, where $x, y \in \mathbf{N}$. By the hypothesis

$$\begin{cases} x^2 + y^2 = 10, \\ (10x + y)(10y + x) = 403. \end{cases}$$

We can rewrite the second equation of the system, with due regard for the first equation, in the form $xy = 3$. The system of equations assumes the form

$$\begin{cases} x^2 + y^2 = 10, \\ xy = 3, \end{cases} \Leftrightarrow \begin{cases} (x + y)^2 = 16, \\ xy = 3. \end{cases}$$

The last system is equivalent to two systems

$$\begin{cases} x + y = 4, \\ xy = 3, \end{cases} \begin{cases} x + y = -4, \\ xy = 3. \end{cases}$$

The first system has the following solutions : (1) $x = 1$, $y = 3$ and (2) $x = 3$, $y = 1$. The second system has the following solutions : (3) $x = -1$, $y = -3$ and (4) $x = -3$, $y = -1$. Only the solutions of the first system serve as solutions of the problem in the set of integral nonnegative number.

**14.** 23. *Hint.* We write the required number in the form $10x + y$, where $x$ is the tens digit and $y$ is the units digit. By the hypothesis

$$\begin{cases} 10x + y = 4(x + y) + 3, \\ 10x + y = 3xy + 5. \end{cases}$$

Solving the system, we get $x = 2$, $y = 3$.

**15.** 91.　　**16.** 64.　　**17.** 72.　　**18.** 71.　　**19.** 32.

**20.** 15, 95.　　　　**21.** 7, 8.

**22.** The denominator of the fraction is equal to 9.

**23.** 3/5.

**24.** 4/15. *Hint.* We write the required fraction in the form $n / (n^2 - 1)$, where $n$ is an integer, $n \neq 0$. By the hypothesis

$$\begin{cases} \dfrac{n + 2}{n^2 + 1} > \dfrac{1}{3}, \\ 0 < \dfrac{n - 3}{n^2 - 4} < \dfrac{1}{10}. \end{cases}$$

Solving this system of inequalities and taking into account that $n$ is an integer, $n \neq 0$, we get $n = 4$.

**25.** 13, 63. *Hint.* Assume that $x$ is the tens digit and $y$ is the units digit. By the hypothesis

$$10x + y = x^2 + xy + y^2, 10(x + 5) + y = (x + 5)^2 + (x + 5)y + y^2,\ \text{or}$$

$$\begin{cases} 10x + y = x^2 + xy + y^2, \\ 25 - 4y = x^2 + xy + y^2. \end{cases}$$

Solving the system of equations and neglecting the extraneous solutions, we get $x = 1$, $y = 3$.

**26.** 1210. *Solution.* We can write the number in the form $4n + 1$, where $n = 3, 4, 5, \ldots , 24$ (there are 22 numbers of this kind). These numbers form an arithmetic progression, whose first term is 13 and the last term is 97. The sum of all numbers is $S = \dfrac{13 + 97}{2} \cdot 22 = 110 \cdot 11 = 1210$.

**27.** 99270.　　　　**28.** 676.　　　　**29.** 49500.　　**30.** 45, 54.

**31.** (7, 2), (9, 6), (23, 22). *Solution.* Assume that $n$ and $m$ are the required natural numbers. By the hypothesis $n^2 - m^2 = 45 = 1 \cdot 3 \cdot 3 \cdot 5$, or

$$(n - m)(n + m) = 1 \cdot 3 \cdot 3 \cdot 5. \tag{1}$$

Since $n$ and $m$ are natural numbers, $n + m > 0$, the right-hand side of the equation is a positive number, it follows that $n - m$ is a natural number, and $n + m > n - m$, $n - m$ and $n + m$ are the divisors of the right-hand side of equation (1). Taking all this into account, we get the following systems of equations for $n$ and $m$ :

$$\begin{cases} n - m = 1, \\ n + m = 45, \end{cases} \begin{cases} n - m = 3, \\ n + m = 15, \end{cases} \begin{cases} n - m = 5, \\ n + m = 9. \end{cases}$$

Solving these systems, we get pairs of numbers (23, 22), (9, 6), (7, 2).

**32.** 156.     **33.** 34, 51.     **34.** 5 and 105, 15 and 35.

**35.** 144 and 864. *Solution.* Assume that $x$ and $y$ are the required numbers. By the hypothesis $x / y = 6k$, where $k \in N$. Since $x$ and $y$ are three-digit numbers, it follows that the equation $x / y = 6k$ is possible only for $k = 1$ and, therefore, $x = 6y$. Next, by the hypothesis, $x + y = 504m$, where $m \in N$, and $m$ can only assume the values 1, 2, 3 since $504m < 2000$. Solving the system of equations $x + y = 504m$, $x = 6y$, we get $y = 72m$. For $m = 1$, $y = 72$ and $x = 432$ do not satisfy the condition of the problem; for $m = 2$, $y = 144$ and $x = 864$ is a solution of the problem ; for $m = 3$, $y = 216$ and $x = 1296$ do not satisfy the condition of the problem.

**36.** $19 = 3^3 - 2^3$. *Hint.* Factoring the left-hand side of the equation $m^3 - n^3 = 19$ and using the fact that 19 is a prime number, we can prove that this representation is unique.

**37.** $(2, \sqrt[3]{3}, \sqrt[3]{9}), (1 / \sqrt[3]{2}, -\sqrt[3]{9/2}, \sqrt[3]{3/2})$. *Hint.* Assume that $x, y, z$ are the required numb0ers. By the hypothesis $x^3 = xyz + 2$, $y^3 = xyz - 3$, $z^3 = xyz + 3$. Multiplying the left-hand and right-hand sides of these equations together and designating $xyz$ as $t$, we get an equation $2t^2 - 9t - 18 = 0$ for $t$.

**38.** 52. *Solution.* We write the required number in the form $10x + y$, where $x$ is the tens digit and $y$ is the units digit. By the hypothesis

$$x + y \geq 7, \tag{1}$$
$$x^2 + y^2 \leq 30, \tag{2}$$
$$10x + y \geq 2 (10y + x),$$

or                                    $8x \geq 19y$. $\qquad\qquad$ (3)

It follows from (3) that $y$ can assume the values 0, 1, 2, 3 (since $x \leq 9$). If $y = 0$, then (1) yields $x \geq 7$. These numbers do not satisfy inequality (2). If $y = 1$, then (1) yields $x \geq 6$. These numbers do not satisfy inequality (2). If $y = 2$, then $x \geq 5$. The numbers $x = 5$ and $y = 2$ satisfy all the inequalities. For $y = 2$ and $x > 5$ inequality (2) is not satisfied. Assume $y = 3$. It follows from (3) that $x \geq 8$. Such numbers do not satisfy inequality (2). Thus there are no more solutions.

**39.** 2573. *Solution.* We write the required number in the form $1000x + 100y + 10z + t$, where $x, y, z, t$ are the thousands, hundreds, tens, and units digits, respectively, with $x \neq 0$. By the hypothesis

$$\begin{cases} x + y + z = 14, \\ y + z + t = 15, \\ z = t + 4. \end{cases}$$

and, therefore, $t = x + 1$, $z = x + 5$, $y = 9 - 2x$. By the hypothesis we must find the values of $x, y, z$, and $t$ such that $x^2 + y^2 + z^2 + t^2 = x^2 + (x+1)^2 + (x+5)^2 + (9-2x)^2 = f(x)$ assumes the least value. Since $x \leq 9$, it follows from the equation $z = x + 5$ that $x$ can only assume the values 1, 2, 3, 4. Calculating $f(1)$, $f(2)$, $f(3)$, $f(4)$ and comparing them, we obtain $f(2) < f(1)$, $f(2) < f(3)$, $f(2) < f(4)$. Therefore, $x = 2$, $y = 5$, $z = 7$, $t = 3$.

**40.** 1738. **41.** *Proof.* Assume that $a = n^2 + (n + m)^2$, where $n, m$ are natural numbers. Then $2a = 2n^2 + 2(n^2 + 2mn + m^2) = (2n + m)^2 + m^2$.

**42.** 300.

**43.** $\left\{\dfrac{25}{9}, \dfrac{26}{10}, \dfrac{27}{11}, \dfrac{28}{12}, \dfrac{29}{13}, \dfrac{30}{14}\right\}$.     **44.** 18.

## Sec. 1.2

**1.** 113/15.         **2.** 2111/990.      **3.** 1.        **4.** $x = 30$.

**5.** 8.              **6.** 6.             **7.** 7.        **8.** 1.

**9.** 0. *Solution.* $1 + \sec 20° - \sqrt{3}\cot 40° = \sec 20° + \dfrac{\sin 40° - \sqrt{3}\cos 40°}{\sin 40°}$

$= \sec 20° - \dfrac{2(\sin 60° \cos 40° - \cos 60° \sin 40°)}{2\sin 20° \cos 20°} = \sec 20° - \dfrac{\sin 20°}{\sin 20° \cos 20°}$

$= \sec 20° - \sec 20° = 0.$

**10.** 100.    **11.** 1.        **12.** 1/4.      **13.** $-3/2$.      **14.** 1.

**15.** 1.      **16.** 0.        **17.** 2.        **18.** $\sqrt{6}/2$      **19.** $\sqrt{3}$.

**20.** 10.     **21.** $\sqrt{2}/2$.   **22.** 1.        **23.** 3/2.         **24.** 1.

**25.** 5.      **26.** $-9\sqrt{2}$.   **27.** 10.       **28.** 4.

**29.** 1. *Hint.* Reduce the given expression to the form $a - b$.

**30.** 0.

**31.** $-10$. *Hint.* If $a \in \mathbf{Z}$, then $a^2 \in \mathbf{Z}$.

**32.** $(2 + \sqrt{2} - \sqrt{6})/4$.

**33.** $\dfrac{\sqrt{10}}{10}\sqrt{2\sqrt{3} + \sqrt{2}}\;\sqrt{\sqrt[3]{3} - \sqrt{2}}\;\sqrt{4 + 2\sqrt[3]{9} + 3\sqrt[3]{3}}$.

**34.** $a < b$.

**35.** *Proof.* Since $b < c$, if follows that

$$2b < b + c. \tag{1}$$

Since                     $$b + c < a + 1, \tag{2}$$

it follows from (1) and (2), in accordance with the law of transitivity, that

$$2b < a + 1. \tag{3}$$

Since $1 < a$, it follows that

$$1 + a < 2a. \tag{4}$$

According to the law of transitivity we have from (3) and (4) that $2b < 2a$, i.e. $b < a$.

**36.** $\log_3 108 > \log_5 375$. *Solution.*

$$\log_3 108 = 3 + \log_3 4 > 4; \tag{1}$$
$$\log_5 375 = 3 + \log_5 3 < 4. \tag{2}$$

From (1) and (2) it follows that $\log_3 108 > \log_5 375$.

**37.** 0, $\ln(7/5)$, $\sqrt{0.8}$, 0.9186.

**38.** 0.37, $\tan 33°$, 1, 65/63, 61/59, $\tan(-314°)$. *Hint.* $\tan(-314°)$

$= \tan 46° = \dfrac{1 + \tan 1°}{1 - \tan 1°} = \dfrac{1 + \tan(\pi/180)}{1 - \tan(\pi/180)} > \dfrac{1 + 1/60}{1 - 1/60} = \dfrac{61}{59}$.

**39.** 0.02, 0.85, $-\cos 571°$, $\sqrt{3}/2$, $\sqrt{0.762}$, 1.

**40.** *Hint.* Prove that the last digits of the decimal notation of the numbers $53^{53}$ and $33^{33}$ are equal to 3.

**41.** $(n^2 + 2n + 2) \times (n^2 - 2n + 2)$.

**42.** $(1 + n + n^2)(1 - n + n^2)(1 + \sqrt{3}n + n^2)(1 - \sqrt{3}n + n^2)$.

**43.** $(1 + x)\left(1 - \dfrac{1 + \sqrt{5}}{2}x + x^2\right)\left(1 + \dfrac{1 + \sqrt{5}}{2}x + x^2\right)$

*Solution.*   $1 + x^5 = (1 + x)(1 - x + x^2 - x^3 + x^4) = (1 + x)x^2 \left( x^2 + \dfrac{1}{x^2} - \right.$

$\left( x + \dfrac{1}{x} \right) + 1 \right) = (1 + x)x^2 \left( \left( x + \dfrac{1}{x} \right)^2 - \left( x + \dfrac{1}{x} \right) - 1 \right) = (1 + x)x^2 \times \left( x + \dfrac{1}{x} - \dfrac{1 + \sqrt{5}}{2} \right)$

$\left( x + \dfrac{1}{x} - \dfrac{1 - \sqrt{5}}{2} \right) = (1 + x)\left( x^2 - \dfrac{1 + \sqrt{5}}{2}x + 1 \right) \times \left( x^2 + \dfrac{\sqrt{5} - 1}{2}x + 1 \right)$. **44.** $ab / (a + b)$.

**45.** 0. **46.** $-a^4 / (a^2 + b^2)$. **47.** $4(a - b) / ab$. **48.** $2 / (a + b)$. **49.** $m - n$. **50.** 1. **51.** $\sqrt{2}$.

**52.** $1/2bc$. **53.** $a(a - b - c)/2$. **54.** 1. **55.** $-1$. **56.** $\dfrac{x^3(x^2 - xy - y^2)}{y^4(x + y)}$. **57.** $12 / (a + 2)$.

**58.** $-3a / (a - 1)$. **59.** $2a + 3$.   **60.** $\sqrt{2}$. **61.** $ab$. **62.** $(a - b)/(a + b)$. **63.** $1/a$.

**64.** $a^{-169/60}b^{-31/30}$. **65.** $1/\sqrt[3]{a^2b}$. **66.** 1. **67.** 1. **68.** $2/(1 - x)$. **69.**   2. **70.** $\sqrt{a}$. **71.** $\sqrt{1 - a}$.

**72.** 2. **73.** $(x - 1)\sqrt{x}$. **74.** 1. **75.** $-1$. **76.** $2/(a - 1)$. **77.** 0. **78.** 1. **79.** $x - 1$.

**80.** $2 - 3\sqrt{2}\,y^{1/5} + 9y^{2/5}$. **81.** $-2y$. **82.** 1, if $a > 0$ and $b > 0$; $-1$, if $a < 0, b < 0$.

**83.** $\sqrt{bx}/(\sqrt{b} + \sqrt{x})$. **84.** 1. **85.** 1. **86.** $2ab$. **87.** $\sqrt{a^2 - 1} + a^2 - 1$. **88.** $1/10$.

**89.** $4/(\sqrt{x} + \sqrt{y})$. **90.** $(1 - a)/\sqrt{a}$. **91.** $\sqrt{a}/(\sqrt{a} - \sqrt{b})$. **92.** 1. **93.** 1. **94.** $2(\sqrt{a} + \sqrt{b})$.

**95.** $-\sqrt{ab}$. **96.** $1 + \dfrac{\sqrt{ab}}{a - b}$. **97.** $1 - \dfrac{2b}{3a}$. **98.** $a(a + 1)$. **99.** $\sqrt[4]{a} + \sqrt[4]{b}$. **100.** $a + 1$. **101.** $a^{2/3}$.

**102.** $2b/\sqrt{a^2 - b^2}$. **103.** $1/a(a^{1/m} - a^{1/m})$. **104.** $2/(1 - a)$. **105.** 0.  **106.** $\sqrt[3]{b} - \sqrt[3]{a}$.

**107.** 2. **108.** $-4$. **109.** $a^2 + ab + b^2$.

**110.** $(x - a)/x$. *Solution.* $\dfrac{(x(x^2 - a^2)^{-1/2} + 1)}{a(x - a)^{-1/2} + (x - a)^{1/2}} \cdot \left( \dfrac{x - (x^2 - a^2)^{1/2}}{a^2(x + a)^{1/2}} \right)^{-1}$

$= \dfrac{(x + \sqrt{x^2 - a^2})\sqrt{x - a}}{\sqrt{x^2 - a^2}(a + (x - a))} \cdot \dfrac{(x - \sqrt{x^2 - a^2})}{a^2\sqrt{x + a}}$

$= \dfrac{(x^2 - (x^2 - a^2))}{(x + a) \cdot x \cdot a^2} = \dfrac{1}{x^2 + ax}$

After this transformation, the initial expression reduces to the form

$\left( \dfrac{2}{x^2 - a^2} \right)^{-1} \left( \dfrac{1}{x^2 + ax} + (x^2 + ax)^{-1} \right)$

$= \dfrac{(x - a)(x + a)}{2} \cdot \left( \dfrac{1}{x(x + a)} + \dfrac{1}{x(x + a)} \right)$

$= \dfrac{(x - a)(x + a) \cdot 2}{2x(x - a)} = \dfrac{x - a}{x}$.

**111.** $\sqrt{a}/(\sqrt{a} - \sqrt{b})$. **112.** $3\sqrt{b}$. *Solution.* $\left( \dfrac{2a + b^{1/2}a^{1/2}}{3a} \right)^{-1} \left( \dfrac{a^{3/2} - b^{3/2}}{a - a^{1/2}b^{1/2}} - \dfrac{a - b}{\sqrt{a} + \sqrt{b}} \right)$

$= \dfrac{3\sqrt{a}}{2\sqrt{a} + \sqrt{b}} \left( \dfrac{(\sqrt{a})^3 - (\sqrt{b})^3}{\sqrt{a}(\sqrt{a} - \sqrt{b})} - (\sqrt{a} + \sqrt{b}) \right)$

$= \dfrac{3\sqrt{a}}{2\sqrt{a} + \sqrt{b}} \left( \dfrac{a + \sqrt{ab} + b}{\sqrt{a}} - \sqrt{a} + \sqrt{b} \right)$

$= \dfrac{3\sqrt{a}}{2\sqrt{a} + \sqrt{b}} \cdot \dfrac{a + \sqrt{ab} + b - a + \sqrt{ab}}{\sqrt{a}}$

$= \dfrac{3\sqrt{a}(2\sqrt{a} + \sqrt{b})\sqrt{b}}{(2\sqrt{a} + \sqrt{b})\sqrt{a}} = 3\sqrt{b}$.

**113.** 0. **114.** 0. **115.** 1, if $a > 0, b > 0, a \neq b; -1$, if $a < 0, b < 0, a \neq b$. **116.** $a^2$.

**117.** 3. *Solution.* $\dfrac{(\sqrt{x} - \sqrt{y})^3 + y\sqrt{y} + 2x^2 : \sqrt{x}}{x\sqrt{x} + y\sqrt{y}} + \dfrac{3\sqrt{xy} - 3y}{x - y}$

$$= \dfrac{x\sqrt{x} - 3x\sqrt{y} + 3\sqrt{xy} - y\sqrt{y} + y\sqrt{y} + 2x\sqrt{x}}{x\sqrt{x} + y\sqrt{y}} + \dfrac{3\sqrt{y}\,(\sqrt{x} - \sqrt{y})}{x - y}$$

$$= \dfrac{3x\sqrt{x} - 3x\sqrt{y} + 3y\sqrt{x}}{x\sqrt{x} + y\sqrt{y}} + \dfrac{3\sqrt{y}}{\sqrt{x} + \sqrt{y}}$$

$$= \dfrac{3x\sqrt{x} - 3x\sqrt{y} + 3y\sqrt{x} + 3\sqrt{y}\,(x - \sqrt{xy} + y)}{x\sqrt{x} + y\sqrt{y}} = \dfrac{3x\sqrt{x} + 3y\sqrt{y}}{x\sqrt{x} + y\sqrt{y}} = 3$$

**118.** $-\sqrt{x}$. **119.** 1. **120.** $(1 + a^{2/3})^2 / (1 - a^{1/3})$. **121.** $1 / (1 - x^2)$. **122.** $xy$. **123.** 1.

**124.** $3(\sqrt{a} - \sqrt{b}) / (\sqrt{a} + \sqrt{b})$. **125.** $\sqrt[4]{a} / (\sqrt{a} + 1)$. **126.** $2\sqrt[4]{b / a^2}$. **127.** 1. **128.** $1 / (x^2 - 1)$.

**129.** $\sqrt[4]{m} - \sqrt[4]{n}$. **130.** 2. **131.** $\sqrt[4]{a}$, if $\sqrt[4]{a} > \sqrt[3]{b} \geq 0; -\sqrt[4]{a}$, if $\sqrt[3]{b} > \sqrt[4]{a} \geq 0$.

**132.** $2 / \sqrt{b^3}$. *Solution.* $\dfrac{b - b^{-2}}{b^{1/2} - b^{-1/2}} - \dfrac{1 - b^{-2}}{\sqrt{b} + b^{-1/2}} - b^{1/2}$

$$= \dfrac{b^3 - 1}{b\sqrt{b}\,(b - 1)} - \dfrac{b^2 - 1}{b\sqrt{b}\,(b + 1)} - \sqrt{b}$$

$$= \dfrac{b^2 + b + 1}{b\sqrt{b}} - \dfrac{b - 1}{b\sqrt{b}} - \sqrt{b}$$

$$= \dfrac{b^2 + 2 - b^2}{b\sqrt{b}} = \dfrac{2}{b\sqrt{b}}.$$

**133.** $2(a + 4) / a$.    **134.** 2.    **135.** 1.    **136.** $\sqrt{a} + \sqrt{5}$.

**137.** 1.    **138.** $-3 / 2$.    **139.** $\sqrt[4]{b}$.    **140.** $1 / \sqrt[3]{a}$.

**141.** $-\sqrt[4]{a^2 b}$.    **142.** $(a + 1) / 4$.    **143.** $x^{1/3}$.

**144.** $\sqrt[4]{a} + \sqrt{5}$.    **145.** $(\sqrt{a} + \sqrt{b}) / 2$.    **146.** 3.

**147.** $\sqrt[3]{x^2}$.    **148.** $\sqrt{ab}$.

**149.** $-25$, if $a > 0, |a| > |b| > 0$; 25, if $a < 0, |a| > |b| > 0$.

**150.** 1.

**151.** $6 - 4a$ for $a \in [0, \sqrt{2}]; 2(a - 1)^2$ for $a \in (\sqrt{2}, + \infty)$.

**152.** $-1$.

**153.** 1.

**154.** $\sqrt{1 - x^2}$. *Solution.* $\left( \dfrac{1 + \sqrt{1 - x}}{1 - x + \sqrt{1 - x}} + \dfrac{1 - \sqrt{1 + x}}{1 + x - \sqrt{1 + x}} \right)^2 \dfrac{x^2 - 1}{2} + 1$

$$= \left( \dfrac{1 + \sqrt{1 - x}}{\sqrt{1 - x}(1 + \sqrt{1 - x})} + \dfrac{1 - \sqrt{1 - x}}{\sqrt{1 + x}\,(\sqrt{1 + x} - 1)} \right)^2 \dfrac{x^2 - 1}{2} + 1$$

$$= \left( \dfrac{1}{\sqrt{1 - x}} - \dfrac{1}{\sqrt{1 + x}} \right)^2 \dfrac{x^2 - 1}{2} + 1 = \dfrac{(\sqrt{1 + x} - \sqrt{1 - x})^2}{(\sqrt{1 - x^2})^2} \dfrac{x^2 - 1}{2} + 1$$

$$= - \dfrac{2 - 2\sqrt{1 - x^2}}{2} + 1 = -1 + \sqrt{1 - x^2} + 1 = \sqrt{1 - x^2}.$$

**155.** $a + b$. **156.** $1 / n$. **157.** $-2b$. **158.** (1) $2b(a - b)$; (2) $2b(a - b)$. **159.** $\sqrt[6]{2}$, if $|a| < 1; -\sqrt[6]{2}$, if $1 < |a| \leq \sqrt{2}$. **160.** $1 + \sqrt{bc}$. **161.** $1 / ab\sqrt[5]{a^3}$. **162.** $-2 / (a + \sqrt{a})$; $(\sqrt{5} - 5) / 10$. **163.** $a^2 + ab + b^2$, 2.52. **164.** $a(b - a) / (a + b)$,15 / 4. **165.** $f(x, y) = y + \sqrt{xy}$, $f(9, 0.04) = 0.64$. **167.** $D(y) = (-4, 0) \cup (0, + \infty)$; $y = -4$.

**168.** *Proof.* $\sqrt{x^2 + \sqrt[3]{x^4 y^2}} + \sqrt{y^2 + \sqrt[3]{x^2 y^4}}$

$$= \sqrt{\sqrt[3]{x^4}\left(\sqrt[3]{x^2} + \sqrt[3]{y^2}\right)} + \sqrt{\sqrt[3]{y^4}\left(\sqrt[3]{y^2} + \sqrt[3]{x^2}\right)}$$

$$= \sqrt{\sqrt[3]{x^2} + \sqrt[3]{y^2}}\left(\sqrt[3]{x^2} + \sqrt[3]{y^2}\right) = \left(\sqrt[3]{x^2} + \sqrt[3]{y^2}\right)^{3/2} = a.$$

Hence it follows that $x^{2/3} + y^{2/3} = a^{2/3}$.

**169.** $n^2 / m^2$.

**170.** The domain of definition is $a > 0$, $b > 0$, $a \neq b$; $A = 0$ or $a > b$; $A = \sqrt{a} - \sqrt{b}$ for $a < b$.

## Sec. 1.3

**2.** *Proof.* (1) Let us verify the truth of the proposition for $n = 1 : 4^1 + 15 \cdot 1 - 1 = 18 = 9 \cdot 2$, and, consequently, $A$ (1) is true. (2) Let us assume that for $n = k$, where $k$ is a natural number, we have

$$4^k + 15k - 1 = 9m, \qquad (1)$$

where $m$ is a natural number, i.e. $A(k)$ is true. For $n = k + 1$, taking (1) into account, we have

$4^{k+1} + 15 (k + 1) - 1$

$$= 4 \cdot 4^k + 60k - 4 - (45k - 18) = 4 \left(4^k + 15k - 1\right) - 9 \left(5k - 2\right)$$

$$= 4 \cdot 9m - 9 \left(5k - 2\right)$$

$$= 9 \cdot (4m - 5k + 2) = 9p,$$

where $p = 4m - 5k + 2$ is a natural number. We have carried out both parts of the proof and, therefore, we can prove by induction that the assumption is true for all natural $n$.

**4.** *Remark.* This problem can be easily proved without resort to induction. Let us factor the given expression : $n^3 - n = (n - 1) n (n + 1)$. By the hypothesis, $n$ is an odd number and, consequently, $n - 1$ and $n + 1$ are even numbers ; out of two successive even numbers one number is divisible by 2 and the other by 4 ; in addition, out of three successive integers $n - 1, n, n + 1$ at least one number is divisible by 3. Thus $n^3 - n$ is divisible by $2 \cdot 4 \cdot 3 = 24$.

**6.** *Proof.* (1) For $n = 1$ the proposition is true $1 - \dfrac{4}{1} = \dfrac{1 + 2 \cdot 1}{1 - 2 \cdot 1}$, i.e. $-3 = -3$.

(2) Let us assume that the equality holds for $n = k$, where $k \in N$, i.e.

$$\left(1 - \frac{4}{1}\right)\left(1 - \frac{4}{9}\right)...\left(1 - \frac{4}{(2k-1)^2}\right) = \frac{1 + 2k}{1 - 2k}. \qquad (1)$$

For $n = k + 1$, taking (1) into account, we have

$$\left(1 - \frac{4}{1}\right)\left(1 - \frac{4}{9}\right)...\left(1 - \frac{4}{(2k-1)^2}\right)\left(1 - \frac{4}{(2k+1)^2}\right)$$

$$= \frac{1 + 2k}{1 - 2k} \cdot \left(1 - \frac{4}{(2k+1)^2}\right) = \frac{1 + 2k}{1 - 2k} \cdot \frac{(2k+1)^2 - 4}{(2k+1)^2}$$

$$= \frac{(2k-1)(2k+3)}{(1-2k)(2k+1)} = \frac{1 + 2(k+1)}{-1 - 2k} = \frac{1 + 2(k+1)}{1 - 2(k+1)}.$$

We have carried out both parts of the proof and, therefore, by induction, the proposition is true for all $n \in N$. *Remark.* The proof can be also carried out with out resort to induction. Indeed, for any $k \in N$, we have

$$1 - \frac{4}{(2k-1)^2} = \frac{(2k-1)^2 - 4}{(2k-1)^2} = \frac{(2k-3)(2k+1)}{(2k-1)^2}. \qquad (2)$$

and therefore, assigning to $k$ the values $1, 2, ..., n$ in formula (2) and multiplying the results, we have

$$\left(1 - \frac{4}{1}\right)\left(1 - \frac{4}{9}\right)\left(1 - \frac{4}{25}\right)...\left(1 - \frac{4}{(2n-1)^2}\right)$$

$$= \frac{(-1 \cdot 3)(1 \cdot 5)(3 \cdot 7)...((2n-3)(2n+1))}{1 \cdot 3^2 \cdot 5^2 ... (2n-1)^2}$$

$$= \frac{-1 \cdot 1 \cdot 3^2 \cdot 5^2 ... (2n-3)^2 \cdot (2n-1)(2n+1)}{1 \cdot 3^2 \cdot 5^2 ... (2n-1)^2} = \frac{-1 \cdot (2n+1)}{2n-1} = \frac{1+2n}{1-2n}.$$

**11.** $x = 4.$ **12.** $x = 11.$ **13.** $x = 5.$ **14.** $x = 7.$ **15.** $x = 8.$ **16.** $x = 4.$ **17.** $x = 5.$ **18.** $x = 10.$
**19.** $x = 8.$ **20.** $x = 9.$ **21.** $x = 3.$ **22.** $n = 6, m = 3.$
**23.** $x = 7.$ *Hint.* Solving the problem, we get an equation $(x+3)(x+2) \times (x+1) = 720.$ On the set of natural numbers the function $f(x) = (x+3) \times (x+2)(x+1)$ is increasing. It is easy to find by selection the natural root of the equation obtained.
**24.** $x = 3.$ *Hint.* Solving the problem, we get an equation

$$6(x+3) - x(x+1) = \frac{120}{x+2}, x \in N.$$ Since for $x \in N$ the right-hand side of the

equation is positive, we have $6(x+3) - x(x+1) \in N,$ $6(x+3) - (x+1)x > 0$ and, consequently, $1 \le x < 8, x \in N.$ In this case 120 must be exactly divisible by $x+2$ and, therefore, $x \ne 5, x \ne 7.$ Therefore we must seek the roots of the equation obtained among the numbers $1, 2, 3, 4, 6.$

**25.** 13.   **26.** 10.
**27.** $(n+2)2^{n-1}.$ *Solution.* We designate the given expression as $S$:

$$S = C_0^n + 2C_1^n + 3C_2^n + ...+ (n+1)C_n^n; \qquad (1)$$

taking into account $C_k^n = C_{n-k}^n,$ we write expression (1) in the form

$$S = (n+1)C_0^n + nC_1^n + (n-1)C_2^n + .... + 2C_{n-1}^n + C_n^n. \qquad (2)$$

Adding equations (1) and (2) together, we get

$$2S = (n+2)(C_0^n + C_1^n + C_2^n + ...+ C_n^n) = (n+2)2^n,$$

whence it follow that $S = (n+2)2^{n-1}.$

**28.** $\{0,1, 2, 3, 4, 5\}.$ **29.** $\{11,12,13,14,15,16,17,18\}.$ **30.** $\{6,7, 8, 9\}.$
**31.** $n > 14, n \in N.$ **32.** $\{5, 6,7, 8, 9,10\}.$
**33.** $x \ge 2, x \in N.$ *Solution.* Since $C_{x-1}^{x+1} - C_2^{x+1} = \frac{(x+1)x}{2},$ the given inequality has

the form $\frac{(x+1)x}{2} > \frac{3}{2}$ or $x^2 + x - 3 > 0,$ i.e.

$x \in (-\infty, -(\sqrt{3}+1)/2) \cup ((\sqrt{13}-1)/2, +\infty),$ but by the hypothesis $x \in N$ and, consequently, $x \ge 2$ is an integer.
**34.** $\{1, 2, 3, 4, 5\}.$ **35.** $\{12,13,14, ....\}.$ **36.** $\{2, 3, 4, 5, 6,7, 8, 9\}.$ **37.** $\{8, 9,10, ...\}.$
**38.** Three. **39.** Four.   **40.** $x_1 = -63/4;$ $x_2 = -23/8.$ **41.** $mnk.$
**42.** In $2(P_5)^2 = 2 \cdot (120)^2$ ways. *Solution.* There is only one way to seat the children alternately girl and boy. The girls can be seated in $P_5$ ways and the boys in $P_5$ ways. In addition, the boys and girls can change places. There are, therefore, $2 \cdot (P_5)^2$ ways.
**43.** In $2 \cdot (P_6)^2 = 2 \cdot 720^2$ ways . **44.** In $A_3^{10}$ ways. **45.** (a) $A_2^{40} = 40 \cdot 39 = 1560;$
(b) $40 C_4^{39}.$ **46.** In $C_5^9 = 126$ ways. **47.** $C_{k-1}^{n+k-1}$ ways. **48.** 26250 distributions. *Hint.* Either there are 3 balls in some urn and one ball in each of the other nine urns, or there are 2

balls in each of some two urns and one ball in each of the other eight urns. **49.** In $(C_3^8 - 6) P_3 P_5 = 36000$ ways. **50.** In 2187 ways. **51.** 31. **52.** $A_2^3 \cdot A_4^5 = 720$. **53.** 125.
**54.** 5274. **55.** $A_5^9 = 15120$. **56.** 576.

**57.** $45 \cdot 10^5$, *Solution.* Let us consider 10 successive seven-digit numbers

$$a_1 a_2 a_3 a_4 a_5 a_6\, 0,$$
$$a_1 a_2 a_3 a_4 a_5 a_6\, 1,$$
$$\cdots \cdots \cdots$$
$$a_1 a_2 a_3 a_4 a_5 a_6\, 9,$$

where $a_1$, $a_2$, $a_3$, $a_4$, $a_5$, $a_6$ are some digits. As we see, half of these 10 numbers, i.e. five numbers have an even sum of digits. The first digit, $a_1$, can assume 9 different values ; each of the digits $a_2, a_3, a_4, a_5, a_6$ can assume 10 different values ; the last digit, $a_7$, can assume only 5 different values for which the sum of all the digits is even. Thus there are $9 \cdot 10^5 \cdot 5 = 45 \cdot 10^5$ seven-digit numbers the sum of whose digits is even.

**58.** $4 \cdot 7^3 = 1372$. *Solution.* By the hypothesis there is always a unity in the notation of a four-digit number and it occurs only once. The other three digits can assume any of the following 7 values : 2, 3, 4, 5, 6, 7, 8. The assigned digit 1 can occupy the first, the second, the third, or the fourth place. Consequently, there are $4 \cdot 7 \cdot 7 \cdot 7 = 4 \cdot 7^3 = 1372$ digits all in all that satisfy the conditions of the problem.

**59.** $C_2^7 \cdot 2^5 = 672$.

**60.** $C_4^9(C_3^6\, P_3 \cdot 4 + C_2^4 C_2^6\, C_2^4 P_2) + C_3^9\, (C_3^5 P_3 + 5 \cdot 3 C_3^5 P_2 + C_2^5 3 C_2^4 P_2 + 5 \cdot 3 C_2^5 C_2^3)$

$= 294840$ numbers. *Hint.* Either there is no zero among the digits or there is a zero. In the first case either some digit occupies three places of there are two "duplicates". In the second case there can be four variants : either the zero occupies three places, or some other digit occupies three places, or, the zero occupies one place and there are two "duplicates", or the zero occupies two places and there is one "duplicate".

**61.** 4373. Thus 12, 21, ... , 199999992 are the required numbers. To make the calculations more convenient, we assume all of them to be nine-digit and designate

$$a_1\ a_2\ a_3\ a_4\ a_5\ a_6\ a_7\ a_8\ a_9,$$

writing the necessary number of zeros in front of the given number. We write three successive nine-digit numbers

$$a_1\ a_2\ a_3\ a_4\ a_5\ a_6\ a_7\ a_8\ 0,$$
$$a_1\ a_2\ a_3\ a_4\ a_5\ a_6\ a_7\ a_8\ 1,$$
$$a_1\ a_2\ a_3\ a_4\ a_5\ a_6\ a_7\ a_8\ 2,$$

where $a_1$ can assume one of the two values 0 or 1 and each of the digits $a_2, a_3, a_4, a_5, a_6, a_7, a_8$ can assume any of the three values 0, 1, 2. We must then eliminate the number for which $0 = a_1 = a_2 = a_3 = a_4 = a_5 = a_6 = a_7 = a_8 = a_9$ from these numbers. For the nine-digit number permissible by the hypothesis to be divisible by 3, the sum of its digits $a_1 + a_2 + a_3 + a_4 + a_5 + a_6 + a_7 + a_8 + a_9$ must be divisible by 3. The sum of the first eight digits can be equal to $3n - 2$, or to $3n - 1$, or to $3n$. In each case $a_9$ can be chosen from 0, 1, 2 in only one way such that the sum of all the nine digits is equal to $3n$. Thus there are $2 \cdot 3^7 \cdot 1 - 1 = 4374 - 1 = 4373$ numbers which satisfy the hypothesis.

**62.** 576. *Solution.* The hypothesis is satisfied in the case when a four-digit number is written by means of one digit. There are $C_1^9$ numbers of that kind since 0 cannot be that digit. The hypothesis is also satisfied in the case when a four-digit number is written by means of two different digits. Let us assume that 0 is one of them. Then the other digit can be chosen in $C_1^9 = 9$ ways. Zero cannot occupy the first place since that place is occupied by the second chosen digit ; consequently, if 0 occurs only once in the notation of the number, then there are $C_1^3$ numbers of that kind, if 0 occurs twice in the notation of the number, then there are $C_2^3$ numbers of that kind, and, finally, if 0 occurs three times, then there are $C_3^3$ numbers. Thus there are $C_1^9 (C_1^3 + C_2^3 + C_3^3) = 9 \cdot (3 + 3 + 1) = 63$ numbers which satisfy the hypothesis and whose notation includes at least one zero. The last possibility : the number is written by means of two different digits with no zero among them. There can $C_2^9$ way of choosing two digits out of line. If the digits have been chosen, then in the notation of the number the first digit can occur once, the second digit, three times, and there are $C_1^4$ numbers of that kind; the first digit can occur twice and the second digit also twice, there are $C_2^4$ numbers of that kind; and, finally, the first digit can occur three times and the second digit can occur once, there are $C_3^4$ digits of that kind. Consequently, there are $C_2^9 (C_1^4 + C_2^4 + C_3^4) = 36 \cdot (4 + 6 + 4) = 36 \cdot 14 = 504$ different four-digit numbers written by means of only two different digits with no zero among them. Thus the hypothesis is satisfied by $C_1^9 + C_1^9 (C_1^3 + C_2^3 + C_3^3) + C_2^9 (C_1^4 + C_2^4 + C_3^4)$ $= 9 + 63 + 506 = 576$ numbers.

63. 261972. *Solution.* The bouquet can be formed from 3, 4, ....,18 different flowers. That can be done in $A = C_3^{18} + C_4^{18} + .... C_{18}^{18}$ different ways. Taking into account that $C_0^{18} + C_1^{18} + C_2^{18} + C_3^{18} + .... + C_{18}^{18} = 2^{18}$, i.e. $1 + 18 + 153 + A = 262144$, we obtain $A = 262144 - 172 = 261972$.

64. 766. *Solution.*Thus, 1, 2,11,12, 21, 22,....,122222222 are the required numbers. Let us calculate this number. There are $C_1^2 = 2$ one-digit numbers written by means of the digits 1 and 2. There are $C_1^2 + C_2^2 C_1^1 = 2 + 1 \cdot 2 = 4$ two-digit numbers written by means of two digits 1 and 2, and there are $C_1^2 + C_2^2 (C_1^3 + C_2^3) = 2 + 1 \cdot (3 + 3) = 8$ three-digit numbers, written by means of two digits 1 and 2. There are $C_1^2 + C_2^2 (C_1^4 + C_2^4 + C_3^4) = C_1^2 + (2^4 - C_0^4 - C_4^4)$ $= 2 + 2^4 - 2 = 2^4 = 16$ four-digit numbers written by means of two digits 1 and 2, and there are $C_1^2 + C_2^2 (C_1^5 + C_2^5 + C_3^5 + C_4^5) = 2 + (2^5 - C_0^5 - C_5^5) = 2 + 2^5 - 2 = 2^5 = 32$ five-digit numbers written by means of two digits 1 and 2. There are $C_1^2 + C_2^2 (C_1^6 + C_2^6 + C_3^6 + C_4^6 + C_5^6) = 2^6 = 64$ six-digit numbers written by means of the digits 1 and 2, and there are $C_1^2 + C_2^2 (C_1^7 + C_2^7 + C_3^7 + C_4^7 + C_5^7 + C_6^7) = 2^7 = 128$ seven-digit numbers written by means of the digits 1 and 2. There are $2^8 = 256$ eight-digit numbers written by means of the digits 1 and 2 . Since all the required numbers are smaller than $2 \cdot 10^8$, all the nine-digit numbers begin with 1, and consequently, there are as many of them as the eight-digit numbers, i.e. 256. Thus there are $2 + 4 + 8 + 16 + 32 + 64 + 128 + 256 + 256 = 766$ numbers which are smaller than $2 \cdot 10^8$ and are written by means of two digits 1 and 2.

65. $(C_1^5 \cdot C_3^5 + C_2^5 \cdot C_3^6 \cdot C_1^3 + C_3^5 \cdot C_3^6 \cdot C_1^3 \cdot C_1^2)$ $(C_1^4 + C_2^4 \cdot C_1^3 + C_3^4 \cdot C_1^3 \cdot C_1^2) + (C_1^5 \cdot C_2^5 + C_2^5 \cdot C_2^5 \cdot C_1^3 + C_3^5 \cdot C_2^5 \cdot C_1^3 \cdot C_1^2)(C_1^4 + C_1^4 \cdot C_1^3 + C_1^4 \cdot C_1^3 + C_2^4 \cdot C_1^3 \cdot C_1^2) + (C_1^5 C_3^5 + C_2^5 C_2^5 C_1^3 + C_3^5 \ C_3^5 \ C_1^3 C_1^2)(C_1^4 + C_1^4 C_1^2 + C_2^4 C_1^2 C_1^2) =$ $1900 \cdot 46 + 950 \cdot 61 + 950 \cdot 36 = 87400 + 57950 + 3420 = 179550$.

*Hint.* The following cases are possible when we choose three odd digits : (1) the three chosen odd digits are the same, (2) two of the chosen odd digits are the same and the third one differs from them, (III) the three chosen odd digits are different. There are two possibilities when we choose three even digits : (1) there is no zero among the even digits, (2) there is a zero among the even digits. Case 1 branches into three possible subcases : (a) the three chosen digits are the same, (b) two of the chosen digits are the same and the third one differs from them, (c) the three chosen digits are different. Case 2 branches into four subcases : (d) the three chosen even digits are zeros, (e) two of the chosen even digits are 0 and the third even digit is different from 0, (f ) one of the chosen even digits is 0 and the other two are the same and different from 0, (g) one of the chosen even digits is 0 and the other two are different even digits different form 0. Each case (1), (2), (III) can be combined with each of the cases (a), (b), (d), (e), (f), (g). In cases (a), (b), (c) we can put the odd digits into any of the 6 places of the six-digit number, and in cases (d), (e), (f), (g) we must either put one of the odd digits in the first place and the other two into any of the remaining five places, or put one even digit, which is different from zero, in the first place and place all the other digits arbitrarily.

**66.** 840.

**67.** $45 \cdot 10^4$. *Hint.* See the solution of problem 50. There is an equal number of six-digit numbers with an even sum of digits and an odd sum of digits. There are $9 \cdot 10^5$ six-digit numbers all in all.

**68.** 13 participants, 156 games. **69.** $C_6^{12} = 924$. **70.** $20x^{-3/2}$.

**71.** $n = 11$. *Solution.* $C_2^n - C_1^n = 44$, i.e. $\dfrac{n\,(n-1)}{2} - n = 44$, whence $n_1 = 11$, $n_2 = -8$.

The hypothesis is satisfied by $n = 11$.

**72.** $T_4 = 70$. *Solution.* $T_k = C_k^8 x^{8-k} (1 / x)^k$ is the required term of the expansion ;

$x^{8-k} x^{-k} = x^0$, whence $8 - 2k = 0$, $k = 4$, $T_4 = \dfrac{8 \cdot 7 \cdot 6 \cdot 5}{1 \cdot 2 \cdot 3 \cdot 4} = 70$. **73.** $T_8 = C_8^{17}$.

**74.** $T_4 = 35x^2$. *Solution.* $T_k = C_k^7 (x^{-2/3})^{7-k} x^k = C_k^7 x^2$ is the required term of the expansion, whence $\dfrac{2k-14}{3} + k = 2$, $k = 4$; $T_4 + C_4^7 x^2 = 35x^2$.

**75.** $T_1 = 14a^{7/2}$. **76.** $T_3 = 165z^{14}$. *Solution.*

$C_0^n + C_1^n + \ldots C_n^n = 2^n = 2048 = 2^{11}$, $n = 11$; $T_3 = C_3^{11} (z^2)^8 (z^{-2/3})^3 = 165z^{14}$.

**77.** $x = 9$. *Solution.* $\dfrac{C_6^x (2^{1/3})^{x-6} (3^{-1/3})^6}{C_{x-6}^x (2^{1/3})^6 (3^{-1/3})^{x-6}} = \dfrac{1}{6}$, or $6^{-4} \cdot 6^{x/3} = 6^{-1}$, whence $x = 9$.

**78.** The sixth. **79.** The third. **80.** $T_6 = 5005$.

**81.** $n = 32$. *Solution.* $C_2^n \cdot \dfrac{1}{16} = 31$, $\dfrac{n\,(n-1)}{32} = 31$, $n_1 = 32$, $n_2 = -31$. The hypothesis is satisfied by $n = 32$.

**82.** $T_4 = 1120x^4$. *Solution.* $C_0^m - 2C_1^m + 4C_2^m = 97, 1 - 2m + 2m + 2m\,(m-1) = 97$, $m_1 = 8$, $m_2 = -6$. The hypothesis is satisfied by $m = 8$. Next, $2\,(8-k) - k = 4$, whence $k = 4$, $T_4 = C_4^8 (x^2)^4 (-2 / x)^4 = 1120x^4$.

**83.** 380.

**84.** $x_1 = -1$, $x_2 = 2$. *Solution.* $C_{m-2}^m + C_{m-1}^m + C_m^m = \dfrac{m\,(m-1)}{2} + m + 1 = 22$, $m = 6$, $T_2 + T_4 = 135$, or $30 \cdot 2^x + 60 \cdot 2^{-x} = 135$, $x_1 = -1$, $x_2 = 2$.

**85.** $x_1 = 10^{-4}$, $x_2 = 10$. *Solution.* $T_3 = C_3^6 x^{3/2(\log x +1)} \cdot x^{1/4} = 200$,

$20x^{\frac{1}{4}+\frac{3}{2(\log x+1)}} = 200, \left(\dfrac{1}{4} + \dfrac{3}{2\,(\log x + 1)}\right) \log x = 1$; $\log x = -4$, $\log x = 1$;

$x_1 = 10^{-4}$, $x_2 = 10$.

**86.** $x = -1/3$.

**87.** $x = 1$. *Solution.* $m\,(m-1)/2 - 20 = m$, $m^2 - 3m - 40 = 0$, $m_1 = 8$, $m_2 = -5$. The hypothesis is satisfied by $m = 8$. Furthermore, $T_3 - T_5 = 56$, or $56 \left(2^x - 2 \cdot 2^{-x}\right) = 56$, or $2^{2x} - 2^x - 2 = 0; 2^{x_1} = 2, x_1 = 1; 2^{x_2} = -1, x_2 \in \phi$.

**88.** $x_1 = 10^{-5/2}, x_2 = 10$.    *Solution.*    $T_2 = C_2^5 x^3 x^{2\log x} = 10^6, x^3 x^{2\log x} = 10^5$;
$(2\log x + 3) \log x = 5, 2\log^2 x + 3 \log x - 5 = 0; \log x = -5/2, x_1 = 10^{-5/2}$;
$\log x = 1, x_2 = 10$.

**89.** $x_1 = 1, x_2 = 2$. *Solution.* The sixth term of the expansion of the power of the binomial is the fifth term of the binomial (see formula (2) on p. 30).

$$T_5 = C_5^7 2^{2\log_2 \sqrt{9^{x-1}} + 7} \cdot 2^{-\log_2 (3^{x-1} + 1)} = 84,$$

$$\frac{7 \cdot 6}{1 \cdot 2} \cdot 2^{\log_2 (9^{x-1} + 7)} \cdot 2^{-\log_2 (3^{x-1} + 1)} = 84,$$

$$\frac{9^{x-1} + 7}{3^{x-1} + 1} = 4, (3^{x-1})^2 - 4 \cdot 3^{x-1} + 3 = 0,$$

$$(3^{x-1} - 1)(3^{x-1} - 3) = 0; 3^{x-1} = 1, x_1 = 1, 3^{x-1} = 3, x_2 = 2.$$

**90.** *Hint.* The given inequality is equivalent to the inequality $\left(1 + \dfrac{1}{n}\right)^n < n$. Prove that

$$\left(1 + \frac{1}{n}\right)^n < 3.$$

**91.** $n = 3$. *Hint.* Since $\left(\dfrac{x}{5} + \dfrac{2}{5}\right)^n = \dfrac{1}{5^n}(x + 2)^n$, the problem must be solved for

$(x + 2)^n$.

**92.** $T_{18} = C_{18}^{50} b^{50} 3^{16}$ is maximum in the absolute value. *Solution.* The ratio

$\left|\dfrac{T_k}{T_{k-1}}\right| = \left(\dfrac{51}{k} - 1\right)\dfrac{1}{\sqrt{3}}$. It is easy to verify that $|T_k / T_{k-1}| > 1$ for $k \le 18$ and

$|T_k / T_{k+1}| > 1$ for $k \ge 18$. Consequently, $k = 18$ and $T_{18}$ is the term of the expansion which is the greatest in absolute value.

**93.** $n = 6; x = \pm 1/2$.

**94.** $(x^{1/3} - x^{-1/2})^{10}, T_4 = C_4^{10}$.

**95.** $T_6 = 84$.

## Sec. 1.4

**1.** $\{4/3\}$.    **2.** $\{-9/2, 13/4\}$.

**3.** All $x \in [2, +\infty)$. *Solution.* Let us consider three cases : (1) $x \le 0$; (2) $0 < x < 2$;
(3) $x \ge 2$.

1st case : $x \le 0$. In this case the equation is equivalent to the system

$$\begin{cases} x \le 0 \\ -x + (x - 2) = 2, \end{cases}$$

whose set of solutions is empty.

2nd case : $0 < x < 2$. In this case the equation is equivalent to the system

$$\begin{cases} 0 < x < 2 \\ x + (x - 2) = 2, \end{cases}$$

whose set of solutions is empty.

3rd case : $x \ge 2$. In this case the equation is equivalent to the system

$$\begin{cases} x \ge 2, \\ x - (x - 2) = 2, \end{cases}$$

whose set of solutions is constituted by all $x \in [2, +\infty)$.

**4.** $\{-1\}$.

**5.** $(-\infty, -3] \cup [-\sqrt{3}/3, 1/2]$. *Hint.* The given equation is equivalent to the equation

$$|a^2 x + 1| + |a^3 + a^2 x| = a^2 x + 1 - (a^3 + a^2 x).$$

The equation $|u| + |v| = u - v$ is equivalent to the system

$$\begin{cases} u \geq 0, \\ v \leq 0. \end{cases}$$

For $a \neq 0$ consider the cases $0 < a < 1, -1 < a^- < 0, a^- = -1$ and $a < -1$.

**6.** $(2, 3)$. *Solution.* The given inequality is equivalent to the inequality $|2x - 5| < 1$, which is equivalent to the two-sided inequality $-1 < 2x - 5 < 1$, or $4 < 2x < 6$, or $2 < x < 3$.

**7.** $(-\infty, 1/6] \cup [3/2, +\infty]$. *Hint.* The given inequality is equivalent to the collection of two inequalities $3x - 2.5 \leq -2$ and $3x - 2.5 \leq 2$.

**8.** $[1, +\infty)$. *Hint.* Consider the following three cases : (1) $x \geq 2$, (2) $-4 < x < 2$; (3) $x \leq -4$. The inequality can also be solved by graphical means, by constructing the graphs of the functions $y = |x + 4|$ and $y = |x - 2|$.

**9.** $(5/3, 3)$. *Hint.* The given inequality is equivalent to the system of inequalities

$$\begin{cases} x - 1 > 0, \\ 1 - x < 2x - 4 < x - 1. \end{cases}$$

**10.** $(-\infty, -2) \cup (2, +\infty)$. *Hint.* Consider the following two cases : (1) $x < -1$, (2) $x \geq -1$.

**11.** $(9/2, +\infty)$. *Hint.* Consider the following three cases : (1) $x \leq -2$; (2) $-2 < x < 1$; (3) $x \geq 1$.

**12.** $(-\infty, -2) \cup (5, +\infty)$, $x = 6$. **13.** $x = -1$. **14.** $b = -50$.

**15.** $\{-4\}$. *Solution.* We write the given equation in the form $|x|^2 - 2|x| - 8 = 0$, or $(|x| - 4)(|x| + 2) = 0$. Taking into account that $|x| + 2 \neq 0$, we obtain $|x| = 4$, whence we find that $x_1 = -4$, $x_2 = 4$. Only $x_1 = -4$ belongs to the domain of definition of the function $y = \sqrt{5 - 2x}$.

**16.** $\{-2, -1/9\}$. **17.** $\{1\}$.

**18.** $\{-5/3\}$. *Hint.* When solving the equation, take into account that $x^2 = |x|^2$.

**19.** $\{-2, 1\}$. *Solution.* (1) If $x^2 + 4x + 2 \geq 0$, then the equation is equivalent to the system

$$\begin{cases} x^2 + 4x + 2 \geq 0, \\ 3x^2 + 7x - 10 = 0, \end{cases}$$

solving which we find $x = 1$.

(2) If $x^2 + 4x + 2 < 0$, then the equation is equivalent to the system

$$\begin{cases} x^2 + 4x + 2 < 0, \\ 3x^2 + 17x + 22 = 0, \end{cases}$$

solving which, we find $x_2 = -2$.

The graphical solution consists in constructing the graphs of the functions $y = |x^2 + 4x + 2|$ and $y = \dfrac{5x + 16}{3}$ and finding the coordinates $x_1$ and $x_2$ of the intersection points of these graphs, which are exactly the roots of the given equation.

**20.** $\{1, 4\}$. **21.** $\{2, 5\}$. **22.** $\{3, 6\}$. **23.** $\{0, 1\}$. *Solution.* Let us consider two cases : (1) $x \geq 1$, (2) $x < 1$.

1st case. For $x \geq 1$ the equation is equivalent to the system

$$\begin{cases} x \geq 1, \\ x^2 + x - 2 = 0, \end{cases}$$

solving which we find $x_1 = 1$.

2nd case. For $x < 1$ the equation is equivalent to the system

$$\begin{cases} x < 1 \\ x^2 - x = 0, \end{cases}$$

solving which we find $x_2 = 0$.

**24.** $\{-20.5, 10\}$. *Hint.* When simplifying the equation, take into account that the coefficients in $x$ in the quadratic trinomials form an arithmetic progression with the common difference equal to 1, and the free terms form an arithmetic progression with the common difference equal to 2.

**25.** $\{2 + \sqrt{4-a}\}$ if $a < -4$; $\{2 - \sqrt{4+a}, 2 + \sqrt{4+a}, 2 + \sqrt{4-a}\}$ if $-4 \leq a < 0$; $\{0, 4\}$ if $a = 0$; $\{2 - \sqrt{4+a}\}$. if $a > 0$.

**26.** $a = 20 \pm 6\sqrt{5}$. **27.** $k_1 = -22/3$, $k_2 = 2$. **28.** $a = 4$. **29.** For all $m \in (1/4, +\infty)$.

**30.** For all $m \in (-\infty, -1/2) \cup (1/2, +\infty)$. **31.** For all $m \in (-\infty, -1/7) \cup (1, +\infty)$.

**32.** For all $c \in (2, 4)$. **33.** $k = 13$. **34.** $a = \pm 2$. **35.** $a = \pm 10$. **36.** $k = \pm 4$.

**37.** $28x^2 - 20x + 1 = 0$. *Solution.* Let us write the required equation in the form

$$x^2 + px + q = 0. \tag{1}$$

Its roots are

$$x_1 = \frac{1}{10 - \sqrt{72}} = \frac{10 + \sqrt{72}}{28} = \frac{5 + 3\sqrt{2}}{14},$$

$$x_2 = \frac{1}{10 + 6\sqrt{2}} = \frac{10 - 6\sqrt{2}}{28} = \frac{5 - 3\sqrt{2}}{14};$$

$$p = -(x_1 + x_2) = -\frac{5}{7}, \quad q = \frac{5 + 3\sqrt{2}}{14} \cdot \frac{5 - 3\sqrt{2}}{14} = \frac{1}{28}.$$

Substituting the values obtained for $p$ and $q$ into equation (1), we find the required equation.

**38.** $(-3, 5)$.　　　　**39.** $[0, 1/2]$.　　　　**40.** $k = 3$.

**41.** $k = 3$.　　　　**42.** $a_1 = -2, a_2 = 1$.

**43.** For all $a \in (-\infty, -2) \cup (0, +\infty)$.

**44.** For all $a \in (-6, 3)$. *Hint.* We must find the values of $a$ for which the equation $(a - 6) x^2 - 2 = 2ax + 1$ does not have real roots.

**45.** $\{-4, -3, 3, 4\}$.

**46.** $a = -4$. *Solution.* By the hypothesis, the roots $x_1$ and $x_2$ of the equation are related as $x_2 = 2x_1$. Applying the Vieta theorem to define $x_1$, we obtain a system of equation

$$\begin{cases} x_1 + 2x_1 = 1 - 2a, \\ 2x_1^2 = a^2 + 2. \end{cases}$$

from which we obtain $a^2 + 8a + 16 = 0$ to define $a$. Hence $a = -4$.

**47.** $a_1 = -3/2, a_2 = 6$. **48.** $a_1 = 2, a_2 = 9/2$. **49.** $\{2, 18\}$ for $a = 6, \{2/19, 18/19\}$ for $a = -6/19$. **50.** $a_1 = -125/8, a_2 = 27/8$. **51.** $p = \pm 7$. **52.** $k = \pm 3\sqrt{5}$. **53.** $a_1 = 3/2, a_2 = 3$. **54.** $b = -13$.

**55.** $p = 0$. *Solution.* Assume that $x_1$ and $x_2$ are roots of the given equation. By the hypothesis and the Vieta theorem

$$\begin{cases} x_1^2 + x_2^2 = 16, \\ x_1 + x_2 = 4. \end{cases}$$

Squaring the second equation of the system and taking the first equation into account, we get $x_1 x_2 = 0$ and $p = x_1 x_2 = 0$.

**56.** $a = 2$. *Solution.* Assume that $x_1$ and $x_2$ are roots of the given equation. Applying the Vieta theorem to define $x_1$ and $x_2$, we obtain a system of equations

$$\begin{cases} x_1 - x_2 = (a-1)/2, \\ x_1 + x_2 = (a+1)/2, \end{cases}$$

whence $x_1 = a/2$, $x_2 = 1/2$. Substituting any one of the roots obtained into the equation, we find the value of $a$.

**57.** $a_1 = 1/2$, $a_2 = 1$.

**58.** $p_1 = 0$, $q_1 = 0$; $p_2 = 1$, $q_2 = -2$.

**59.** $a = -2$. *Solution.* The first equation has real roots if $a^2 - 4 \geq 0$, i.e. if $a \in (-\infty, -2) \cup [2, +\infty]$. The second equation has real roots if $1 - 4a \geq 0$, i.e. if $a \in (-\infty, 1/4]$. Thus the given equations may have a common root if $a \in (-\infty, -2]$. Assume that $x_0$ is a common root. Then the numerical equations

$$x_0^2 + ax_0 + 1 = 0 \text{ and } x_0^2 + x_0 + a = 0 \qquad (1)$$

hold true. Subtracting the second equation from the first, we obtain $(a-1)x_0 + (1-a) = 0$, or $x_0 = (a-1)/(a-1) = 1$ (in the domain $[-\infty, -2]$ the division by $a-1$ is legitimate). Substituting $x_0$ into any equation (1), we find $a = -2$.

**60.** $m = -2$. *Solution.* Let us find the roots of the first equation $x_1 = \dfrac{1 - \sqrt{1-4m}}{2}$, $x_2 = \dfrac{1 + \sqrt{1-4m}}{2}$, and the roots of the second equation $x_3 = \dfrac{1 - \sqrt{1-12m}}{2}$, $x_4 = \dfrac{1 + \sqrt{1-12m}}{2}$. The roots are real under the condition $m \leq 1/12$. If $0 < m \leq 1/12$, then $\sqrt{1-12m} < \sqrt{1-4m} < 1$, and, consequently, all the four roots are positive, with $(m \neq 0) 0 < x_1 < x_3 < x_4 < x_2$. It can be seen that $x_3 \neq 2x_2$ and $x_4 \neq 2x_2$. We must investigate the two remaining possibilities : (a) $x_3 = 2x_1$, or (b) $x_4 = 2x_1$, i.e.

$$\dfrac{1 \pm \sqrt{1-12m}}{2} = 1 - \sqrt{1-4m}$$

$$\Rightarrow \qquad 1 \pm \sqrt{1-12m} = 2 - 2\sqrt{1-4m}$$
$$\Rightarrow \qquad 2\sqrt{1-4m} = 1 \mp \sqrt{1-12m}$$
$$\Rightarrow \qquad 4(1-4m) = 1 \mp 2\sqrt{1-12m} + 1 - 12m$$
$$\Rightarrow \qquad 1 - 2m = \mp \sqrt{1-12m}.$$

The left-hand side of the equation includes $1 - 2m > 0$, and, consequently, the right-hand side must contain a positive number, i.e. $+\sqrt{1-12m}$, i.e. possibility (b) must be excluded. Hence $x_4 \neq 2x_1$. Let us square the left-hand and right-hand sides of the equation $1 - 2m = \sqrt{1-12m}$. Then, after tansformations, we get $m^2 + 2m = 0$, whence $m = -2$ since $m \neq 0$. But we consider the case $0 < m < 1/12$, i.e. we have arrived at a contradiction in case (a) as well, i.e. $x_3 \neq 2x_1$.

Let us consider the case $m < 0$; then $1 < \sqrt{1-4m} < \sqrt{1-12m}$, and, consequently, $x_1 < 0$, $x_3 < 0$, $x_4 > 0$, $x_2 > 0$, with $x_3 < x_1 < 0 < x_2 < x_4$. It can be seen that $x_3 \neq 2x_2$ and $x_4 \neq 2x_1$. We must investigate the two remaining possibilities : (a) $x_4 = 2x_2$ and (b) $x_3 = 2x_1$. Then

$$\dfrac{1 \pm \sqrt{1-12m}}{2} = 1 \pm \sqrt{1-4m}$$

$$\Rightarrow \qquad 1 \pm \sqrt{1-12m} = 2 \pm 2\sqrt{1-4m}$$
$$\Rightarrow \qquad \pm\sqrt{1-12m} = 1 \pm 2\sqrt{1-4m}. \qquad (1)$$

In the case of the upper signs (case (a)), we square the left-hand and right-hand sides : $1-12m = 1+ 4\sqrt{1-4m} + 4\,(1-4m) \Rightarrow m-1 = \sqrt{1-4m}$. Here $m-1 < 0$ is on the left and $\sqrt{1-4m} > 0$ on the right ; we have arrived at a contradiction. Consequently, $x_4 \neq 2x_2$. In the case of the lower signs in (1) (case (b)), we carry out the following transformation :

$$2\sqrt{1-4m} = 1 + \sqrt{1-12m}$$
$$\Rightarrow \qquad 4\,(1-4m) = 1 + 2\sqrt{1-12m} + 1 - 12m$$
$$\Rightarrow \qquad 1 - 2m = \sqrt{1-12m}$$
$$\Rightarrow \qquad 1 - 4m + 4m^2 = 1 - 12m$$
$$\Rightarrow \qquad m^2 = -2m$$

and, since $m \neq 0$, it follows that $m = -2$. Consequently, $x_3 = 2x_1$ for $m = -2$.

**61.** $c < 0$. *Solution.* The trinomial $f(x) = ax^2 + bx + c$ does not possess real roots and, consequently, $f(x)$ has the same sign for any $x \in \mathbf{R}$, but $f(1) = a + b + c < 0$; therefore, $0 > f(0) = c$.

**62.** $x_1^3 + x_2^3 = 3pq - p^3$. **63.** $x_1^3 + x_2^3 = a\,(a^2 - 18a + 9)/27$. **64.** $215/27$.

**65.** $\dfrac{1}{x_1^3} + \dfrac{1}{x_2^3} = -\dfrac{27a^3 + 38a}{8}$. *Solution.* On the basis of the Vieta theorem we have

$x_1 + x_2 = 3a/2, \; x_1 x_2 = -1;$ then
$$(x_1 + x_2)^3 = 27a^3/8 \Rightarrow x_1^3 + x_2^3 + 3x_1 x_2\,(x_1 + x_2) = 27a^3/8$$
$$\Rightarrow x_1^3 + x_3^3 = \frac{27a^3}{8} - 3\,(-1) \cdot \frac{3a}{2}$$
$$\Rightarrow x_1^3 + x_2^3 = \frac{27a^3 + 36a}{8}.$$

We calculate $\dfrac{1}{x_1^3} + \dfrac{1}{x_2^3} = \dfrac{x_1^3 + x_2^3}{x_1^3 x_2^3} = -\dfrac{27a^3 + 36a}{8}.$

**66.** For all $a \in (2,17/4)$. **67.** For all $p \in [3,15/4]$. **68.** For all $a \in (-\infty, -6)$. **69.** For all $a \in (6, +\infty)$. **70.** For all $a \in (5/3, +\infty)$. **71.** $k = 5$.

**72.** $(-\infty, -(a+\sqrt{a^2-4a})/2) \cup (\sqrt{a^2-4a}-a)/2, +\infty)$ for all $a \in (-\infty, 0]$ $\cup [4, +\infty); (-\infty, +\infty)$ for all $a \in (0, 4)$.

**73.** For all $m \in (1, +\infty)$. **74.** For all $a \in (11/9, +\infty)$. **75.** $x^2 - (a^2+1)\,x + 4 = 0$.

**76.** $4x^2 - 8x - 4a^2 - 11 = 0$. **77.** For all $a \in [1/2, +\infty)$. **78.** For all $a \in (0,1) \cup (1, 6/5)$.

**79.** For all $a \in (-1/2, 0) \cup (0, \sqrt{5}/5)$.

**80.** For all $a \in (0, 1/3)$. *Solution.* Let us consider two possibilities : (1) $3a \leq a + 3$, (2) $3a > a + 3$.

(1) $3a \leq a + 3 \Rightarrow a \leq 3/2$. In this case the given inequality is satisfied on the interval $(3a, \; a+3)$. For this inequality to hold for all $x \in [1, 3]$, it is necessary and sufficient that the system of inequalities

$$\begin{cases} 3a < 1, \\ a + 3 > 3, \end{cases} \Leftrightarrow \begin{cases} a < 1/3 \\ a > 0 \end{cases} \Leftrightarrow 0 < a < 1/3,$$

should have a nonempty set of solutions. Since in this case $a < 3/2$, the problem has a solution for all $a \in (0, 1/3)$.

(2) $3a > a + 3 \Rightarrow a > 3/2$. In this case the given inequality is satisfied on the interval $(a + 3, 3a)$. For the inequality to hold for all $x \in [1, 3]$, it is necessary and sufficient that the system of inequalities

$$\begin{cases} a + 3 < 1, \\ 3a > 3, \end{cases} \Leftrightarrow \begin{cases} a < -2, \\ a > 1, \end{cases}$$

should have a nonempty set of solutions. We have arrived at a contradictory system and, therefore, in this case the problem does not have a solution.

**81.** $k \in (-\infty, 0] \cup \{1\}$. *Solution.* Let us rewrite the inequalities in the form $(x - 3k)^2 > 4(k-1)^2$ and $(x - 2k)^2 \geq k$. We can see from the second inequality that for $k \in (-\infty, 0]$ it holds for any real $x$. It remains to investigate the case where $k \in (0, +\infty)$. From the first inequality we obtain
$$x \in (-\infty, 3k - 2|k-1|) \cup (3k + 2|k-1|, +\infty).$$
From the second inequality we obtain
$$x \in (-\infty, 2k - \sqrt{k}] \cup [2k + \sqrt{k}, +\infty).$$
If there are positive values of $k$ such that the inequality
$$3k + 2|k-1| \leq 2k - \sqrt{k}, \tag{1}$$
or the inequality
$$2k + \sqrt{k} \leq 3k - 2|k-1| \tag{2}$$
is satisfied, then for these values of $k$ any real $x$ is a solution of at least one of the given inequalities. Inequality (1) can be rewritten as $k + \sqrt{k} + 2|k-1| \leq 0$, where $k > 0, \sqrt{k} > 0, |k-1| \geq 0$, and, consequently, inequality (1) is contradictory. We rewrite inequality (2) in the form
$$\sqrt{k} \leq k - 2|k-1|. \tag{3}$$
Let us consider two cases : $k \in [1, +\infty)$ and $k \in (0,1)$. In the first case inequality (3) can be rewritten as $\sqrt{k} \leq 2 - k$, and, consequently, if it is satisfied, then it holds only for some $k \in [1, 2)$. For these $k$, after squaring the left-hand and right-hand sides, we obtain, after transformations, an inequality $k^2 - 5k + 4 \geq 0$. This inequality is valid only for $k(-\infty, 1] \cup [4, +\infty)$. Out of all these values, only the value $k = 1$ satisfies the condition $k \in [1,2)$. Let us now consider the case $k \in (0,1)$. In this case we can rewrite inequality (3) as $\sqrt{k} \leq 3k - 2$, and, consequently, if it is satisfied, then it holds only for some $k \in (2/3, 1)$. For these $k$, after squaring the left-hand and right-hand sides, we obtain, after transformations, an inequality $9k^2 - 13k + 4 \geq 0$. It is valid only for $k \in (-\infty, 4/9] \cup [1, +\infty)$. Since $4/9 < 2/3$, none of these values of $k$ satisfies the condition $k \in (2/3,1)$. Combining all the results obtained, we find that the problem has a solution only for $k \in (-\infty, 0] \cup \{1\}$.

**82.** $a \in (-1/4, 1)$. *Solution.* We find the roots of the equation
$$x^2 - 2x - a^2 + 1 = 0; \tag{1}$$
they are $x_1 = 1 - |a|, x_2 = 1 + |a|, x_1 \leq x_2$. Next we find the roots of the equation
$$x^2 - 2(a+1)x + a(a-1) = 0; \tag{2}$$
they are $x_3 = a + 1 - \sqrt{1+3a}, x_4 = a + 1 + \sqrt{1+3a}, x_3 \leq x_4$. The roots of equation (2) are real provided that $-1/3 \leq a$. We must also find $a$ such that the inequalities
$$a + 1 - \sqrt{1+3a} < 1 - |a| \leq 1 + |a| < a + 1 + \sqrt{1+3a} \tag{3}$$
hold. Let us consider the case $a < 0$; then inequalities (3) assume the form $a + 1 - \sqrt{1+3a} < 1 + a \leq 1 - a < a + 1 + \sqrt{1+3a}$. In this case we must only satisfy the inequality $1 - a < a + 1 + \sqrt{1+3a}$, or $-2a < \sqrt{1+3a}$, where $-2a > 0$; squaring the right-hand and left-hand sides of the inequality, we get, after the requisite transformations, an inequality $4a^2 - 3a - 1 < 0$, which is satisfied for $a \in (-1/4, 1)$. Since we have $a < 0$, the problem has a solution for $a \in (-1/4, 0)$. Assume now that $a \geq 0$; then inequality (3) can be rewritten as $a + 1 - \sqrt{1+3a} < 1 - a \leq 1 + a < a + 1 + \sqrt{1+3a}$. In this case we must only satisfy

the inequality $a + 1 - \sqrt{1 + 3a} < 1 - a$, or $2a < \sqrt{1 + 3a}$, where $2a \geq 0$; squaring the right-hand and left-hand sides of the inequality, we get, after the requisite transformation, an inequality $4a^2 - 3a - 1 < 0$, which holds for $a \in (-1/4, 1)$. Since in this case $a \geq 0$, the problem has a solution for $a \in [0, 1)$. Thus the roots of the first equation lie between the roots of the second equation if $a \in (1/4, 1)$.

**83.** $\phi$.   **84.** $[1, 4/3]$.   **85.** $\phi$.

**86.** $(-2/3, 3)$.   **87.** $(-\infty, +\infty)$.

**88.** $(-\infty, -1) \cup (15, +\infty)$.   **89.** $[-2, 1]$.

**90.** $(-3, 2) \cup (2, 3)$. *Solution.* The given inequality is equivalent to the inequality $(|x| - 2)(|x| - 3) < 0$, whence $2 < |x| < 3$. For $x > 0$ we get $2 < x < 3$, and for $x < 0$ we get $2 < -x < 3$, or $-3 < x < -2$.

**91.** $(-\infty, -2] \cup [2, +\infty)$. *Solution.* The given inequality is equivalent to the inequality $(|x| + 1)(|x| - 2) \geq 0$, which is equivalent to the inequality $|x| \geq 2$ (since $|x| + 1 > 0$ for all $x \in R$). Finding the absolute value, we get $x \geq 2$ for $x \geq 0$ and $-x \geq 2$ or $x \leq -2$ for $x < 0$.

**92.** $(-1, 5)$. *Hint.* The inequality $|x^2 - 2x| < 5$ is equivalent to the two-sided inequality $-5 < x^2 - 4x < 5$, which is equivalent to the system of inequalities

$$\begin{cases} x^2 - 4x - 5 < 0, \\ x^2 - 4x + 5 > 0, \end{cases}$$

solving which we get $-1 < x < 5$.

**93.** $(-(1 + \sqrt{21})/2, (\sqrt{21} - 1)/2)$.   **94.** $(-1, 2) \cup (3, 6)$.

**95.** $(1, 3)$. *Hint.* The given inequality is equivalent to the system of inequalities.

$$\begin{cases} x > 0, \\ -x < x^2 - 2x < x. \end{cases}$$

**96.** $(2, 5)$.  **97.** $(1 - \sqrt{3}, 2 - \sqrt{2})$.

**98.** $(2, 4)$. *Solution.* We must consider two cases : (1) $x \geq 4$, (2) $x < 4$

(1) For $x \geq 4$ the given inequality is equivalent to the system of inequalities

$$\begin{cases} x \geq 4, \\ (x - 4)^2 < 0, \end{cases}$$

which has an empty set of solutions.

(2) For $x < 4$ the given inequality is equivalent to the system of inequalities

$$\begin{cases} x < 4, \\ x^2 - 6x + 8 < 0, \end{cases}$$

solving which we get $2 < x < 4$.

**99.** $(-5, 3 + 2\sqrt{2})$.

**100.** $(1, 3)$.

**101.** $(-\infty, 1) \cup (3, +\infty)$.

**102.** $(-\infty, (4 - \sqrt{2})/2] \cup [(5 + \sqrt{3})/2, +\infty)$.

**103.** $(-\infty, (4 - \sqrt{19})/3) \cup ((4 + \sqrt{19})/3, +\infty)$.

**104.** $(-\infty, -5 - \sqrt{19}) \cup (\sqrt{2} - 2, +\infty)$.

**105.** $(-\infty, (5 - \sqrt{57})/2) \cup ((5 + \sqrt{57})/2, +\infty)$.

**106.** $(-\infty, -1) \cup (2, +\infty)$.

**107.** $(1, 3)$. *Hint.* The given inequality is equivalent to the inequality $|x - 6| > x^2 - 5x + 9$ since $x^2 - 5x + 9 > 0$ for all $x \in R$.

**108.** $(-7, -2) \cup (3, 4)$. *Hint.* Consider three cases : (1) $x \leq -2$; (2) $-2 < x < 1$; (3) $x \geq 1$.

**109.** $(-\infty, 2\sqrt{2}) \cup (2 + 2\sqrt{3}, + \infty)$. *Solution.* For $x \le 0$ the inequality evidently holds true. Since $x^2 - 2x - 8 = (x - 4)(x + 2)$, we must consider two cases for $x > 0$:
(1) $0 < x < 4$, (2) $x \ge 4$.
(1) For $0 < x < 4$ the given inequality is equivalent to the system of inequalities
$$\begin{cases} 0 < x < 4, \\ x^2 - 8 < 0, \end{cases}$$
solving which we get $0 < x < 2\sqrt{2}$.
(2) For $x \ge 4$ the given inequality is equivalent to the system of inequalities
$$\begin{cases} x \ge 4, \\ x^2 - 4x - 8 > 0, \end{cases}$$
solving which we get $x > 2 + 2\sqrt{3}$.
Combining the solutions obtained, we find the expression given above.
**110.** $k = 15$. **111.** $(-\sqrt{2}, 1) \cup [\sqrt{2}, + \infty)$. **112.** $a_1 = 1 - \sqrt{2}, a_2 = 5 + \sqrt{10}$. **113.** $a < 0$.

## Sec. 1.5

**1.** $\{\sqrt{3} - 2, \sqrt{3} - 1, \sqrt{3} + 1, \sqrt{3} + 2\}$.
**2.** $\{-3, 3\}$.
**4.** $\{(5 - \sqrt{21})/2, (5 + \sqrt{21})/2\}$. *Hint.* Represent the equation in the form $(x^2 - 5x + 4)(x^2 - 5x + 6) = 15$ and set $x^2 - 5x + 4 = y$.
**5.** $\{(-5a \pm \sqrt{5a^2 + 4\sqrt{a^4 + b^4}})/2\}$    if    $|a| \ge 2|b|/\sqrt{3}$;    $\{(-5a \pm \sqrt{5a^2 + 4\sqrt{a^4 + b^4}})/2\}$ if $|a| < 2|b|/\sqrt{3}$. *Hint.* Represent the equation in the form $(x^2 + 5ax + 4a^2)(x^2 + 5ax + 6a^2) = b^4$ and set $x^2 + 5ax + 4a^2 = y$.
**6.** $\phi$.       **7.** $\{-1, 1/2\}$.        **8.** $\{1/2\}$.        **9.** $\{2\}$.
**10.** $x_1 = a, x_2 = -1$ for all $a \in (-\infty, -1)$; $x_1 = a, x_3 = -2a^2$ for all $a \in (-1/2, 0)$. *Solution.* Solving this equation, we find three values $x_1 = a, x_2 = -1, x_3 = -2a^2$ which turn either the first or the second factor of the given equations into zero. Note that $x_2 - x_1 < 0$ and $x_3 - x_1 < 0$, i.e. $x_2 - a < 0, x_3 - a < 0$ for $a \ge 0$, that is if for $a \ge 0$ we substitute $x_2$ or $x_3$ for $x$ into the equations, then the factor $\sqrt{x - a}$ is meaningless. Consequently, for $a \ge 0$ the given equation has only one root $x_1 = a$. We assume $a < 0$. For $a = -1$, we have $x_1 = x_2 > x_3$, i.e. $x_3 - x_1 = x_3 - a < 0$; thus $x_3 = -2a^2$ is not a root of the equation for $a = -1$; the equation has two coincident roots $x_1 = x_2$, which fact does not satisfy the hypothesis. For $a = -2a^2$, i.e. for $a = -1/2$, we have $x_2 - a = -1 + 1/2 = -1/2 < 0$, i.e. $x_2 = -1$ is not a root of the equation; the equation has two coincident roots $x_1 = x_3$ which does not satisfy the hypothesis. Let us now consider three cases : (1) $a \in (-\infty, -1)$, (2) $a \in (-1, -1/2)$ (3) $a \in (-1/2, 0)$. In the first case we have $-2a^2 < a < -1$, i.e., $x_3 < x_1 < x_2$. Consequently, $x_3 - x_1 = x_3 - a < 0$ and $x_3$ is not a root of the equation, but $x_2 - x_1 = x_2 - a > 0$ and consequently, $x_2$ is a root of the equation, i.e. for $a \in (-\infty, -1)$ the equation has only two distinct roots $x_1$ and $x_2$. In the second case we have $-1 < a$ and $a < -1/2$, i.e. $a^2 > -a/2$ or $-2a^2 < a$.    Consequently, in this case $x_2 - x_1 = x_2 - a < 0$ and $x_3 - x_1 = x_3 - a < 0$, i.e., neither $x_2$ nor $x_3$ is a root of the given equation ; for $a \in (-1, -1/2)$ the equation has only one root $x_1 = a$. In the third case we have $a > -1/2$; consequently, $-2a^2 > a > -1$ i.e. $x_2 < x_1 < x_3$ ; in this case $x_3 - x_1 = x_3 - a > 0$, and $x_2 - x_1 = x_2 - a < 0$. Thus $x_3$ is a root of the equation and $x_2$ is not . For $a \in (-1/2, 0)$ the equation has two distinct real root $x_1 = a$ and $x_3 = -2a^2$. Thus the given equation has only two distinct real root

$x_1 = a$ and $x_2 = -1$ for $a \in (-\infty, -1)$ and two distinct real roots $x_1 = a$ and $x_3 = -2a^2$ for $a \in (-1/2, 0)$. **11.** $\{3, 4, 5, 6, 7, 10\}$.

**12.** (a) The equation has no solutions for $a \in (-\infty, -1) \cup (5/4, +\infty)$;
    (b) the equation has one solution for $a = -1$;
    (c) the equation has two solutions for $a \in (-1,1) \cup \{5/4\}$;
    (d) the equation has three solutions for $a = 1$. *Solution.* The roots of the given equation have the form

$$x = \pm\sqrt{a - \frac{1}{2} \pm \sqrt{\frac{5}{4} - a}}. \tag{1}$$

(a) If we consider expression (1), we can see that the equation does not have a single root if $5/4 < a$, i.e. $a \in (5/4, +\infty)$ or $a \le 5/4$, but $a - \frac{1}{2} + \sqrt{\frac{5}{4} - a} < 0$, i.e. $\sqrt{\frac{5}{4} - a} < \frac{1}{2} - a$; this inequality can only be satisfied for $a < 1/2$; squaring the left-hand and right-hand sides of the inequality, we get (after the requisite transformations), $a^2 > 1$ or $|a| > 1$; taking into account that $a < 1/2$, we get $a < -1$, i.e. $a \in (-\infty, -1)$. Thus the equation has no solutions if $a \in (-\infty, -1) \cup (5/4, +\infty)$.

(b) The equation has only one solution $x = 0$ if $a - \frac{1}{2} + \sqrt{\frac{5}{4} - a} = 0$, i.e. $\sqrt{\frac{5}{4} - a} = \frac{1}{2} - a$. This equality is possible only under the condition that $a < 1/2$. Squaring the right-hand and left-hand sides of this equation, we get (after the requisite transformations) $a^2 = 1$, i.e. $a = -1$ ($a = 1 > 1/2$, and we must have $a < 1/2$). Thus, for $a = -1$ the equation has only one solution.

(c) If $a - \frac{1}{2} + \sqrt{\frac{5}{4} - a} > 0$ and $a - \frac{1}{2} - \sqrt{\frac{5}{4} - a} < 0$, or $a = 5/4$ (2) in expression (1), then the equation has only two solutions. Let us transform inequality (2) as follows : $\frac{1}{2} - a < \sqrt{\frac{5}{4} - a}$, $a - \frac{1}{2} < \sqrt{\frac{5}{4} - a}$. The second inequality holds for all $a \le 1/2$ and for $a > 1/2$ for which $a^2 < 1$, i.e. $1/2 < a < 1$, i.e. the second inequality (2) holds for all $a \in (-\infty, 1)$. The first inequality (2) holds for all $5/4 \ge a \ge 1/2$ and for $a < 1/2$ for which $a^2 < 1$ or $-1 < a < 1/2$, i.e. the first inequality (2) holds for all $a \in (-1, 5/4)$. Thus the equation has two solutions if the two inequalities (2) hold true simultaneously, i.e. for all $a \in (-1, 1)$ or for $a = 5/4$.

(d) The equation has three solutions provided that

$$a - \frac{1}{2} + \sqrt{\frac{5}{4} - a} > 0 \text{ and } a - \frac{1}{2} - \sqrt{\frac{5}{4} - a} = 0. \tag{3}$$

The first of these conditions is satisfied for all $a \in (-1, 5/4)$ (sec (c)) The second condition is equivalent to the equality $a - \frac{1}{2} = \sqrt{\frac{5}{4} - a}$, i.e. $a \ge 1/2$.

Squaring the right-hand and left-hand sides of the equation, we get, after the requisite transformation, $a^2 = 1$, i.e. $|a| = 1$; since $a > 1/2$, we get $a = 1$; this value satisfies the first of the condition (3) as well. Consequently, the equation has only three solutions for $a = 1$.

(e) If $a \in (1, 5/4)$, then the equation has four solutions.

**13.** For all $a \in (0, 1) \cup (1, 5 - \sqrt{15}) \cup (5 + \sqrt{15}, +\infty)$.
**14.** $x = b/2$. The solution exists and is unique for $b \ne 0$.

**15.** $a \in (-1/2, 3)$. *Solution.* $\dfrac{x - 2a - 3}{x - a + 2} < 0 \Leftrightarrow 2 - \dfrac{x + 7}{x + 2 - a} < 0 \Leftrightarrow 2 < \dfrac{x + 7}{x + 2 - a}$. The

last inequality must hold for all $1 \le x \le 2$; for these values of $x$ the numerator $x + 7 < 0$, consequently, for the inequality

$$2 < \frac{x + 7}{x + 2 - a} \tag{1}$$

to hold true, the denominator $x + 2 - a$ must be positive, i.e. $x + 2 - a > 0$, or $a < x + 2$ for all $1 \le x \le 2$, consequently, $a < 3$. We rewrite inequality (1) as follows (taking into consideration that $1 \le x \le 2$):

$$\frac{1}{2} > \frac{x + 2 - a}{x + 7} \Leftrightarrow \frac{1}{2} > 1 - \frac{a + 5}{x + 7} \quad \Leftrightarrow \quad \frac{a + 5}{x + 7} > \frac{1}{2}$$

$$\Leftrightarrow a + 5 > \frac{x + 7}{2} \qquad \Leftrightarrow a > \frac{x - 3}{2} \Rightarrow a > -\frac{1}{2}.$$

Thus inequality (1) and, hence, that given in the hypothesis, hold true for all $a \in (-1/2, 3)$.

**16.** $x = -6$.        **17.** $x = 2$.      **18.** $x = -4$.      **19.** $x = 1$.

**20.** $x = 2$.         **21.** $x = 2$.      **22.** $x = -2$.

**23.** $\{11, 12, 14, 15\}$.

**24.** (b) For all $a \in [10/3, +\infty)$, (a) follows from (b).

**26.** $k \in [(7 + 3\sqrt{5})/2, +\infty)$. *Solution.* We write the inequality in the form

$$\frac{x^2 - kx + k^2 - 6k}{k(6 + x)} \ge 0. \tag{1}$$

The discriminant of the quadratic trinomial

$$x^2 - kx + k^2 - 6k \tag{2}$$

is equal to $3k(8 - k)$. It is negative for $k < 0$ and $k > 8$. For these values of $k$ the numerator of the fraction (1) is positive and, therefore, for $k < 0$, the set of solutions of inequality (1) is $(-\infty, -6)$, and for $k > 8$ it is $(-6, +\infty)$. Consequently, all $k > 8$ satisfy the hypothesis and all $k < 0$ do not.

For $k = 0$ inequality (1) is indeterminate.

For $k = 8$ inequality (1) has the form $\dfrac{(x - 4)^2}{8(6 + x)} \ge 0$ and hold for all $1 < x < 1$.

Let us consider $0 < k < 8$. In this case the quadratic trinomial (2) has two roots

$$x_1 = \frac{k - \sqrt{3k(8 - k)}}{2} \text{ and } x_2 = \frac{k + \sqrt{3k(8 - k)}}{2} \text{ with } x_1 < x_2 \text{ for all } 0 < k < \infty.$$

Indeed, for $0 < k < 8$ the inequality $x_1 > -6$ is equivalent to the inequality $k + 12 > \sqrt{3k(8 - k)}$, which is equivalent to the inequality $k^2 + 24k + 144 > 24k - 3k^2$, or $4k^2 + 144 > 0$. Since $x_2 > 0$ for all $0 < k < 8$, for the given values of $k$ the problem has a solution only for $x_1 \ge 1$, i.e. if $\dfrac{k - \sqrt{3k(8 - k)}}{2} \ge 1$, or $k - 2 \ge \sqrt{3k(8 - k)}$, which can only be satisfied for $2 < k \le 8$, and for $2 < k < 8$ is equivalent to the inequality $(k - 2)^2 \ge 3k(8 - k)$. Solving the system of inequalities

$$\begin{cases} 2 < k < 8, \\ (k - 2)^2 \ge 3k(8 - k), \end{cases}$$

we obtain a set of solutions $[(7 + 3\sqrt{5})/2, 8]$. Combining the results obtained, we find that $k \in [(7 + 3\sqrt{5})/2, +\infty)$.

**27.** $(1, 2) \cup (2, 3)$. **28.** $(-1/4, 5/6)$. **29.** $(-\infty; 32) \cup (7/3, +\infty)$. **30.** $(-\infty, +\infty)$. **31.** $(2, 3)$. **32.** $(-3, 1)$. **33.** $(-\infty, -2) \cup (-2, -1) \cup (1, +\infty)$. **34.** $(-\infty, -2) \cup (-2, -1/2) \cup (1, +\infty)$. **35.** $(-2, -1) \cup (1, 2)$. **36.** $[-3, 3]$. **37.** $(-\infty, 2) \cup (5, +\infty)$. **38.** $(-\infty, 1) \cup (3/2, +\infty)$. **39.** $(-\infty, 5/2) \cup (33/8, +\infty)$. **40.** $(-6, 3)$. **41.** $(-\infty, 0) \cup (4, +\infty)$. **42.** $(-7, -3)$. **43.** $(-17/25, -3/8)$. **44.** $(-\infty, -5) \cup (5, +\infty)$. **45.** $(-\infty, 1) \cup (4, +\infty)$. **46.** $(-3, 1)$. **47.** $(-\infty, 0) \cup (3, +\infty)$. **48.** $(-\infty, 3) \cup (4, +\infty)$. **49.** $(-\infty, +\infty)$. **50.** $(-1, 5)$. **51.** $(5/2, 8)$. **52.** $(2, 3)$. **53.** $(-\infty, -1) \cup (5, +\infty)$. **54.** $(-\infty, -1) \cup (1/3, +\infty)$. **55.** $(-1/2, 2)$. **56.** $(-\infty, -20) \cup (23, +\infty)$. **57.** $(-2, +\infty)$. **58.** $(-1, 0) \cup (4, +\infty)$. **59.** $(-5, -2) \cup (-2/3, +\infty)$. **60.** $(-\infty, (2-\sqrt{22})/3) \cup ((2+\sqrt{22})/3, 5/2)$. **61.** $(-17/2, -3) \cup (1, +\infty)$. **62.** $(-\infty, -3) \cup (3, +\infty)$. **63.** $[1, 3] \cup (5, +\infty)$. **64.** $(-\infty, -1/\sqrt[3]{2}) \cup (0, +\infty)$. **65.** $(-\infty, 6) \cup [-2, 0) \cup (3, +\infty)$. **66.** $(2, 3) \cup (5, 6)$. **67.** $(1, 7)$. **68.** $(-6, 3)$. **69.** $(-5, 1) \cup \{3\}$. **70.** $(-1, +\infty)$. **71.** $(-9/2, -2) \cup (3, +\infty)$. **72.** $(-1, 1) \cup (4, 6)$. **73.** $(-(1+\sqrt{21})/2, (\sqrt{21}-1)/2)$. **74.** $(1/2, 3)$. **75.** $(-\infty, -1) \cup (4, +\infty)$. **76.** $(-\infty, -5/2) \cup (-2, 8)$. **77.** $(3/4, 1) \cup (7, +\infty)$. **78.** $(-1, 0) \cup (0, 1)$. **79.** $(-5, 1)$. **80.** $(-\infty, -4) \cup (-3, 3) \cup (6, +\infty)$. **81.** $(-\infty, -\sqrt{7}) \cup (-2, \sqrt{7}) \cup [8/3, +\infty)$.

**82.** $\left(-\infty, -\dfrac{\sqrt{63}}{2}\right) \cup \left(-3, \dfrac{\sqrt{63}}{2}\right) \cup (4, +\infty)$. **83.** $[1, 2) \cup (3, 4]$.

**84.** $(-\sqrt{2}, -1) \cup (-1, \sqrt{2}) \cup [3, 4)$.

**85.** $(-\infty, -\sqrt{7/2}) \cup (-1, \sqrt{7}/2) \cup (4/3, +\infty)$.

**86.** $(-(11+\sqrt{737})/28, 4/7) \cup ((\sqrt{737}-11)/28, 1)$.

**87.** $(-\infty, -\sqrt{10}) \cup (-1/2, 2/5) \cup (\sqrt{10}, +\infty)$. **88.** $(-\infty, -2) \cup (-1, 4)$. **89.** $(-\infty, -1) \cup (1, +\infty)$. **90.** $(-\infty, -2) \cup (-1, 0) \cup (1/2, +\infty)$. **91.** $(-\infty, 0) \cup (1, 2) \cup (3, +\infty)$. **92.** $(-\infty, -5) \cup (1, 2) \cup (6, +\infty)$. **93.** $(-\infty, -1) \cup (0, 1/2) \cup (1, +\infty)$. **94.** $(-\infty, 0) \cup (1, 6)$. **95.** $(1, 2) \cup (7, +\infty)$. **96.** $(-\infty, -3) \cup (-2, 3)$. **97.** $(-\sqrt{2}, 0) \cup (1, \sqrt{2}) \cup (2, +\infty)$. **98.** $(-5, 1) \cup (2, 3)$. **99.** $(-\infty, -2) \cup (-1, 3) \cup (4, +\infty)$. **100.** $(-\infty, -7) \cup (-4, -2)$. **101.** $(-\infty, -3) \cup (-2, -1)$.

**102.** $(-\infty, -4] \cup [-2, -1] \cup [1, +\infty)$.     **103.** $(-2, -1) \cup (2, 3)$.

**104.** $(-\infty, -1] \cup (0, 1] \cup (2, 3]$.     **105.** $[-4, -3) \cup [-3/2, 0) \cup [1, +\infty)$.

**106.** $(-\infty, -|a|) \cup (|a|, +\infty)$ if $a \neq 0$, $(-\infty, 0) \cup (0, 1) \cup (1, +\infty)$ if $a = 0$.

**107.** $(-\infty, -1) \cup (0, +\infty)$.     **108.** $(2, 3)$.

**109.** $(-7, -4) \cup (-4, 1)$.     **110.** $(2, +\infty)$.

**111.** $(2, 4) \cup (4, 6)$. *Hint.* The given inequality is equivalent to the system of inequalities $|x - 4| < 2, x \neq 4$.

**112.** $(3/4, 1) \cup (1, +\infty)$. *Hint.* The given inequality is equivalent to the collection of two inequalities $\dfrac{2x-1}{x-1} > 2$ and $\dfrac{2x-1}{x-1} < -2$.

**113.** $(-\infty, -2) \cup (-1, +\infty)$. *Hint.* Since $x^2 + x + 1 > 0$ for all $x \in R$, the given inequality is equivalent to the inequality $|x^2 - 3x - 1| < 3(x^2 + x + 1)$, or to $-3(x^2 + x + 1) < x^2 - 3x - 1 < 3(x^2 + x + 1)$, or to

$$\begin{cases} -3(x^2 + x + 1) < x^2 - 3x - 1, \\ x^2 - 3x - 1 < 3(x^2 + x + 1). \end{cases}$$

**114.** $(-5, -2) \cup (2, 3) \cup (3, 5)$.

**115.** $(-5, -2) \cup (-1, +\infty)$.

**116.** $(-\infty, -2) \cup (-1/2, +\infty)$

**117.** $(-\infty, 0) \cup (1, +\infty)$. *Hint.* We can write the inequality in the form $|x + 2|/x < 3$.

**118.** $(-\infty, -5) \cup (-3, 3) \cup (5, \infty)$

**119.** $(-\infty, -4] \cup [-1, 1] \cup [4, +\infty)$. *Hint.* The given inequality is equivalent to two-sided inequality $-1 \le \dfrac{3x}{x^2-4} \le 1$, or to the system

$$\begin{cases} \dfrac{3x}{x^2-4}+1 \ge 0, \\ \dfrac{3x}{x^2-4}-1 \le 0. \end{cases}$$

**120.** $[0, 8/5] \cup [5/2, +\infty)$.
**121.** $[3/2, 2)$. *Hint.* Since $x^2 - 5x + 6 = (x-2)(x-3)$, for $x > 3$ the given inequality is equivalent to the inequality $1/(x-2) \ge 2$, and for $x < 3$ it is equivalent to the inequality $1/(2-x) \ge 2$.
  **122.** $(-\infty, 3)$.   **123.** $(-\sqrt{-a}, 0)$ for $a < 0$; $(0, \sqrt{a})$ for $a > 0$, $\phi$ for $a = 0$. **124.** $(3, 4)$.
**125.** $[-3, 0) \cup [20, +\infty)$. **126.** $(-\infty, -3) \cup (-2, +\infty)$. **127.** $(-\infty, -1) \cup (1,3)$. **128.** $(-\infty, -1] \cup (5, 6]$. **129.** $(-\infty, -3) \cup [1, 4)$. **130.** $(-\infty, -1.4] \cup (2, 2.6]$. **131.** $[1, 6]$. **132.** $[2, 3]$.
**133.** $[-1, 7/3]$.  **134.** $(-2, -1) \cup (2, +\infty)$.  **135.** $(-2, (3-\sqrt{17})/2] \cup (0, 2) \cup [(3+\sqrt{17})/2, +\infty)$.

## Sec. 1.6

  **1.** $\{1/2, 1\}$. **2.** $\{-1, 2\}$. **3.** $\{-3, 2\}$. **4.** $\{-4, 3\}$. **5.** $\{6\}$. **6.** $\{-27/8, 1\}$. **7.** $\{-8, 27\}$.
**8.** $\{8, 27\}$. **9.** $\{3\}$. **10.** $\{1\}$. **11.** $\{3\}$. **12.** $\{9\}$. **13.** $\{-1, 8\}$.
  **14.** $\{2\}$. *Solution.* Setting $\sqrt[6]{\log_2 x} = t$, we get an equation $t^3 + t^2 - 2 = 0$, or $t^3 - 1 + t^2 - 1 = 0$, or $(t-1)(t^2+t+1)+(t-1)(t+1) = 0$, or $(t-1)(t^2+2t+2) = 0$, whence it follows that $t = 1$, i.e. $\log_2 x = 1$, $x = 2$.
  **15.** $\{-1\}$.
  **16.** $\{-3/2, 1/2\}$. *Hint.* Set $\sqrt{(3-x)/(2+x)} = t$.

| | | | |
|---|---|---|---|
| **17.** $\{5/2\}$. | **18.** $\{5/3\}$. | **19.** $\{3\}$. | **20.** $\{3\}$. |
| **21.** $\{8\}$. | **22.** $\{28\}$. | **23.** $\{0\}$. | **24.** $\{4\}$. |
| **25.** $\{19\}$. | **26.** $\{3\}$. | **27.** $\{6\}$. | **28.** $\{-1\}$. |
| **29.** $\{3\}$. | **30.** $\{2\}$. | **31.** $\{1\}$. | **32.** $\{5\}$. |

**33.** $\{0, 4/3\}$.   **34.** $\left\{3 + \dfrac{2\sqrt{5}}{5}\right\}$.   **35.** $\{1\}$.

**36.** $\{4\}$.   **37.** $\{2\}$.   **38.** $\{5\}$.

**39.** $\left\{\arctan\dfrac{2}{3} + \pi n \mid n \in \mathbf{Z}\right\}$.

**40.** $\{2\}$. *Hint.* Set $x^2 - 4x + 6 = y$.
**41.** $\{-9/2, 3\}$. *Hint.* Set $\sqrt{2x^2 - 4x + 9} = y$.
**42.** $\{-4, 2\}$. *Hint.* Set $\sqrt{2x^2 - 4x + 12} = y$.
  **43.** $\{1 - \sqrt{3}, 1 + \sqrt{3}\}$. **44.** $\{-5, 0\}$. **45.** $\{2\}$. **46.** $\{2/3\}$. **47.** $\{-3\}$. **48.** $\{-1, 9/16\}$. **49.** $\{2\}$.
**50.** $\{5\}$. **51.** $\{-1, 15\}$. **52.** $\{4\}$. **53.** $\{20\}$. **54.** $\{10\}$. **55.** $\{6\}$. **56.** $\{-1, 3\}$. **57.** $\{-5, 4\}$.
**58.** $\{-1\}$. *Solution.* Note that

$$\sqrt{3x^2 + 6x + 7} = \sqrt{3(x+1)^2 + 4} \ge \sqrt{4} = 2,$$

$$\sqrt{5x^2 + 10x + 14} = \sqrt{5(x+1)^2 + 9} \ge \sqrt{9} = 3,\ 4 - 2x - x^2 = 5 - (x+1)^2 \le 5.$$

Consequently, the left-hand side of the equation $\ge 5$ for any $x \in \mathbf{R}$ and the right-hand side $\le 5$ for any $x \in \mathbf{R}$. Their equality is only possible under the condition that both sides of the equation are equal to 5 for the same value of $x$. In our case the right-hand and left-hand sides of the equation are equal to 5 for $x = -1$.

**59.** $\{3\}$. **60.** $\{-1, 2\}$. **61.** $\{2, 34\}$. **62.** $\{-1\}$. **63.** $\{5\}$. **64.** $\{8 + 4\sqrt{2}\}$. **65.** $\{-1\}$. **66.** $\{4\}$.
**67.** $\{1\}$. **68.** $\{5\}$.

**69.** $\{-15, 13\}$. *Solution.* $\sqrt[3]{12 - x} + \sqrt[3]{14 + x} = 2 \Rightarrow \sqrt[3]{12 - x} = 2 - \sqrt[3]{14 + x} \Rightarrow$
$12 - x = 8 - 12\sqrt[3]{14 + x} + 6\sqrt[3]{(14 + x)^2} - 14 - x \Rightarrow 6(\sqrt[3]{14 + x})^2 - 12.$
$\sqrt[3]{14 + x} - 18 = 0 \Rightarrow (\sqrt[3]{14 + x})^2 - 2\sqrt[3]{14 + x} - 3 = 0 \Rightarrow \sqrt[3]{14 + x} = 1 \pm \sqrt{1 + 3} =$
$1 \pm 2 \Rightarrow \sqrt[3]{14 + x} = -1$ or $\sqrt[3]{14 + x} = 3$; if $\sqrt[3]{14 + x} = -1$, then $14 + x = -1$ and
$x = -15$; if $\sqrt[3]{14 + x} = 3$, then $14 + x = 27$ and $x = 13$.

**70.** $\{(3 - \sqrt{73}) / 4, 0, 3 / 2, (3 + \sqrt{73}) / 4\}$. *Hint.* Set $\sqrt{2x^2 - 3x + 1} = y$.

**71.** $\{1 / 4\}$. **72.** $\{0, 16 / 9\}$. **73.** $\{0, 5\}$. **74.** $\{7\}$. **75.** $\{16 / 25\}$. **76.** $\{4\}$. **77.** $\{-1\}$.
**78.** $\{2\}$. **79.** $\{4\}$. **80.** $\{12 / 5, 4\}$. **81.** $\{4\}$. **82.** $\{-1\}$. **83.** $\{0\}$. **84.** $\{4\}$. **85.** $\{2\}$. **86.** $\{1\}$. **87.** $\{8\}$.
**88.** $\{22/9, 6\}$. **89.** $\{12 / 127\}$. **90.** $\{12\}$. **91.** $\{3, 5\}$. **92.** $\{0, -1\}$. *Hint.* Set $x^2 + x + 4 = y$.

**93.** $\{1\}$. *Hint.* Find the domain of definition of the equation.

**94.** $\{-47 / 24\}$.

**95.** $\{73 / 32\}$.

**96.** $\{[-1, 1]\}$. *Hint.* Write the equation in the form $\sqrt{(x - 1)^2} + \sqrt{(x + 1)^2} = 2$, or $|x - 1| + |x + 1| = 2$, and then consider three cases : (1) $x < -1$; (2) $-1 \le x \le 1$; (3) $x > 1$. Make sure that the equation is identically satisfied for $-1 \le x \le 1$.

**97.** $\{[2, +\infty)\}$

**98.** $\{[2, +\infty)\}$

**99.** $\{[5, 10]\}$

**100.** $\{(-1 - \sqrt{5}) / 2, (\sqrt{5} - 1) / 2, 1\}$. *Solution.* Set $\sqrt[3]{2x - 1} = y$; then $2x - 1 = y^3$, $y^3 + 1 = 2x$. We obtain a system of equations
$$\begin{cases} x^3 + 1 = 2y, \\ y^3 + 1 = 2x, \end{cases}$$
whence it follows that $x^3 - y^3 = 2(y - x)$, or $(x - y)(x^2 + xy + y^2) + 2(x - y) = 0$. Hence, we get (1) $x - y = 0$; [(2) $x^2 + xy + y^2 = -2$. The last equation can be written in the form $\left(x + \dfrac{1}{2}y\right)^2 + \dfrac{3}{4}y^2 + 2 = 0$. This equation has no solutions.
Returning to the unknown $x$, We write the first equation as $\sqrt[3]{2x - 1} = x$, or $x^3 - 2x + 1 = 0$, or $x^3 - 1 - 2(x - 1) = 0$, or $(x - 1)(x^2 + x - 1) = 0$. Solving this equation, we get $x = 1$, $x = (-1 \pm \sqrt{5}) / 2$.

**101.** $\{-2, 1 + \sqrt{5}\}$.

**102.** $x \in \phi$ if $a \in (-\infty, 0) \cup (0, 1)$; $x = \{0\}$ if $a = 0$; $x = \{(2a - 1 - \sqrt{4a - 3}) / 3\}$ if $a \in [1, +\infty)$. *Solution.* Since $x$ is under the radical sign, we have $x \ge 0$ and the left-hand side of the equation $x + \sqrt{a + \sqrt{x}} \ge 0$, and, consequently, $a \ge 0$; in addition, since $\boldsymbol{a}$ is the sum of $x$ and the nonnegative quantity $\sqrt{a + \sqrt{x}}$, we have $x \le a$. Thus, considering the equation itself, we infer that the equation has no real solutions for $a \in (-\infty, 0)$. We assume that $a \ge 0$. Then
$$x + \sqrt{a + \sqrt{x}} = a \Rightarrow \sqrt{a + \sqrt{x}} = a - x$$
$$\Rightarrow a + \sqrt{x} = a^2 - 2ax + x^2$$
$$\Rightarrow a^2 - (2x + 1)a + x^2 - \sqrt{x} = 0$$
$$\Rightarrow a = (2x + 1 \pm \sqrt{4x^2 + 4x + 1 - 4x^2 + 4\sqrt{x}}) / 2$$
$$\Rightarrow a = (2x + 1 \pm \sqrt{4x + 4\sqrt{x} + 1}) / 2$$
$$\Rightarrow a = (2x + 1 \pm (2\sqrt{x} + 1)) / 2$$
$$\Rightarrow a = x + \sqrt{x} + 1 \text{ or } a = x - \sqrt{x}.$$

Let us consider the first possibility $a = x + \sqrt{x} + 1$; taking into account that the right-hand side is the sum of three nonnegative terms, one of which is equal to 1, we find that if $0 \le a < 1$, then the equation we have written has no solutions. We shall consider the case $a \ge 1$; then

$$a = x + \sqrt{x} + 1 \Rightarrow x + \sqrt{x} + 1 - a = 0 \Rightarrow \sqrt{x} = (-1 \pm \sqrt{1 - 4 + 4a}) / 2$$

$$\Rightarrow \sqrt{x} = (-1 + \sqrt{4a - 3}) / 2, \text{ or } \sqrt{x} = -(1 + \sqrt{4a - 3}) / 2,$$

but $\sqrt{x} \ge 0, a \ge 1$, and, therefore, the second case is impossible ; thus we have

$$\sqrt{x} = (\sqrt{4a - 3} - 1) / 2 \Rightarrow x = (2a - 1 - \sqrt{4a} - 3) / 2 \text{ for } a \ge 1.$$

Let us now consider the case $a = x - \sqrt{x}$; here $a$ can be any non negative number, and then

$$a = x - \sqrt{x} \Rightarrow x - \sqrt{x} - a = 0 \Rightarrow \sqrt{x} = (1 \pm \sqrt{1 + 4a}) / 2.$$

Note that the relation $\sqrt{x} = (1 - \sqrt{1 + 4a}) / 2 \le 0$, and, consequently, is possible only for $a = 0$; then $x = 0$. In the case $\sqrt{x} = (1 + \sqrt{1 + 4a}) / 2 \Rightarrow$ $x = (2a + 1 + \sqrt{4a + 1}) / 2 > a$, which is impossible. Combining all the results, we find that $x \in \phi$ if $a \in (-\infty, 0) \cup (0, 1), x = \{0\}$ for $a = 0$ and $x = \{(2a - 1 - \sqrt{4a - 3}) / 2\}$ for $a \in [1, + \infty)$.

103. $x \in \phi$ if $a \in (-\infty, -1 / 4), x = \{(-1 \pm \sqrt{4a + 1}) / 2\}$ if $a \in [-1 / 4, 0]$; $x = \{(-1 - \sqrt{4a + 1}) / 2\}$ if $a \in (0, 1); x = \{(-1 - \sqrt{4a + 1}) / 2, (1 + \sqrt{4a - 3}) / 2\}$ if $a \in [1, + \infty)$. *Solution.* *1st technique.* We shall use this technique to carry out the solution in detail. Considering the equation, we see that $x \le a \le x^2$. Then

$$x^2 - \sqrt{a - x} = a \Rightarrow x^2 - a = \sqrt{a - x} \Rightarrow x^4 - 2x^2 a + a^2 = a - x$$

$$\Rightarrow a^2 - (2x^2 + 1) a + x + x^4 = 0$$

$$\Rightarrow a = (2x^2 + 1 \pm \sqrt{4x^4 + 4x^2 + 1 - 4x - 4x^4}) / 2$$

$$\Rightarrow a = (2x^2 + 1 \pm (2x - 1)) / 2$$

$$\Rightarrow a = x^2 + x \text{ or } a = x^2 - x + 1.$$

Thus

$$a = x^2 + x \text{ or } a = x^2 - x + 1 \tag{1}$$

In the case $a = x^2 + x$ only $x \le 0$ are suitable since $a \le x^2$. We solve the equation $x^2 + x - a = 0 \Rightarrow x = (-1 \pm \sqrt{4a + 1}) / 2$. Consequently, for $a < -1 / 4$ the given equation has no solutions, for $a \in [-1 / 4, 0]$ it has a set of solutions $x = \{(-1 \pm \sqrt{4a + 1}) / 2\}$, for $a \in (0, + \infty)$ only the set $x = \{(-1 - \sqrt{4a + 1}) / 2\}$ is a solution since for $a > 0$ the expression $x = (-1 + \sqrt{4a + 1}) / 2 > 0$ which is impossible in the given case. In the case $a = x^2 - x + 1$ only $x \ge 1$ are suitable since $a \le x^2$, and in that case $a \ge 1$ since $a = 1 + x(x - 1)$. We solve the equation

$$x^2 - x + 1 - a = 0 \Rightarrow x = (1 \pm \sqrt{1 - 4 + 4a}) / 2$$

$$\Rightarrow \qquad x = (1 \pm \sqrt{4a - 3} / 2).$$

For $a \ge 1$ the expression $(1 - \sqrt{4a - 3}) / 2 \le 0$ whereas in this case $x \ge 1$. Consequently, in this case the only solution is $x = (1 + \sqrt{4a - 3}) / 2$ which is equal to or greater than unity for $a \ge 1$. Thus, combining the results, we get

$x \in \phi$ if $a \in (-\infty, -1/4)$;

$x = \{(-1 \pm \sqrt{4a+1})/2\}$, if $a \in [-1/4, 0]$;

$x = \{(-1 - \sqrt{4a+1})/2\}$, if $a \in (0, 1)$;

$x = \{(-1 - \sqrt{4a+1})/2 ; (1 + \sqrt{4a-3})/2\}$, if $a \in [1, +\infty)$.

*2nd technique.* We write the initial equation in the form

$$x_2 - x + \frac{1}{4} = (a - x) + \sqrt{a-x} + \frac{1}{4} \Leftrightarrow \left(x - \frac{1}{2}\right)^2 = \left(\sqrt{a-x} + \frac{1}{2}\right)^2$$

$$\Rightarrow x - \frac{1}{2} = \pm \left(\sqrt{a-x} + \frac{1}{2}\right). \tag{2}$$

Rationalizing equations (2), we obtain equations (1).

*3rd technique.* Setting $\sqrt{a-x} = y$ and taking into account that $a - x = y^2$, we get the following system of equations for $x$ and $y$:

$$\begin{cases} x^2 - y = a, \\ y^2 + x = a. \end{cases} \tag{3}$$

Subtracting the second equation of system (3) from the first, we obtain

$$x^2 - y^2 - y - x = 0 \Leftrightarrow (x - y - 1)(x + y) = 0$$

$$\Rightarrow \begin{bmatrix} x - y - 1 = 0, \\ x + y = 0, \end{bmatrix} \Leftrightarrow \begin{bmatrix} x - 1 = \sqrt{a-x}, \\ x = -\sqrt{a-x}. \end{bmatrix}$$

We solve the equation $x - 1 = \sqrt{a-x}$. This equation is equivalent to a system consisting of one equation and one inequality

$$\begin{cases} x^2 - x + 1 - a = 0, \\ x - 1 \geq 0, \end{cases} \Leftrightarrow \begin{cases} x = \dfrac{1 \pm \sqrt{4a-3}}{2}, \\ x \geq 1, \end{cases}$$

$$\Leftrightarrow \begin{cases} x = \dfrac{1 + \sqrt{4a-3}}{2} \\ x \geq 1, \end{cases} \Rightarrow x = \frac{1 + \sqrt{4a-3}}{2} \text{ for } a \geq 1.$$

Let us now solve the equation

$$x = -\sqrt{a-x}. \tag{4}$$

This equation is equivalent to a system consisting of one equation and one inequality

$$\begin{cases} x^2 + x - a = 0, \\ x \leq 0, \end{cases} \Leftrightarrow \begin{cases} x = \dfrac{-1 \pm \sqrt{4a+1}}{2}, \\ x \leq 0. \end{cases}$$

The root $x = -(\sqrt{4a+1}+1)/2$ satisfies equation (4) for all $a \geq -1/4$ (and only for those $a$).

A system consisting of one equation and one inequality

$$\begin{cases} x = \dfrac{\sqrt{4a+1}-1}{2}, \\ x \le 0, \end{cases}$$

has a solution only for $-1/4 \le a \le 0$ and, therefore, the root $x = (\sqrt{4a+1}-1)/2$ satisfies equation (4) for all $-1/4 \le a \le 0$ and only for those values of $a$. Combining all the results, we find all the roots of the initial equation.

104. $x = \{0\}$. *Solution.* If $a < 0$, then the left-hand side of the equation includes two expressions $a\sqrt{x} \le 0$ and $-\sqrt{x} + 2ax\sqrt{x^2 + 7a^2} \le 0$, whose sum is equal to zero. This is possible only under the condition that each of them is equal to zero, and that is possible only if $x = 0$. If $a = 0$, the equation assumes the form $-\sqrt{x} = 0$, which is satisfied only in the case $x = 0$. If $a > 0$, then we can transfer the second radical into the right-hand side of the equation and square both sides of the equation, to obtain, after the requisite transformations, an equation $x (a^2 - 1 - 2a$ $\sqrt{x^2 + 7a^2}) = 0$, whose root is $x = 0$. If there are some other roots, they satisfy the equation $a^2 - 1 = 2a\sqrt{x^2 + 7a^2}$, from which we can see that $a^2 > 1$. Eliminating the radical and transforming, we obtain $4a^2x^2 + 27a^4 + 2a^2 - 1 = 0$, which is impossible. Thus, for any $a$ the equation has the only root $x = 0$.

105. $\{4\}$. *Hint.* Reduce the equation to the form $(x - 4)^2 + (2\sqrt{x} - 4)^2 = 0$.

106. $(a^2 - 4a)$ for $a > 2$; $\phi$ for $a \le 2$.

107. $\phi$. 108. $\{-2\}$. 109. $\{-4/3\}$. 110. $\{1/3\}$. 111. $\{2\}$. 112. $\{-3\}$. 113. $\{2^{10}\}$.

114. $\{76\}$. 115. $\{1\}$. 116. $\{2\}$. 117. $\{1\}$. 118. $\{5\}$.

119. One ; $x = -4/3$. *Solution.* Assume that the equation has roots ; then (taking into account that $x^2 > 5/3$)

$$\sqrt{x^2+1} - \frac{1}{\sqrt{x^2-5/3}} = x \Rightarrow \sqrt{x^2+1} = x + \frac{1}{\sqrt{x^2-5/3}}$$

$$\Rightarrow x^2 + 1 = x^2 + \frac{2x}{\sqrt{x^2-5/3}} + \frac{1}{x^2-5/3} \Rightarrow x^2 - \frac{8}{3} = 2x\sqrt{x^2 - \frac{5}{3}}$$

$$\Rightarrow x^4 - \frac{4}{9}x^2 - \frac{64}{27} = 0 \Rightarrow x^2 = \frac{2}{9} \pm \frac{14}{9}, x^2 = \frac{16}{9} \text{ or } x^2 = -\frac{4}{3}.$$

Only the first result, $x^2 = 16/9$, is possible, whence $x = 4/3$ or $x = -4/3$. Verification shows that the only root is $x = -4/3$.

120. $\{11/6\}$. *Hint.* The domain of definition of the left-hand side of the equation is the interval $[7/5, 5/2]$, and, therefore, $7/5 < a/6 < 5/2$, i.e. $9 < a < 15$.

121. $[2, +\infty)$.                    122. $(-\infty, -1] \cup [2, +\infty)$.

123. $(1/2, 2]$.                    124. $(3/4, 2)$.

125. $[0, 2) \cup (9, +\infty)$.          126. $(4, 6]$.

127. $(-1 - \sqrt{5}, -3] \cup [1, \sqrt{5} - 1)$.                    128. $(6/5, -1] \cup [2, 3)$.

129. $(-\infty, 1)$. *Hint.* Set $\sqrt{2-x} = t$ and take into account that $t \ge 0$.

130. $(-\infty, -17/2] \cup [1, 10)$.                    131. $(6, 8]$.

132. $(-1/2, +\infty)$.                    133. $[1, 2) \cup (2, +\infty)$.

134. $[-18, -2)$. *Hint.* The inequality $\sqrt{f(x)} < \varphi(x)$ is equivalent to the system of inequalities

$$\begin{cases} f(x) \ge 0, \\ \varphi(x) > 0, \\ f(x) < \varphi^2(x). \end{cases}$$

**135.** $(3, 24/5]$. **136.** $[20/9, 4) \cup (5, +\infty)$. **137.** $((1 + \sqrt{29})/2, +\infty)$. **138.** $(5, +\infty]$.
**139.** $(3, +\infty)$. **140.** $(1/2, 5/2)$. **141.** $(3, +\infty)$. **142.** $[1, 6]$. **143.** $[0, 2]$. **144.** $[5/2, 3)$.
**145.** $(2/3, +\infty)$. **146.** $(-\infty, -2] \cup [5, 74/13)$. **147.** $[-2, -8/5) \cup (0, 2]$. **148.** $[0, 3]$.
**149.** $(-1, 0) \cup (3/5, 1]$. **150.** $\left(-1, \dfrac{5\sqrt{13}}{2}\right] \cup (2, +\infty)$. **151.** $(-\infty, -2)$.

**152.** $(-\infty, 1)$. *Hint.* The inequality $\varphi(x) < \sqrt{f(x)}$ is equivalent to the collection of two systems of inequalities

$$\begin{cases} \varphi(x) < 0, \\ f(x) \geq 0 \end{cases} \text{ and } \begin{cases} \varphi(x) \geq 0, \\ f(x) \geq 0, \\ \varphi^2(x) < f(x). \end{cases}$$

**153.** $(-33, 3)$. **154.** $(-\infty, -1]$. **155.** $[-7, 1)$. **156.** $[2, (7 + \sqrt{5})/2)$. **157.** $[-14, 2)$.
**158.** $(-\infty, 3)$. **159.** $(4, 5)$. **160.** $(-\infty, 2)$. **161.** $[-2, 2)$. **162.** $(-\infty, +\infty)$. **163.** $(24/19, +\infty)$.
**164.** $(-\infty, -3]$. **165.** $[2, +\infty)$. **166.** $(-\sqrt{5}, 2)$. **167.** $[-1, (\sqrt{5}-1)/2)$. **168.** $(8/3, +\infty)$.
**169.** $(-\infty, -2] \cup (2, +\infty)$. **170.** $\phi$. **171.** $(-\infty, -5] \cup [1, +\infty)$. **172.** $(-\infty, -3] \cup (-2,$
$(\sqrt{13}-7)/6)$. **173.** $(-\infty, 1] \cup [2, (17 + \sqrt{13})/6)$. **174.** $((5 - \sqrt{13})/6, +\infty)$. **175.** $(1, 4]$.
**176.** $[-2, (\sqrt{5}-15)/10)$. **177.** $[-6, \sqrt{2}-4)$. **178.** $(3, 5]$. **179.** $(1, 2\sqrt{3}/3]$. **180.** $[-1$
$-\sqrt{13}, 0] \cup [(1 + \sqrt{17})/2, \sqrt{13} - 1]$. **181.** $[-\sqrt{3}, 0) \cup (0, 2]$. **182.** $[0, 1] \cup [4, 16]$.
**183.** $[3, 4) \cup (7, +\infty)$.      **184.** $[-4, (3 - \sqrt{21})/2) \cup (1, +\infty)$.
**185.** $[1/2, 2) \cup (5, +\infty)$.      **186.** $[-20, 0) \cup (5, +\infty)$.
**187.** $(-\infty, -4) \cup (1/2, 8/7)$.      **188.** $(-2 - 2\sqrt{6}, -1) \cup [-2 + 2\sqrt{6}, 3]$.
**189.** $((\sqrt{13} - 5)/2, 1]$.      **190.** $[-1, -\sqrt{3}/2) \cup (\sqrt{3}/2, 1]$.
**191.** $(1, +\infty)$.      **192.** $(4, +\infty)$.
**193.** $[-3, 1)$.      **194.** $[-1, -\sqrt{15}/4) \cup (\sqrt{15}/4, 1]$.
**195.** $[-2, +\infty)$.     **196.** $[(16 + \sqrt{7})/2, 10]$.     **197.** $[1, 3/2)$.
**198.** $(2\sqrt{21}/3, +\infty)$.     **199.** $(-2, -1] \cup [-2/3, 1/3)$.
**200.** $[1, +\infty)$.     **201.** $[-5, 2\sqrt{\sqrt{5}-2}-4)$.

**202.** $[-1/2, 0) \cup (0, 1/2]$. *Hint.* The domain of definition of the inequality is the set of numbers $[-1/2, 0) \cup (0, 1/2]$. Reduce the inequality to the form

$$\frac{1 - (1 - 4x)^2}{x(1 + \sqrt{1 - 4x^2})} < 3, \text{ or } \frac{4x}{1 + \sqrt{1 - 4x^2}} < 3,$$

or to $4x < 3 + 3\sqrt{1 - 4x^2}$, and prove that this resulting inequality holds in the whole domain of definition.

**203.** $[\sqrt[3]{5/4}, +\infty)$.

**204.** $[-1, +\infty]$ if $a \in (-\infty, 0)$; $\left[-1, \dfrac{1}{a^2} - 1\right)$ if $a \in (0, +\infty)$.

**205.** $(-\infty, 2]$ if $a \in (-\infty, -1]$; $\left(2 - \dfrac{1}{(a+1)^2}, 2\right]$ if $a \in (-1, +\infty)$.

**206.** $(-\infty, 0) \cup [1, +\infty)$. **207.** $(-\infty, 0] \cup (9/2, +\infty)$. **208.** $(-(3 + \sqrt{29})/10, 2]$.
**209.** $\{5\} \cup (4 + \sqrt{2}, +\infty)$. **210.** $[-1, \sqrt[3]{4})$. **211.** $[-2, 2)$. **212.** $(1, 5/4) \cup (5/3, +\infty)$. **213.** 2.
**214.** 13. **215.** $x = 2$.

## Sec. 1.7

**1.** $\{(0,1)\}$.

**2.** $\{(-5, 2), (-2, -1)\}$. *Solution.* We must consider two possible cases : (1) $y \le 0$, (2) $y > 0$.

(1) For $y \le 0$, the given system of equations is equivalent to a system consisting of two equations and one inequality

$$\begin{cases} x - 3y = 1, \\ x + y = -3, \\ y \le 0. \end{cases} \tag{1}$$

Solving the system of equations

$$\begin{cases} x - 3y = 1, \\ x + y = -3. \end{cases}$$

we get $x = -2$, $y = -1$. This solution satisfies the inequality $y \ge 0$ and, thus is a solution of system (1).

(2) For $y > 0$ the given system of equations is equivalent to a system consisting of two equations and one inequality

$$\begin{cases} x + 3y = 1, \\ x + y = -3, \\ y > 0. \end{cases} \tag{2}$$

Solving the system of equations

$$\begin{cases} x + 3y = 1, \\ x + y = -3, \end{cases}$$

we get $x = -5$, $y = 2$. This solution satisfies the inequality $y > 0$ and is thus a solution of system (2).

**3.** $\{(2, 3)\}$. **4.** $\{(0, -1)\}$. **5.** $\{(0, 3), (4,1)\}$. **6.** $\{(1 / 4, 7 / 4), (3 / 4, 5 / 4)\}$.

**7.** $\{(4 /7, 5 / 7), (8 / 7, 3 / 7)\}$. **8.** $\{(-11 /19, 23 /19), (1, -1)\}$.

**9.** $\{(24 / 25, 23 / 25)\}$.

**10.** $\{(c, 4 - c)\}$ if $c \in [1, 2]$; $\{(c, c + 2)\}$ if $c \in [0, 1]$; $\phi$ for the other values of $c$. *Solution.* For $x \le 1$ the given system of equations is equivalent to a system of two equations and one inequality

$$\begin{cases} |y - 2| = x, \\ y = 2 + x, \\ x < 1. \end{cases} \tag{1}$$

If $y < 2$, then system of equations (1) has the form

$$\begin{cases} y = 2 - x, \\ y = 2 + x, \\ x < 1, \\ y < 2. \end{cases} \tag{2}$$

System (2) has no solutions since the solution of a system of the first two equations $x = 0$ and $y = 2$ does not satisfy the inequality $y < 2$.

If $y \ge 2$, then system (1) has the form

$$\begin{cases} y = 2 + x, \\ y = 2 + x, \\ x < 1, \\ y \ge 2. \end{cases} \tag{3}$$

The pair of numbers $x = c$, $y = 2 + c$ is a solution of system (1) for any $c < 1$ and, consequently, a solution of the given system of equations.

For $x > 1$ the given system of equations is equivalent to a system of two equations and one inequality

$$\begin{cases} |y - 2| = 2 - x, \\ y = 4 - x, \\ y \geq 1. \end{cases} \tag{4}$$

For $y < 2$ system (4) can be written in the form

$$\begin{cases} y = x, \\ y = 4 - x, \\ x \geq 1, \\ y < 2. \end{cases} \tag{5}$$

System (5) does not have solutions since the solution of a system of the first two equations $x = 2$ and $y = 2$ does not satisfy the inequality $y < 2$.

If $y \geq 2$, then system (4) has the form

$$\begin{cases} y = 4 - x, \\ y = 4 - x, \\ x \geq 1, \\ y \geq 2. \end{cases} \tag{6}$$

The pair of numbers $x = c$ and $y = 4 - c$ is a solution of system (6) for any $c \in [1, 2]$ and only for those $c$, and, consequently, a solution of the given system of equations.

**11.** $\{(1, 1, 0)\}$. **12.** $\{(2, 1, 1)\}$. **13.** $\{(2, 3, 1)\}$.

**14.** $\{(1, 5), (5, 1), (2, 3), (3, 2)\}$. *Solution.* We set $x + y = u$, $xy = v$. Then the system of equations assumes the form

$$\begin{cases} u + v = 11, \\ uv = 30. \end{cases}$$

whose solutions are $u = 5$, $v = 6$ and $u = 6$, $v = 5$. Then, we get two systems for $x$ and $y$:

$$\begin{cases} x + y = 6, \\ xy = 5, \end{cases} \qquad \begin{cases} x + y = 5, \\ xy = 6. \end{cases}$$

The solutions of the first system are $(5, 1)$ and $(1, 5)$, the solutions of the second system are $(2, 3)$ and $(3, 2)$. We have found the solutions using the Vieta theorem.

**15.** $\{(4, -1); (-1, 4)\}$. **16.** $\{(5, 3)\}$. **17.** $\{(-5, -4); (4, 5)\}$. **18.** $\{(4, 5), (5, 4)\}$.
**19.** $\{(1, 6), (6, 1)\}$. **20.** $\{(-6, -2), (-4, -4)\}$. **21.** $\{(3, 2), (2, 3)\}$.
**22.** $\{(-3, -2), (3, 2)\}$. **23.** $\{(2, 3), (3, 2)\}$. **24.** $\phi$. **25.** $\{(5, 1)\}$.

**26.** $\{(-5, -1), (5, 1)\}$. *Hint.* Set $(x - y)(x + y) = t$. Then $(x + y)/(x - y) = 1/t$.

**27.** $\{(-2, -1), (2, 1), (-\sqrt{6}, -\sqrt{6}/3), (\sqrt{6}, \sqrt{6}/3)\}$.

**28.** $\{(8/13, 12/5)\}$. *Hint.* Introduce new variables setting $1/3x = u$, $1/2y = v$.

**29.** $\{(11/13, -24/5)\}$. **30.** $\{(-7/4, -3), (-7/6, -2)\}$. **31.** $\{(-1, 9/4), (4, -9)\}$.
**32.** $\{(-7, 3), (7/3, -1)\}$. **33.** $\{(-1, -3/2), (2, 3)\}$. **34.** $\{(4, -3), (4, 3)\}$.
**35.** $\{(-3\sqrt{2}, -\sqrt{2}), (-3\sqrt{2}, \sqrt{2}), (3\sqrt{2}, -\sqrt{2}), (3\sqrt{2}, \sqrt{2})\}$. **36.** $\{(-6, -2), (6, 2)\}$.
**37.** $\{(-2, -3), (2, 3)\}$.     **38.** $\{(-2, -4), (-4, -2), (2, 4), (4, 2)\}$.
**39.** $\{(2, 8), (8, 2), (-2, -8) (-8, -2)\}$.
**40.** $\{(-9/5, -16/5), (9/5, 16/5)\}$.

**41.** $\{(-3, -2)(3, 2)\}$. *Hint.* Adding the equations together, we get $(x + y)^2 = 25$ and so on. **42.** $\{(-10, -11)(10, 11)\}$. **43.** $\{(-7, -3), (7, 3)\}$.

**44.** $\{(7\sqrt{2/5}, -3\sqrt{2/5}), (-7\sqrt{2/5}, 3\sqrt{2/5})\}$. *Hint.* Multiplying the first equation by 7 and the second by –3 and adding the resulting equations together, we find $7y^2 - 4xy - 3x^2 = 0$, whence $y = x$, and $y = -3x/7$ and so on.

**45.** $\{(-3, -2), (3, 2)\}$. **46.** $\{(-3, 4), (4, -3)\}$. **47.** $\{(-3, 4), (4, -3)\}$.

**48.** $\{(-1, -3), (3/2, 2), (-\sqrt{7}/2, 0), (\sqrt{7}/2, 0)\}$.

**49.** $\{(1, 2), (2, 1)\}$. *Hint.* Adding the two equations together, we get a quadratic equation $(x + y)^2 + (x + y) - 12 = 0$ with respect to $x + y$.

**50.** $\{(-5 - \sqrt{55}, -5 + \sqrt{55}), (-5 + \sqrt{55}, -5 - \sqrt{55}), (4, 12), (12, 4)\}$.

**51.** $\{(-3, -5), (-5/3, -13/3), (5/3, 13/3), (3, 5)\}$. *Hint.* Subtract the second equation of the system from the first.

**52.** $\{(-4, -5), (-3\sqrt{3}, -\sqrt{3}), (3\sqrt{3}, \sqrt{3}), (4, 5)\}$. *Hint.* Multiplying the first equation by 5 and the second by 7, we get after addition, an equation $5x^2 - 19xy + 12y^2 = 0$, Solving which, we get $x = 3y$ and $x = \dfrac{4}{5}y$.

**53.** $\{(-1, -2), (-\sqrt{2}, -\sqrt{2}), (1, 2), (\sqrt{2}, \sqrt{2})\}$.

**54.** $\{(1, 4), (4, 1), (2 - \sqrt{3}, 2 + \sqrt{3}), (2 + \sqrt{3}, 2 - \sqrt{3})\}$. *Hint.* Set $x + y = u$, $\sqrt{xy} = v$.

**55.** $\{(1, -1), (3, -3), (\sqrt{157} - 13, (\sqrt{157} - 13)/2), (-13 - \sqrt{157}), -(13 + \sqrt{157})/2\}$. *Hint.* Solving the first equation of the system, we get $y = -x$ and $x = 2y$.

**56.** $\{(2, -3), (c, 1) \mid c \in \mathbf{R}\}$. *Solution.* Multiplying the first equation by –2 and adding the resulting equation to the second equation of the system, we get, after the requisite simplifications, an equation $y^2 + 2y - 3 = 0$, whence $y_1 = 1$, $y_2 = -3$. For $y = 1$ the two equations of the system are identically satisfied for all $x \in \mathbf{R}$. For $y = -3$ we have $x = 2$.

**57.** $\{(-1, 3), (c, 2) \mid c \in \mathbf{R}\}$. **58.** $\{(2, -1), (-1, c) \mid c \in \mathbf{R}\}$.

**59.** $\{(2, 3), (3, 2)\}$. *Hint.* Taking into account that $x + y = 5$, we reduce the first equation of the system $(x + y)(x^2 - xy + y^2) = 35$ to the form $x^2 - xy + y^2 = 7$. As a result, we get the following system of equations :
$$\begin{cases} x^2 - xy + y^2 = 7, \\ x + y = 5. \end{cases}$$

**60.** $\{(-1, -2), (2, 1)\}$. **61.** $\{(-1, 2), (2, -1)\}$. *Hint.* Multiply the second equation by 3 and add it to the first equation of the system.

**62.** $\{(4, 8), (8, 4)\}$. **63.** $\{(-3, -1), (-1, -3), (1, 3), (3, 1)\}$. *Hint.* From the second equation of the system, we find that $y = 3/x$; excluding $y$ from the first equation of the system, we get a quadratic equation for $x^4$.

**64.** $\{(2, -1), (-1, 2)\}$. *Hint.* Set $x^3 = u$, $y^3 = v$.

**65.** $\{(-3, -2), (-2, -3), (2, 3), (3, 2)\}$. *Hint.* From the first equation of the system we find that $x^2 + y^2 = 78/xy$. Squaring both sides of this equation and taking the second equation of the system into account, we get $97 + 2(xy^2) = 78^2/(xy)^2$.

**66.** $\{(-3, -2), (-2, -3), (2, 3), (3, 2)\}$. **67.** $\{(1, 3, 9), (9, 3, 1)\}$.

**68.** $\{(12/7, 12/5, -12)\}$.

**69.** $\{(3, 3, 3)\}$. *Solution.* Adding the equations of the system together, we obtain
$$(x - 3)^3 + (y - 3)^3 + (z - 3)^3 = 0. \tag{1}$$
The triple $(3, 3, 3)$ is a solution of equation (1). Verification shows that it is also a solution of the system. (Note that the verification of this statement is obligatory). Let us show that the system has no other solutions. From the first equation of

the system, we have $y^3 = 9x^2 - 27x + 27$. The discriminant of the quadratic trinomial $9x^2 - 27x + 27$ is negative. Therefore, if $(x_0, y_0, z_0)$ is a solution of the system, then $y_0^3 > 0$. Consequently, $y_0 > 0$. Similarly, from the second and third equations of the system, we get $z_0 > 0$ and $x_0 > 0$. From the first equation of the system, we have

$$(y_0 - 3)(y_0^2 + 3y_0 + 9) = 9x_0(x_0 - 3). \tag{2}$$

Since $x_0 > 0$ and $y_0^2 + 3y_0 + 9 > 0$, it follows from (2) that the numbers $y_0 - 3$ and $x_0 - 3$ cannot have different signs. Similarly, it follows from the second equation that the numbers $z_0 - 3$ and $y_0 - 3$ cannot have different signs either. Thus, if $x_0 > 3$, then $y_0 > 3$ and $z_0 > 3$; now if $x_0 < 3$, then $y_0 < 3$ and $z_0 < 3$. The solution $(x_0, y_0, z_0)$ of the system is also a solution of equation (1), and for (1) neither $x_0 > 3$, $y_0 > 3$, $z_0 > 3$ nor $x_0 < 3$, $y_0 < 3$, $z_0 < 3$ are suitable. Hence $x_0 = 3$, $y_0 = 3$, $z_0 = 3$.

**70.** $\{(-1, -1, -1)\}$.     **71.** $\{(2, 2, 2)\}$.     **72.**   $\{(-1, 2)\}$.

**73.** $\{(1, 0, -1)\}$.

**74.** $\{(1, -1)\}$. *Solution.* The given system of equations is equivalent to the system

$$\begin{cases} y^2 = 2x / (1 + x^2), \\ 2(x-1)^2 + 1 + y^2 = 0. \end{cases} \tag{1}$$

Since $2x / (1 + x^2) \leq 1$ for all $x \in R$, it follows from the first equation of system (1) that $-1 \leq y \leq 1$: From the second equation of the system it follows that $y \leq -1$. Hence $y = -1$, $x = 1$.

**75.** $\{(1,1,1), ((-5, \pm \sqrt{11}) / 2, (10 \mp \sqrt{11}) / 2; (4 \mp \sqrt{11}) / 2)\}$. *Hint.* Add the first equation multiplied by 2 and the second equation multiplied by 3 to the third equation.

**76.** $\{(c, c)\}, c \in R$ for $a = 0$; $\{(3a / 2, 9a / 2), (3a / 4, -9a / 4)\}$ for $a \neq 0$.

**77.** $\{(3, 0), (-1 / 3, -5 / 3)\}$.

**78.** $\{((4a - a^2) / (2a - 4), (a - 4) / (a - 2))\}$, if $a \in [2 / 3, 3 - \sqrt{5}]$;
$\{((a^2 - 12a + 8) / (6a - 4); a / (3a - 2))\}$, if $a \in (3 - \sqrt{5}, 2]$.

**79.** $\{(0, 1 - 2\sqrt{3})\}$ for $a = 7 - 4\sqrt{3}$, $\{(0, 1 + 2\sqrt{3})\}$ for $a = 7 + 4\sqrt{3}$. *Solution.* For $x \geq 0$ the given system has the form

$$\begin{cases} ax + (a-1)y = 2 + 4a, \\ 3x + 2y = a - 5, \end{cases} \tag{1}$$

its solution is $x = \dfrac{a^2 - 14a + 1}{a - 3}$, $y = -\dfrac{a^2 - 17a - 6}{a - 3}$. Since $x \geq 0$, we have $\dfrac{a^2 - 14a + 1}{a - 3} \geq 0$ for $\dfrac{(a - 7 + 4\sqrt{3})(a - 7 - 4\sqrt{3})}{a - 3} \geq 0$; applying the method of intervals, we obtain $a \in [7 - 4\sqrt{3}; 3) \cup [7 + 4\sqrt{3} + \infty)$. For $x \leq 0$ the given system has the form

$$\begin{cases} ax + (a-1)y = 2 + 4a, \\ -3x + 2y = a - 5, \end{cases} \tag{2}$$

its solution is $x = -\dfrac{a^2 - 14a + 1}{5a - 3}$, $y = \dfrac{a^2 + 7a + 6}{5a - 3}$. Since $x \leq 0$, we have $\dfrac{a^2 - 14a + 1}{5a - 3} \geq 0$; applying again the method of intervals, we obtain $a \in [7 - 4\sqrt{3}; 3/5) \cup [7 + 4\sqrt{3}; +\infty)$. We have to find $a$'s for which the given system has a unique solution independent of whether $x \geq 0$ or $x \leq 0$. Consequently, we have to find $a$'s for which the following equalities are satisfied :

$$\begin{cases} \dfrac{a^2-14a+1}{a-3} = -\dfrac{a^2-14a+1}{5a-3}, \\ -\dfrac{a^2-17a-6}{a-3} = \dfrac{a^2+7a+6}{5a-3}. \end{cases} \qquad (3)$$

From the first equation, we obtain

$$(a^2-14a+1)\left(\dfrac{1}{a-3}+\dfrac{1}{5a-3}\right)=0$$

$$\Rightarrow \qquad \dfrac{(a-7-4\sqrt{3})\,(a-7+4\sqrt{3})\,6\,(a-1)}{(a-3)\,(5a-3)}=0$$

i.e. the values of $x$ we have found from system (1) and (2) are the same for $a=1$, $a=7-4\sqrt{3}$, $a=7+4\sqrt{3}$. Let us substitute the values of $a$ we have obtained into the second equation of system (3). For $a=1$, we get $-11=7$, which is impossible, i.e. for $a=1$ the given system has two solutions $(6,-11)$ and $(6,7)$.

For $a=7-4\sqrt{3}$ we get $\dfrac{(7-4\sqrt{3})^2-17\,(7-4\sqrt{3})-6}{7-4\sqrt{3}-3}=1-2\sqrt{3}$ on the left-hand

side and $\dfrac{(7-4\sqrt{3})^2+7\,(7-4\sqrt{3})+6}{5\,(7-4\sqrt{3})-3}=1-2\sqrt{3}$ on the right-hand side, i.e. for

$a=7-4\sqrt{3}$ the given system has a unique solution $(0,1-2\sqrt{3})$. For $a=7+4\sqrt{3}$

we get $-\dfrac{(7+4\sqrt{3})^2-17(7+4\sqrt{3})-6}{7+4\sqrt{3}-3}=1+2\sqrt{3}$ on the left-hand side and

$\dfrac{(7+4\sqrt{3})^2+7\,(7+4\sqrt{3})+6}{5\,(7+4\sqrt{3})-3}=1+2\sqrt{3}$ on the right-hand side, i.e. for

$a=7+4\sqrt{3}$ the given system has a unique solution $(0,\,1+2\sqrt{3})$.

80. $\{((1-2a^2)/(1-a^2),a/(1-a^2))\}$, if $a\neq\pm1$; $\phi$ if $a=\pm1$.

81. $\{((a^2+1)/a^2,(a+1)/a^2)\}$, if $a\neq0$; $\phi$ if $a=0$.

82. $\{(0,a)\}$ if $a\neq\pm1$; $\{(c+1,c)\,|\,c\in R\}$ if $a=-1$; $\{(c,1-c)\,|\,c\in R\}$ if $a=1$.

83. $\{((1+a+a^2)/(1+a);-a/(1+a))\}$ for $a\neq\pm1$; $\{(c,1-c)\,|\,c\in R\}$ for $a=1$; $\phi$ for $a=-1$.

84. $a=3$. 85. $a=-7$. 86. $a=1/17$.

87. $\{(11/5,1/5)\}$ if $a=-3$, $\phi$ if $a\neq3$.

88. $\{((a-3)^2;\cos(a-2);3)\}$ if $a\in[3,2+\pi]$; $\phi$, if $a\notin[3,2+\pi]$.

89. $\{\log_2(a+1);5;\sin(a+2))\}$ if $a\in[-1,(\pi-4)/2]$; $\phi$ if $a\notin(-1,(\pi-4)/2]$.

90. $\{(\arctan(a-3)^2+\pi n;\pm\sqrt{a+1},6)\,|\,n\in Z\}$ if $a\in[-1,3]$; $\phi$ if $a\notin[-1,\ 3]$.

91. $a\in[-1,0)$. *Solution.* For $b\neq2$ and $b\neq-2$ and for any $a$ and $c$ the given system has a unique solution

$$x=\dfrac{2ac^2+2c-bc+b}{4-b^2},\ y=\dfrac{2c-2-abc^2-bc}{4-b^2}.$$

If $b=2$, then the given system has the form

$$\begin{cases} 2x+2y=ac^2+c, \\ 2x+2y=c-1. \end{cases}$$

For this system to have solutions, the condition $ac^2 + c = c - 1$ must be satisfied, i.e. $ac^2 = -1$. Considering the last relation as an equation with respect to $c$, we note that it is solvable for any $a \in (-\infty, 0)$. If $b = -2$, then the system has the form

$$\begin{cases} 2x - 2y = ac^2 + c, \\ -2x + 2y = c - 1. \end{cases}$$

For this system to have solutions, the condition $ac^2 + c = -c + 1$ must be satisfied, i.e. $ac^2 + 2c - 1 = 0$. Considering the last relation as an equation with respect to $c$, we note that it has solutions if $1 + a \geq 0$, i.e. $a \in [-1, +\infty)$. Thus, for $a \in [-1, 0)$ there is always a $c$ such that the given system has at least one solution for any value of $b$.

**92.** $a \in [-(3\sqrt{2} + 4)/8, (3\sqrt{2} - 4)/8]$. *Hint.* For $b \neq \pm\sqrt{2}$ the system has a unique solution. For $b = \sqrt{2}$ the given system has solutions when the condition $ac^2\sqrt{2} + (\sqrt{2} - 1)c + 1 = 0$ is satisfied. With respect to $c$ this equation has solutions if $a \in (-\infty, (3\sqrt{2} - 4)/8]$. For $b = -\sqrt{2}$ the system has solutions when the condition $ac^2\sqrt{2} + (\sqrt{2} + 1)c - 1 = 0$ is satisfied. With respect to $c$ this equation has a solution if $a \in [-(3\sqrt{2} + 4)/8, +\infty)$. Compare with the solution of problem 91.

**93.** For any $a \in (-\infty, -2] \cup [2, +\infty)$. *Hint.* See the solution of problem 91.

**94.** For any $a \in (-\infty, -4] \cup [4, +\infty)$. *Hint.* See the solution of problem 91.

**95.** For all $k \in (-2, 2) \cup (2, 4)$.

**96.** For all $b \in ((35 - \sqrt{1009})/6, (35 + \sqrt{1009})/6)$.

**97.** For all $n \in (-(2 + \sqrt{22})/6, 0) \cup (\sqrt{22} - 2)/6, 9/4)$.

**98.** For all $a \in (0, 5)$.

**99.** $\{(9a + 54; a^3 - 36a)\}$ for $a \neq 6, a \neq 12$; $\{(8c, c) \mid c \in R\}$ for $a = 12$; $\phi$ for $a \neq 6$; $x + y > 0$, where $x = 9(a + 6)$, $y = a(a - 6)(a + 6)$ for $a \in (-6, 3) \cup (3, 6) \cup (6, 12) \cup (12, +\infty)$.

**100.** $x_{gr} = 3$. **101.** (a) $\begin{pmatrix} 4 & 7 \\ 7 & 11 \end{pmatrix}$; (b) $X = \begin{pmatrix} 2 & 1 \\ 1 & 3 \end{pmatrix}$. **102.** $\{(6, 10), (10, 6)\}$.

**103.** $\{(1, 4), (4, 1)\}$. *Hint.* Setting $x/y = t$, solve the first equation of the system.

**104.** $\{(5/3, 1/3)\}$.

**105.** $\{(-\sqrt{15/17}); -4\sqrt{15/17}), (-4\sqrt{15/17}, -\sqrt{15/17}), \sqrt{15/17}, 4\sqrt{15/17}), (4\sqrt{15/17}, \sqrt{15/17})\}$.

**106.** $\{(-9, -9/4), (4, 1)\}$. **107.** $\{(-6, -1), (-3, 2), (9, -4), (2, 3)\}$. **108.** $\{(12, 4)\}$.

**109.** $\{(-1, -27), (27, 1)\}$. **110.** $\{(1, 8), (8, 1)\}$. **111.** $\{(1, 27), (27, 1)\}$.

**112.** $\{(1, 81), (81, 1)\}$. **113.** $\{(-9, 25), (5, 4)\}$. **114.** $\{(1, 4), (4, 1)\}$.

**115.** $\{(5, 4)\}$. *Hint.* Multiply the left-hand and right-hand sides of the equations of the system. **116.** $\{(-2, -1), (-1, -2), (1, 2), (2, 1), (0, c) \mid c \in R\}$.

**117.** $\{(-\sqrt{59}/2, -3/2), (-\sqrt{59}/2, 3/2), (5/\sqrt{2}, 3/\sqrt{2}), (5/\sqrt{2}, -3/\sqrt{2})\}$. *Hint.* Consider two cases : (1) $x + y > 0, x - y \geq 0$. (2) $x + y < 0, x - y \leq 0$.

**118.** $\{(1, 8), (8, 1)\}$. **119.** $\{(2, 2)\}$. **120.** $\{(4, 2), (4/3, -2/3)\}$. **121.** For $a = 1/3$ and all $a < 0$. **122.** $(-7/2, 0)$. **123.** $x = 8$. **124.** $x = 1$. **125.** $0, 1$. **126.** $[-1, 7/3]$.

**127.** $\alpha_1 = 0, \alpha_2 = 1$.

**128.** $a \in [-2/3, 0)$. *Solution.* Since $a < 0$ by the hypothesis, and $ax > 0$, it follows that $x < 0$. The right-hand side of the first inequality includes $3a - x > 2\sqrt{ax} > 0$, consequently, $x < 3a$. Similarly, from the second inequality $x - (6/a) > \sqrt{x/a}$ we see that $6/a < x$. We solve the first inequality

$$2\sqrt{ax} < 3a - x \Rightarrow x^2 - 10ax + 9a^2 > 0 \Rightarrow x \in (-\infty, 9a) \cup (a, 0);$$

but we have found that $x < 3a$, consequently, $x \in (-\infty, 9a)$. We solve the second inequality :

$$x - \sqrt{\frac{x}{a}} > \frac{6}{a} \Rightarrow x - \frac{6}{a} > \sqrt{\frac{x}{a}}$$

$$\Rightarrow \quad x^2 - \frac{13}{a}x + \frac{36}{a^2} > 0; \; \Rightarrow \; x \in (-\infty, \; 9/a) \cup (4/a, \; 0);$$

but we have found that $x > 6/a$ for the second inequality, consequently, $x \in (4/a, 0)$. For the two inequalities to have solutions in common, the intervals $(-\infty, 9a)$ and $(4/a, 0)$ must have points in common, i.e. $4/a \le 9a \Rightarrow 4/9 \ge a^2 \Rightarrow$ $-2/3 \le a < 0$. Thus, for $a \in [-2/3, 0)$ the two given inequalities have solutions in common for $a < 0$.

**129.** All $k \in (-\infty, 1/2) \cup (3/2, +\infty)$. **130.** All $k \in [-1, +\infty)$. **131.** $(1/3, 6)$.
**132.** $[0, 1/2]$. **133.** $(-\infty, -3) \cup (2, +\infty)$. **134.** $[3, 4)$. **135.** $(-1, 1/2)$.
**136.** $(-\infty, -(3+\sqrt{5})/2] \cup (0, 6]$. **137.** $[2/3, (3+\sqrt{5})/6]$. **138.** $(-2, -1) \cup (1, 4)$.
**139.** $(-1, 2)$. **140.** $(-2, 0]$. **141.** $(-2, 1)$. **142.** (a) $x = 1; y = 1;$ (b) yes; (c) no.
**143.** For all $a \in (-(1+\sqrt{13})/2, -2)$.
**144.** $a < -1$. *Solution.* Let us first assume that the given system has solutions. Then there is always a solution for the inequality resulting from the summation of the first inequality of the system multiplied by $-2$ and the second inequality of the system : $(x + 3y)^2 \le -4/(a+1)$. Hence, $a + 1 < 0, a < -1$. Assume now that $a_0 < 1$. Then, $(1 - a_0)/(a_0 + 1) < -1$. Therefore, if the system of equations

$$\begin{cases} x^2 + 2xy - 7y^2 = -1, \\ 3x^2 + 10xy - 5y^2 = -2, \end{cases} \quad (1)$$

has solutions, then the given system of inequalities also has solutions. But system (1) has a solution, say, $(-3/2, 1/2)$. (It is possible to find the indicated solution by multiplying the first equation by $-2$ and summing it up with the second equation.)

**145.**

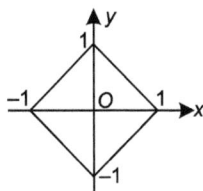

**146.**

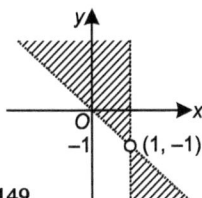

**147.**

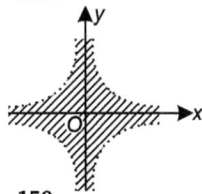

**148.**

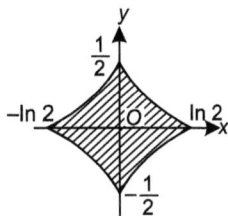

**149.**

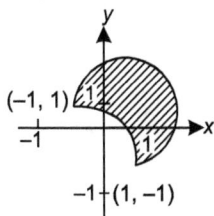

**150.**

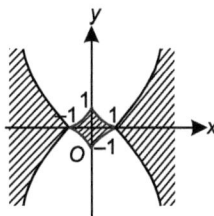

**151.**

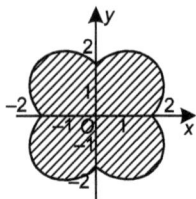

**152.**

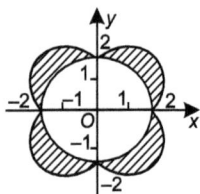

**153.**

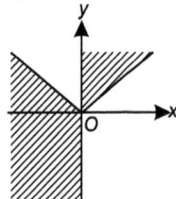

**154.**

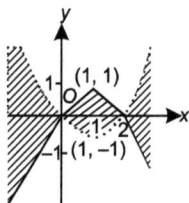

**155.**

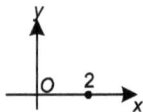

**156.**

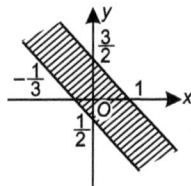

**157.**

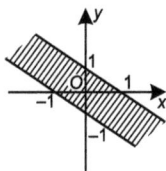

**158.**

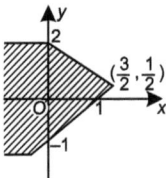

**159.**

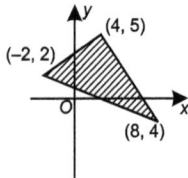

**160.**

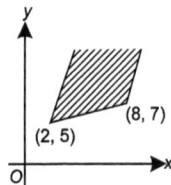

**161.**

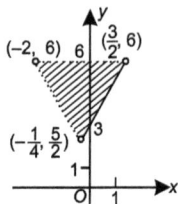

**162.**

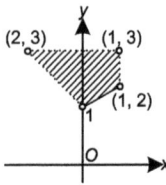

**163.**

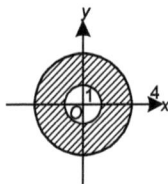

**164.**

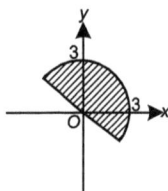

165.

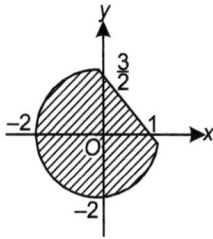

166.

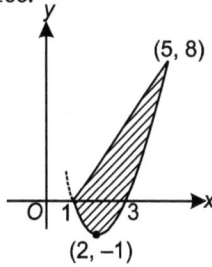

(5, 8)

(2, −1)

167.

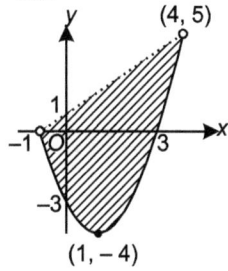

(4, 5)

(1, − 4)

168.

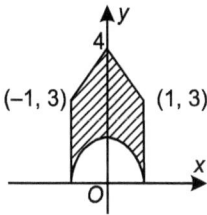

(−1, 3)   (1, 3)

169.

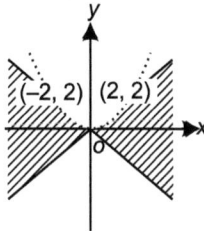

(−2, 2)   (2, 2)

170.

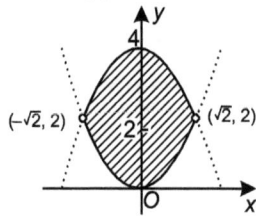

$(-\sqrt{2}, 2)$   $(\sqrt{2}, 2)$

171.

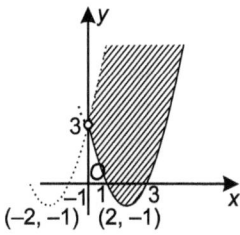

(−2, −1)   (2, −1)

172.

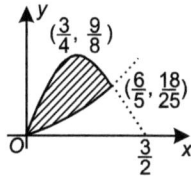

$(\frac{3}{4}, \frac{9}{8})$

$(\frac{6}{5}, \frac{18}{25})$

$\frac{3}{2}$

173.

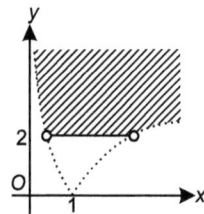

174.

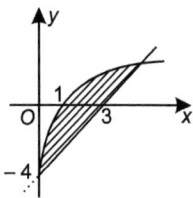

175.

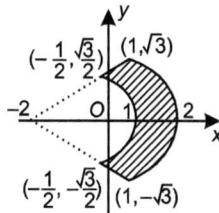

$(-\frac{1}{2}, \frac{\sqrt{3}}{2})$   $(1, \sqrt{3})$

$(-\frac{1}{2}, -\frac{\sqrt{3}}{2})$   $(1, -\sqrt{3})$

176.

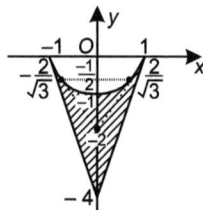

$-\frac{2}{\sqrt{3}}$   $\frac{2}{\sqrt{3}}$

**177.**

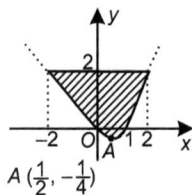

$A\left(\frac{1}{2}, -\frac{1}{4}\right)$

**178.**

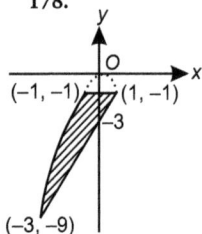

**179.**

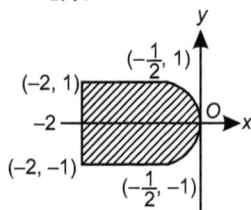

**180.**

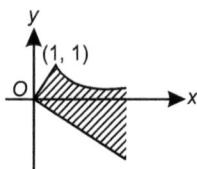

**181.**

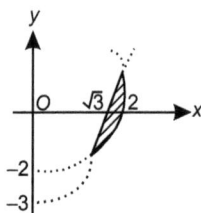

**182.**

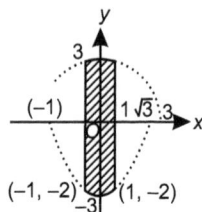

**183.**

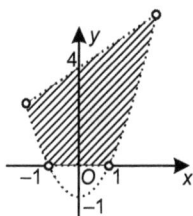

**184.**

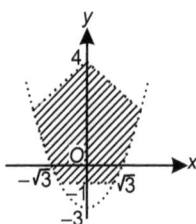

**185.**

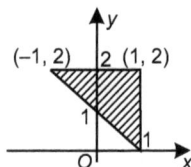

## Sec. 1.8

**1.** $D(y) = [0, 2]$.  **2.** $D(y) = [-1, 1]$.  **3.** $D(y) = [1, 6]$.

**4.** $D(y) = (-\infty, 2] \cup [3, +\infty)$.  **5.** $D(y) = [-3, 5)$.

**6.** $D(f) = [-1, 2]$.  **7.** $D(y) = [-1/2, 3/2]$.

**8.** $D(y) = [-2/3, 3]$.  **9.** $D(y) = [-2, 1] \cup (1, +\infty)$.

**10.** $D(y) = (-\infty, -1/2) \cup (3, +\infty)$.  **11.** $D(f) = (-\infty, -2] \cup [0, 2]$.

**12.** $D(f) = (-\infty, -\sqrt{3}) \cup [0, \sqrt{3}]$.

**13.** $D(y) = (-\infty, 1) \cup (1, +\infty)$. **14.** $D(y) = (-4, 1]$. **15.** $D(y) = \left[\dfrac{7}{3}, 63\right) \cup (63, +\infty)$.

**16.** $(-3, 0) \cup (0, 2) \cup (2, 4]$. **17.** $D(y) = (-\infty, 0) \cup [2, 3]$. **18.** $D(f) = (-\infty, -4] \cup [5, 6]$.
**19.** $D(f) = (-\infty, -3] \cup (2, +\infty)$. **20.** $D(y) = (-3, 3) \cup (3, 4]$. **21.** $D(y) = [-2, 1) \cup (1, 2]$.
**22.** $D(y) = (-3, 1]$. **23.** $D(y) = (-\infty, 3/4) \cup (4, 7]$. **24.** $D(y) = (-\infty, -1) \cup (4, +\infty)$.
**25.** $D(y) = (-\infty, -4) \cup (-2, 2] \cup (2, +\infty)$. **26.** $D(y) = \{1\}$. **27.** $D(y) = (-\infty, -4] \cup$
$(7, +\infty)$. **28.** $D(y) = [5, 7)$. **29.** $D(y) = [-2, -1) \cup (-1, 1) \cup (1, 3]$. **30.** $D(f) = (-\infty,$
$+\infty)$. **31.** $D(y) = (-\infty, 0]$. **32.** $D(f) = \{[\pi n, \pi(2n+1)/2] \mid n \in \mathbf{Z}\}$. **33.** $D(y) = (-1$
$/6, \pi/3] \cup [5\pi/3, 6)$. **34.** $D(y) = \left\{\left(\dfrac{\pi}{6} + 2\pi n, \dfrac{5\pi}{6} + 2\pi n\right) \cup \left(\dfrac{5\pi}{6} + 2\pi n; \dfrac{13}{6}\pi + 2\pi n\right)\right.$

$\left. \mid n \in \mathbf{Z}\right\}$. **35.** $D(y) = (3, 30) \cup (30, +\infty)$. **36.** $D(y) = [-5, 8) \cup (8, 9)$. **37.** $D(f) =$
$[4, 11/2) \cup (11/2, +\infty)$. **38.** $D(y) = (-\infty, -2) \cup (2, +\infty]$. **39.** $D(y) = (-\infty, -3)$
$\cup (-1, +\infty)$. **40.** $D(y) = [0, +\infty)$. **41.** $D(y) = (-\infty, -7) \cup (-1, +\infty)$. **42.** $D(y) = (-1, 1)$.
**43.** $D(y) = [-1, 1)$. **44.** $D(y) = (-5, 0) \cup (3, +\infty)$. **45.** $D(y) = (2, 4]$. **46.** $D(y)$
$= [1, +\infty)$. **47.** $D(f) = (2, +\infty)$. **48.** $D(y) = [1, +\infty)$. **49.** $D(y) = (0, 1)$. **50.** $D(y)$
$= (0, 3/2]$. **51.** $D(y) = (-3, -2/3]$. **52.** $D(f) = [0, +\infty)$. **53.** $D(y) = (-\infty, -11)$
$\cup (-11, -10) \cup (-10, -9) \cup (-9, 2] \cup [3, +\infty)$. **54.** $D(y) = (9, -\infty)$. **55.** $D(y) = [1, 4]$.
**56.** $D(y) = \{4\} \cup [5, +\infty)$. **57.** $D(f) = [-2, -\sqrt{3}) \cup (\sqrt{3}, 2]$. **58.** $D(y) = (-3, -2$
$/5) \cup (2, +\infty)$. **59.** $D(y) = (-\infty, -1) \cup [0, +\infty)$. **60.** $D(y) = (1, 4]$. **61.** $D(y) = (1, 4)$.
**62.** $D(y) = (1, +\infty)$. **63.** $D(y) = (0, 1)$. **64.** $D(y) = [2, 3]$. **65.** $D(y) = [-5,$
$-3] \cup (3, 5)$. **66.** $D(y) = (1, 6) \cup (6, +\infty)$. **67.** $D(f) = (-2, (1 - \sqrt{21})/2] \cup [1 + \sqrt{21})$
$/2, 3)$. **68.** $D(y) = [2, 4)$. **69.** $D(f) = (-\infty, -2)$. **70.** $D(y) = [(1 - \sqrt{5})/2, 0) \cup$
$[(1 + \sqrt{5})/2, +\infty)$. **71.** $D(f) = (-\infty, -7] \cup (-3, +\infty)$. **72.** $D(y) = (8/3, +\infty)$. **73.** $D$
$(f) = (-\infty, -2] \cup (1, 2]$. **74.** $D(y) = [-1, 1) \cup (3, 5]$. **75.** $D(f) = [0, 1)$. **76.** $(-\infty;$
$(1 - \sqrt{65})/4] \cup [0, (1 + \sqrt{65})/4)$. **77.** $D(f) = (0, 16]$. **78.** $D(y) = [1, 2) \cup (2, 5]$. **79.** $D$
$(y) = (-\infty, 2) \cup (3, +\infty)$. **80.** $D(y) = (1, 3]$. **81.** $D(y) = [3, +\infty)$. **82.** $D(f) = (2, +\infty)$.
**83.** $D(f) = (5, +\infty)$. **84.** $D(f) = [3, +\infty)$. **85.** $D(y) = (-\infty, -1) \cup (-1, 4)$. **86.** $D(y) =$
$(1, 2) \cup (2, 3]$. **87.** $D(y) = (-\infty, -1/2] \cup (0, 1)$. **88.** $D(y) = [4, 5) \cup (5, 39)$. **89.** $D(y)$
$= (2, 3)$. **90.** $D(y) = (1/3, +\infty)$. **91.** $D(y) = [10^3, 10^4)$. **92.** $D(y) = [4 - \sqrt{2}, 3) \cup [4$
$+ \sqrt{2}, +\infty)$. **93.** $D(y) = (100, +\infty)$. **94.** $D(y) = (0, 10^{-2}) \cup (10^{-2}, 10^{-1/2})$. **95.** $D(y)$
$= (0, 1)$. **96.** $D(y) = (-\infty, -5) \cup (-5, -4) \cup (4, 5) \cup (5, +\infty)$. **97.** $D(y) = [-4, -\pi]$
$\cup [0, \pi]$. **98.** $D(y) = (0, 10^2) \cup (10^3, +\infty)$. **99.** $D(y) = (-\infty, -5/3) \cup (1, +\infty)$.
**100.** $D(y) = (3 - 2\pi, 3 - \pi) \cup (3, 4]$. **101.** $D(y) = (9, +\infty)$. **102.** $D(y) = [1, 4)$.
**103.** $D(y) = [-1, 3]$. **104.** $D(y) = (-1/2, 1/2) \cup (1/2, 1) \cup (3/2, +\infty)$. **105.** $D(y)$
$= (0, 10^{-2}) \cup (10^{-2}, 10^{-1})$. **106.** $D(y) = \{1/2\}$. **107.** $D(y) = \{[2\pi n, \pi(2n+1)] \mid n \in \mathbf{Z}\}$.
**108.** $D(y) = \{[2\pi n + \arcsin(1/3); \pi(2n+1) - \arcsin(1/3)] \mid n \in \mathbf{Z}\}$. **109.**
$D(y) = \{4\pi n, 2\pi(2n+1)] \mid n \in \mathbf{Z}\}$. **110.** $D(y) = \{(\pi(2n-1) + \arccos(1/4),$
$\pi(2n+1) - \arccos(1/4) \mid n \in \mathbf{Z}\}$. **111.** $D(y) = \{[\pi(6n-1)/3, \pi(6n+1)/3] \mid n \in \mathbf{Z}\}$.
**112.** $D(y) = \{\pi(4n+1)/2 \mid n \in \mathbf{Z}\} \cup \{[\pi(2m+1), 2\pi(m+1)] \mid m \in \mathbf{Z}\}$. **113.** $D(y)$
$= (-\infty, 0) \cup (0, +\infty)$, $E(y) = \{-1, 1\}$. **114.** $D(f) = [0, 1]$, $E(f) = [0, 1/2]$. **115.**
$D(y) = (-\infty, +\infty)$, $E(y) = [\sqrt{11}/3, +\infty)$. **116.** $D(y) = (-\infty, +\infty)$, $E(y)$
$= [\log(11/3); +\infty)$. **117.** $D(y) = (-\infty, +\infty)$, $E(y) = [\log(4/5), +\infty)$. **118.** $D(f)$
$= [1, 3]$, $E(f) = [\sqrt{2}, \sqrt{10}]$. **119.** $D(f) = (-\infty, +\infty)$, $E(f) = [1, 2]$. **120.** $D(f) = [-1, 2]$,
$E(f) = [\sqrt{3}, \sqrt{6}]$. **121.** $x \in (-3, 2]$.

**122.** $a \in (-\infty, -1/4)$. *Solution.* Assume that $-k \in [-1, -1/3]$. If the function $y = -k \, (1/3 \le k \le 1)$ at a certain point $x$, then

$$\frac{x-1}{a-x^2+1} = -k \Rightarrow \frac{x}{k} - \frac{1}{k} = -a + x^2 - 1 \Rightarrow x^2 - \frac{1}{k}x - a - 1 + \frac{1}{k} = 0$$

$$\Rightarrow \qquad x = \frac{1}{2k} \pm \sqrt{\frac{1}{4k^2} - \frac{1}{k} + 1 + a} \Rightarrow x = \frac{1}{2k} \pm \sqrt{\left(1 - \frac{1}{2k}\right)^2 + a}.$$

Consequently, for the function to contain no values from the interval $[-1, -1/3]$, it is necessary that $((2k-1)/2k)^2 + a < 0$ for any $k \in [1/3, 1]$, i.e. $a < -((2k-1)/2k)^2$. For $2k - 1 = 0$, i.e. for $k = 1/2$, the function $-((2k-1)/2k)^2$ assumes its greatest values $0$; the function assumes its least values on the interval $1/3 \le k \le 1$ at the terminal points : for $k = 1$ and $k = 1/3$ we have $-((2k-1)/2k)^2 = -1/4$. Consequently, for any $a \in (-\infty, -1/4)$ the set of the values of the given function does not contain any value from the interval $[-1, -1/3]$.

**123.** $a \in (-\infty, -1) \cup (-1, 5/4]$. *Solution.* For $a = -1$ the function assumes the form $y = (x-1)/(x^2-1)$. It is defined everywhere except for the points $x = \pm 1$. If we exclude these points from the consideration, we can rewrite the function as $y = 1/(x+1)$. This function can assume any values except for $y = 0$. Consequently, for $a = -1$ the function $y = (x+1)/(a+x^2)$ cannot assume the value $y = 0$. If $a < 0$ and $a \ne -1$, then for $x \to \pm \sqrt{|a|}$ the value $y \to \pm \infty$. Let us consider the function $y = (x+1)/(a+x^2)$ on the interval $(-1, |a|)$ if $-1 < -|a|$; on this interval $x + 1 \ge 0$, $x^2 + a > 0$ and, consequently, $y \ge 0$. At the point $x = -1$ the function $y = 0$. When $x$ increases from $-1$ to $-|a|$, the function $y$ increases from $0$ to $+\infty$ and the set of values of the function includes the interval $[0, 1]$. If $a < 0$ and $-|a| < -1$, then the inequalities $x + 1 \le 0$, $x^2 + a < 0$ are satisfied on the interval $(-|a|, -1)$, and, consequently, $y \ge 0$, and when $x$ varies from $-1$ to $-|a|$, the function $y$ increases from $0$ to $+\infty$. Therefore, in this case as well the set of values of the function includes the interval $[0, 1]$. If $a \ge 0$, then the function $y = (x+1)/(x^2+a)$ assume the values $\ge 1$ not for any $a$. Let us find the permissible values of $a$. We shall require that

$$(x+1)/(x^2+a) \ge 1 \Rightarrow x + 1 \ge x^2 + a$$

$$\Rightarrow \qquad x^2 - x + a - 1 \le 0$$

$$\Rightarrow \qquad x \in [(1 - \sqrt{5-4a})/2, (1 + \sqrt{5-4a})/2].$$

Consequently, to find the points $x$ at which $y \ge 1$, it is necessary that $5 - 4a \ge 0 \Rightarrow a \le 5/4$. Thus, for $x = -1$ the function $y = 0$ ; upon a further increase of $x$ the function $y$ will increase, and if $a \le 5/4$, the vertex of the graph of the function will be higher than $y = 1$, i.e. the set of values of the function contains an interval $[0, 1]$.

**124.** $a \in \phi$. *Solution.* $y(1) = 0$ for all $a$ except for $a$'s for which the denominator is zero. For $x = 1$ the denominator vanishes only for $a = 0$. But for $a = 0$, we have $y = -(x-1)/(x^2-1) = -1/(x+1)$. This function has values belonging to the interval $[-1, 1]$. Thus, for all $a$ the given function has values belonging to the interval $[-1, 1]$.

## Sec. 1.9

**1.** $\{2\}$. **2.** $\{-1\}$. **3.** $\{-7/2, 2\}$. **4.** $\{2/7\}$. **5.** $\{2^{-1/2}\}$. **6.** $\{1\}$. **7.** $\{2\}$. **8.** $\{3\}$. **9.** $\{-1, 1\}$.

**10.** $\{9/4, 3\}$. **11.** $\{1, 2\}$. **12.** $\{-2\}$. **13.** $\{0\}$. **14.** $\{3\}$. **15.** $\{0\}$. **16.** $\{-2\}$. **17.** $\{2\}$. **18.** $\{0\}$.

**19.** $\phi$. **20.** $\{-2, 3\}$. **21.** $\{\log_{(\sqrt{5}-1)/2}(3/2)\}$. **22.** $\{10\}$. **23.** $\{2 - \sqrt{7/2}, 2 + \sqrt{7/2}\}$. **24.** $\{3/2\}$. **25.** $\{-1\}$. **26.** $\{0, 1/2\}$. **27.** $\{3 \log_6 2, 3\}$.

**28.** {2}. **29.** {3/2}. **30.** {1/5, 6}. **31.** {1/4}. **32.** {1}. **33.** {$\pi\,(6n - (-1)^n)/18 \mid n \in \mathbf{Z}$}.

**34.** {25}. **35.** {100}. **36.** {−1, 2}. **37.** {3/2}. **38.** {9}. **39.** {−2}. **40.** {1}. **41.** {−3/2, 4}.

**42.** {$-\sqrt{3}, \sqrt{3}$}. **43.** {$-\sqrt{2}, -1, 1, \sqrt{2}$}. **44.** {$-3/2, -1$}. **45.** {1}. **46.** {arctan 10 + $\pi n$

$\mid n \in \mathbf{Z}$]. **47.** {$\log_3 (2 + \sqrt{11/3})$}. **48.** {−2}. **49.** $[3, + \infty] \cup \{-1/2, 1/2\}$.

**50.** {3}. *Solution.* Since the given equation contains a factor $\sqrt[x]{8x - 1}$, it follows that
$x \in \mathbf{N}$. We raise the right-hand and left-hand sides of the equation to the degree $x$;
then, after transformations, we get $5x^2 \cdot 2^{3x-3} = 5^{3x} \cdot 2^{2x}$. We take logarithms of
the right-hand and left-hand sides of the given relation to the base 5 ; after a
transformation, the equation will have the form $(x - 3) \times (x + \log_5 2) = 0$. Hence,
$x = 3$ and $x = -\log_5 2$. The second solution does not satisfy the condition
$-\log_5 2 \in \mathbf{N}$ and, consequently, the only solution is $x = 3$.
*Remark.* If the given equation were written in the form $5^x \cdot 8^{(x-1)/x} = 500$, then
$x \in \mathbf{R}$. In that case the roots of the equation would be $x = 3$ and $x = -\log_5 2$.

**51.** {$\pi/9, 1, \pi\,(6k \pm 1)/9 \mid k \in \mathbf{N}$}. **52.** {10}. **53.** {4}.

**54.** {1}. **55.** {$\pi\,(3k \pm 1) \mid 3 \mid k \in \mathbf{Z}$}. **56.** {−2, 2}. **57.** {1, 3}.

**58.** {1/9, 9}. *Solution.* $3^{\log_3^2 x} + x^{\log_3 x} = 162 \Rightarrow 3^{\log_3^2 x} + (3^{\log_3 x})^{\log_3 x} = 162 \Rightarrow$
$3^{\log_3^2 x} + 3^{\log_3^2 x} = 162 \Rightarrow 3^{\log_3^2 x} = 3^4 \Rightarrow \log_3^2 x = 4 \Rightarrow \log_3 x = 2,$
or $\log_3 x = -2 \Rightarrow x = 9$, or $x = 1/9$.

**59.** {−2, 4}. **60.** {−1}. **61.** {2}. **62.** {6}. **63.** {$\pi\,(2k + 1)/2 \mid k \in \mathbf{Z}$}. **64.** {9}. **65.** {1
/16, 1}. **66.** −1. **67.** {3}. *Remark.* In this problem $x \in N$ and, therefore, the second value
$x = 3/\log_2 6$ is a root of the equation $64^{1/x} - 2^{(3x+2)/x} + 12 = 0$ but not of the given
equation. **68.** {$10^{-2}$}. **69.** {1}.

**70.** {$\log_3 2 - 1, 2$}. *Solution.* $3^{x+1} + 18 \cdot 3^{-x} = 29 \Rightarrow 3 \cdot (3^x)^2 - 29 \cdot 3^x + 18 = 0 \Rightarrow$
$3\,(3^x - 9)\left(3^x - \dfrac{2}{3}\right) = 0.$ Consequently, $3^x = 9 \Rightarrow x = 2$ and
$3^x = 2/3 \Rightarrow x = \log_3 2 - 1.$

**71.** {100}. **72.** {1, 3}. **73.** {8}.

**74.** {$10^{-1/2}, 10^2$}. *Solution.* Since $x$ is the base of an exponential function, we have
$x > 0$. Then
$$x^{1 - \frac{1}{3}\log x^2} = \frac{1}{\sqrt[3]{100}} \Rightarrow 10^{\log x - \frac{2}{3}(\log x)^2} = 10^{-2/3}$$
$$\Rightarrow 2\log^2 x - 3\log x - 2 = 0 \Rightarrow 2\,(\log x - 2)\left(\log x + \frac{1}{2}\right) = 0$$
$$\Rightarrow \log x = 2 \text{ or } \log x = -1/2 \Rightarrow x = 10^2 \text{ or } x = 10^{-1/2}.$$

**75.** {1/3, 9}. **76.** {1/10, 1000}. **77.** {10, $10^5$}. **78.** {−5/2, 3}. **79.** {−1, 1}. **80.** {$10^{-5}$,
$10^3$}. **81.** {9}. **82.** {5}. **83.** $p \in \{0, 25/4\}$. **84.** $(-\infty, 1/2)$. **85.** $(-3/4, +\infty)$. **86.** $(-\infty,$
$1/2) \cup (1, +\infty)$. **87.** $(2, +\infty)$. **88.** $x \in \mathbf{R}$.

**89.** $x \in \phi$. *Solution.* Since $x + 1 \in \mathbf{N}$, we take this remark into account and have
$$\sqrt[x+1]{3} > 9 \Rightarrow 3^{1/(x+1)} > 3^2 \Rightarrow 1/(x+1) > 2$$
$$\Rightarrow \frac{1}{x+1} - 2 > 0 \Rightarrow \frac{x + 0.5}{x+1} < 0 \Rightarrow x \in (-1; -1/2).$$

Thus, if the problem has solutions, then $x \in (-1, -1/2)$, but $x$ can only be equal
to the numbers 0, 1, 2,..., none of these numbers belong to the interval $(-1, -1/2)$
and, consequently, the given inequality does not hold for any $x \in \mathbf{N}_0$.

*Remark.* If the initial inequality were given in the form $3^{1/(x+1)} > 9$, then $x \in R$ and the interval obtained above would be a solution of the inequality.

**90.** $(4/3, +\infty)$. **91.** $(-\infty, -2) \cup (-2/5, +\infty)$. **92.** $(0, 1)$. **93.** $(1/2, 1)$. **94.** $(-2, +\infty)$. **95.** $(-5, 5)$. **96.** $(-\infty, 0) \cup (\log_4 3, +\infty)$. **97.** $x \in R$. **98.** $(-\infty, 0) \cup (1, +\infty)$. **99.** $(3, +\infty)$. **100.** $\{1, 2, 3, 4, 5, 6, 7\}$. **101.** $[(\log_2 3 - 2)/2, +\infty)$. **102.** $(-\infty, 1 - (\log_3 5)/3)$. **103.** $(10^3, +\infty)$. **104.** $(-\infty, 2)$. **105.** $(0, +\infty)$. **106.** $(0, +\infty)$. **107.** $[0, 64]$. **108.** $(-\infty, \log_{0.4} 2)$. **109.** $(-\infty, -\log_5 10]$. **110.** $(10^{-2}, +\infty)$. **111.** $(2, +\infty)$. **112.** $(-1/2, 0)$. **113.** $(0, \log_{2/3}(1/3))$. **114.** $(-\sqrt{2}, -1) \cup (1, \sqrt{2})$. **115.** $(2, 6)$. **116.** $[-1, 0] \cup [2, 3]$. **117.** $(\log_2 3, +\infty)$. **118.** $(-\infty, \log_2 3]$. **119.** $(-\infty, -1) \cup (0, 2)$. **120.** $(2, +\infty)$. **121.** $x \in (a^2 \log_3^2 2, +\infty)$ for $a \in [0, +\infty)$, $x \in [0, +\infty)$ for $a \in (-\infty, 0)$. **122.** $(-\infty, \log_2(\sqrt{2}-1) \cup [1/2, +\infty)$. **123.** $(5/3, 2)$. **124.** $(1, 4)$. **125.** $\{1 + \sqrt{3}\}$. **126.** $\{2\}$. **127.** $\{4\}$. **128.** $\{2\}$. **129.** $\{0\}$. **130.** $\{-5\}$. **131.** $\{1/3\}$. **132.** $\{0\}$. **133.** $\{3\}$. **134.** $\{3/2, 10\}$. **135.** $\{16/\sqrt[3]{5}\}$. **136.** $\{1\}$. **137.** $\{100\}$. **138.** $\{2 + 10^{-7}, 3, 102\}$. **139.** $\{-2 - \sqrt{10}\}$. **140.** $\{10^{-1}, 10\}$. **141.** $\{\sqrt{10^{1-\sqrt{3}}}, \sqrt{10^{1+\sqrt{3}}}\}$. **142.** $\{10^{-5}, 10^3\}$. **143.** $\{3^{-\sqrt{2}}, 3^{\sqrt{2}}\}$. **144.** $\{1/5, 25\}$. **145.** $\{10^{-3}, 10^2\}$. **146.** $\{10^{-4}, 10\}$. **147.** $\{5\}$. **148.** $\{10^{-3}, 10\}$. **149.** $\{2, 16\}$. **150.** $\{10, 10^4\}$. **151.** $\{1/81, 1/3\}$. **152.** $\{\sqrt{5}, 5\}$. **153.** $\{1/4, 2\}$. **154.** $\{2^{\log_a b}, 3^{\log_a b}\}, a > 0, a \neq 1, b > 0, b \neq 1$. **155.** $\{\sqrt{10^{-9}}, 10\}$. **156.** $\{5\}$. **157.** $\{3 + \sqrt{6}\}$. **158.** $\{-5\}$. **159.** $\{-4\}$. **160.** $\{-(3 - \sqrt{3})/3, 8\}$. **161.** $\{1/9, 3\}$. **162.** $\{2^{-7}, 2\}$. **163.** $\{1/\sqrt{2}, 1, 4\}$. **164.** $\{7\}$. **165.** $\{-1, 0\}$. **166.** $\{2\}$. **167.** $\{1\}$. **168.** $\{1/16, 2\}$. **169.** $\{3\}$. **170.** $\{-9/5, 23\}$. **171.** $\{2\}$. **172.** $\{1/2, 128\}$. **173.** $\{1/\sqrt{10}, \sqrt[3]{10}\}$. **174.** $\{1\}$. **175.** $\{8\}$. **176.** $\{2\}$. **177.** $\{\sqrt[5]{5}, 5\}$. **178.** $\{10\}$. **179.** $\{9\}$. **180.** $\{4\}$. **181.** $\{2\}$. **182.** $\{3\}$. **183.** $\{3, 3+\sqrt{2}\}$. **184.** $\{16\}$. **185.** $\{10^{-5/3}, 10^2\}$. **186.** $\{48\}$. **187.** $\{10^{-\sqrt{3}}, 10^{\sqrt{3}}\}$. **188.** $\{13\}$. **189.** $\{29\}$. **190.** $\{1, 9\}$. **191.** $\{-3\}$. **192.** $\{\log_3 4\}$. **193.** $x \in \varnothing$. **194.** $\{0\}$. **195.** $\{98\}$. **196.** $\{0\}$. **197.** $\{3\}$. **198.** $\{-1, 2\}$. **199.** $\{10\}$. **200.** $\{2\}$. **201.** $\{2\}$. **202.** $\{1/5\}$. **203.** $\{3\}$. **204.** $\{10^{-2}, 10^2\}$. **205.** $\{10^{-1}, 10\}$. **206.** $\{-2\}$. **207.** $\{2\}$. **208.** $\{2, 4\}$. **209.** $\{0\}$. **210.** $\{2^{-2/5}, 16\}$. **211.** $\{1/4, 1/2\}$. **212.** $\{1, 2\}$. **213.** $\{100\}$. **214.** $\{10^{-1}, 2, 10^3\}$. **215.** $\{10^{-2/3}, 10^{2/3}\}$. **216.** $\{13/4, 10\}$. **217.** $\{16\}$. **218.** $\{-13/20, 13/6\}$. **219.** $\{3^{(5+3\sqrt{5})/10}\}$. **220.** $\{10^{-3}, 10\}$. **221.** $\{0, 2\}$. **222.** $\{0, 7/4, (3 + 2\sqrt{6})/2\}$. **223.** $\{-17\}$.

**224.** $\{-1/4\}$. *Solution.* The bases of the logarithms $3x + 7$ and $2x + 3$ must exceed zero and be different from 1. Then, taking this into account, we have

$\log_{3x+7}(9 + 12x + 14x^2) + \log_{2x+3}(6x^2 + 23x + 21) = 4$

$\Rightarrow \quad \log_{3x+7}(2x+3)^2 + \log_{2x+3}(2x+3)(3x+7) = 4$

$\Rightarrow \quad 2\log_{3x+7}(2x+3) + 1 + \log_{2x+3}(3x+7) = 4$

$\Rightarrow \quad 2\log_{3x+7}(2x+3) + \dfrac{1}{\log_{3x+7}(2x+3)} - 3 = 0$

$\Rightarrow \quad 2(\log_{3x+7}(2x+3))^2 - 3\log_{3x+7}(2x+3) + 1 = 0$

$\Rightarrow \quad (\log_{3x+7}(2x+3) - 1)(2\log_{3x+7}(2x+3) - 1) = 0$

$\Rightarrow \quad (\log_{3x+7}(2x+3) = 1$ or $2\log_{3x+7}(2x+3) = 1$

$\Rightarrow \quad 2x + 3 = 3x + 7$ or $(2x+3)^2 = 3x + 7$.

The first possibility leads to $x = -4$, but for this value $2x + 3 = -5 < 0$, which contradicts the condition $2x + 3 > 0$. The second possibility yields $4x^2 + 9x + 2 = 0$, whence $x = -2$ or $x = -1/4$, but for $x = -2$ the bases of the logarithms $2x + 3 = -1 < 0$, and this contradicts the condition too. The values $x = -1/4$ satisfies all the conditions.

**225.** $\{0\}$. **226.** $\{2, 3\}$. **227.** $\{2^{-2/3}, 8\}$. **228.** $\{\log_2(3/5), \log_2(2/5)\}$. **229.** $\{2\}$.

**230.** $\{-\log_2 3\}$. **231.** $\{14\}$. **232.** $\{3, 10\}$. **233.** $\{1/12\}$.

**234.** $\{(-1)^n \times \arcsin 2^{\sqrt{-(\log_2 a)/2}} + \pi n \mid n \in \mathbf{Z}, 0 < a < 1\}$.

**235.** $x \in \phi$ for $a \in (-\infty, 0] \cup \{1, 2\}$, $x = a + 2$ for $a \in (0, 1) \cup (1,2) \cup \{3\}$, $x = a \pm 2$ for $a \in (2, 3) \cup (3, +\infty)$. *Solution.* Since $\sqrt{x}$ and $a^2$ are the bases of the logarithms, its follows that $x \neq 1$, $a \neq 1$; since the unknown $x$ enters into the expression $\sqrt{x} x$, it follows that $x > 0$, $a > 0$; since we can take logarithms only of positive numbers; in addition $0 < (a^2 - 4)/(2a - x) < +\infty$, and, consequently, $a \neq 2$, $x \neq 2a$ and $x > 2a$ for $0 < a < 2$ and $x < 2a$ for $a > 2$. No that under these conditions
$$\log_{\sqrt{x}} a = 2/\log_a x$$
and $$\log_{a^2}((a^2 - 4)/(2a - x)) = 0.5 \log_a((a^2 - 4)/(2a - x))$$
and we can write the given equation in the form
$$\log_a \frac{a^2 - 4}{2a - x} = \log_a x \Rightarrow \frac{a^2 - 4}{2a - x} = x \Rightarrow x^2 - 2ax + a^2 - 4 = 0 \Rightarrow x = a \pm 2.$$
Since $x > 2a$ for $a \in (0, 1) \cup (1, 2)$, only $x = a + 2$ of the two solutions obtained satisfies this conditions. For $a > 2$ the quantity $x < 2a$ and, consequently, the condition is satisfied by both solutions. However for $a = 3$ we have $x = a - 2 = 1$, i.e. this value of $x$ does not belong to the domain of definition of the equation and, consequently, for $a = 3$ we have only one solution, $x = a + 2 = 5$, for $x$. Combining these results, we obtain the solution written above.

**236.** $\{25\}$. **237.** $\{3^{-2}, 1, 3\}$.

**238.** $x \in \phi$ for $a \in (-\infty, 1]$; $x = a \pm 1$ for $a \in (1, \sqrt{2}) \cup (\sqrt{2}, 2) \cup (2, +\infty)$, $x = a + 1 = 3$ for $a = 2$. *Solution.* Considering the equation we infer that $x > 0$, $x \neq 1$, $a > 0, a \neq 1$. $a^2 - 1 > 0$, i.e. $|a| > 1$, and since $a > 0$, it follows that $a > 1$; $a^2 - 1 \neq 1$, ie, $a^2 \neq 2$ or $a \neq \sqrt{2}$, and, finally, $2a - x > 0$, i.e. $x < 2a$. Consequently, we get the boundaries $0 < x < 2a, x \neq 1$, for the unknown $x$. Writing the given equation for the base of all the logarithms which is equal to 2, we get
$$\log_2(2a - x) + \log_2 x = \log_2(a^2 - 1)$$
$\Rightarrow \quad (2a - x) x = a^2 - 1$
$\Rightarrow \quad x^2 - 2ax + a^2 - 1 = 0$
$\Rightarrow \quad (x - a)^2 = 1$
$\Rightarrow \quad x = a \pm 1.$
The solutions obtained satisfy all the conditions enumerated above except for the condition $x \neq 1$; indeed, for $a = 2$ we get two values for $x$, one of which, $x = a + 1 = 3$, satisfies all the conditions, and the second, $x = 1$, is not a solution. Combining all the results, we obtain the solution written above.

**239.** $\{a^2 \mid a > 0, a \neq 1/\sqrt{2}, a \neq 1\}$. **240.** $\{4\}$. **241.** $\{(2 + \sqrt{3})^{\log_2(2+\sqrt{3})}\}$. **242.** $\{\arcsin (1/14) + 2\pi n \mid n \in \mathbf{Z}\}$. **243.** $\{1\}$. **244.** $\{1\}$. **245.** $\{-1\}$. **246.** $\{-1, 3\}$. **247.** $\{16\}$. **248.** $\{9\}$. **249.** $\{2\}$. **250.** $\{1, \log_2 3\}$. **251.** $\{2, 64\}$. **252.** $\{1/\sqrt[3]{3}\}$. **253.** $\{4, 36\}$. **254.** $\{0, 3\}$. **255.** $\{9\}$. **256.** $\{1/3, 9\}$. **257.** $\{10, 10^4\}$. **258.** $\{1\}$. **259.** $\{5^{-4}, 5\}$. **260.** $\{-1/2, 1/2\}$. **261.** $\{-1/2, 1/2\}$. **262.** For all $a \in (0, 1/8)$. **263.** $(1/5, 2/5)$. **264.** $(1/3, 2)$. **265.** $(-1/2, 1/2)$. **266.** $(1, 2) \cup (3, 4)$. **267.** $[-1, 1) \cup (3, 5]$. **268.** $(2, 3)$. **269.** $(-\infty, 1/2)$. **270.** $(4, 6)$. **271.** $[1/3, 1/2]$. **272.** $[1/3, 2/3)$. **273.** $[-7, -\sqrt{35}) \cup [5, \sqrt{35})$. **274.** $(-\infty, -3/2) \cup (3, +\infty)$. **275.** $[1/2, 4]$. **276.** $(1/\sqrt{27}, +\infty)$.

**277.** $(0, 1/2] \cup (2, 4]$. **278.** $(0, 10)$. **279.** $(0.1, 1) \cup (1, 10)$. **280.** $(-1, 0) \cup (1, 2)$.
**281.** $[\log_3 0.9, 2)$. **282.** $(-1, +\infty)$. **283.** $(4/3, -17/22)$.
**284.** $(-2, 2 - \sqrt{15})$. *Solution.* Since the denominator of the fraction is always positive, the numerator must be $\geq 0$ for t,he inequality to be satisfied. The trinomial $2 - 5x - x^2 > 0$ for all $x \in (-2, 3)$, and the denominator is meaningful only for these values. The trinomial $x^2 - 4x + 11 = (x - 2)^3 + 7 \geq 7$ for any $x \in \textbf{R}$, and in that case $\log_5 (x^2 - 4x + 11)^2 > 0$. The trinomial $x^2 - 4x - 11$ assumes nonnegative values for $x \in (-\infty, 2 - \sqrt{15}) \cup (2 + \sqrt{15}, +\infty)$, and only for these values $\log_{11} (x^2 - 4x - 11)^3$ is meaningful. Thus, for the left-hand side of the inequality to have sense, it is necessary that both inequalities be satisfied, i.e. that $x \in (-2, 2 - \sqrt{15})$. On that interval $f(x) = \log_{11} (x^2 - 4x - 11) < 0$. Indeed, $f(-2) = \log_{11} 1^3 = 0$,
$f(2 - \sqrt{15}) = \log_{11} 0 = -\infty$. The trinomial $x^2 - 4x - 11$ attains its minimum value for $x = 2$. When varies to the left of $x = 2$, the values of the trinomial increase continuously, for $x = 2 - \sqrt{15}$ its value is equal to zero, for $x = \pm 2$ it is equal to 1; thus $f(x)$ increasea monotonically when $x$ varies from $2 - \sqrt{15}$ to $-2$, remaining negative all the time. Consequently, on the left-hand side of the inequality the numerator of the fraction is positive for $x \in (-2, 2 - \sqrt{15})$ and the inequality is valid for these values.

**285.** $(-\infty, -2) \cup (-1/2, +\infty)$. **286.** $(6, +\infty)$. **287.** $(2, 5/2)$.
**288.** $(-\infty, -1) \cup (4, +\infty)$. **289.** $(-\infty, -2) \cup (4, +\infty)$.
**290.** $(-4/3, (3 - \sqrt{17})/2) \cup (3 + \sqrt{17})/2, +\infty)$.
**291.** $(-\sqrt{2}, -1) \cup (1, \sqrt{2})$. **292.** $(-1, 0) \cup (1 + \infty)$.
**293.** $(-16/3, -3)$. **294.** $(1, 2) \cup (3, +\infty)$. **295.** $(2, +\infty)$.
**296.** $(1, 2] \cup [3, 4)$. **297.** $(-\sqrt{5}, -2) \cup (1, \sqrt{5})$. **298.** $(2, 3)$.
**299.** $(-\infty, 2)$. **300.** $(-\infty, 0] \cup [\log_6 5, 1)$.
**301.** $(-3, -\sqrt{6}) \cup (\sqrt{6}, 3)$. **302.** $(-4, -3) \cup (8, +\infty)$.
**303.** $(-\infty, -1 - \sqrt{2}) \cup (-1 + \sqrt{2}, 1) \cup (1, +\infty)$.
**304.** $(\log_3 10, +\infty)$. **305.** $(3, 4) \cup (6, +\infty)$. **306.** $(-1/2, 0]$.
**307.** $(-1/2, -1/3)$. **308.** $(0, 1/2) \cup (2, 3)$.
**309.** $(-\infty, -2) \cup (-\sqrt{2}, -1) \cup (1, \sqrt{2}) \cup (2, +\infty)$.
**310.** $(-\infty, -2/3] \cup [1/2, 2]$. **311.** $[\sqrt{6} - 1, 2) \cup (2, 5]$.
**312.** $(\log_2 (2/3), 0) \cup (0, +\infty)$.
**313.** $(-1, (1 - \sqrt{5})/2) \cup (1 + \sqrt{5})/2, 2)$. **314.** $[4, +\infty)$.
**315.** $(0, 1) \cup (\sqrt{3}, 9)$. **316.** $(\log_2 (5/4), \log_2 3)$. **317.** $[\log_2 14, 4]$. **318.** $(3, 10)$.
**319.** $(5/8, +\infty)$. **320.** $[0, 27/16]$. **321.** $(0, 1/5) \cup (1, 5\sqrt{5})$. **322.** $(1, 4)$. **323.** $(0, 27)$.
**324.** $[0, 1/3) \cup (3, 10/3)$. **325.** $(0, 1/8] \cup [1/4, +\infty)$. **326.** $[2 + \sqrt{2}, 4)$. **327.** $(3/4, 4 /3)$. **328.** $(0, 1/2) \cup (1, 2) \cup (3, 6)$. **329.** $(2, +\infty)$. **330.** $(-3/2, -1) \cup (-1, 0) \cup (0, 3)$.
**331.** $(10 - \sqrt{43}, 4) \cup (10 + \sqrt{43}, +\infty)$. **332.** $(2, \sqrt{2} + 1) \cup (3.5, +\infty)$. **333.** $(5, 5.5)$
$\cup (6.5, +\infty)$. **334.** $[-\sqrt{8}, -1) \cup (1, (\sqrt{41} - 1)/5]$. **335.** $(0, 1) \cup (2, +\infty)$. **336.** $(3, 5 - \sqrt{3}) \cup (7, +\infty)$. **337.** $(-6, -5) \cup (-3, -2)$. **338.** $(0, (3 - \sqrt{5})/2) \cup [5/2, (3 + \sqrt{5})/2)$.
**339.** $(-2, -1) \cup (-1, 0) \cup (0, 1) \cup (2, +\infty)$. **340.** $(1 - \sqrt{7}, -1) \cup (-1/3, 0) \cup (0, 1/3)$
$\cup (2, 1 + \sqrt{7})$. **341.** $[1/2, 1) \cup [2, +\infty)$. **342.** $(3, 4.5) \cup (8, +\infty)$. **343.** $(-\sqrt{2}, -1) \cup (1, \sqrt{2})$.
**344.** $[-4, -1]$. **345.** $(-4 - \sqrt{2}, -5) \cup (-3, -4 + \sqrt{2}) \cup (1, 2)$. **346.** $(0.7, 1)$. **347.** $(-1, 2)$.
**348.** $(-1, \sqrt{5}) \cup (2/\sqrt{5}, 1)$. **349.** $(4\pi n, \pi (12n + 1)/3) \cup (\pi (12n + 5)/3, 2\pi + 4\pi n)$. $(n \in \textbf{Z})$ **350.** $(0, 1) \cup (2, +\infty)$. **351.** $(0, 1/\sqrt{5}) \cup (1, 3)$. **352.** $(0, 10^{(\log 0.5 \cdot \log 3)/\log 1.5})$.

**353.** $(0, 1/2) \cup (\sqrt{2}, +\infty)$. **354.** $(0, 1/4] \cup [1, 4)$. **355.** $(3, 9)$. **356.** $[2, +\infty)$. **357.** $(0, 1/4] \cup (4, +\infty)$. **358.** $(\log_5 (1 + \sqrt{2}), \log_5 3)$. **359.** $(\log_2 5, 3) \cup (\log_2 14, +\infty)$. **360.** $(1, 10^3)$.

**361.** $(-1, 0)$.      **362.** $(0, 3) \cup (7, +\infty)$. **363.** $x \in (1, (1 + \sqrt{1 + 4a^2})/2)$ for $a \in (0, 1)$;

$x \in ((1 + \sqrt{1 + 4a^2})/2, +\infty)$ for $a \in (1, +\infty)$. **364.** $x \in (a^4, a^{-1})$ for $a \in (0, 1)$; $x \in (a^{-1}, a^4)$ for $x \in (1, +\infty)$. **365.** $x \in ((202 - 53d)/24, +\infty)$ for $d \in (-\infty, 2)$, $x \in (4, +\infty)$ for $d = 2$; $x \in (3d - 2, +\infty)$ for $d \in (2, +\infty)$. **366.** $x \in (0, a^5) \cup (a^3, a^2) \cup (a^{-1}, +\infty)$ for $a \in (0, 1)$; $x \in (0, a^{-1}) \cup (a^2, a^3) \cup (a^5, +\infty)$ for $a \in (1, +\infty)$. **367.** $(0, 10^{-3}) \cup (10^{-2}, 10^2) \cup (10^3, +\infty)$. **368.** $(-\infty, -2)$. **369.** $(1, 2)$. **370.** $(0, 10^{-1}) \cup (10^2, +\infty)$.

**371.** $[3, +\infty)$. *Solution.* The expressions on the left-hand side have sense for $x \in (1, +\infty)$. Then, taking into account that $\log_{1/3} y$ is a decreasing function, we have $\log_{1/3} x + 2\log_{1/9} (x - 1) \leq \log_{1/3} 6 \Rightarrow \log_{1/3} (x^2 - x) \leq \log_{1/3} 6 \Rightarrow x^2 - x \geq 6 \Rightarrow x \leq -2$, or $x \geq 3$. Since, in addition, $x \in (1, +\infty)$, the first inequality obtained is not suitable and the given inequality only holds for $x \in [3, +\infty)$.

**372.** $(3, 4)$.      **373.** $x \in [\log_a (4 + \sqrt{16 + a^2}), 3\log_a 2)$      for $a \in (0, 1)$, $x \in [\log_a (4 + \sqrt{16 + a^2}), +\infty)$ for $a \in (1, +\infty)$. **374.** $[2, +\infty)$.

**375.** $(-1 + \log_2 (5 + \sqrt{33}), +\infty)$.

**376.** $(-1 - \sqrt{5}, -3) \cup (\sqrt{5} - 1, 5)$. **377.** $(0, 1/6) \cup (3/2, +\infty)$.

**378.** $(-2, 2)$. **379.** $(-1, +\infty)$. **380.** $[2, +\infty)$.

**381.** $(-3, -\sqrt{5}) \cup (\sqrt{5}, 3)$. **382.** $(1, 2)$. **383.** $(1/3, 1)$.

**384.** $(-\infty, -5) \cup (-5, -1) \cup (3, +\infty)$.

**385.** $(2, 5/2)$. **386.** $\{1\}$. **387.** $\{(2, 3/2)\}$. **388.** $\{(1, 2), (2, 1)\}$. **389.** $\{(2, 2)\}$.

**390.** $\{(2, 3), (3, 2)\}$. **391.** $\{(4, 2), (1, 1)\}$. **392.** $\{(-2, 0)\}$. **393.** $\{(-2, 1/784)\}$. **394.** $\{(12, 4)\}$. **395.** $\{(5, 2)\}$. **396.** $\{(8, 2)\}$. **397.** $\{(27, 4)(1/81, -3)\}$. **398.** $\{(2, 10), (10, 2)\}$.

**399.** $\{(2\sqrt{2}, \sqrt[4]{8})\}$. **400.** $\{(3, 5)\}$. **401.** $\{(5, 1/2)\}$. **402.** $\{(5, 1)\}$. **403.** $\{(4, 16)\}$. **404.** $\{(9, 7)\}$. **405.** $\{(4, 2), (2, 4)\}$. **406.** $\{(10/3, 20/3), (-10, 20)\}$. **407.** $\{(2, 1/2)\}$. **408.** $\{(2, 2)\}$.

**409.** $\{(625, 3), (125, 4)\}$. **410.** $\{(2^{\sqrt[3]{\log_2^2 7/\log_3 5}}, 3^{\sqrt[3]{\log_3^2 5/\log_2 7}})\}$. **411.** $\{(a^3, a^{-1}), (a^{-1}, a^3)$ $| a > 0\}$. **412.** $\{(2, 1/6)\}$. **413.** $\{(20, 16)\}$. **414.** $\{(4, 2)\}$. **415.** $\{(2, 5)\}$. **416.** $\{(6, 2)\}$.

**417.** $\{(4, 16)\}$. **418.** $\{(2, 4), (4, 2)\}$. **419.** $\{(512, 1)\}$. **420.** $\{(3, 9), (9, 3)\}$. **421.** $\{(5, 5)\}$.

**422.** $\{(64, 1/4)\}$. **423.** $\{(2, 1/4), (2 + \sqrt{7}, 2 + \sqrt{7})\}$. **424.** $\{(4, 1), (16, 2)\}$.

**425.** $\{2, 1\}$. *Solution.* We can rewrite the first equation of the system in the form $(xy \neq 0)$:

$$2^{2(x^2 + y^2)/xy} = 2^5 \Rightarrow \frac{2(x^2 + y^2)}{xy} = 5$$

$$\Rightarrow \quad 2\frac{x}{y} + \frac{2}{x/y} = 5 \Rightarrow 2\left(\frac{x}{y}\right)^2 - 5\left(\frac{x}{y}\right) + 2$$

$$= 0 \Rightarrow \left(\frac{x}{y} - 2\right)\left(\frac{2x}{y} - 1\right) = 0 \Rightarrow x = 2y \text{ or } y = 2x.$$

Thus, the initial system of equations reduces to two systems

$$\begin{cases} x = 2y \\ \log_3 (x - y) = 1 - \log_3 (x + y), \\ xy \neq 0, \end{cases} \quad \begin{cases} y = 2x, \\ \log_3 (x - y) = 1 - \log_3 (x + y), \\ xy \neq 0 \end{cases}$$

We infer from the second equation of the system that $x - y > 0$ and $x + y > 0$. We can rewrite the systems as follows :

$$\begin{cases} x = 2y, \\ x^2 - y^2 = 3, \\ x - y > 0, \\ x + y > 0, \\ xy \neq 0; \end{cases} \quad \begin{cases} y = 2x, \\ x^2 - y^2 = 3, \\ x - y > 0, \\ x + y > 0, \\ xy \neq 0. \end{cases}$$

Solving two equations of the first system simultaneously, we get two solutions $(2, 1)$ and $(-2, -1)$. The first solution satisfies all the conditions and the second solution does not satisfy the third and the fourth conditions. Substituting $y = 2x$ from the first equation of the second system into the second equation, we obtain a contradictory equality $-3y^2 = 3$; consequently, the second system has no solutions. Thus, the given system has a unique solution $(2, 1)$.

**426.** $\{(1, 1)\}$.    **427.** $\{(2, 6)\}$.

## Sec. 1.10

**61.** $-1$. **62.** $1 / \sin\alpha = \mathrm{cosec}\,\alpha$. **63.** $2\sqrt{2} \sin\left(2x + \dfrac{\pi}{4}\right)$. **64.** $32\sin^2\alpha \times \cos^4\alpha$. **65.** $1$.

**66.** $\tan\alpha$. **67.** $2$. **68.** $\tan 5\alpha$. **69.** $3/2$. **70.** $1$. **71.** $\cos 2\alpha$. **72.** $0$. **73.** $\tan 2\alpha$. **74.** $1/2$.

**75.** $(\sin 4\alpha)/2$. **76.** $\sin^2\alpha$. **77.** $1$. **78.** $\sin 2\alpha$. **79.** $1/4$. **80.** $\sqrt{2-\sqrt{2}}/2$, $(\sqrt{6}-\sqrt{2})/4$.

**81.** $2$. **82.** $42$. **83.** $4.5$. **84.** $\sqrt{3}$. **85.** $(3-\sqrt{3})/2$. **86.** $\sqrt{3}/8$. **87.** $0$. **88.** $4$. **89.** $9$. **90.** $\sqrt{2}/8$.

**91.** $-1$. **95.** $225/128$. **96.** $\cot(\alpha/2) = -3$  or  $\cot(\alpha/2) = -\sqrt{15}/3$. **97.** $65/113$.

**98.** $-3\sqrt{7}/8$. **99.** $(1-a^2-2a)/(1-a^2+2a)$. **100.** $31/49$. **101.** $-2/\sqrt{13}$. **102.** $5/\sqrt{26}$.

**103.** $1/\sqrt{5}$. **104.** $10/11$. **105.** $(4-3(m^2-1)^2)/4, |m| \leq \sqrt{2}$. **106.** $\sqrt{2}\tan\alpha \sin\left(\dfrac{\pi}{4}+\alpha\right)$.

**107.** $2\sqrt{3}\cos\dfrac{\alpha}{2}\sin\left(\dfrac{\alpha}{2}+\dfrac{\pi}{6}\right)$. **108.** $4\cos\alpha \sin\dfrac{5\alpha+\beta}{2}\cos\dfrac{3\alpha+\beta}{2}$. **109.** $\sin(\alpha-15°)$

$\cos(\alpha+15°)$.    **110.** $4\sqrt{2}\sin\left(\dfrac{\pi}{4}-11\alpha\right)$.    **111.** $2\sin(\alpha-\beta) \times \sin\dfrac{\alpha+\beta-30°}{2}$

$\cos\dfrac{\alpha+\beta+30°}{2}\sec^2\alpha \sec^2\beta$. **112.** $0.5\cos(\alpha-\beta)\sec\alpha \sec\beta$. **113.** $3\pi/4$. **114.** $17\pi/12$

**115.** $127°$. **116.** $\sqrt{3}/2$. **117.** $\pi/2$. **118.** $24/25$. **119.** $17/25$. **120.** $3/5$. **121.** $-3/5$.

**122.** $2/3$. **123.** $3/4$. **124.** $14/15$. **125.** $-7/25$. **126.** $-3/4$. **133.** $4/5$. **134.** $3/5$.

**135.** $2/\sqrt{5}$. **136.** $1/\sqrt{5}$. **137.** $-\pi/7$. **138.** $3\pi/7$. **139.** $\pi/7$. **140.** $-\pi/7$. **141.** $6\pi/7$.

**142.** $\arccos(12/13) = \arctan(5/12) = \mathrm{arccot}(12/5)$. **143.** $\{2, 4\}$. **144.** $1/2$. **145.** $1$.

**146.**

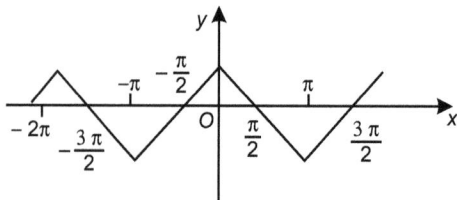

**147.** $x \in (2 - \sqrt{9-2\pi}, 2 + \sqrt{9-2\pi})$.

## Sec. 1.11

**1.** $x = (-1)^n \dfrac{\pi}{6} + \pi n \ (n \in \mathbf{Z})$. **2.** $x = (-1)^n \arcsin (-1/3) + \pi n = (-1)^{n+1} \arcsin$

$(1/3) + \pi n \ (x \in \mathbf{Z})$. **3.** $x = (-1)^n \arcsin 0 + \pi n = \pi n \ (n \in \mathbf{Z})$. **4.** $x = (-1)^n \arcsin 1 + \pi n$

$= (-1)^n \dfrac{\pi}{2} + \pi n \ (n \in \mathbf{Z})$. Setting $n = 2k$ and $n = 2k+1$, we get in both cases

$x = \dfrac{\pi}{2} + 2\pi k \ (k \in \mathbf{Z})$. **5.** $x = \pm \dfrac{\pi}{3} + 2\pi n \ (n \in \mathbf{Z})$. **6.** $x = \pi + 2\pi n \ (n \in \mathbf{Z})$. **7.** $x = \dfrac{\pi}{2} + \pi n$

$(n \in \mathbf{Z})$. **8.** $x = \dfrac{\pi}{3} + \pi n \ (n \in \mathbf{Z})$. **9.** $x = -\dfrac{\pi}{4} + \pi n \ (n \in \mathbf{Z})$. **10.** $\pi (4n + (-1)^n)/4 \ (n \in \mathbf{Z})$.

**11.** $\pi (3n - (-1)^n)/3 \ (n \in \mathbf{Z})$. **12.** $\pi (4n-1)/2 \ (n \in \mathbf{Z})$. **13.** $[2\pi n \ (n \in \mathbf{Z})$. **14.** $\pi n$

$(n \in \mathbf{Z})$. **15.** $\phi$. **16.** $\pi (4n-1)/4 \ (n \in \mathbf{Z})$. **17.** $\arctan 7 + 1 + \pi n \ (n \in \mathbf{Z})$. **18.** $(\pi (3n$

$+1) - 9)/6 \ (n \in \mathbf{Z})$. **19.** $\pi (12n-1)/4 \ (n \in \mathbf{Z})$ **20.** $(\pi (4n-1)+4)/3 \ (n \in \mathbf{Z})$.

**21.** $\pi (8n \pm 1)/28, \pi (2k+1)/14 \ (n, k \in \mathbf{Z})$. **22.** $\pi n, 2\pi (3k \pm 1)/3 \ (n, k \in \mathbf{Z})$. **23.** $\pi +$

$2\pi n, \pi (6k + (-1)^k)/6 \ (n, \ k \in \mathbf{Z})$. **24.** $2\pi n \ (n \in \mathbf{Z})$. **25.** $\pi (4n+1)/2, \pi k \ (n, k \in \mathbf{Z})$.

**26.** $\pi (4n-1)/4, \pi (3k \pm 1)/3 \ (n, \ k \in \mathbf{Z})$. **27.** $\pi (3n-1)/6 \ (n \in \mathbf{Z})$. **28.** $\pi n /2$

$(n \in \mathbf{Z})$. **29.** $\pi (4n+1)/2, \pi (3k-1)/3 \ (n, \ k \in \mathbf{Z})$. **30.** $\pi (4n+1)/2, \pi (6k + (-1))/6$

$(n, \ k \in \mathbf{Z})$. **31.** $2\pi n \ (n \in \mathbf{Z})$. **32.** $\pi (2n+1)/2, 2\pi (3m \pm 1)/3 \ (n, \ m \in \mathbf{Z})$. **33.** $(-1)^n$

$\arcsin (1/3) + \pi n, \pi (6k + (-1)^n)/6, \ (n, \ k \in \mathbf{Z})$. **34.** $\pi + 2\pi n, \pm \arccos(3/4) + 2\pi k$

$(n, \ k \in \mathbf{Z})$. **35.** $\pm \arccos((\sqrt{19}-2)/5) + 2\pi n \ (n \in \mathbf{Z})$. **36.** $\pi (2n+1)/4 \ (n \in \mathbf{Z})$.

**37.** $\pi (3n \pm 1)/3 \ (n \in \mathbf{Z})$. **38.** $\pi (2n+1)/4, \pi (3m \pm 1)/3 \ (n, m \in \mathbf{Z})$. **39.** $\pi n /2$

$(n \in \mathbf{Z})$. **40.** $\pi (4n-1)/4 \ (n \in \mathbf{Z})$. **41.** $\arctan 2 + \pi n, [\pi (4k+1)/4 \ (n, \ k \in \mathbf{Z})$.

**42.** $\pi (4n+1)/4, \arctan (-1/3) + \pi k \ (n, \ k \in \mathbf{Z})$. **43.** $\arctan (-\sqrt{2}) + \pi n \ (n \in \mathbf{Z})$.

**44.** $\pi (4k-1)/4, \ \pi (3n \pm 1)/3 \ (n, \ k \in \mathbf{Z})$. **45.** $\pi n /2, \pi (6k \pm 1)/3 \ (n, k \in \mathbf{Z})$. **46.**

$2\pi n, \pi (4m-1)/2, \ \pi (4k+1) /4 \ (n, m, k \in \mathbf{Z})$. **47.** $-\dfrac{\pi}{4} + (-1)^n \arcsin \dfrac{1+\sqrt{3}}{2\sqrt{2}} 3 + \pi n$

$(n \in \mathbf{Z}) \equiv \pi (12n -3 + 5 (-1)^n)/12 \ (n \in \mathbf{Z})$. **48.** $(\pi (4n+1) - 2 \arccos (4/5))/4$

$(n \in \mathbf{Z})$. **49.** $\pi (12n-5)/12 \ (n \in \mathbf{Z})$. **50.** $\pi n /3, \pi (6k+1)/18 \ (n, \ k \in \mathbf{Z})$. **51.** $\pi (2n+1)$

$/5, 2\pi k \ (n, \ k \in \mathbf{Z})$. **52.** $\pi n /3, \pi (2k+1)/8 \ (n, \ k \in \mathbf{Z})$. **53.** $\pi k /2, 2\pi (3n \pm 1)$

$/3 \ (n, \ k \in \mathbf{Z})$. **54.** $2\pi n /5, \pi (2k+1) /2, \pi (2m+1) \ (n, k, m \in \mathbf{Z})$. **55.** $\pi n \ (n \in \mathbf{Z})$.

**56.** $\pi n /2, \pi (3k - (-1)^k)/21 \ (n, \ k \in \mathbf{Z})$. **57.** $\pi (2n+1)/6, \pi (4k-1)/4 \ (n, \ k \in \mathbf{Z})$.

**58.** $2\pi (3n \pm 1)/3, \pi (4k+1)/8 \ (n, \ k \in \mathbf{Z})$. **59.** $\pi (4n-1)/4, \pi (4k+3)/16 \ (n,$

$k \in \mathbf{Z})$. *Hint.* $\sin 5x = \cos \left( \dfrac{\pi}{2} - 5x \right)$. **60.** $((4n-1) \pi - 4)/12, ((2k-1) \pi - 4)/24$

$(n, \ k \in \mathbf{Z})$. **61.** $\pi (12n-1)/24, \pi (12k+1)/12 \ (n, \ k \in \mathbf{Z})$. **62.** $\pi (8n + 3)/24, \ \pi (8k$

$+1)/16 \ (n, \ k \in \mathbf{Z})$. **63.** $\arctan (2 \pm \sqrt{3}) + \pi n \equiv \pi (6n + (-1)^n)/12 \ (n \in \mathbf{Z})$. **64.** $\pi (2n$

$+1)/16, \pi (2k+1)/10 \ (n, \ k \in \mathbf{Z})$. **65.** $\pi (2n+1)/4, \pi(2k+1)/2 \ (n, \ k \in \mathbf{Z})$. **66.**

$\pi (2n+1)/4 \ (n \in \mathbf{Z})$. **67.** $\pi (4n+1)/4 \ (n \in \mathbf{Z})$. **68.** $\pi n /4 \ (n \in \mathbf{Z})$. **69.** $\pi n /8 \ (n \in \mathbf{Z})$.

**70.** $\pi (2n+1)/4, \pi (6k \pm 1)/6 \ (n, k \in \mathbf{Z})$. **71.** $\pi (2n+1)/10, \pi (2k+1)/4 \ (n, \ k \in \mathbf{Z})$.

**72.** $\pi n, \pi (2k+1)/20 \ (n, k \in \mathbf{Z})$. **73.** $\pi (4n+1)/6, 2\pi k /3 \ (n, k \in \mathbf{Z})$. **74.** $\pi (2n$

$+1)/10, \pi (2m+1)/6 \ (n, \ m \in \mathbf{Z})$. **75.** $\pi (6n + (-1)^n)/18, \pi (2k+1)/4 \ (n, \ k \in \mathbf{Z})$.

**76.** $\pi (2k+1)/10 \ (n, \ k \in \mathbf{Z})$. **77.** $\pi (3n \pm 1)/2 \ (n \in \mathbf{Z})$. **78.** $2\pi n \pm \arccos$

$(-1/4) \ (n \in \mathbf{Z})$. **79.** $\pi (4n+1)/4 \ (n \in \mathbf{Z})$. *Solution.* $\sin^4 x + \cos^4 x = \sin x \cos x$

$\Rightarrow \left( \dfrac{1 - \cos 2x}{2} \right)^2 \left( \dfrac{1 + \cos 2x}{2} \right)^2 = \dfrac{1}{2} \sin 2x \Rightarrow 1 + \cos^2 2x = \sin 2x \Rightarrow 1 + 1 =$

$\sin^2 2x = \sin 2x \Rightarrow \sin^2 2x + \sin 2x = 2$ This relation is possible only for $\sin 2x = 1$, whence $2x = \dfrac{\pi}{2} + 2\pi n \Rightarrow x = \dfrac{\pi}{4} + \pi n$. Thus $x = \pi\,(4n+1)\,/\,4$.

**80.** $\pi\,(4n+1)/4$, $2\pi k\,(n,\ k \in \mathbf{Z})$. **81.** $\pi n\,/\,5$, $\pi\,(2k+1)\,/\,2\,(n,\ k \in \mathbf{Z})$. **82.** $\pi\,(2n+1)/8$, $\pi\,(3k\pm 1)\,/\,3\,(n,\ k \in \mathbf{Z})$. **83.** $\pi n\,/\,3$, $2\pi k$, $\pi\,(2m+1)/11\,(n, k, m \in \mathbf{Z})$. **84.** $\pi\,(2k+1)$ $/4$, $\pi\,(2m+1)/10\,(m, k \in \mathbf{Z})$. **85.** $\pi\,(2n+1)/16$, $(\pi(4k+1)-8)/4$, $(\pi\,(4m+1)+8)$ $/12\,(n, k,\ m \in \mathbf{Z})$. **86.** $\pi(2n+1)/10\,(n \in \mathbf{Z})$. **87.** $\pi\,(3n\pm 1)/3\,(n \in \mathbf{Z})$. **88.** $\pi\,(4k$ $+1)/10\,(k \in \mathbf{Z})$. **89.** $\pi\,(4n+1)\,/\,2$, $2\pi k\,(n, k \in \mathbf{Z})$. **90.** $\pi\,(2n+1)\,/\,2(n \in \mathbf{Z})$. **91.** $\pi\,(6n-(-1)^n)/12\,(n \in \mathbf{Z})$. **92.** $\pi\,(2n+1)\,/\,4$, $\pi\,(6k+(-1)^k)12\,(n, k \in \mathbf{Z})$. **93.** $\pi\,(4n-1)$ $/2$, $(-1)^k \arcsin\,(3\,/\,4) + \pi k\,(n,\ k \in \mathbf{Z})$. **94.** $\pi\,(2n+1)/4$, $\pi\,(6k+(-1)^k)\,/\,24\,(n,\ k \in \mathbf{Z})$. **95.** $2\pi\,(3n\pm 1)\,/\,3\,(n \in \mathbf{Z})$. **96.** $2\pi n$, $\pi\,(2k+1)\,/\,2\,(n, k \in \mathbf{Z})$. **97.** $\pi n\,/\,2$, $(12k\pm 1)$ $/6\,(n, k \in \mathbf{Z})$. **98.** $\pi\,(4n+1+(-1)^n)/4\,(n \in \mathbf{Z})$. **99.** $\pi\,(4n+1)\,/\,2\,(n \in \mathbf{Z})$. **100.** $\pi\,(4n+$ $(-1)^n)/16\,(n \in \mathbf{Z})$. **101.** $\pi n + \arctan\,(-1 \pm \sqrt{3})\ (n \in \mathbf{Z})$. **102.** $\pi\,(6n+(-1)^n)/6\,(n \in \mathbf{Z})$. **103.** $\pi\,(2n+1)/14\,(n \in \mathbf{Z})$. **104.** $\pi\,(2n+1)\,/\,4$, $\pi\,(4k+1)\,/\,2\ (n, k \in \mathbf{Z})$. **105.** $\pi\,(4n-1)\,/\,4$, $2\pi m$, $\pi\,(4k-1)\,/2$, $\pi\,(4l+1)\,/\,4 + (-1)^l \arcsin\,(\sqrt{2}\,/\,4)\,(n, m, k, l \in \mathbf{Z})$. *Solution.* Since $\qquad \cos 3x = \cos^3 x - 3\cos x \sin^2 x, \sin 3x \qquad = 3\cos^2 x \sin x - \sin^3 x$, $\cos 2x = \cos^2 x - \sin^2 x$, we can rewrite the given equation as

$$\cos^3 x + \sin^3 x - 3\cos x \sin x\,(\cos x + \sin x)$$

$$-(\cos x + \sin x)\,(\cos x - \sin x) = 0$$

or as

$$(\cos x + \sin x)\,(\cos^2 x + \sin^2 x - 4\sin x \cos x + \sin x - \cos x) = 0.$$

Then either

(1) $\cos x + \sin x = 0$, or $\qquad$ (2) $1 - 2\sin 2x + \sin x - \cos x = 0$.

In the first case $\tan x = -1$ and $x = \pi\,(4n-1)\,/\,4\,(n \in \mathbf{Z})$. In the second case we set

$$\sin x - \cos x = y; \qquad\qquad\qquad (1)$$

then $y^2 = (\sin x - \cos x)^2 = 1 - \sin 2x$ and, consequently,

$$\sin 2x = 1 - y^2. \qquad\qquad\qquad (2)$$

Substituting these values into the equation, we obtain $1 - 2\,(1 - y^2) + y = 0$ or $2y^2 + y - 1 = 0$. The roots of this equation are $-1$ and $1\,/\,2$. Then we get the following equations for $x$ :

(a) $\sin x - \cos x = -1$ and (b) $\sin x - \cos x = 1\,/\,2$.

(Note that if we use relation (2), then it is simple to find the value of $x$, but in that case extraneous roots may appear since equations (1) and (2) are not equivalent. Equation (2) follows from equation (1), but not viceversa.) In case (a) we can rewrite the equation as follows :

$$\sin x + 1 - \cos x = 0 \Rightarrow 2\sin\,(x\,/\,2)\cos(x\,/\,2) + 2\sin^2(x\,/\,2)$$

$$\Rightarrow 2\sin^2\,(x\,/\,2)\,(\cot\,(x\,/\,2) + 1) = 0 \Rightarrow \sin(x\,/\,2) = 0 \ \text{ or } \cot x = -1$$

$$\Rightarrow x = 2\pi m, \ \ x = \pi\,(4k-1)\,/\,2 \ (m, k \in \mathbf{Z}).$$

In case (b) we multiply the left-hand and right-hand sides by $\sqrt{2}\,/\,2$ and rewrite it in the form

$$\sin x \cos\frac{\pi}{4} - \cos x \sin\frac{\pi}{4} = \frac{\sqrt{2}}{4},$$

or in the form

$$\sin\left(x - \frac{\pi}{4}\right) = \frac{\sqrt{2}}{4},$$

whence

$$x = \frac{\pi}{4} + (-1)^l \arcsin\left(\frac{\sqrt{2}}{4}\right) + \pi l \ (l \in \mathbf{Z}).$$

**106.** $\pi (2n+1)/44,\ \pi (6k+(-1)^k)/48\ (n, k \in \mathbf{Z})$.

**107.** $\pi n,\ \pi (6k \pm 1)/6 (n, k \in \mathbf{Z})$. *Solution.* $\sin 3x - 4\sin x \cos 2x = 0$

$\Rightarrow 3\sin x \cos^2 x - \sin^3 x - 4\sin x \cos 2x = 0$

$\Rightarrow \sin x \left(3\dfrac{1+\cos 2x}{2} - \dfrac{1-\cos 2x}{2} - 4\cos 2x\right) = 0$

$\Rightarrow \sin x (1 - 2\cos 2x) = 0 \Rightarrow \sin x = 0$ or $1 - 2\cos 2x = 0 \Rightarrow x = \pi n,$

$2x = \pm\dfrac{\pi}{3} + 2\pi k (n, k \in \mathbf{Z})$. Thus the roots of the equation are

$\pi n,\ \pi (6k \pm 1)/6 (n,\ k \in \mathbf{Z})$.

**108.** $\pi (6n + (-1)^n)/12(n \in \mathbf{Z})$.

*Solution.* $\sin^2 x \tan x + \cos^2 x \cot x - \sin 2x = 1 + \tan x + \cot x$

$\Rightarrow \dfrac{\sin^4 x + \cos^4 x}{\sin x \cos x} - \sin 2x = 1 + \dfrac{\sin^2 x + \cos^2 x}{\sin x \cos x}$

$\Rightarrow \dfrac{2}{\sin 2x} - \sin 2x - \sin 2x = 1 + \dfrac{2}{\sin 2x} \Rightarrow -2\sin 2x = 1$

$\Rightarrow \sin 2x = -1/2$. Whence, $x = \pi (6n + (-1)^n)/12\ (n \in \mathbf{Z})$.

**109.** $\pi n,\ \pi (4k+1)/4\ (n, k \in \mathbf{Z})$. **110.** $\dfrac{(-1)^n}{2} \times \arcsin (4 - 2\sqrt{3}) + \dfrac{\pi}{2}n\ (n \in \mathbf{Z})$.

**111.** $2\pi (3n \pm 1)/9 (n \in \mathbf{Z})$. **112.** $\pi (4n+1)/2,\ \pi + 2\pi k (n,\ k \in \mathbf{Z})$. **113.** $\pi (4n+1)/2, 2\pi k, \pi (4m+1)/4 (n,\ k, m \in \mathbf{Z})$. **114.** $\pi (4n+1)/4 (n \in \mathbf{Z})$. **115.** $\pi (8k+1)/4,\ \pi(8n+3)/12\ (n,\ k \in \mathbf{Z})$. **116.** $2\pi n,\ \pi (4k-1)/4, \pi(4m-1)/2(n, k, m \in \mathbf{Z})$. **117.** $\pi (12n - 3 + 2 (-1)^n)/60 (n \in \mathbf{Z})$. **118.** $\pi (4n+1)/4 (n \in \mathbf{Z})$. **119.** $\pi (4n+1)/2,$ $\pi (2k+1)\ (n, k \in \mathbf{Z})$. **120.** $2\pi n,\ 2\arctan(3/2) + 2\pi k\ (n,\ k \in \mathbf{Z})$.

**121.** $\pi n - \arctan(1/2) (n \in \mathbf{Z})$. *Solution.* We rewrite the given equation in the form

$$\cos^2 x - 2 (1 - \sin x) \cos x - 4\sin x = 0.$$

This is a quadratic equation with respect to $\cos x$. Let us solve it :

$\cos x = 1 - \sin x \pm \sqrt{1 - 2\sin x + \sin^2 x + 4\sin x}$

$= 1 - \sin x \pm \sqrt{1 + 2\sin x + \sin^2 x} = 1 - \sin x \pm (1 + \sin x).$

In the case of the upper sign, we get $\cos x = 2$. We must discard this case. In the case of the lower sign, we get $\cos x = -2\sin x$ or $\tan x = -1/2$, whence $x = -\arctan(1/2) + \pi n\ (n \in \mathbf{Z})$.

If, we solve the given equation, expressing all the functions in terms of $\tan (x/2)$, we arrive at an equation

$$3\tan^4 (x/2) - 12\tan^3 (x/2) - 2\tan^2 (x/2) - 4\tan(x/2) - 1 = 0,$$

which can be factored into two factors :

$$(\tan^2(x/2) - 4 \tan (x/2) - 1) (3\tan^2(x/2) + 1) = 0.$$

The second factor is not equal to zero for any value of $x \in \mathbf{R}$, and the first factor vanishes for $\tan (x/2) = 2 \pm \sqrt{5}$, i.e. $x = 2\arctan (2 \pm \sqrt{5}) + 2\pi n (n \in \mathbf{Z})$.

This is another form of notation for the solution. It is easy to show that if $\tan(x/2) = 2 \pm \sqrt{5}$, then it follows that $\tan x = -1/2$. The converse is also true. Thus the two notations for the solution are equivalent.

**122.** $\pi n + \arctan(3 - 2\sqrt{2})$ $(n \in \mathbf{Z})$.

**123.** $\pi n / 3$, $\pi k \pm \arctan(1/\sqrt{2}) \equiv \pi k \pm 0.5 \arccos(1/3)$ $(n, k \in \mathbf{Z})$.

*Solution.* Note that $\tan 2x = 2\tan x / (1 - \tan^2 x)$,

and $\tan 3x = (3\tan x - \tan^3 x) / (1 - 3\tan^2 x)$, with $x \neq \pi (2l+1)/4$,

$x \neq \pi ((2m+1)/6$ $(l, m \in \mathbf{Z})$. After the substitution of these values and transformations, the given equation assumes the form

$$\tan x (2\tan^4 x - 7\tan^2 x + 3) = 0,$$

whence follow two possibilities :

$$\tan x = 0, \quad 2\tan^4 x - 7\tan^2 x + 3 = 0.$$

The first possibility yields $x = \pi p$ $(p \in \mathbf{Z})$. The second possibility yields $\tan x = \pm\sqrt{3}$ and $\tan x = \pm 1/\sqrt{2}$ and, consequently,

$$x = \pm\frac{\pi}{3} + \pi m \ (m \in \mathbf{Z}), \ x = \pm \arctan\frac{1}{\sqrt{2}} + \pi k \ (k \in \mathbf{Z}).$$

The solution $x = \pi p$ obtained earlier and the solution $x = \pi (3m \pm 1)/3$ can be combined and then we can write $x = \pi n / 3$ $(n \in \mathbf{Z})$. (For $n = 3p$ we get $\pi p$, and for $n = 3m \pm 1$ we get $\pi (3m \pm 1)/3$). We can thus write all the solutions as follows : $\pi n / 3$, $\pi k \pm \arctan(1/\sqrt{2})$ $(n, k \in \mathbf{Z})$. It is easy to prove that this notation is equivalent to the other notation presented above.

**124.** $\pi (2n+1)/6$, $2\pi (3k \pm 1)/9$ $(n, k \in \mathbf{Z})$. *Hint.* Introduce a new variable $3x = y$.

**125.** $4\pi n$ $(n \in \mathbf{Z})$.

**126.** $\pi (24n + 1)/12$, $\pi (24k - 7)/12$ $(n, k \in \mathbf{Z})$. *Solution.* For the given equation to have a solution, conditions (1) $\sin\left(3x + \dfrac{\pi}{4}\right) \geq 0$ and (2) $1 + 8\sin 2x \cos^2 2x \geq 0$ must be satisfied. Under these conditions we square the right-hand and left-hand sides.

$$4\sin^2\left(3x + \frac{\pi}{4}\right) = 1 + 8\sin 2x (1 - \sin^2 2x) \Rightarrow 2\left(1 - \cos\left(6x + \frac{\pi}{2}\right)\right)$$

$$= 1 + 8\sin 2x - 8\sin^3 2x \Rightarrow 2 + 2\sin 6x = 1 + 8\sin 2x - 8\sin^3 2x$$

$$\Rightarrow 2 + 2(3\sin 2x - 4\sin^3 2x) = 1 + 8\sin 2x - 8\sin^3 2x$$

$$\Rightarrow 1 = 2\sin 2x \Rightarrow x = \frac{\pi}{12} + \pi l \text{ and } x = \frac{5\pi}{12} + \pi m \ (l, m \in \mathbf{Z}).$$

Let us verify whether all the values of $x$ we have obtained satisfy conditions (1) and (2). We substitute $x = \dfrac{\pi}{12} + \pi l$ into the first condition

$$\sin\left(3x + \frac{\pi}{4}\right) = \sin\left(\frac{\pi}{4} + 3\pi l + \frac{\pi}{4}\right) = \sin\left(3\pi l + \frac{\pi}{2}\right) = \cos 3\pi l,$$

$\cos 3\pi l = 1 > 0$ for even $l$ and $\cos 3\pi l = -1 < 0$ for odd $l$. To fulfil the first condition we set $l = 2n$ and then $x = -\dfrac{\pi}{12} + 2\pi n$ or $\sin x = 1/2$. We substitute this value of $x$ into the second condition. Then we have

$$1 + 8\sin 2x \cos^2 2x = 1 + 8 \cdot \frac{1}{2} \cdot \frac{3}{4} = 4 > 0,$$

i.e. the second condition is also fulfilled, i.e. $x = \pi\,(24n+1)/12$ is a solution of the given equation. Let us substitute now, $x = \dfrac{5\pi}{12} + \pi m$ into the first condition ;

we obtain $\sin\left(3x + \dfrac{\pi}{4}\right) = \sin\left(\dfrac{5\pi}{4} + 3\pi m + \dfrac{\pi}{4}\right) = \sin\left(\dfrac{3\pi}{2} + 3\pi m\right)$

$= -\cos 3\pi m, -\cos 3\pi m = -1 < 0$ for even $m$ and $-\cos 3\pi m = 1 > 0$ for odd $m$. To satisfy the first condition, we set in this case $m = 2k-1$; then we have $x = -\dfrac{7\pi}{12} + 2\pi k$. Since in this case, as well we have $\sin 2x = 1/2$, the second condition is also fulfilled and, consequently, $x = \pi\,(24k-7)/12$ is a solution of the given equation.

**127.** $\pi\,(2n+1)/2,\ \pi\,(6k+(-1)^k)/24\,(n,\ k \in \mathbf{Z})$. **128.** $\pi\,(6n-(-1))^n/12$ $(n \in \mathbf{Z})$. **129.** $\arctan(2/3) + \pi n\,(n \in \mathbf{Z})$. **130.** $\pi\,(6n-(-1)^n)/18\ (n \in \mathbf{Z})$. **131.** $\pi n - (-1)^n \arcsin(1/3)\ (n \in \mathbf{Z})$. **132.** $\pi n + \arctan\dfrac{1\pm\sqrt5}{2},\ \pi k + \arctan\dfrac{1\pm\sqrt{13}}{2}(n,\ k \in \mathbf{Z})$.
**133.** $\pi\,(8n-1)/4\,(n \in \mathbf{Z})$. **134.** 4. **135.** $\phi$. **136.** $\pi n \pm \arccos(1/3)\,(n \in \mathbf{Z})$. **137.** $\pi\,(2n+1)/6,\ \pi(4k+(-1)^k/20\,(n,\ k \in \mathbf{Z})$. **138.** $2\pi\,(3n\pm1)/3\,(n \in \mathbf{Z})$. **139.** $\pi\,(6n\pm1)/18\,(n \in \mathbf{Z})$. **140.** $\pi n/5, (2\pi k \pm \arccos(1/4))/5\,(n,\ k \in \mathbf{Z})$. **141.** $\pi n, \pi(6k\pm1)/3\,(n,\ k \in \mathbf{Z})$. **142.** $\pi\,(2n+1)/2, \pi\,(4k+1)/4\,(n,\ k \in \mathbf{Z})$. **143.** $\pi\,(2n+1)/4\,(n \in \mathbf{Z})$. **144.** $\pi\,(2k+1)/4,\ \pi\,(6n\pm1)/6,(n,\ k \in \mathbf{Z})$. **145.** $\pi\,(6n\pm1)/6\,(n \in \mathbf{Z})$. **146.** $\pi n/2, \pi\,(2k+1)/8\,(n,\ k \in \mathbf{Z})$. **147.** $\pi n, \pi\,(3k+1)/3\,(n, k \in \mathbf{Z})$. **148.** $\pi\,(2n+1)/4, \pi\,(6k\pm1)/6\,(n,\ k \in \mathbf{Z})$. **149.** $\pi\,(2n+1)/8, \pi\,(3k\pm2)/9(n,\ k \in \mathbf{Z})$. **150.** $\pi\,(2n+1)/6,\ \pi\,(6k-(-1)^k)/12\,(n,\ k \in \mathbf{Z})$. **151.** $5\pi n, 5\pi\,(2k+1)/2(n,\ k \in \mathbf{Z})$. **152.** $2\pi n/3, \pi\,(4k+1)/4, \pi\,(4m-1)/4\,(n, k, m \in \mathbf{Z})$. **153.** $\arctan(-1/3) + \pi n, \pi\,(4k+1)/4, \pi\,(2m+1)/2\ (n, k, m \in \mathbf{Z})$. **154.** $\pi n, \pi(4k-1)/4(n,\ k \in \mathbf{Z})$. **155.** $\pi n,\ \pi\,(6k\pm1)/3\,(n,\ k \in \mathbf{Z})$. **156.** $\pi n/3,\ \pi\,(2k+1)/7\ (n,\ k \in \mathbf{Z})$. **157.** $2\pi n, 2\pi\,(3k\pm1)/3\ (n,\ k \in \mathbf{Z})$. **158.** $\pi n/3, \pi(4k+1)/12\ (n,\ k \in \mathbf{Z})$. **159.** $\pi\,(2n+1)/8\ (n \in \mathbf{Z})$. **160.** $\pi\,(2n+1)/2, \pi\,(6k\pm1)/15\ (n,\ k \in \mathbf{Z})$. **161.** $\pi\,(2n+1)/11, 2\pi k/5, \pi\,(2m+1)/2\ (n, k, m \in \mathbf{Z})$. **162.** $\pi n, \pi(6k+1)/48\ (n,\ k \in \mathbf{Z})$. **163.** $\pi\,(4n+1)/4, 2\pi\,(3k\pm1)/3\ (n,\ k \in \mathbf{Z})$. **164.** $\pi n,\ \pi\,(6k+(-1)^k)/6\ (n,\ k \in \mathbf{Z})$. **165.** $\pi\,(4n+(-1)^n)/4\ (n \in \mathbf{Z})$. **166.** $\pi\,(3n\pm1)/6\ (n \in \mathbf{Z})$. **167.** $\pi\,(2n+1)/4, \pi\,(3k-1)/6,\ (n,\ k \in \mathbf{Z})$. **168.** $\pi\,(2n+1)/6(n \in \mathbf{Z})$. **169.** $\pi\,(12k+3+4\,(-1)^k)/36\ (k \in \mathbf{Z})$. **170.** $\pi\,(6n+(-1)^n/6\ (n \in \mathbf{Z})$. **171.** $\pi\,(4n-1)/4,\ \pi\,(4m+1)/8\ (n,\ m \in \mathbf{Z})$. **172.** $\pi\,(4n+1)/4\ (n \in \mathbf{Z})$. **173.** $(\pi n+3)/2\ (n \in \mathbf{Z})$. **174.** $2\pi\,(3n+1)/3\ (n \in \mathbf{Z})$. **175.** $\pi n,\ \pi\,(3k+(-1)^k)/6\ (n,\ k \in \mathbf{Z})$. **176.** $\pi\,(4n-1)/4, \pi k + \arctan 3\ (n,\ k \in \mathbf{Z})$. **177.** $\pi\,(2n+1)/4\ (n \in \mathbf{Z})$. **178.** $\pi\,(2n+1)/4\ (n \in \mathbf{Z})$. **179.** $2\pi n/3, \pi(4k+1)/6\ (n,\ k \in \mathbf{Z})$. **180.** $\pi\,(3n+1)/3, \pi + 2\pi k\ (n,\ k \in \mathbf{Z})$. **181.** $\pm\arccos(1/4) + 2\pi n\ (n \in \mathbf{Z})$. **182.** $\pi\,(6n+1)/6,\ \pi\,(6k-1)/12\ (n,\ k \in \mathbf{Z})$. **183.** $\pi + 2\pi n, \pi\,(4k+1)/2\ (n,\ k \in \mathbf{Z})$. **184.** $\pi n/12\ (n \in \mathbf{Z})$. **185.** $\pi n/4, \pi\,(4k+1)/32\ (n,\ k \in \mathbf{Z})$. **186.** $\dfrac{(-1)^n}{2}\arcsin\dfrac{\sqrt5-1}{2} + \dfrac{\pi}{2}n, \pi\,(4k+1)/4\ (n,\ k \in \mathbf{Z})$. **187.** $\pi n, \pi\,(3k+(-1)^k)/6\ (n,\ k \in \mathbf{Z})$. **188.** $\pi\,(4n+1)/4\ (n \in \mathbf{Z})$. **189.** $\pi\,(4n+1)/2, 2\arctan(3/5) + 2\pi k\,(n, k \in \mathbf{Z})$. **190.** $\pi n/11, \pi m\ (n \in N_0, m \in N)$ **191.** $\pi n/3, \pi k/11\ (n = 0, 1, 2, \ldots; -k \in N)$ **192.** $\pi\,(2n+1)/4\ (n \in \mathbf{Z})$. **193.** $\pi\,(4n+1)/4, \pi\,(4k-(-1)^k)/8\ (n,\ k \in \mathbf{Z})$. **194.** $\pm\arccos(-1/4) + 2\pi n\ (n \in \mathbf{Z})$. **195.** $\pi\,(2n+1)/20, \pi\,(2k+1)/8\ (n,\ k \in \mathbf{Z})$. **196.** $\pi\,(4n-1)/2, (\pi(4k-1)+4\,(-1)^k \arcsin((\sqrt2-2)/2)/4(n,\ k \in \mathbf{Z})$. *Hint.* The given equation is equivalent to the equation
$$\sin x\,(1+\sin x) + \cos x\,(1-\sin^2 x) = 0.$$

The equation $\sin x + \cos x - \sin x \cos x = 0$ is equivalent to the equation $(\sin x + \cos x)^2 - 2(\sin x + \cos x) - 1 = 0$.

**197.** $\pi(6n + (-1)^n)/12$, $\pi(4k+1)/14$, $\pi(4m+1)/6$ $(n,\ k,\ m \in \mathbf{Z})$. **198.** $\pi(6n - (-1)^n)/6$ $(n \in \mathbf{Z})$. **199.** $\pi(6k + (-1)^k)/6$ $(k \in \mathbf{Z})$. **200.** $2\pi n/5, 2\pi k/3$ $(n,\ k \in \mathbf{Z})$ **201.** $\pi(4n+1)/6$, $\pi(6k - (-1)^k)/18$ $(n,\ k \in \mathbf{Z})$. **202.** $\pi(3n \pm 1)/3$ $(n \in \mathbf{Z})$. **203.** $\pi(4n+1)/4$ $(n \in \mathbf{Z})$. **204.** $\pi(4n+1)/2$ $(n \in \mathbf{Z})$. **205.** $\pi(4n+1)/4$ $(n \in \mathbf{Z})$. **206.** $\pi(6n \pm 1)/12$ $(n \in \mathbf{Z})$. **207.** $\pi n/2$, $\pi(6k - (-1)^k)/12(n,\ k \in \mathbf{Z})$. **208.** $\pi(2n+1)/4$, $\pi k$ $(n,\ k \in \mathbf{Z})$. **209.** $\pi n/4$, $\pi(3k \pm 1)/12(n,\ k \in \mathbf{Z})$. **210.** $\pi n/2$, $\pi(2k+1)/12$ $(n,\ k \in \mathbf{Z})$. **211.** $\pi n/4$ $(n \neq 4m,\ m \in \mathbf{Z})$. **212.** $\pi(8n \pm 1)/8$ $(n \in \mathbf{Z})$. **213.** $\pi(4n+1)/8$, $\pi(4k+3)/4$ $(n,\ k \in \mathbf{Z})$. **214.** $\pi(3n \pm 1)/3$ $(n \in \mathbf{Z})$. *Hint.* Since, $\cos x \neq 0$, divide the equation by $\cos x$ and apply formula (2) of Sec. 1.10. **215.** $\pi(3n + (-1)^n)/6$ $(n \in \mathbf{Z})$. **216.** $\pi(3n+1)/6$ $(n \in \mathbf{Z})$. **217.** $\pi n$ $(n \in \mathbf{Z})$. **218.** $\pi(3n \pm 1)/12$ $(n \in \mathbf{Z})$. **219.** $\pi(6n-1)/3$ $(n \in \mathbf{Z})$. **220.** $2\pi n$ $(n \in \mathbf{Z})$. **221.** $\dfrac{\pi}{2} + \pi n$, $2\pi k \pm \arccos\sqrt[4]{2/3}$, $2\pi m + \pi \pm \arccos\sqrt[4]{2/3}$ $(n,\ k,\ m \in \mathbf{Z})$. **222.** $\pi n$, $2\pi k \pm \arccos((-3+\sqrt{5})/6)$, $2\pi m \pm \arccos((-3-\sqrt{5})/6)$ $(m,\ k,\ n \in \mathbf{Z})$. **223.** If $a = 2\pi n$ $(n \in \mathbf{Z})$, then

$$\{[\pi(4k-1)/2, \pi(4k+1)/2] \mid k \in \mathbf{Z}\};$$

if $a = (2n+1)\pi$ $(n \in \mathbf{Z})$, then

$$\{[\pi(4k+1)/2; \pi(4k+3)/2] \mid k \in \mathbf{Z}\};$$

if $a \in (4\pi n, (4n+1)\pi)$ $(n \in \mathbf{Z})$, then

$$\{(\pi(8k+7\pm1)-2a)/4 \mid k \in \mathbf{Z}\};$$

if $a \in ((4n+1)\pi, (4n+2)\pi)$ $(n \in \mathbf{Z})$, then

$$\{(\pi(8k+5\pm1)-2a)/4 \mid k \in \mathbf{Z}\};$$

if $a \in ((4n+2)\pi, (4n+3)\pi)$ $(n \in \mathbf{Z})$, then

$$\{(\pi(8k+3\pm1)-2a)/4 \mid k \in \mathbf{Z}\};$$

if $a \in ((4n+3)\pi, (4n+4)\pi)$ $(n \in \mathbf{Z})$, then

$$\{(\pi(8k+1\pm1)-2a)/4 \mid k \in \mathbf{Z}\}.$$

**224.** $2/(3\pi(2n+1)-4), 1/(3(\pi k + (-1)^k \arcsin(3/4))-2)$ $(k, n \in \mathbf{Z})$. **225.** $2\pi n; \pm 2\arccos((-1+\sqrt{4a+5})/4)+4\pi k$ for $a \in [-5/4, 5]$; $\pm 2 \arccos((-1-\sqrt{4a+5})/4)+4\pi m$ for $a \in [-5/4, 1]$ $(n,\ k,\ m \in \mathbf{Z})$. **226.** $(-1)^{n+1}\arcsin(\pi/8)+\pi(3n-1)/3$, $(-1)^k\arcsin(\pi/4\sqrt{2})+\pi(4k-1)/4$ $(n, k \in \mathbf{Z})$. **227.** $\pi(24n+13)/6$ $(n \in \mathbf{Z})$. *Hint.* $\sqrt{3}\cos x + \sin x - 2 = 2\left(\cos\left(x - \dfrac{\pi}{6}\right)-1\right)$. **228.** $x = [(-1)^k \arcsin(1-\sqrt{2a+3})+\pi k]/2\,(k \in \mathbf{Z})$ for $a \in [-3/2, 1/2]$. **229.** $\pi(2n+1)/4$ $(n \in \mathbf{Z})$. **230.** $\pi n/2$, $\pi(4k \pm 1)/8$ $(n,\ k \in \mathbf{Z})$. **231.** $\pi(2n+1)$, $\pi(4k+1)/4(n,\ k \in \mathbf{Z})$. **232.** $\pi(4n+1)/4$ $(n \in \mathbf{Z})$. **233.** $\pi(12n \pm 5)/6$ $(n \in \mathbf{Z})$. **234.** $\pi(6n+1)/6$ $(n \in \mathbf{Z})$. *Solution.* To solve the given equation, it is necessary that $4\cos 2x > -1$, i.e. $\cos 2x > -1/4\,(2\cos 2x + 2 \geq 0$ for any $x \in \mathbf{R})$. Then, after multiplication by $\sqrt{4\cos 2x + 1}$, squaring, and transformation, we obtain an equation

$$8\cos^2 2x + 10\cos 2x - 7 = 0$$

or

$$(4\cos 2x + 7)(2\cos 2x - 1) = 0.$$

In this expression $4\cos 2x + 7 > 0$ for any $x \in \mathbf{R}$ and therefore, $2\cos 2x = 1$ and $x = \pi\,(6n \pm 1)/6\,(n \in \mathbf{Z})$. **235.** $\sqrt{3}$. **236.** $\pi\,(4n+1)/4\,(n \in \mathbf{Z})$. *Solution.*

$$\sin 2x + \tan x = 2 \Rightarrow \frac{2\tan x}{1 + \tan^2 x} + \tan x = 2 \Rightarrow \tan^3 x - 2\tan^2 x + 3\tan x - 2 = 0$$

$\Rightarrow (\tan x - 1)(\tan^2 x - \tan x + 2) = 0$. The second factor cannot vanish for any $x \in \mathbf{R}$. Therefore, provided that $x \neq \pi\,(2l+1)/2\,(l \in \mathbf{Z})$, the given equation is equivalent to the equation $\tan x - 1 = 0$, whence $x = \pi\,(4n+1)/4\,(n \in \mathbf{Z})$.
**237.** $\pi n$, $\pi\,(8k \pm 1)/4\;(n, k \in \mathbf{Z})$. **238.** $\pi n + \arctan(3/4)$, $\pi\,(4k-1)/4(n, k \in \mathbf{Z})$.
**239.** $\pi\,(3n \pm 1)/3\;(n \in \mathbf{Z})$. **240.** $\pi(2n+1)$, $\pi m + \arctan 4\;(n, m \in \mathbf{Z})$. **241.** $\pi\,(4n+1)/4$, $\pi\,(2k+1)/2\;(n, k \in \mathbf{Z})$. **242.** $\pi\,(4n-1)/4$, $\pi k\,(n, k \in \mathbf{Z})$. *Solution.* The given equation has sense under the condition $x \neq \pi\,(2l+1)/2\,(l \in \mathbf{Z})$. Since $1 - \tan x = (\cos x - \sin x)/\cos x, 1 + \sin 2x = (\cos x + \sin x)^2, 1 + \tan x$

$= (\cos x + \sin x)/\cos x$, with $\cos x \neq 0$, the equation assumes the form
$$\frac{(\cos x - \sin x)(\cos x + \sin x)^2}{\cos x} = 1 + \tan x$$

$\Rightarrow (\cos^2 x - \sin^2 x)(\cos x + \sin x)/\cos x = 1 + \tan x$

$\Rightarrow \cos 2x\,(1 + \tan x) = (1 + \tan x)$

$\Rightarrow (t + \tan x)(\cos 2x - 1) = 0$

$\Rightarrow \tan x = -1$ or $\cos 2x = 1$

$\Rightarrow x = \pi\,(4n-1)/4$, or $x = \pi k\,(n, k \in \mathbf{Z})$.
**243.** $\pi n\,(n \in \mathbf{Z})$.
**244.** $2\pi n\,(n \in \mathbf{Z})$. *Solution.* $2\tan^2 x + 3 = \dfrac{3}{\cos x} \Rightarrow 2\sec^2 x + 1 - \dfrac{3}{\cos x} = 0$

$\Rightarrow \qquad 2\left(\dfrac{1}{\cos x}\right)^2 - 3\left(\dfrac{1}{\cos x}\right) + 1 = 0$

$\Rightarrow \qquad \left(\dfrac{1}{\cos x} - 1\right)\left(\dfrac{2}{\cos x} - 1\right) = 0.$

Note that the equation has sense under the condition that $\cos x = 0$. The second factor in the product obtained is not equal to zero for any $x \in \mathbf{R}$. The first factor is zero if $\cos x = 1$, i.e. $x = 2\pi n\,(n \in \mathbf{Z})$.

**245.** $\pi\,(6n \pm 1)/6$, $\pi\,(3m \pm 1)/3\,(m,\;n \in \mathbf{Z})$. *Solution.* We designate $81^{\sin^2 x} = y$; then $81^{\cos^2 x} = 81^{1 - \sin^2 x} = 81 \cdot y^{-1}$ and, after transformations, the equation assumes the form $y^2 - 30y + 81 = 0$, whose roots are $y = 3$ and $y = 27$. Then $81^{\sin^2 x} = 3$ or $81^{\sin^2 x} = 27$. The first equation can be rewritten as $3^{4\sin^2 x} = 3$
$\Rightarrow 4\sin^2 x = 1 \Rightarrow \sin x = \pm 1/2 \Rightarrow x = \pm\,\pi/6 + \pi n\,(n \in \mathbf{Z})$. The second equation yields $81^{\sin^2 x} = 27 \Rightarrow 3^{4\sin^2 x} = 3^3 \Rightarrow 4\sin^2 x = 3 \Rightarrow \sin x = \pm\sqrt{3}/2 \Rightarrow x = \pm\pi/3 + \pi m$
$(m \in \mathbf{Z})$. Consequently, the solutions of the given equation are the numbers $\pi\,(6n \pm 1)/6$, $\pi\,(3m \pm 1)/3\,(n, m \in \mathbf{Z})$. This set of solutions can be written in another form : $\pi\,(30n + 15 \pm 4)/60\,(n \in \mathbf{Z})$. **246.** $p \in [\sqrt{5}-1, 2]$. **247.** $a \in [2, 6]$. **248.** $7\pi/18$.
**249.** $0°$. **250.** $-31\pi/24$, $\pi/24, 17\pi/24$. **251.** $\pi\,(2m+1)/2$, $\pi\,(6n \pm 1)/3$, $2\pi k \pm \arccos(-2/3)\,(m, n, k \in \mathbf{Z})$. The least distance between the positive roots is equal to $\pi/6$. **252.** $-5\pi/6$, $-2\pi/3$, $0$, $\pi/6$, $\pi/3$. **253.** $0$, $\pi/8, \pi/4, \pi/2, 5\pi/8, 3\pi/4$.
**254.** $\pi\,(2k+1) - \arctan 3\,(k \in \mathbf{Z})$, i.e. all numbers for which $n$ is an odd number.
**255.** $-7\pi/12$, $\pi/12$, $5\pi/12$. **256.** $-\pi$, $-\dfrac{5\pi}{6}$, $-2\pi/3, 0, \pi/6$, $\pi/3$. **257.** $\pi\,(3n-1)/6\,(n = 3, 4, ...,)$. **258.** $n, (-1 \pm \sqrt{3 + 4k})/2\;(n \in \mathbf{Z}, k = 0, 1, 2, ....), x_7 = (-1 + \sqrt{23})/2$. **259.** $\pi n\,(n \in \mathbf{Z})$, $4950\pi$. **260.** $\pi n + (-1)^n \arcsin(1/3)\,(n \in \mathbf{Z})$. **261.** $\pi n\,(n = -1, 0, 1)$, $\pi(\,6k+1)/6\,(k = -2, -1, 0, 1)$, $\pi\,(6m-1)/6\,(m = -1, 0, 1, 2)$.

**262.** $\pi\,(8n+5)/8\,(n\in\mathbf{Z})$. **263.** $(2\pi n,\ \pi+2\pi n)\,(n\in\mathbf{Z})$. **264.** $[\pi\,(12n+1)/6,\!\pi\,(12n+5)/6)\,(n\in\mathbf{Z})$. **265.** $(4\pi n,\ \pi\,(12n+1)/3)\cup(\pi\,(12n+1)\ /3,2\pi+4\pi n)(n\in\mathbf{Z})$. **266.** $[\pi\,(12n-7)/6,\pi\,(12n+1)/6]\,(n\in\mathbf{Z})$. **267.** $\pi\,(4n-1)/2\,(n\in\mathbf{Z})$. **268.** $(\pi\,(8n+3)/4,\ \pi\,(8n+5)/4\,(n\in\mathbf{Z})$. **269.** $(\pi n,\ \pi\,(2n+1)/2\,(n\in\mathbf{Z})$. **270.** $(\pi\,(2n-1)/2,\ \pi\,(3n-1)/3\,(n\in\mathbf{Z})$. **271.** $(\pi\,(8n-3)/4,\pi\,(8n+1)/4)\ (n\in\mathbf{Z})$. **272.** $(\pi\,(8n-1)4,\ \pi\,(8n+3)/4)\cup(\pi+2\pi n2\pi n)\cup(\pi\,(8n+1)/4,\ \pi\,(8\pi+5)/4)\,(n\in\mathbf{Z})$. **273.** $(\pi\,(6n-1)/3,\ 2\pi\,(3n+1)/3)\,(n\in\mathbf{Z})$. **274.** $(\pi\,(4n+1)/4,\pi\,(3n+1)/3\ (n\in\mathbf{Z})$. **275.** $[2\pi\,(3n+1)/3,\pi+2\pi n]\cup[2\pi\,(3n-1)/3,\ 2\pi n]\ (n\in\mathbf{Z})$. **276.** $[2\pi\,(3n-1)/3,\ 2\pi\,(3n+1)/3]\ (n\in\mathbf{Z})$. **277.** $[\pi\,(12n-7)/6,\ \pi\,(12n+1)/6]\ (n\in\mathbf{Z})$. **278.** $\{(\pi n,1/2)\mid n\in\mathbf{Z}\}$. **279.** $\pi+2\pi n,\ \pi\,(12k\pm5)/6\ (n,\ k\in\mathbf{Z})$. **280.** $0,\ \pi/4,\ \pi/2,5\pi/4$. **281.** $/2,-\pi-\pi/6,\ 0,\ \pi/6,\pi/2,\ 5\pi/6,\pi$. **282.** $\pi\,(8n+3)/4\ (n\in\mathbf{Z})$. **283.** $\{(\pi\,(6n+6k-1)/6,\ \pi\,(6n-6k-1)/6),\ (\pi\,(6n+6k+1)/6,\ \pi\,(6n-6k+1)/6\}\ (n,\ k\in\mathbf{Z})$. **284.** $\{(\pi\,(6k+6m-1)/6,\pi\,(6k-6m-1)/6,\ (\pi\,(6k+6l+1)/6,\ \pi\,(6k-6l+1)/6\}\ (k,m,\ l\in\mathbf{Z})$. **285.** $\{(\pi/6,2\pi/3),(5\pi/6,2\pi/3)\}$. **286.** $\{(n,\ (4n+1)/4,\ ((4n-1)/4,n\ )\}\,(n\in\mathbf{Z})$. **287.** $\{(\pi\,(6k+6n\pm1)/6,\ \pi\,(2k-2n\pm1)/2),\ (\pi\,(2k+2n\pm1)/2,\ \pi\,(6k-6n\pm1)/6\}\ (k,\ n\in\mathbf{Z})$. *Remark.* We have briefly written here four groups of values; the upper sign in the formula for $x$ corresponds to the upper sign in the formula for $y$ and the lower sign in the formula for $x$ corresponds to the lower sign in the formula for $y$. It is convenient to use this rule is various calculations.

**288.** $\{(-\pi\,(12n-5)/12,\ \pi\,(3n+1)/3),\ (-\pi\,(12n-1)/12,\pi\,(3n+2)/3)\}(n\in\mathbf{Z})$.

**289.** $\left\{((-1)^n\arcsin\sqrt{2-\sqrt3}-\dfrac{\pi}{12}+\pi n,\ \dfrac{5\pi}{12}-(-1)^n\arcsin\sqrt{2-\sqrt3}-\pi n)\right\}(n\in\mathbf{Z})$.

**290.** $\{(\pi\,(2n+1)/2,\ -\pi\,(6n-1)/6\}\,(n\in\mathbf{Z})$.

**291.** $\{(\pi\,(4m+1)/4,\ \pi\,(8n-4m+1)/4)\},(n,m\in\mathbf{Z})$.

**292.** $\left\{\left(\dfrac{13}{2}\pi+\arccos((\sqrt{57}-6)/3)+2\pi n,\ \arccos((\sqrt{57}-6)/3)+2\pi n\right),\right.$
$\left.\left(\dfrac{13}{2}\pi-\arccos((\sqrt{57}-6)/3)+2\pi n,\ \arccos((\sqrt{57}-6)/3)+2\pi n\right)\right\}(n\in\mathbf{Z})$.

**293.** $\{(\pi\,(6n+(-1)^n)/6,\ (-1)^k\arcsin(3/4)+\pi k\}\ (n,k\in\mathbf{Z})$.

**294.** $\{(\pi\,(8n-1)/4,\arctan3+\pi k\}\ (n,\ k\in\mathbf{Z})$.

**295.** $\{(\pi\,(8n+3)/4,\ \pi\,(6k+(-1)^k)/6\},\ (n,\ k\in\mathbf{Z})$.

**296.** $\{(\arccos(3/4)+2\pi k,\ -\arccos(1/8)+2\pi n),\ (-\arccos(3/4)+2\pi k,\ \arccos(1/8)+2\pi n)\}\,(n,k\in\mathbf{Z})$. *Solution.* The first equation yields $6\cos x=5-4\cos y$ and the second equation yields $6\sin x=-4\sin y$. Squaring and adding, we get $\cos y=1/8$ or $y=\pm\arccos(1/8)+2\pi n\ (n\in\mathbf{Z})$. Then it follows from the first equation of the system that $\cos x=3/4$, or $x=\pm\arccos(3/4)+2\pi k\ (k\in\mathbf{Z})$. Since, when squaring, we could obtain extraneous roots, we must verify, by a direct substitution of the values obtained into the second equation, that only two out of four combinations of signs are suitable :
$$x_1=\arccos(3/4)+2\pi k,\quad y_1=-\arccos(1/8)+2\pi n,$$
$$x_2=-\arccos(3/4)+2\pi k,\quad y_2=\arccos(1/8)+2\pi n.$$

**297.** $\{(\pi m,\ 2\pi n),(2\pi m-\arccos(1/7),\ 2\pi\,(3n+1)/3),(2\pi m+\arccos2\pi\,(3n-1)/3)\}(1/7),(m,n\in\mathbf{Z})$. *Hint.* Multiplying the equation term-by-term, we get, after the necessary transformations, an equation
$$\tan^2(y/2)=4\sin^2 y.$$
It has three series of solutions

$$y_1 = 2\pi n, \quad y_2 = \frac{2\pi}{3} + 2\pi n, \quad y_3 = -\frac{2\pi}{3} + 2\pi n \ (n \in \mathbf{Z}).$$

We substitute each value of $y$, we have obtained into the two equations of the initial system.

For $y_1 = 2\pi n$ we get

$$\sin x = 0 \text{ and } x_1 = \pi m.$$

For $y_2 = \frac{2\pi}{3} + 2\pi n$ we get a system

$$\begin{cases} \sqrt{3} \cos x - 5 \sin x = 3\sqrt{3}, \\ 3\sqrt{3} \cos x - \sin x = \sqrt{3}, \end{cases}$$

whence

$\cos x = 1/7, \quad \sin x = -4\sqrt{3}/7$ i.e. $-\pi/2 < x < 0, x_2 = 2\pi m - \arccos(1/7)$.

For $y_3 = -2\pi/3 + 2\pi l$ we get, by analogy,

$$\cos x = 1/7, \quad \sin x = 4\sqrt{3}/7,$$

i.e. $0 < x < \pi/2$, and consequently, $x_3 = 2\pi m + \arccos(1/7)$.

**298.** $\{(\pi (6n+1)/3; \pi (6m-1)/3), (\pi (6n-1)/3, (6m+1)/3)\}$ $(m, \ n \in \mathbf{Z})$.

**299.** $\{(\pi (4m+1)/4, \pi (6n \pm 1)/3); ((\pi m + (-1)^m \arcsin\sqrt{2/3})/2, 2\pi n \pm \arccos\sqrt{2/3}); \quad ((\pi m - (-1)^m \arcsin \sqrt{2/3})/2, 2\pi n \pm (\pi - \arccos\sqrt{2/3}))\}$

$(m, \ n \in \mathbf{Z})$.

*Solution.* Since $\cos 2y = 2\cos^2 y - 1$, we rewrite the first equation of the given system as follows :

$$(4\cos^2 y - 1) = (2\cos y - 1)(1 + 2\sin 2x)$$

$\Rightarrow \qquad (2\cos y - 1)(2\cos y + 1 - 1 - 2\sin 2x) = 0$

$\Rightarrow \qquad 2\cos y - 1 = 0 \text{ or } \cos y - \sin 2x = 0.$

In the first case $\cos y = 1/2 \Rightarrow y = \pi (6n \pm 1)/3 \ (n \in \mathbf{Z})$. Substituting this value of $y$ into the second equation of the system, we obtain

$$\pm \frac{\sqrt{3}}{2} \left( \tan^3 x + \frac{1}{\tan^3 x} \right) = \pm \frac{3}{\sqrt{3}} \Rightarrow \tan^3 x + \frac{1}{\tan^3 x} = 2$$

$\Rightarrow \qquad (\tan^3 x)^2 - 2(\tan^3 x) + 1 = 0$

$\Rightarrow \qquad \tan^3 x = 1$

$\Rightarrow \qquad \tan x = 1$

$\Rightarrow \qquad x = \pi (4m+1)/4 \ (m \in \mathbf{Z}).$

Thus, the solution of the system is the set

$$(\pi (4m+1)/4, \ \pi (6n \pm 1)/3 \ (n, \ m \in \mathbf{Z}).$$

In the second case $\cos y = \sin 2x$. Then the second equation of the system assumes the form

$$\sin y \left( \frac{\sin^6 x + \cos^6 x}{\sin^3 x \cos^3 x} \right) = 3 \frac{\cos y}{\sin y}$$

$\Rightarrow \qquad \sin^2 y ((\sin^6 x + 3\sin^4 x \cos^2 x + 2\sin^2 x \cos^4 x + \cos^6 x)$

$- 3\sin^2 x \cos^2 x (\sin^2 x + \cos^2 x)) = 3\cos y (\sin x \cos x)^3$

$\Rightarrow \qquad (1 - \cos^2 y)((\sin^2 x + \cos^2 x)^3 - 3(\sin x \cos x)^2)$

$\Rightarrow \qquad 3\cos y (\sin x \cos x)^3$

$\Rightarrow \qquad (1 - \sin^2 2x)\left(1 - \frac{3\sin^2 2x}{4}\right) = 3\sin 2x \frac{\sin^3 2x}{8}$

$\Rightarrow \qquad 3\sin^4 2x - 14\sin^2 2x + 8 = 0$

$\Rightarrow \qquad (\sin^2 2x - 4)(3\sin^2 2x - 2) = 0.$

The first factor $\sin^2 2x - 4 \neq 0$ for $x \in \mathbf{R}$. We equate the second factor to zero and then

$$3\sin^2 2x = 2 \Rightarrow \sin 2x = \pm\sqrt{2/3}.$$

In the case $\sin 2x = \sqrt{2/3} \Rightarrow x = (\pi m + (-1)^m \arcsin \sqrt{2/3})/2$ $(m \in \mathbf{Z})$, and since $\cos y = \sin 2x = \sqrt{2/3}$, we have $y = \pi n \pm \arccos \sqrt{2/3}$ $(n \in \mathbf{Z})$. Similarly, in the case $\sin 2x = -\sqrt{2/3}$, we get $x = (\pi m - (-1)^m \arcsin \sqrt{2/3})/2$ and $y = \pi n \pm (\pi - \arccos \sqrt{2/3})$ $(m, n \in \mathbf{Z})$.

300. $\{x = \pi (4k \pm 1)/4,\ y = \pm \arctan 2 + \pi n,\ z = \pi - x - y\}$ $(n, k \in \mathbf{Z})$. *Remark.* The upper sign in the formula for $x$ corresponds to the upper sign in the formula for $y$ and the lower sign corresponds to the lower sign.

301. $(x, y) \in \{(1/2, -1/2), (-1/2, 1/2), (1, 0), (-1, 0), (0, 1), (0, -1)\}$.

302. $\{(\pi m,\ \pi l)\}$ $(m, l \in \mathbf{Z})$.

303. $\left\{\left(\dfrac{\pi}{8} - \arccos\dfrac{a}{\sqrt{2+\sqrt{2}}} + 2\pi n, \dfrac{\pi}{8} + \arccos\dfrac{a}{\sqrt{2+\sqrt{2}}} - 2\pi n\right),\right.$

$\left.\left(\dfrac{\pi}{8} + \arccos\dfrac{a}{\sqrt{2+\sqrt{2}}} + 2\pi n, \dfrac{\pi}{8} - \arccos\dfrac{a}{\sqrt{2+\sqrt{2}}} - 2\pi n\right)\right\}$

$\hfill (n \in \mathbf{Z}),\ a \in [-\sqrt{2}, \sqrt{2}].$

304. $\{(\arctan (a-1) + \pi n, (-1)^k \arcsin (a+1) + \pi k)\}$ $(n, k \in \mathbf{Z})$ for $a \in [-2, 0]$, $\{(\arctan (a+1) + \pi n, (-1)^k \arcsin (a-1) + \pi k\}$ $(n, k \in \mathbf{Z})$ for $a \in [0, 2]$.

305. $\{(\pi (6n \pm 1)/3, \pi (6m + 1)/6, (\pi (6n \pm 1)/6, \pi (6m - 1)/3)\}$, $(n, m \in \mathbf{Z})$.

306. $\{(\pi (1 - 2(n - m))/2, \pi n/2, (\pi (6(m - 2l) + 3 \pm 4)/6, \pi (3l \mp 1)/3)\}$ $(n, m, l \in \mathbf{Z})$. *Remark.* We must take either the upper signs or the lower signs in the formulas.

307. $\left\{\left(\dfrac{1}{2}(-1)^n \arcsin\dfrac{2}{5} + (-1)^m \arcsin\dfrac{4}{5} + \pi (n + m)\right),\right.$

$\dfrac{1}{2}\left.\left((-1)^n \arcsin\dfrac{2}{5} - (-1)^m \arcsin\dfrac{4}{5} + \pi (n - m)\right)\right\}$ $(n, m \in \mathbf{Z})$.

308. $\{(\pi (8n + 1)/4, \pi (8k + 5)/4\}$ $(n,\ k \in \mathbf{Z})$.

309. $\{(\pi (8n - 3)/4)^4, n \in \mathbf{N}\}$.

310. $\{(\pi (8k + 1)/2, \pi (4k + 1)/2)\}$ $(k \in \mathbf{Z})$.

311. $\{4\pi k\}$ $(k \in \mathbf{Z})$.

312. $\{(\pi (4m - (-1)^n)/8, (\pi n - \arctan (1/\sqrt{2}))/5\}$ $(n, m \in \mathbf{Z})$.

313. $\{((\pi (8n + 1) - 6)/4, 1\}$ $(n \in \mathbf{Z})$.

314. $\{(\pi/3, \pi/6), (5\pi/12, \pi/6), (\pi/12, 11\pi/6)\}$.

## Sec. 1.12

1. $a_1 = 13$, $d = -1$. *Solution.* For an arithmetic progression $a_n = a_1 + d(n-1)$. Expressing the summands in terms of $a_1$ and $d$, we obtain

$$\begin{cases} a_1 + 3d = 10, \\ 2a_1 + 9d = 17. \end{cases}$$

From the first equation, we find that $a_1 = 10 - 3d$. Then $2(10 - 3d) + 9d = 17 \Rightarrow d = -1, a_1 = 13$.

2. *Proof.* $S_n = 2n^2 - 3n$, $S_{n-1} = 2(n-1)^2 + 3(n-1) = 2n^2 - n - 1$; $a_n = S_n$ $-S_{n-1} = 4n + 1$, $a_{n-1} = 4(n-1) + 1 = 4n - 3$; $d = a_n - a_{n-1} = 4n + 1 - (4n - 3)$ $= 4$ is constant and, consequently, this sequence is an arithmetic progression.

3. 1, 9, 17. 4. 610. 5. $a_1 = -2$, $d = 1$. 6. 69, 87. 7. 44. 8. $(116k - 39)/90$.

9. $a_1 = -1, d = 2$. 10. 7. 11. 1.05, 1.1, 1.15, 1.2, 1.25. 12. 11.2, 18.4, 25.6, 32.8.

**13.** 164850. **14.** 7. **15.** $a_1 = 5, d = 4.$ **16.** 01, 0.2, 0.3, 0.4. **17.** 5. **18.** $a_1 = 4, d = 5$ or $a_1 = -79/7, d = -37/14.$ **19.** $d = -5/4.$ **20.** $d = 33/20.$ **21.** 26. **22.** $a_1 = 8, d = 4.$

**23.** They can; obtuse ; $\angle A = \arccos(-29/48)\,\pi - \arccos(29/48),$ $\angle B = \arccos(61/72), \angle C = \arccos(101/108).$ *Solution.* By the hypothesis we have $(a, b, c > 0)$

$$\begin{cases} 2\log b = \log a + \log c, \\ 2\,(\log 2b - \log 3c) = \log 3c - \log 2b \end{cases} \text{ or } \begin{cases} b^2 = ac, \\ 2b = 3c. \end{cases}$$

Solving the last system for $a$ and $b$, we get $a = 9c/4, b = 3c/2.$ Thus the triple of numbers that satisfies the hypothesis is $9c/4, 3c/2$ and $c\,(c \neq 0).$ For the segments of lengths $a = 9c/4,$ $b = 3c/2,$ and $c$ to form a triangle, it is sufficient to verify the fulfilment of the conditions (1) $a + b > c,$ (2) $b + c > a,$ (3) $a + c > b.$ But since $a + b = 15c/4 > c, b + c = 5c/2 > 9c/4 = a,$ and $a + c = 13c/4 > 3c/2 = b$ $(c > 0),$ a triangle with sides equal to $a, b,$ and $c$ exists, and since $a^2 > b^2 + c^2,$ it is obtuse. To find the angles of this triangle, we apply the cosine rule, whence we get

$$\cos A = \frac{b^2 + c^2 - a^2}{2bc} = -\frac{29}{48}, \qquad \cos B = \frac{a^2 + c^2 - b^2}{2ac} = \frac{61}{72},$$

$$\cos C = \frac{a^2 + b^2 - c^2}{2ab} = \frac{101}{108}.$$

Thus, bearing in mind that $\angle A, \angle B,$ and $\angle C$ are angles of a triangle, we obtain
$$\angle A = \arccos(-29/48) = \pi - \arccos(29/48),$$
$$\angle B = \arccos(61/72), \quad \angle C = \arccos(101/108).$$

**24.** $d = 24/11.$ **25.** $d = 12/5.$      **26.** 101100.

**27.** $(6 - \sqrt{6})/6, 1, (6 + \sqrt{6})/6.$      **29.** $a \in [12, +\infty).$

**30.** 19680.

**31.** $1/5, 1, 5, 25.$

**32.** 728. *Solution.* From the condition

$$\begin{cases} b_4 - b_1 = 52, \\ b_1 + b_2 + b_3 = 26, \end{cases} \Rightarrow \begin{cases} b_1 q^3 - b_1 = 52, \\ b_1 + b_1 q + b_1 q^2 = 26, \end{cases}$$

$$\Rightarrow \begin{cases} b_1(q^3 - 1) = 52, \\ b_1\,(1 + q + q^2) = 26, \end{cases} \Rightarrow \begin{cases} b_1\,(q-1)\,(q^2 + q + 1) = 52, \\ b_1\,(1 + q + q^2) = 26. \end{cases}$$

We divide the left-hand and right-hand sides of the equations and obtain $q - 1 = 2, q = 3.$ Substituting the value $q$ into the first equation, we get $b_1 = 2.$ Since $S_n = b_1\,(q^n - 1)/(q - 1),$ it follows that $S_6 = 2 \cdot (3^6 - 1)/(3 - 1) = 728.$

**33.** 8190.      **34.** 8/3 or 5000/3. **35.** 5, 10, 20 .

**36.** 12 or 108/7.      **37.** 121.      **38.** $b_1 = 3.5, q = -2.$

**39.** 0.      **40.** 2 , 6, 18 or 18, 6, 2.      **41.** 31.

**42.** 4, 12, 36, 108.      **43.** $-2/5.$

**44.** 1, 3, 9. *Solution.* By the hypothesis $b_1 + b_2 + b_3 = 13$ and $b_1^2 + b_2^2 + b_3^2 = 91,$ i.e. we have a system of equations

$$\begin{cases} b_1\,(1 + q + q^2) = 13, \\ b_1^2\,(1 + q^2 + q^4) = 91. \end{cases}$$

Squaring the left-hand and right-hand sides of the first equation of the system, we obtain
$$b_1^2\,(1 + q + q^2)^2 = 169 \Rightarrow b_1^2\,(1 + q^2 + q^4) + 2b_1^2 q\,(1 + q + q^2)$$
$$= 169 \Rightarrow b_1^2\,(1 + q^2 + q^4) + 2b_1 q \cdot b_1\,(1 + q + q^2) = 169$$
$$\Rightarrow \qquad 91 + 2b_1 q \cdot 13 = 169 \Rightarrow b_1 q = 3 \Rightarrow b_1 = 3/q$$

Substituting $b_1 = 3/q$ into the first equation of the system, we get after transformations, $3q^2 - 10q + 3 = 0$. Solving this equation, we find that $q = 3$ or $q = 1/3$. Then the corresponding values of $b_1$ are $b_1 = 1$ or $b_1 = 9$. Consequently, in the first case the required progression is 1, 3, 9 and in the second case, 9, 3, 1. The required numbers are 1, 3, 9 in both cases.

**45.** 0, 0 or 10/3, 4/3 or – 3/4, – 3/10. **46.** 6. **47.** 1/625 or 15625. **48.** 2. **49.** 5 or 20. **50.** 4, 8, 16. **51.** 10, 6, 2 or – 6, 6, 18. **52.** –2. **53.** 22.5 or 2.5. **54.** 2, 4, 8, 12 or 12.5, 7.5, 4.5, 1.5. **55.** 75/4, 45/4, 27/4, 9/4 or 3, 6, 12, 18. **56.** 12.2 (the given numbers are 27.8, 20, 12.2 or 17, 20, 23). **57.** $b_1 = 8$, $q = -0.5$ or $b_1 = 24/29$, $q = 3/2$. **58.** 186. **59.** 2, 6, 18. **60.** 27 or 3. **61.** 1, $3 - 2\sqrt{2}$, $3 + 2\sqrt{2}$. **62.** – 2 or 1. **63.** – 2. **64.** 1, 3, 9 or 1/9, 7/9, 49/9. **65.** 931. **66.** 8 problems, 127.5 min. **67.** $1 - 1/\sqrt{3}$, 3. **68.** 100/3. **69.** $(3 \pm \sqrt{5})/2$.

**70.** 2, –1/2, 1/8. **71.** 1/16. **72.** $b_1 = 6$, $q = -1/2$. **73.** If it is a geometric progression, then $x = \pm 2\sqrt{6}/11$; $S_4 = (3 \pm 2\sqrt{6})/3$; if it is an arithmetic progression, then $x = 1/2$; $S_4 = 27/11$. **74.** 4. **75.** 486. **76.** 4. **77.** $S^2/(2S - 1)$. **78.** $b_1 = 1$, $q = 1/3$.

**79.** 9/8. **80.** 2, 4, 8. **81.** $a = 3$, $b = 6$ or $a = 27$, $b = 18$. **82.** $x = 1 - \log_2 5$. **83.** 217/30. **84.** 4. **85.** 8/3. **86.** $b_1 = 405$, $q = -2/3$. **87.** $b_1 = 3/19$, $q = 17/19$. **88.** $b_1 = 2$, $q = 1/3$.

## Sec. 1.13

1. 50 km/h. *Solution.* Assume that $v$ km/h is the speed of the train before it stops. Then $(v + 10)$ km/h is its speed for the next 80 km. According to the time-table, the train should have covered this distance in $80/v$ h but it actually covered it in $80/(v + 10)$ h. It follows that
$$\frac{80}{v} - \frac{80}{v + 10} = \frac{16}{60}.$$
Transforming the last equation, we obtain $v^2 + 10v - 3000 = 0$, whence $v = 50$ km/h.

2. 25 km/h.    3. 16 km/h.    4. $(-c + \sqrt{c^2 + 16ac - 16bc})/2$ if $a > b$.

5. 6 km/h. *Solution.* By the hypothesis, the pedestrians met in 5 hours. Consequently, they covered $50:5 = 10$ km in a hour. If $v_1$ and $v_2$ are the speeds of the pedestrians, then $v_1 + v_2 = 10$. The first pedestrians travelled $5v_1$ km before they met and the second pedestrians covered that distance in $5v_1/(v_2 + 1)$ h. The second pedestrians travelled $5v_2$ km before they met and the first pedestrians covered that distance in $5v_2/(v_1 - 1)$ h. Since, after their meeting, the second pedestrian walked 2 h longer, it follows that
$$\frac{5v_1}{v_2 + 1} - \frac{5v_2}{v_1 - 1} = 2.$$
We have a system of equations
$$\begin{cases} v_1 + v_2 = 10, \\ \dfrac{5v_1}{v_2 + 1} - \dfrac{5v_2}{v_1 - 1} = 2. \end{cases}$$
Solving it, we get $v_1 = 6$.

6. 75.6 km/h, 147 m. *Hint.* If $l$ m is the length of the train, then $l/7 = (378 + l)/25$.

7. 720 km.

8. 30 km/h, 40 km/h. *Solution.* If $v_1$ and $v_2$ are the speeds of the trains, then $v_1 + v_2 = 70$, at 2 p.m. the distance between the trains is $2v_1$ km. Each

four this distance decreases by $(v_2 - v_1)$ km $(v_2 > v_1)$. Since the second train needed $20 - 14 = 6$ h to overtake the first train, it follows that $2v_1 = 6(v_2 - v_1)$. We have a system of equations

$$\begin{cases} v_1 + v_2 = 70, \\ 2v_1 = 6(v_2 - v_1), \end{cases}$$

whence we have $v_1 = 30$ km/h, $v_2 = 40$ km/h.

9. 4 km/h, 16km/h.

10. 18 km/h, 24 km/h.

11. $v_1 = 10$ km/h, $t_1 = 3.5$ h, $v_2 = 8$ km/h, $t_2 = 3$ h, or $v_1 = 14$ km/h, $t_1 = 2.5$ h, $v_2 = 12$ km/h, $t_2 = 2$ h.

12. 12 km/h, 16 km/h.

13. 30 km/h. *Solution.* Assume that $v_1, v_2,$ and $v_3$ are the speeds of the cars. Then, proceeding from the condition that three cars started from point $A$ in equal time intervals, we can write

$$\frac{|AB|}{v_1} - \frac{|AB|}{v_2} = \frac{|AB|}{v_2} - \frac{|AB|}{v_3}.$$

Since the second car arrived at $C$ an hour earlier than the first, we get a second equation :

$$\frac{120}{v_1} - \frac{120}{v_2} = 1.$$

It follows from the hypothesis that the third car covered $120 + 40 = 160$ km during the same time that the first car needed to cover $120 - 40 = 80$ km. Therefore, we obtain a third equation :

$$\frac{160}{v_3} = \frac{80}{v_1}.$$

Thus, we obtain a system of equations

$$\begin{cases} \dfrac{|AB|}{v_1} - \dfrac{|AB|}{v_2} = \dfrac{|AB|}{v_2} - \dfrac{|AB|}{v_3}, \\ \dfrac{120}{v_1} - \dfrac{120}{v_2} = 1, \\ \dfrac{160}{v_3} = \dfrac{80}{v_1}. \end{cases}$$

Solving it, we get $v_1 = 30$ km/h.

14. 18 km/h, 24 km/h.    15. 240 km.    16. $20\sqrt{3}$ km.

17. 35 km/h.    18. 4 h.    19. 24 km.

20. 0.03 m/s, 0.05 m/s. *Hint.* When solving problems on the motion of bodies along a circle, bear in mind the following. If two bodies begin moving from the same point in the same direction with different speeds, then the body moving with the higher speed can overtake the other body if the difference between the distances covered is equal to the length of the circle. In the given problem we assume that $v_1$ and $v_2$ are the speeds of the dots and $v_1 > v_2$. Then

$$\begin{cases} 60v_1 - 60v_2 = 1.2, \\ 15v_1 + 15v_2 = 1.2. \end{cases}$$

Solving the system, we get $v_1 = 0.03$ m/s, $v_2 = 0.05$ m/s.

21. 180 km.    22. 1 m/s.    23. 56 km.    24. 60 km/h.

25. $t_p = 21$ h, $t_g = 28$ h.

26. He walked for 2 hours and sailed for 6 hours.

27. The first car covered 200 km and the second car 100 km.

28. 4 km/h, 6 km/h.    29. 15 km.    30. 20 m/min, 15 m/min.

**31.** $(1 + \sqrt{3})/2$, $(2 \mp \sqrt{2})/2$. *Hint.* Designating $|KM|$ as $s$, and
$t_c / t_b = v_b / v_c = x$, we get from the hypothesis

$$\begin{cases} | s - (0.4s + 0.4s \cdot x) | = 4, \\ \left| s - \left( 0.5s + \dfrac{0.5s}{x} \right) \right| = 10. \end{cases}$$

After transformations, the system reduces to an equation

$$2|1.5 - x| = \left| 1 - \frac{1}{x} \right|.$$

Solving it, we get

$$x = (1 + \sqrt{3})/2, \quad x = (2 \mp \sqrt{2})/2.$$

**32.** $50 \text{ m}^3/\text{min.}$

**33.** The first mason would need $(5a - 2 - \sqrt{25a^2 + 4a + 4})/4$ h, the second would need $(5a + 2 + \sqrt{25a^2 + 4a + 4})/4$ h, and the third would need $(5a - 2 + \sqrt{25a^2 + 4a + 4})/2$ h.

**34.** 20 days, 30 days.

**35.** 30 h. *Solution.* Assume that the reservoir is filled through the first pipe in $x$ h. Then a $1/x$th part of the reservoir is filled through that pipe in an hour and a $1/(x+10)$th part of the reservoir is filled through the second pipe in an hour. Since, by the hypothesis, a $1/12$th part of the reservoir is filled through both pipes in an hour, we have

$$\frac{1}{x} + \frac{1}{x+10} = \frac{1}{12}.$$

Hence $x = 30$ h.

**36.** 20 h, 25 h.  **37.** 12 h, 15 h.  **38.** 11 h, 14 h.  **39.** 5, 7.

**40.** The productivity of the second worker is 1.5 times higher than that of the first worker.

**41.** 6 h, 9 h or 9 h , 6 h.

**42.** The third team. *Solution.* We designate the productivities of the first, the second, and the third, team as $x$, $y$, and $z$, respectively. By the hypothesis

$$\begin{cases} x + z = 2y, \\ y + z = 3x, \end{cases} \text{ or } \begin{cases} \dfrac{x}{z} + 1 = 2 \cdot \dfrac{y}{z}, \\ \dfrac{y}{z} + 1 = 3 \cdot \dfrac{x}{z}. \end{cases}$$

Solving this system, we get $x/z = 3/5$, $y/z = 4/5$, i.e.
$x : y : z = 3 : 4 : 5$. Thus the third team is the winner.

**43.** 8.  **44.** 10 days.

**45.** 3 h. *Solution.* Assume that $x$, $y$, $z$, $u$, and $v$ are the productivities of the first, the second, the third, the fourth, and the fifth worker respectively. It follows from the hypothesis that

$$\begin{cases} \dfrac{1}{x} + \dfrac{1}{y} + \dfrac{1}{z} = \dfrac{2}{15}, \\ \dfrac{1}{x} + \dfrac{1}{z} + \dfrac{1}{v} = \dfrac{1}{5}, \\ \dfrac{1}{x} + \dfrac{1}{z} + \dfrac{1}{u} = \dfrac{1}{6}, \\ \dfrac{1}{y} + \dfrac{1}{u} + \dfrac{1}{v} = \dfrac{1}{4}. \end{cases}$$

There are five unknowns in the system obtained, and four equations. We must not necessarily seek each unknown. According to the requirement of the problem, we must find the sum $\left(\dfrac{1}{x}+\dfrac{1}{y}+\dfrac{1}{z}+\dfrac{1}{u}+\dfrac{1}{v}\right)$. There are many ways of solving the problem. Here is one of them. We multiply the left-hand and right-hand sides of the fourth equation by 2 and then sum up the left-hand and right-hand sides of all the equations

$$3\left(\frac{1}{x}+\frac{1}{y}+\frac{1}{z}+\frac{1}{u}+\frac{1}{v}\right)=1, \quad \text{i.e.} \quad \frac{1}{x}+\frac{1}{y}+\frac{1}{z}+\frac{1}{u}+\frac{1}{v}=\frac{1}{3}.$$

**46** 8 h, 6 h.  **47.** 27.5 $(3-\sqrt{5})$ km/h, 27.5 $(\sqrt{5}-1)$ km/h. **48.** 9 days. **49.** 6 days, 12 days. **50.** 3 km/h. **51.** 10 km/h, 15 km/h. **52.** 4 km/h.
**53.** $\sqrt{3}/2$. *Solution.* Assume that $v_1$ is the speed of the launch in still water and $v_f$ is the speed of the river flow. From the hypothesis we have

$$\frac{|AB|}{2v_1+v_f}+\frac{|AB|}{2v_1-v_f}=\frac{1}{5}\left(\frac{|AB|}{v_1+v_f}+\frac{|AB|}{v_1-v_f}\right)$$

or

$$\frac{|AB|}{v_f}\left(\frac{1}{2\frac{v_1}{v_f}+1}+\frac{1}{2\frac{v_1}{v_f}-1}\right)=\frac{|AB|}{5v_f}\left(\frac{1}{\frac{v_1}{v_f}+1}+\frac{1}{\frac{v_1}{v_f}-1}\right).$$

Dividing both sides by $|AB|/v_f$, we obtain an equation in the unknowns $v_1/v_f$, whence $v_1/v_f=\sqrt{3/2}$.

**54.** 12.5 km/h. **55.** $v_f=3$ km/h, $v_1=9$ km/h. **56.** 20. **57.** 100. **58.** 3150, 3450.
**59.** 24. **60.**    20 of type $A$ and 30 of type $B$. **61.** 20 rows and 25 seats. **62.** 54, 75. **63.** 10. **64.** 40 kettles 2.5 roubles each and 24 kettles 1.5 roubles each. **65.** 18. **66.** 32.
**67.** 49. **68.** $(3+\sqrt{5})/2$ and $(1+\sqrt{5})/2$ or $(3-\sqrt{5})/2$ and $(1-\sqrt{5})/2$. **69.** 863 **70.**
6 and 54. **71.** 5 and 105 or 15 and 35. **72.** 3 / 5. **73.** 72. **74.** 54 and 45. **75.** 90 and 24. **76.** 78 and 13 or 26 and 39.
**77.** 51 and 34. *Solution.* Assume that $a$ and $b$ are the numbers and $d$ is their greatest common divisor, then $a=a_1d$, $b=b_1d$. The least common multiple of the numbers $a$ and $b$ is $a_1b_1d$, $a_1$ and $b_1$ having no common factors. Then $a_1b_1d=102=1\cdot2\cdot3\cdot17$, $(a_1+b_1)d=85=1\cdot5\cdot17$. Consequently, $d=1$ or $d=17$. In the first case, $ab=102$, $a+b=85$ and the numbers $a$ and $b$ are not integral. In the second case, $a_1b_1=6$, $a_1+b_1=5$, $a_1=3$, $b_1=2$ or $a_1=2$, $b_1=3$, then the given numbers are equal to 51 and 34.
**78.** 28 and 27 or 8 and 3. **79.** 137. **80.** 813. *Solution.* Assume that $x$, $y$, and $z$ are the digits constituting the given three-digit number. Then, $100x+10y+z$ is the given number, and $100z+10y+x$ is the number written by the same digits in the reverse order. By the hypothesis

$$\begin{cases} x^2+y^2+z^2=74, \\ x+z=2y, \\ (100x+10y+z)-(100z+10y+x)=495. \end{cases}$$

Solving the system, we find that $x=8$, $y=1$, $z=3$.
**81.** 13 or 31. **82.** 24. **83.** 3 and 7. **84.** 25%. **85.** 2 kg for $p>60, a$ kg, where $a\in[0,\ 2]$, for $p=60$, 0 kg for $0<p<60$. **86.** 6, 8. **87.** [0, 20] m/s. **88.** Two times.
**89.** 2.4 kg or 80%. **90.** 4 kg, 6 kg. **91.** 243 litres. **92.** 7 kg of the first alloy and 21 kg of the second. **93.** 99. **94.** 35.
**95.** 0.5 litre of glycerine and 3.5 litres of water. *Solution.* If we designate the capacity of the vessel as $v1$, then after the first operation (replacement of

two litres of glycerine by water) glycerine will occupy a $(v - 2)/v$ th part of the vessel. Pouring off 2 litres of the mixture, we get $(v - 2)\dfrac{v - 2}{v}$ litres of glycerine in the vessel, and after the addition of water, glycerine will occupy a $\left(\dfrac{v - 2}{v}\right)^2$ th part of the vessel. After the third operation glycerine will occupy a $\left(\dfrac{v - 2}{v}\right)^3$ th part of the vessel and the quantity of glycerine in the vessel will be equal to $v\left(\dfrac{v - 2}{v}\right)^3$ litres. By the hypothesis, there are 3 litres more of water in the vessel, i.e.

$$v\left(\frac{v - 2}{v}\right)^3 + 3.$$

Adding up these quantities, we get an equation

$$2v\left(\frac{v - 2}{v}\right)^3 + 3 = v,$$

whence

$$v^3 - 9v^2 + 24v - 16 = 0,$$

next we factor the left-hand side of this equation :

$$(v - 1)(v - 4)^2 = 0.$$

Hence $v = 4$ litres ($v = 1$ litre does not suit the hypothesis) and this means that the volume of glycerine is

$$v\left(\frac{v - 2}{v}\right)^3 = 0.5 \text{ litre}$$

**96** 170 kg. **97.** 4 kg. **98.** 25%. **99.** 166 roubles. **100.** 12%. **101.** $(12\sqrt{78} - 100)$%. **102.** 20%. **103.** 4 s. **104.** The altitude is $h \in [1, (5 - \sqrt{5})/2]$. **105.** $v_b \in [4, (8 + \sqrt{61})/3]$. **106.** $(a\sqrt{(u + w)^2 - v^2}/uv)$ s. **107.** $t = 2$ hours if $|BC| \geq 120$ km, $t = |BC|/60$ if $0 < |BC| < 120$ km. **108.** 2 h. **109.** $v \in (5, 15)$ (km/h). **110.** ln 3.6 h. **111.** 25 km. **112.** 3.75 km. **113.** 18 km/h. **114.** $x_1 = 1.25$, $x_2 = 2$, $x_3 = 0.5$. **115.** $-0.5$. **116.** 0.25. **117.** $x_1 = x_2 = 10$. **118.** 9 km away from $A$. **119.** 12 m/s, 28m.

**120.** 7/9. *Solution.* Let us introduce designations. The train which started from $A$ had an acceleration $a_1$, in $t_1$ h it attained the speed $v_1 = a_1 t_1$ with which it continued travelling until it reached point $B$; it took the first train, as well as the second, $t_4$ h to cover the distance. The second train which started from point $B$, had an acceleration $a_2 \neq a_1$ (we assume for definiteness that $a_2 < a_1$, otherwise the solution will be the same with the only difference that the first and second trains will change parts), in $t_2$ h it attained the speed $v_2 = a_2 t_2$ with which it continued travelling until it reached point $A$. We designate $a_2/a_1 = x < 1$. By the hypothesis the trains travelled with the same speed when they met (we designate the time of meeting as $t_3$). This is possible only under the condition that by the moment of their meeting one of the trains (the train whose acceleration was greater, the first train in our case) had moved uniformly with constant speed ($v_1 = a_1 t_1$ in our case) and the other train still had travelled with uniform acceleration (and had the speed $a_2 t_3$ at the moment $t_3$). These speeds are equal by the hypothesis, i.e. $a_2 t_3 = a_1 t_1$, whence

$$t_3 = t_1/x \tag{1}$$

The speed of the uniform motion of the first train is $v_1 = a_1 t_1$ and that of the second train is $v_2 = a_2 t_2 > v_1$, by the hypothesis $v_2 = 4v_1 / 3 \Rightarrow a_2 t_2 = 4a_1 t_1 / 3$ and, consequently,

$$t_2 = 4t_1 / 3x. \tag{2}$$

The distance covered by the second train till the moment of meeting, $s_1 = a_2 t_3^2 / 2$, is the distance traversed by the first train after the moment of meeting, i.e. $s_1 = v_1 (t_4 - t_3) = a_1 t_1 (t_4 - t_3)$. Consequently

$$\frac{a_2 t_3^2}{2} = a_1 t_1 (t_4 - t_3),$$

or, taking into account relation (1) and the fact that $a_2 = a_1 x$, we obtain

$$\frac{a_2 t_1^2}{2x^2} = a_1 t_1 \left( t_4 - \frac{t_1}{x} \right) \Rightarrow \frac{t_1^2}{2x} = t_1 t_4 - \frac{t_1^2}{x} \Rightarrow t_1 t_4 = \frac{3t_1^2}{2x};$$

since $t_1 \neq 0$, this equation yields

$$t_4 = 3t_1 / 2x. \tag{3}$$

By the hypothesis the trains covered the distance from $A$ to $B$ in the same time $t_4$. Let us calculate that distance for the first and the second train separately. For the first train

$$s = \frac{a_1 t_1^2}{2} + v_1 (t_4 - t_1) = \frac{a_1 t_1^2}{2} + a_1 t_1 \left( \frac{3t_1}{2x} - t_1 \right) = \frac{3 a_1 t_1^2}{2x} - \frac{a_1 t_1^2}{2}.$$

For the second train

$$s = \frac{a_2 t_2^2}{2} + v_2 (t_4 - t_2) = \frac{a_2 \cdot 16 t_1^2}{18 \cdot x^2} + a_2 t_2 \left( \frac{3t_1}{2x} - \frac{4t_1}{3x} \right) = \frac{8 a_2 t_1^2}{9x^2}$$

$$+ a_2 \cdot \frac{4t_1}{3x} \left( \frac{3t_1}{2x} - \frac{4t_1}{3x} \right) = \frac{2 a_2 t_1^2}{x^2} - \frac{8 a_2 t_1^2}{9x^2} = \frac{10 a_2 t_1^2}{9x^2} = \frac{10 a_1 t_1^2}{9x}.$$

Equating these values to each other and dividing the right-hand and left-hand sides by $a_1 t_1^2$, we get

$$\frac{3}{2x} - \frac{1}{2} = \frac{10}{9x} \Rightarrow \frac{27}{18x} - \frac{20}{18x} = \frac{1}{2} \Rightarrow x = \frac{7}{9} = \frac{a_2}{a_1}.$$

**121.** 10 km/h. *Solution.* If $|AC| = x$ km, then

$$\frac{x}{v_0} + \frac{120 - x}{v_1} = \frac{x}{v_1} + \frac{120 - x}{v_0} \Rightarrow x = 60.$$

If $t$ is the time it takes the motor-cyclist, travelling with the speed $v_0$ km/h, to cover the distance $|AC|$, then

$$v_0 t = 60, \, v_0 (8 - t) + \frac{a (8 - t)^2}{2} = 60.$$

Taking into account that $a = 2v_0$, we get

$$v_0^2 - 15v_0 + 50 = 0, \quad v_0 = 5 \text{ or } v_0 = 10.$$

The value $v_0 = 5$ does not satisfy the hypothesis since in that case 12 hours are needed to cover the distance $|AC|$. Thus $v_0 = 10$ km/h.

## Sec. 1.14

**1.** $\cos (\pi + 2\pi k) + i \sin (\pi + 2\pi k), \, k \in \mathbf{Z}$. **2.** $\cos \left( 2\pi k - \frac{\pi}{2} \right) + i \sin \left( 2\pi k - \frac{\pi}{2} \right), \, k \in \mathbf{Z}$.

**3.** $2 \left( \cos \left( 2\pi k + \frac{5}{6}\pi \right) + i \sin \left( 2\pi k + \frac{5}{6}\pi \right) \right), \, k \in \mathbf{Z}$.  **4.** $2 \left( \cos \left( 2\pi k + \frac{7}{6}\pi \right) + \right.$

$i \sin \left( 2\pi k + \frac{7}{6}\pi \right), \, k \in \mathbf{Z}$.  **5.** $5 [\cos (2\pi k + \pi + \arctan (4/3)) + i \sin (2\pi k + \pi +$

arctan $(4/3))$, $k \in \mathbf{Z}$.

**6.** $5(\cos(2\pi k - \arctan(4/3)) + i \sin(2\pi k - \arctan(4/3)))$, $k \in \mathbf{Z}$. **7.** $5(\cos 2\pi k + i \sin 2\pi k)$, $k \in \mathbf{Z}$. **8.** $\sqrt{3} \times \left(\cos\left(2\pi k - \dfrac{\pi}{6}\right) + i \sin\left(2\pi k - \dfrac{\pi}{6}\right)\right)$, $k \in \mathbf{Z}$. **9.** $\cos(360° k$

$+58°) + i \sin(360° k + 58°)$, $k \in \mathbf{Z}$. **10.** $\cos(360° k - 12°) + i \sin(360° k - 12°)$, $k \in \mathbf{Z}$.

**11.** $\cos(360° k + 200°) + i \sin(360° k + 200°)$, $k \in \mathbf{Z}$. **12.** $\cos\left(2\pi k - \dfrac{\pi}{2} + \alpha\right)$

$+ i\sin\left(2\pi k - \dfrac{\pi}{2} + \alpha\right)$, $k \in \mathbf{Z}$. **13.** $\cos\left(2\pi k + \dfrac{\pi}{2} - \alpha\right) + i \sin\left(2\pi k + \dfrac{\pi}{2} - \alpha\right)$, $k \in \mathbf{Z}$.

**14.** $\cos\left(2\pi k + \dfrac{3\pi}{2} - \alpha\right) + i \sin\left(2\pi k + \dfrac{3\pi}{2} - \alpha\right)$, $k \in \mathbf{Z}$. **15.** $\dfrac{1}{\cos\alpha}(\cos(2\pi k + \alpha)$

$+ i \sin(2\pi k + \alpha))$, if $2\pi n - \dfrac{\pi}{2} < \alpha < 2\pi n + \dfrac{\pi}{2}; \dfrac{1}{|\cos\alpha|}(\cos(2\pi k + \pi + \alpha)$

$+ i \sin(2\pi k + \alpha))$, if $2\pi n + \dfrac{\pi}{2} < \alpha < 2\pi n + \dfrac{3\pi}{2}$; $k, n \in \mathbf{Z}$. **16.** $\dfrac{1}{\sin\alpha}\left(\cos\left(2\pi k + \dfrac{\pi}{2} - \alpha\right)\right.$

$+ i \sin\left(2\pi k + \dfrac{\pi}{2} - \alpha\right)$, if $2\pi n < \alpha < \pi(2n+1)$; $\dfrac{1}{|\sin\alpha|}\left(\cos\left(2\pi k + \dfrac{3\pi}{2} - \alpha\right)\right) + i \sin$

$\left(2\pi k + \dfrac{3\pi}{2} - \alpha\right)$ if $\pi(2n+1) < \alpha < 2\pi(n+1)$; $n, k \in \mathbf{Z}$. **17.** $(1 + i\sqrt{3})/2$, **18.** $-16$. **9.** $8i$.

**20.** $8$. **21.** $-2^{10}$. **22.** $2^n \cos^n(\varphi/2)(\cos(n\varphi/2) + i \sin(n\varphi/2))$. **23.** $2^n \sin^n$

$(\varphi/2)(\cos(n(\pi - \varphi)/2 + i \sin(n(\pi - \varphi)/2))$. **25.** $\cos 3\alpha = 4\cos^3\alpha - 3\cos\alpha$,

$\cos 4\alpha = 8\cos^4\alpha - 8\cos^2\alpha + 1$, $\sin 3\alpha = 3\sin\alpha - 4\sin^3\alpha$. **26.** $\dfrac{\sqrt{2}}{2} + i\dfrac{\sqrt{2}}{2}, -\dfrac{\sqrt{2}}{2} - i\dfrac{\sqrt{2}}{2}$.

**27.** $i; -\dfrac{\sqrt{3}}{2} - \dfrac{i}{2}; \dfrac{\sqrt{3}}{2} - \dfrac{i}{2}$. **28.** $\dfrac{1}{2} + \dfrac{\sqrt{3}}{2}i; -1; \dfrac{1}{2} - \dfrac{\sqrt{3}}{2}i$. **29.** $\dfrac{1}{2} + i\dfrac{\sqrt{3}}{2}; -$

$\dfrac{\sqrt{3}}{2} + \dfrac{i}{2}; -\dfrac{1}{2} - i\dfrac{\sqrt{3}}{2}; \dfrac{\sqrt{3}}{2} - \dfrac{i}{2}$. **30.** $\cos(2\pi k/n) + i \sin(2\pi k/n)$, $k = 0, 1, 2, \dots, n-1$.

**31.** $\{-(1+i)/2; (3-5i)/2\}$. **32.** $\{\cot(k\pi/n) \mid k = 1, 2, \dots, n-1\}$. **33.** *Hint.* It is

sufficient to prove that the roots of the polynomial $x^2 + x + 1$ are the roots of the

polynomial $x^{3n} + x^{3m+1} + x^{3k+2}$. **34.** The straight lines $y = \pm x$. **35.** The straight lines

$y = 0$ and $x = 0$. **36.** The ray $y = 0, x \le 0$. **37.** $\{-1, 0\}$. **38.** $\{0, -i, i\}$. **39.** $\{0, -i, i\}$. **44.** A

point of the straight line $y = 2x + (3/2)$. **45.** Points of the straight line $y = -x$. **46.** Points

of the straight line $x = -1/2$. **47.** Points of the straight line $y = 1$. **48.** Points of the plane

which lie above the straight line $y = x$ (which satisfy the inequality $y > x$). **49.** All the

points of the plane which lie under the straight line $2x + 4y + 3 = 0$. **50.** The points

$x > 0$, i.e. all the points of the right–hand half–plane. **51.** Points of an open circle with

centre at the point $(0, -1)$ and of radius 1. **52.** Points of an open circle with centre at the

point $(0, 1/2)$ of radius 1. **53.** Points of an open ring bounded by circles with radii 1 and

2 and with centre at the point $(-1, 1)$. **54.** Points of an open ring bounded by circles with

radii $1/3$ and 1 and with centre at the point $(0, -1/3)$.

55. Points of an open ring bounded by circles with radii 1/2 and 1 and with centre at the point $(-1/2, -1/2)$.

56. All the points of the plane which lie under the parabola $y = (x^2 + 1)/2$.

57. All the points of the plane outside the straight line $x = 1$ which lie under the parabola $y = -(x^2 + 1)/2$.

58. $-1/8$

# Part-2

## Sec. 2.1

**1.** $q = 2/3$. **2.** $\sqrt{2} - 1$. **3.** $1/2$. **4.** $7/8$. **5.** $1/2$. **6.** $1$. **7.** $4/3$. **8.** $0$. **9.** $0$. **10.** $0$. **11.** $0$. **12.** $-2/5$. **13.** $1/2$. **14.** $-4/7$. **15.** $48$. **16.** $6$. **17.** $10$. **18.** $4/3$. **19.** $1/8$. **20.** $-5/3$. **21.** $8/13$. **22.** $-1/2$. **23.** $3/4$. **24.** $0$. **25.** $1$. **26.** $1/3$. **27.** $m/n$. **28.** $n(n+1)/2$. **29.** $1/4$. **30.** $1/6$. **31.** $1/2$. **32.** $1/4$. **33.** $1/16$. **34.** $4/3$. **35.** $-2$. **36.** $2/3$. **37.** $1/12$. **38.** $1/4$. **39.** $3/2$. **40.** $-2/9$. **41.** $3$. **42.** $1/2$. **43.** $1/3$. **44.** $1$. **45.** $5$. **46.** $1/2$. **47.** $2$. **48.** $-\sin a$. **49.** $-3$. **50.** $\sqrt{2}$. **51.** $-1/2$. **52.** $-1/5$. **53.** $6/5$. **54.** $0$. **55.** $5^{-5}$. **56.** $0$. **57.** $-5/2$. **58.** $-1/8$. **59.** $1/2$.

## Sec. 2.2

**1.** $2\sin 2x$. **2.** $2x\cos 2x + (1+x)\sin 2x + 2$. **3.** $f(x) = 4/(x^2 - 4)$, $f'(x) = -8x/(x^2 - 4)^2$. **4.** $f(x) \equiv 3, f'(x) \equiv 0$. **9.** $\pi(4n+1)/4, n \in \mathbf{Z}$. **10.** $x = m\pi/6$, $m \in \mathbf{Z}, x = \{(\pm\arccos(2/3) + 2\pi n)/6 \mid n \in \mathbf{Z}\}$. **11.** $x = \{\pi(2n+1)/4, (\pi m + (-1)^m$ $\arcsin(1/3))/2 \mid n, m \in \mathbf{Z}\}$. **12.** $x = \{0, 2\pi(3m \pm 1)/3 \mid m \in \mathbf{Z}\}$. **13.** $x = \{2\pi(6n \pm 1)/3 \mid n \in \mathbf{Z}\}$. **14.** $x = \{2\pi(2n+1)/3 \mid n \in \mathbf{Z}\}$. **15.** $x = \{\pi(4m+1)/8, \pi(12k+1)/12 \mid m, k \in \mathbf{Z}\}$. **16.** $x = \{\pi m/4, \pi(6n \pm 1)/3 \mid m, n \in \mathbf{Z}\}$. **17.** $x \in (-\infty, 0) \cup (2, +\infty)$. **18.** $x \in (-\infty, 0) \cup (1, +\infty)$. **19.** $x = [-1, 0)$. **20.** $x \in (5, +\infty)$. **21.** $x \in (0, +\infty)$. **22.** $y_1 = C_1 \cdot e^{(\sqrt{3}-2)x}, C_1 \in \mathbf{R}, y_2 = C_2 \cdot e^{-(\sqrt{3}+2)x}, C_2 \in \mathbf{R}$. **23.** $(x+2)(3x+4)$. **24.** $-\dfrac{3\tan^2 x}{\cos^2 x}$ $x)\cos(\cos^2(\tan^3 x))$. **25.** $-(x^2 + 4x + 9)/x^4$. **26.** $2(1+x^2)/(1-x^2)^2$. **27.** $x^2 e^x$. **28.** $(6\log^2(x^2))/x \ln 10$. **29.** $-(6 + 3\sqrt{x} + 2\sqrt[3]{x^2})/6x^2$. **30.** $x/(x^4 - 1)$. **31.** $a^2/(a^2 - x^2)^{3/2}$. **32.** $1/\sin x$. **33.** $x^2/(\cos x + x\sin x)^2$. **34.** $-1/\cos x$. **35.** $(1 + 2x^2)/\sqrt{1 + x^2}$. **36.** $\cos x\cos(\sin x)\cos(\sin(\sin x))$. **37.** $x^2\sin x$. **38.** *Proof.* Let us consider the function $f(x) = e^x - x - 1$. Since $f(0) = 0, f(x)$ is a continuous function, $f'(x) = e^x - 1 > 0$ for $x > 0$, it follows that the function $f(x)$ increases for all $x > 0$; therefore, $f(x) > f(0)$ for all $x > 0$, i.e. $e^x - x - 1 > 0$ for all $x > 0$.

39. The function decreases for $x \in (-\infty, 2/3) \cup (2, +\infty)$ and increases for $x \in (2/3, 2)$.

40. The function decreases for $x \in (-1, 0)$ and increases for $x \in (-\infty, -1) \cup (0, +\infty)$.

41. The function decreases for $x \in (-1, 2)$ and increases for $x \in (-\infty, -1) \cup (2, +\infty)$.

42. The function decreases for $x \in (-\infty, 0)$ and increases for $x \in (0, +\infty)$.

43. The function decreases for $x \in (\log_2(4/3), 1)$ and increases for $x \in (-\infty, \log_2(4/3)) \cup (1, +\infty)$.

44. A local maximum for $x = -1$, a local minimum for $x = 3$. The function decreases for $x \in (-1, 3)$ and increases for $x \in (-\infty, -1) \cup (3, +\infty)$.

45. Local minima for $x = -4$ and $x = 1$, a local maximum for $x = 0$. The function decreases for $x \in (-\infty, -4) \cup (0, 1)$ and increases for $x \in (-4, 0) \cup (1, +\infty)$.

**46.** Local minima for $x = -1$ and $x = 4$, a local maximum for $x = 0$. The function decreases for $x \in (-\infty, -1) \cup (0, 4)$ and increases for $x \in (-1, 0) \cup (4, +\infty)$.

**47.** Local minima for $x = -1/2$ and $x = 1/2$, a local maximum for $x = 0$. The function decreases for $x \in (-\infty, -1/2) \cup (0,1/2)$ and increases for $x \in (-1/2, 0) \cup (1/2, +\infty)$.

**48.** The function increases for $x \in (-\infty, +\infty)$.

**49.** A local maximum for $x = 16/5$ a point of discontinuity for $x = 0$. The function decreases for $x \in (-\infty, 0) \cup (16/5, +\infty)$ and increases for $x \in (0,16/5)$.

**50.** A minimum for $x = 2/3$. The function decreases for $x \in (-\infty, 2/3)$ and **increases** for $x \in (2/3, +\infty)$.

**51.** A maximum for $x = 1/3$. The function decreases for $x \in (1/3, +\infty)$ and increases for $x \in (-\infty, 1/3)$.

**52.** A minimum for $x = 1$. The function decreases for $x \in (0,1)$ and increases for $x \in (1, \infty)$.

**53.** A minimum for $x = -1/4$.

**54.** Critical points $x = 0$ and $x = 1$; a minimum for $x = 1$.

**55.** A minimum for $x = \pm\sqrt{5}$, a local maximum for $x = 0$.

**56.** A minimum for $x = 2$ and $x = 3$, a local maximum for $x = 5/2$.

**57.** A local maximum for $x = 0$ and a local minimum for $x = 1$.

**58.** A local maximum for $x = 0$ and a local minimum for $x = \sqrt[3]{2/5}$.

**59.** A maximum for $x = -3$, a minimum for $x = 3$.

**60.** A maximum for $x = -\sqrt{2}$, a minimum for $x = \sqrt{2}$.

**61.** Minima for $x = \pm 5$.

**62.** A minimum for $x = -1$.

**63.** A minimum for $x = -1/\sqrt{2}$ and a maximum for $x = 1/\sqrt{2}$.

**64.** $y = -9/4$ for $x \in \{\pi(12n-1)/6, \pi(12m-5)/6 \mid n,m \in \mathbf{Z}\}$.

**65.** For $a = -9/5$ and $b \in (36/5, +\infty)$ and for $a = 81/25$ and $b \in (400/243, +\infty)$.

*Solution.* Since, the coefficient in $x^3$ is positive, the maximum, provided that it exists, lies to the left of the minimum. To find the extremal points, we seek the derivative of the given function, equate it to zero, and find the roots of the quadratic equation; they are $x_1 = -9/(5a)$ and $x_2 = 1/a$. If $a < 0$, then $x_2 < x_1$ and, consequently, $x_2 = x_0$, i.e. $1/a = -5/9 \Rightarrow a = -9/5$. Then $x_1 = -9/(5a) = 1$. At that point $f(x)$ has a local minimum which must be positive. We have $f(1) = -\dfrac{36}{5} + b > 0$, i.e. $b > 36/5$. If $a > 0$, then $x_1 < x_2$ and, consequently, $x_1 = x_0$, i.e.

$-9/(5a) = -5/9 \Rightarrow a = 81/25$. Then, the point of minimum is $x_2 = 1/a = 25/81$. At that point $f(25/81) = -\dfrac{400}{243} + b > 0$, i.e. $b > 400/243$.

Thus, the extrema of $f(x)$ are positive for $a = -9/5$ and $b \in (36/5, +\infty)$ and for $a = 81/25$ and $b \in (400/243, +\infty)$.

**66.** For $a = -2$ and $b \in (-\infty, -11/27)$ and for $a = 3$ and $b \in (-\infty, -1/2)$.

**67.** For $a = -1/3$ and $b \in (-\infty, -5/9)$ and for $a = 1$ and $b \in (-\infty, -1)$.

**69.** If $p > 0$, then $a \in (p,(32p^3 + 27p)/27)$; if $p = 0$, then $a \in \phi$; if $p < 0$, then $a \in ((32p^3 + 27p)/27, p)$. *Solution.* For the equation $x^3 + 2px^2 + p = a$ to have three real roots, the function $f(x) = x^3 + 2px^2 + p - a$ must have a local maximum and a local minimum with $f(x_{\max}) > 0$ and $f(x_{\min}) < 0$. Since, the coefficient in $x^3$ is positive, we have $x_{\max} < x_{\min}$. Let us find the extrema of that function. The derivative $f'(x) = 3x^2 + 4px$. It vanishes for $x_1 = 0$ and $x_2 = -4p/3$. Three cases are possible here. If $p > 0$, then $x_2 < x_1$ and, consequently, $f(x_2) = \dfrac{32}{27}p^3 + p - a > 0$, and $f(x_1) = p - a < 0$,

whence $p < a < (32p^3 + 27p)/27$. If $p = 0$, then $x_1 = x_2$, and the function has no extrema; $f'(x)$ is positive everywhere except for $x = 0$ and the function increases monotonically. Thus, in this case there are no values of $a$ for which the given equation would have three real roots. If $p < 0$, then $x_1 < 0$ and, consequently, $f(x_1) = p - a > 0$, and $f(x_2) = \dfrac{32}{27}p^3 + p - a < 0$, whence $(32p^3 + 27p) / 27 < a < p$.

**70.** $y = 24 \ln 3$ for $x = 3$. **71.** $y = 1 + 8 \ln 5$ for $x = 2$. **72.** $y = 6(1 - \ln 3)$ for $x = 2$.
**73.** For $x = 0$. **74.** $y(1) = -8/3 = \min\limits_{[0,2]} f(x)$, $y(2) = 8/3 = \max\limits_{[0,2]} f(x)$. **75.** $\min\limits_{[0,3]}$
$f(x) = f(0) = f(3) = 5$, $\max\limits_{[0,3]} f(x) = f(2) = 9$. **76.** $\min\limits_{[0,3]} f(x) = f(0) = 0$, $\max\limits_{[0,3]}$
$f(x) = f(3) = 9$. **77.** $\min\limits_{[-1,4]} f(x) = f(-1) = -10$, $\max\limits_{[-1,4]} f(x) = f(4) = 50$. **78.** $\min\limits_{[-3,3]}$
$f(x) = f(1) = 23$, $\max\limits_{[-3,3]} f(x) = f(3) = 75$. **79.** $\min\limits_{[-1,5]} f(x) = f(1) = -6$, $\max\limits_{[-1,5]} f(x) = f(5)$
$= 266$. **80.** $\min\limits_{[-1,3]} f(x) = f(-1) = -9$, $\max\limits_{[-1,3]} f(x) = f(3) = 6$. **81.** $\min\limits_{[-2,2]} f(x) = f(-2) = -24$,
$\max\limits_{[-2,2]} f(x) = f(2) = 4$. **82.** $\min\limits_{[-3,6]} f(x) = f(3) = -57$, $\max\limits_{[-3,6]} f(x) = f(6) = 132$. **83.** $\min\limits_{[0,3]} f(x)$
$= f(2) = 1$, $\max\limits_{[0,3]} f(x) = f(0) = f(3) = 5$. **84.** $\min\limits_{[0,3]} f(x) = f(2) = -25$, $\max\limits_{[0,3]} f(x) = f(3)$
$= 0$. **85.** $\min\limits_{[-1,1]} f(x) = f(-1) = f(1) = -16$, $\max\limits_{[-1,1]} f(x) = f(0) = -9$. **86.** $\min\limits_{[-1,1]} f(x)$
$= f(-1) = 1$, $\max\limits_{[-1,1]} f(x) = f(1) = 3$. **87.** $\min\limits_{[3/4,2]} f(x) = f(1) = 1$, $\max\limits_{[3/4,2]} f(x) = f(2) = \sqrt[3]{4/3}$.
**88.** $\min\limits_{[-1,2]} f(x) = f(0) = 2 / \ln 2$, $\max\limits_{[-1,2]} f(x) = f(2) = 17 / (4 \ln 2)$. **89.** $\min\limits_{[-1,1]} f(x) = f(0)$
$= 0$, $\max\limits_{[-1,1]} f(x) = f(1) = 24$. **90.** $\min\limits_{[-1,1]} f(x) = f(-1) = f(1) = 4$, $\max\limits_{[-1,1]} f(x) = f(0) = 5$. **91.**

$\min\limits_{[0,\pi]} f(x) = f(\pi / 2) = 0$, $\max\limits_{[0,\pi]} f(x) = f(\arcsin(1/4)) = f(\pi - \arcsin(1/4)) = 9/8$. **92.**

$\min\limits_{[\pi/3,\,3\pi/2]} f(x) = f(\pi / 3) = -1/2$, $\max\limits_{[\pi/3,\,3\pi/2]} f(x) = f(-\pi) = 22$. **93.** $\min\limits_{[0,\,p/2]} f(x) = f(\pi/2)$
$= 3\pi/2$, $\max\limits_{[0,\,p/2]} f(x) = f(\arcsin((\sqrt{57} - 5)/4))) = (15 + \sqrt{57})\sqrt{10\sqrt{57} - 66}/16 + 3$
$\arcsin((\sqrt{57} - 5)/4)$.
**94.** $\max\limits_{[-2,1]} f(x) = f(-2) = -1 / (3b^2 - 8b + 16)$      for      $b \in (-\infty, 2]$,      $\max\limits_{[-2,1]} f(x)$.
$= f(0) = -1/3b^2$    for    $b \in [2; +\infty]$.      *Solution.* We note that
$f(+a) = f(-a)$. To find the greatest value of $f(x)$ on the interval $[-2,1]$, we seek critical points. To do that, we find

$$f'(x) = \frac{4x(x^2 - b)}{(2b^2 x^2 - x^4 - 3b^2)^2} \quad (1)$$

If $b < 0$, then $x^2 - b > 0$ for any $x \in \mathbf{R}$. Therefore, $f'(x) = 0$ only for $x = 0$, with the derivative $f'(x) < 0$ for $x < 0$ and the derivative $f'(x) > 0$ for $x > 0$. Consequently, at the point $x = 0$ the function $f(x)$ has a minimum and attains the greatest value at the left–hand terminal point (taking into account the symmetry and the fact that $|-2| > 1$). Consequently, in this case the greatest value of the function is $f(-2) = -1 / (3b^2 - 8b + 16)$. For $b = 0$ we have $f(x) = -1 / x^4$, for $x < 0$ the function decreases and for $x > 0$ it increases. Consequently, in this case the function attains the greatest value for $x = -2$, namely, $f(-2) = -1 / (3b^2 - 8b + 16) = -1/16$. For $b > 0$ the derivative has three critical points $x_1 = -\sqrt{b}, x_2 = 0, x_3 = \sqrt{b}$. Verification shows that $f'(x) < 0$ for $x \in (-\infty, -\sqrt{b}) \cup (0, \sqrt{b})$ and $f'(x) > 0$ for $x \in (-\sqrt{b}, 0) \cup (\sqrt{b}, +\infty)$.

Thus, for $x = 0$ we have a local maximum $f(0) = -1/3b^2$. For this value to be the greatest, we must require that

$$-\frac{1}{3b^2 - 8b + 16} \le -\frac{1}{3b^2} \Rightarrow \frac{1}{3b^2 - 8b + 16} \ge \frac{1}{3b^2}$$
$$\Rightarrow 3b^2 - 8b + 16 \le 3b^2 \Rightarrow -8b + 16 \le 0 \Rightarrow b \ge 2.$$

Thus, $\max_{[-2,1]} f(x) = f(0) = -1/3b^2$ for $b \in [2, +\infty)$,

$\max_{[-2,1]} f(x) = f(-2) = -1/(3b^2 - 8b + 16)$ for $b \in (-\infty, 2]$.

95. $\max_{[-2,1]} f(x) = f(-2) = 16 - 24b + b^2$ for $b \in (-\infty, 2/3]$,

$\max_{[-2,1]} f(x) = f(0) = b^2$ for $b \in [2/3, +\infty)$.

96. $\min_{[-\pi/2, 3\pi/8]} f(x) = f(-\pi/2) = -1 - \sqrt{3}$,

$\min_{[-\pi/2, 3\pi/8]} f(x) = f(3\pi/8) = ((3\pi - 4)\sqrt{2} - 4\sqrt{3})/4.$

97. $\min_{[-\pi,\pi]} f(x) = -2\sqrt{3}/9 > -7/18.$

98. $\max_{[-\pi,\pi]} f(x) = 4\sqrt{3}/9 < 0.77.$

99. $y = -2x$, $y = 2x - 4$. 100. $y = -4x + 3$. 101. $y = 4x + 1$. 102. $y = 4x$, $y = -4x + 16$. 103. $y = -2x + 5$. 104. $y = (3x + 10 - 6\ln 2)/8$. 105. $y = -(2x - 7)/16$. 106. $y = ((25\ln 3)x + 28 - 25\ln 3)/9$. 107. $y = 1$. 108. $y = 4x + 2$. 109. $(-(2 + \sqrt{3})/2, (9 - 4\sqrt{3})/4)$. 110. $k = \pm 4$. 111. (2, 3). 112. (1, 1).
113. (2, 4); the equation of the tangent is $y = 4x - 4$; the equation of the secant is $y = 4x - 3$.
114. On the second curve $y = \varphi(x)$ at the point $((k + 4)/6, (k^2 - 4)/12)$ the tangent $y = kx - (k^2 + 8k + 2)/12$ is parallel to the tangent $y = kx \mp 2((k+1)/3)^{3/2}$ at the points $(\pm\sqrt{(k+1)/3}, -1 \pm (k-2)\sqrt{3(k+1)}/9)$ of the first curve $y = f(x)$ for any $k \in [-1, +\infty)$.
Solution. We seek the equation of the tangent in the form $y - y_1 = k(x - x_1)$, where $(x_1, y_1)$ is a point on the curve at which the tangent is drawn with the slope $k$. The coordinates of the point of tangency have been found from the equations $y_1 = f(x_1), f'(x_1) = k$ for the first curve and from the equations $y_2 = \varphi(x_2), \varphi'(x_2) = k$ for the second curve. For the curve $y = f(x)$ we have $y_1' = f'(x_1) = 3x_1^2 - 1 = k$, whence we see that $k \ge -1$, i.e. if we assign the value $k \in (-\infty, -1)$, then there is no point on the curve $y = f(x)$ at which the tangent would have such a slope. For the second curve $y = \varphi(x)$ we have $y'_2 = \varphi'(x_2) = 6x_2 - 4 = k$ and $k$ can assume any values. Thus we assume that $k \in [-1, +\infty)$. Then for the curve $y = f(x)$ we get $x_1 = \pm\sqrt{(k+1)/3}$ for the given $k$, and from the equation of the curve $y_1 = -1 \pm (k - 2)\sqrt{3(k+1)}/9$ and the equation of the tangent at the points $(\pm\sqrt{k+1}/3, -1 \pm (k-2)\sqrt{3(k+1)}/9)$ is $y = kx \mp 2(\sqrt{(k+1)}/3)^3$. For the curve $y = \varphi(x)$, we get $x_2 = (k + 4)/6$ and $y_2 = (k^2 - 4)/12$ for the same value of $k$, and the equation of the tangent at the point $((k + 4)/4, (k^2 - 4)/12)$ is $y = kx - (k^2 + 8k + 2)/12$. At the points we have found the tangents are parallel for any $k \in [-1, +\infty)$.
115. The acute angle between the tangents is equal to arctan $(1/2)$. 116. $\pi - 2$ arctan $\sqrt{8}$. 117. $\pi - 2$ arctan $\sqrt{24}$. 118. $y = \dfrac{3 + \sqrt{7}}{2}x - \dfrac{8 + 3\sqrt{7}}{8},$

$$y = \frac{3-\sqrt{7}}{2}x - \frac{8-3\sqrt{7}}{8}.$$

**119.** $y = 2 - 5x$. **120.** $(0,-1),(4,3)$. **121.** $(k\pi,(2k\pi -1+ 32(-1)^k)/2)k \in \mathbf{Z}$. **122.** $0°$.

**123.** $5\sqrt{5}$. **124.** $V = 2\pi a^3 / 9\sqrt{3}$. **125.** $\alpha = 2\pi(1 - \sqrt{2/3})$. **126.** $\alpha = \pi/3, S = 3l^2\sqrt{3}/16$.

**127.** $\dfrac{R^3 \tan\varphi \tan\alpha}{3\tan^3(\varphi/2)\tan(\alpha/2)}, \alpha = \dfrac{\pi}{3}$. **128.** $\dfrac{R^2}{\tan^2\left(\dfrac{\pi}{4} - \dfrac{\alpha}{2}\right)\tan\alpha} = \dfrac{2R^2(1+\sin\alpha)^2}{\sin 2\alpha}, \alpha = \dfrac{\pi}{6}$.

**129.** $(\pi V \sin^2\alpha\, \cos\alpha)/2, \alpha = \arctan\sqrt{2} = \arccos(1/\sqrt{3})$. **130.** $y = -(3x-5)/9$. **131.**

$1/\sqrt{5}$. **132.** $\sqrt{5S/3}, \sqrt{4S/15}$. **133.** 21 cm and 28 cm. **134.** 16/3 and 128/3. **135.** $a = 4$

m and $h = 2$ m.

**136.** Trivial solution: $\alpha = \arctan(y_0/x_0)$ if the point $(x_0, y_0)$ lies in the first or third quadrant and $\alpha = \pi + \arctan(y_0/x_0)$ if that point lies in the second or fourth quadrant; in that case the length of the segment is 0. Non trivial solution: $\alpha = \pi -$ arctan $\sqrt[3]{y_0/x_0}$ if the point lies in the first or third quadrant and $\alpha = -$ arctan $\sqrt[3]{y_0/x_0}$ if that point lies in the second or fourth quadrant. *Solution.* We assume for definiteness that $x_0 > 0, y_0 > 0$. Other cases can be treated by analogy. The equation of the straight line is $y = y_0 + k(x-x_0)$. This straight line cuts the coordinate axes at points $(0, y_0 - kx_0)$ and $\left(x_0 - \dfrac{y_0}{k}, 0\right)$. The square of the length

of the segment connecting those points is

$$s^2 = \left(x_0 - \frac{y_0}{k}\right)^2 + (y_0 - x_0 k)^2 = (y_0 - x_0 k)^2\left(1 + \frac{1}{k^2}\right).$$

It is evident that $s$ and $s^2$ attain their minimum values under the same conditions. Therefore, we find $(s^2)'$ and equate the derivative to zero. Then we obtain

$$(s^2)' = -2(y_0 - x_0 k)\left(x_0 + \frac{y_0}{k^3}\right) = 0.$$

Consequently, there are two critical points (not counting the values $k = 0$ and $k = \infty$, these are the cases of a horizontal and a vertical straight line when $s^2 = \infty$); $k = \tan\alpha = y_0/x_0$ and $k = \tan\alpha = -\sqrt[3]{y_0/x_0}$. The first value, $\alpha = \arctan (y_0/x_0)$, corresponds to the trivial case $s = 0$. The second value, $\alpha = \pi - \arctan$ $\sqrt[3]{y_0/x_0}$, corresponds to the local minimum.

**137.** $\dfrac{S\sqrt{2S}}{6\sqrt{\sin\alpha}\,\cos(\alpha/2)\tan\beta}, \alpha = \dfrac{\pi}{3}$. **138.** $\sqrt{2/3}$. **139.** $2\pi/3$. **140.** A cylinder the

diameter of whose base is equal to the altitude. **141.** $\sqrt[3]{4V}$. **142.** $\alpha = \arctan(\sqrt{2}/2)$.

**143.** $(4V)^{1/3}/\sqrt[6]{3}$. **144.** A parallelepiped in which the length of a side of the base is equal to the radius $R$ of the circumscribed sphere, and the altitude is $R\sqrt{2}$. **145.** $R/3$.

**146.** $(2\pi)^{-1/3}$dm. **147.** $\sqrt{2R}$. **148.** $\sqrt{S(\sqrt{2}-1)/\pi}$. **149.** $8\sqrt{3\pi}R^2/9$. The ratio of the radius of the base of the cone to the radius of the sphere is $2\sqrt{2}:3$. **150.** $l/\sqrt{3}, l^3/2$.

**151.** 3. **152.** $V = \pi(4R^2 - H^2)H/4, V_{max} = 4\pi R^3/3\sqrt{3}$ for $H = 2R/\sqrt{3}$. **153.** $\sqrt{2}$.

**154.** $\sqrt{3}$ times. **155.** $\pi R^3/\sqrt{2}$. **156.** 3/4. **157.** $3a/2$.

**158.** $\pi \cot^2 \varphi / \cos 2\varphi$. The area of the lateral surface assumes the greatest value for $\varphi = 0.5 \arccos(\sqrt{2} - 1)$, i.e., $\cos 2\varphi = \sqrt{2} - 1$. That value is equal to $\pi(\sqrt{2} + 1)^2$.

**159.** $4R / 3$, $64R^3 / 81$. **160.** $H / 3, 2R^2 H \sqrt{3} / 9$. **161.** 2. **162.** 3/2. **163.** $H / 3$, $4\pi R^2 H / 27$. **164.** At the midpoint of the edge $[AC]$. **165.** 8. **166.** 4.5 km. **167.** $a = 12$ km/h.

## Sec. 2.3

1.

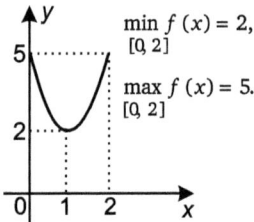

min $f(x) = 2$, [0, 2]

max $f(x) = 5$. [0, 2]

2.

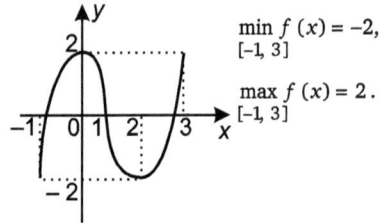

min $f(x) = -2$, [-1, 3]

max $f(x) = 2$. [-1, 3]

3.

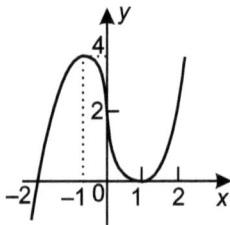

4.

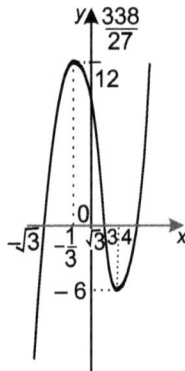

5.

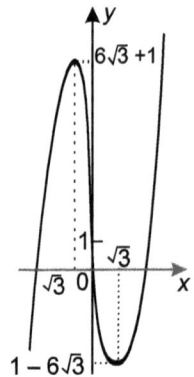

**6.**

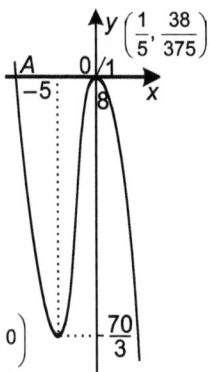

$A\left(\dfrac{-\sqrt{399}+18}{5}, 0\right)$

$B\left(\dfrac{\sqrt{399}-18}{5}, 0\right)$

**7.**

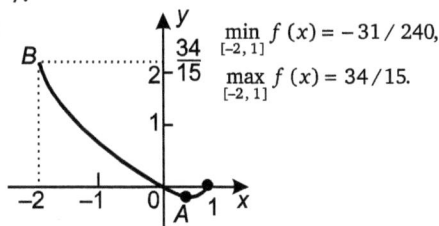

$\displaystyle\min_{[-2,\,1]} f(x) = -31/240,$

$\displaystyle\max_{[-2,\,1]} f(x) = 34/15.$

**8.**

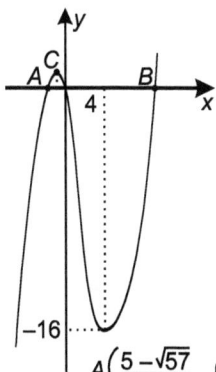

$A\left(\dfrac{5-\sqrt{57}}{2}, 0\right)$

$B\left(\dfrac{5+\sqrt{57}}{2}, 0\right)$

$C\left(-\dfrac{2}{3}, \dfrac{76}{81}\right)$

**9.**

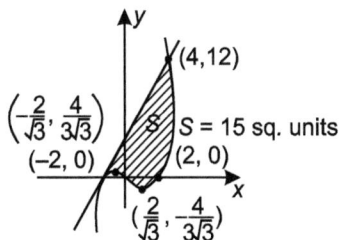

$S = 15$ sq. units

**10.**

$A\left(-1-\dfrac{1}{\sqrt{3}}, -\dfrac{2\sqrt{3}}{9}\right)$

$B\left(-1+\dfrac{1}{\sqrt{3}}, -\dfrac{2\sqrt{3}}{9}\right)$

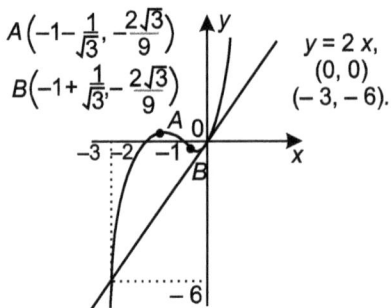

$y = 2x,$
$(0, 0)$
$(-3, -6).$

**11.**

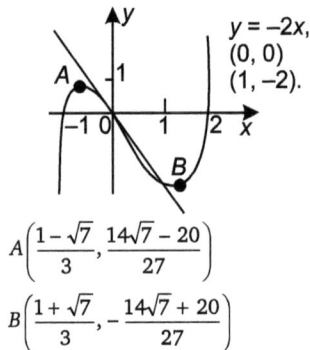

$y = -2x,$
$(0, 0)$
$(1, -2).$

$A\left(\dfrac{1-\sqrt{7}}{3}, \dfrac{14\sqrt{7}-20}{27}\right)$

$B\left(\dfrac{1+\sqrt{7}}{3}, -\dfrac{14\sqrt{7}+20}{27}\right)$

**12.**

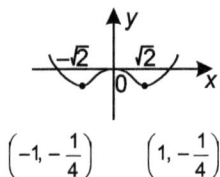

$\left(-1, -\dfrac{1}{4}\right)$     $\left(1, -\dfrac{1}{4}\right)$

**13.**

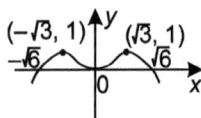

**14.**

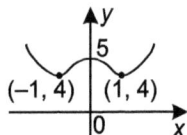

**15.**

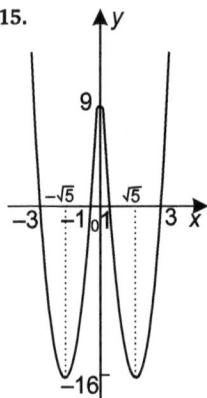

**16.**

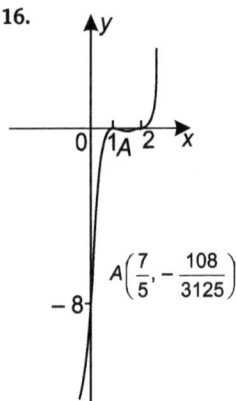

$A\left(\dfrac{7}{5}, -\dfrac{108}{3125}\right)$

**17.**

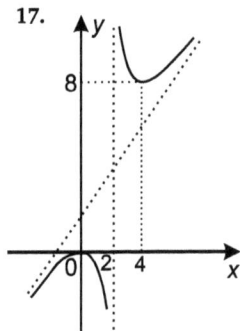

**18.**

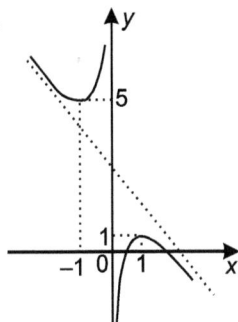

**19.**

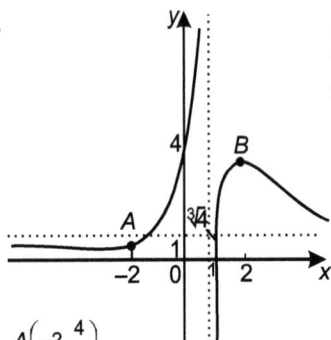

$A\left(-2, \dfrac{4}{9}\right)$

$B\,(2,\,4)$

One root for $c \in (-\infty, 4/9) \cup (4, +\infty)$; two roots for $c \in \{4/9, 1, 4\}$; three roots for $c \in (4/9, 1) \cup (1, 4)$.

**20.**

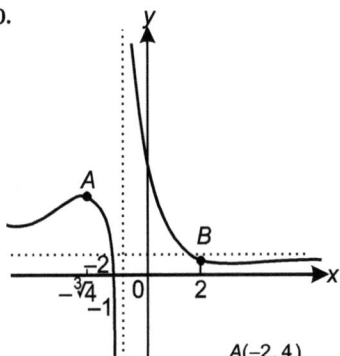

$A(-2, 4)$

$B\left(2, \dfrac{4}{9}\right)$

One root for $c \in (-\infty, 4/9) \cup (4, +\infty)$; two roots for $c \in \{4/9, 1, 4\}$; three roots for $c \in (4/9, 1) \cup (1, 4)$.

**21.**

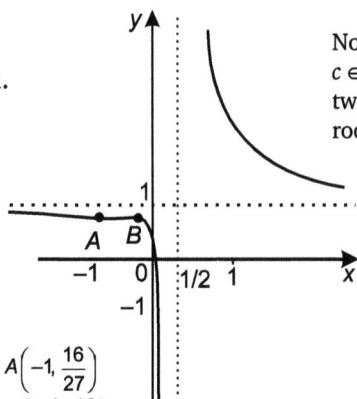

$$A\left(-1, \frac{16}{27}\right)$$
$$B\left(-\frac{1}{3}, \frac{16}{25}\right)$$

No roots for $c = 1$; one root for $c \in (-\infty, 16/27) \cup (16/25, 1) \cup (1, +\infty)$; two roots for $c \in \{16/27, 16/25\}$; three roots for $c \in (16/27, 16/25)$.

**22.**

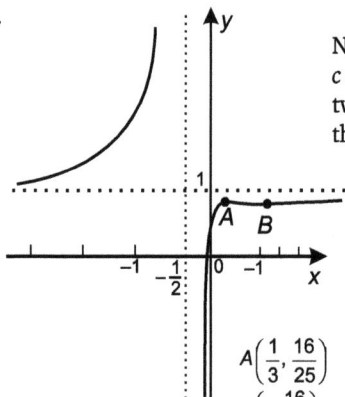

$$A\left(\frac{1}{3}, \frac{16}{25}\right)$$
$$B\left(1, \frac{16}{27}\right)$$

No roots for $c = 1$; one root for $c \in (-\infty, 16/27) \cup (16/25, 1) \cup (1, +\infty)$; two roots for $c \in \{16/27, 16/25\}$; three roots for $c \in [16/27, 16/25)$.

**23.**

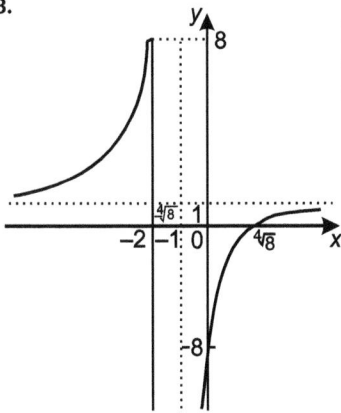

No roots for $c \in (8, +\infty)$; one root for $c \in \{1, 8\}$; two roots for $c \in (-\infty, 1) \cup (1, 8)$.

**24.**

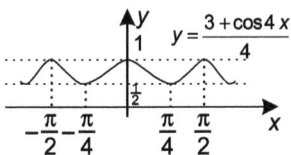

$$y = \frac{3 + \cos 4x}{4}$$

**25.**

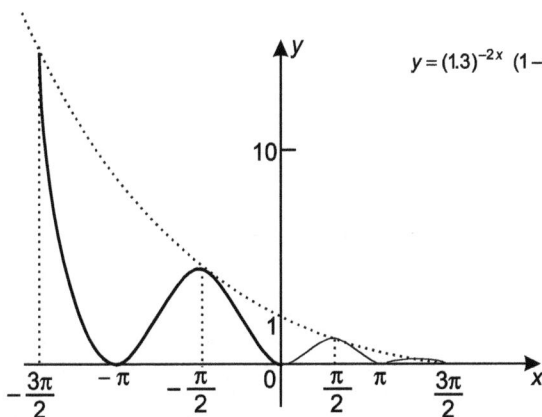

$$y = (1.3)^{-2x} (1 - \cos 2x) / 2.$$

**26.**

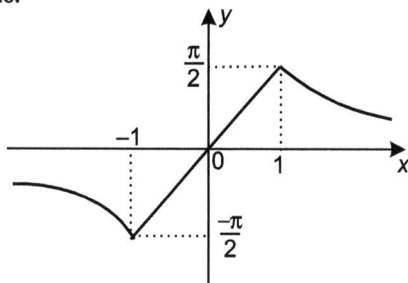

$y = 2 \arctan x$ for $x \in [-1,1]$; $y = -\pi - 2 \arctan x$ for $x \in (-\infty, -1]$ ; $y = \pi - 2 \arctan x$ for $x \in [1, +\infty)$.

**27.**          **28.**                              **29.**

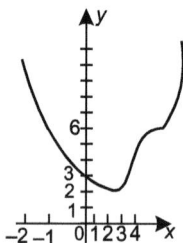          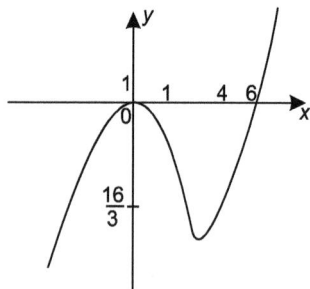          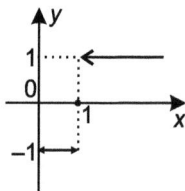

## Sec. 2.4

**1.** $F(x) = 2x^2 + x + 1$. **2.** $F(x) = x^3 - 2x^2 + 1$. **3.** $F(x) = \dfrac{7}{3}x^3 - x^2 + 3x + \dfrac{2}{3}$. **4.** $F(x) = x + \dfrac{x^2}{2} + \dfrac{1}{2}\sin 2x + 1$. **5.** $F(x) = -0.5\cos 2x + x^3 + 2.5$. **6.** $\sqrt{3} - \cot x$. **7.** $6\sin(x/2) - \dfrac{2}{5}\cos 5x - \dfrac{14}{5}$. **8.** $5\tan(\pi/5)$. **9.** $(4-\pi)/8$. **10.** $1066/243 \ln 3$. **11.** $A = 1/\ln 2$, $B = 7(\ln^2 2 - 1)/(3\ln^2 2)$. **12.** $a > 0$. **13.** $b = 2$. **14.** $\alpha = 3$. **15.** $\alpha > \ln 2$. **16.** $a \le -1$. **17.** $4/3$.

**18.** 121.5. **19.** 32/3. **20.** 4/3. **21.** 10/3. **22.** 9/2. **23.** 32/3. **24.** 9/2. **25.** 81/2. **26.** 1/6. **27.** 9. **28.** 9. **29.** 4/3. **30.** 1/3. **31.** 4.5. **32.** 2/3. **33.** 7/3. **34.** $(8\sqrt{2} - 3)/6$. **35.** 8/3.

**36.** 1/3. **37.** 9. **38.** 5/3. **39.** 7/4. **40.** $16\sqrt{3}/15$. **41.** 88/15. **42.** 64/3. **43.** 4/3. **44.** 2. **45.** $(15 - 16\ln 2)/4$. **46.** $(6e - 5)/3$. **47.** 125/6. **48.** 3/2. **49.** 4/3. **50.** $e^2 - 1$. **51.** 5. **52.** 1/3. **53.** $\ln 2$. **54.** 8/5. **55.** $(6 - \pi)/3\pi$. **56.** 53/15. **57.** $12 - 5\ln 5$. **58.** $4 - 3\ln 3$. **59.** $2.5 - 6\ln(3/2)$. **60.** $5\ln(6/5) - 0.5$. **61.** $1.5 - \ln 2$. **62.** $4.5 + 9\ln 3$. **63.** $14.5 + 4\ln 2$. **64.** $1.5 - \ln 2$. **65.** 18. **66.** 7/6. **67.** 8/3. **68.** 5/12. **69.** 8/9. **70.** 2. **71.** 1/3.

**72.** $(\pi - 2) / 4$. **73.** $(3\pi - 4)/4$. **74.** 1. **75.** $5\pi$. **76.** $5\pi / 2$. **77.** $3\pi$. **78.** 16. **79.** $(4 - 3 \ln 3)/2$.
**80.** $11 - 6 \ln 3$. **81.** $90.5 - 6 \ln 2$. **82.** 16/3. **83.** 9/8. **84.** 16/3. **85.** 128/3. **86.** 18. **87.** 1/3.
**88.** 4/3. **89.** 61/24. **90.** 7/6. **91.** $\ln 2 - 5/8$. **92.** $4 \ln 3 - 2$. **93.** 1/8. **94.** 4/3. **95.** $(2e^9$
$-101)/6$. **96.** $(\ln 10 - \pi + 4)/2$. **97.** $(4\sqrt{2} - 16 + \pi\sqrt{2})/16$. **98.** 5/6. **99.** $2 - 2 \ln 2$. **100.**
$2\pi + 4$. **101.** 4/3. **102.** $4\pi + 8$. **103.** 8. **104.** 9/8. **105.** $S(-1) = 125/6$, min
$S(k) = S(2) = 32/3$. **106.** $S = 1$. **107.** $x_0 = 1/2, y_0 = 5/4$. **108.** $x_0 = 3/2, y_0 = 2/3$.
**109.** $p = -1$. **110.** $\alpha_1 = \arctan(27/2), \alpha_2 = \arctan(27/4)$. **111.** $\alpha_1 = \pi + \arctan(4/3\pi)$,
$\alpha_2 = \pi + \arctan(8/3\pi)$. **112.** $625\pi$ cm$^2$. **113.** $18\sqrt{5\pi}$ cm$^2$.

$$\boxed{\textbf{Part-3}}$$

## Sec. 3.1

**1.** 33. **2.** – 33. **3.** – 8. **4.** –10. **5.** $\mathbf{d} = \dfrac{2}{5}\mathbf{a} + \dfrac{3}{5}\mathbf{b} + \dfrac{3}{5}\mathbf{c}$. **6.** – 4. **7.** –2. **8.** 40. **9.** 2. **10.** 7.
**11.** arccos (4/9). **12.** arccos (1/11). **13.** 13. **14.** – 6. **16.** $\mathbf{d} = \{5, 12, -16\}$, $|\mathbf{d}| = 5\sqrt{17}$.
**17.** 90°. **18.** $3\pi/4$. **19.** – 61. **20.** Yes, they are. **21.** $\mathbf{a} = \{1, 2, 2,\}$. **22.** $\mathbf{a} = \{4, -2, 0\}$.
    **23.** $\{1, 1/2, -1/2\}$. **24.** $\mathbf{a} = \{2, -6, 2\}$. **25.** $\mathbf{b}_1 = \{4\sqrt{2}, -2, 8\}, \mathbf{b}_2 = \{-4\sqrt{2}, 2, -8\}$
**26.** $\mathbf{c} = \{-3, 3, 3\}$. **27.** $\mathbf{b} = \{8/3, -10/3, 13/3\}, 3.66...$ **28.** $\mathbf{b} = \{9/16, 3/16, 21/16\}$.
**29.** $\cos\alpha = -4/5$. **30.** $\mathbf{p} = 2a - 3b$. **31.** $-22/\sqrt{133}$ **32.** $\mathbf{x} = \{-4, -6, 12\}$. **33.** $\mathbf{x} = \{-24,$
$32, 30\}$. **34.** $\mathbf{x} = \{-3/2, 5/2, 3\}$.
    **35.** $\{(1 \pm \sqrt{2})\sqrt{3}/6, (5 \mp 7\sqrt{2})\sqrt{3}/30, (10 \pm \sqrt{2})\sqrt{3}/30\}$. In this notation we must
take either the upper or lower signs everywhere. *Solution.* Assume that the required vector
is $\mathbf{c} = \{x, y, z\}$. Then, $|\mathbf{c}|^2 = x^2 + y^2 + z^2 = 1$; furthermore, cos $(\widehat{\mathbf{a}, \mathbf{c}}) = \cos\pi/4$
$= 1/\sqrt{2} \Rightarrow a \cdot c/(|\mathbf{a}| \cdot |\mathbf{c}|) = (x + y + 2z)/\sqrt{6} = 1/\sqrt{2} \Rightarrow x + y + 2z = \sqrt{3}$. Since the
vector $\mathbf{c}$ lies in the plane of the vectors $\mathbf{a}$ and $\mathbf{b}$, the vector $\mathbf{c}$ can be resolved into
components with respect to the vectors $\mathbf{a}$ and $\mathbf{b}$ : $\mathbf{c} = k\mathbf{a} + l\mathbf{b} \Rightarrow x = k - l, y = k + 3l, z$
$= 2k + l$. Removing $k$ and $l$ from the last relations, we get $5x + 3y - 4z = 0$. Thus, we must
simultaneously solve the system of equations

$$\begin{cases} x^2 + y^2 + z^2 = 1, \\ 5x + 3y - 4z = 0, \\ x + y + 2z = \sqrt{3}. \end{cases}$$

From the second and third equations, we find that

$$x = 5z - \frac{3\sqrt{3}}{2}, y = \frac{5\sqrt{3}}{2} - 7z. \qquad (1)$$

Substituting these values of $x$ and $y$ into the first equation, we get, after
transformations, $150z^2 - 100\sqrt{3}z + 49 = 0$, whence we find that

$$z = \frac{1}{\sqrt{3}} \pm \frac{1}{5\sqrt{6}} = \frac{(10 \pm \sqrt{2})\sqrt{3}}{30}. \qquad (2)$$

Substituting these two values of $z$ into relation (1), we get, correspondingly, two values for $x$ and $y$ each

$$x = \frac{1}{2\sqrt{3}} \pm \frac{1}{\sqrt{6}} = \frac{(1 \pm \sqrt{2})\sqrt{3}}{6},$$

$$y = \frac{1}{2\sqrt{3}} \mp \frac{7}{5\sqrt{6}} = \frac{(5 \mp 7\sqrt{2})\sqrt{3}}{30}. \tag{3}$$

Note that in the notations for relations (2) and (3) the upper sign of $z$ corresponds to the upper sign of $x$ and $y$, and the lower sign corresponds to the lower sign.

**36.** 0. **37.** arccos (4/9). **38.** $3\sqrt{34}/2$.

**39.** $\pi - \arccos(4/9)$. $\mathbf{a} = 2\overrightarrow{AB} = \{4, -2, 4\}$.

**40.** $D(-1, 1, 1), (\widehat{\overrightarrow{AC}, \overrightarrow{BD}}) = 2\pi/3$.

**41.** $\cos \hat{A} = \dfrac{\sqrt{3} + \sqrt{2} - 1}{\sqrt{9 + 2\sqrt{6} - 2\sqrt{3}}}$, $\cos \hat{B} = \dfrac{1 - \sqrt{2}}{\sqrt{6}}$,

$\cos \hat{C} = \dfrac{2\sqrt{3} + \sqrt{2} - 1}{\sqrt{18 + 4\sqrt{6} - 4\sqrt{3}}}$, $m = 2\sqrt{9 + 2\sqrt{6} - 2\sqrt{3}}$.

**42.** $B((2\sqrt{3} - 1)/3, (7 - 2\sqrt{3})/3, (7 + \sqrt{3})/3)$ $C(-(3 + 2\sqrt{3})/9, (21 + 2\sqrt{3})/9,$ $(21 - \sqrt{3})/9)$, or $B(-(1 + 2\sqrt{3})/3, (7 + 2\sqrt{3})/3, (7 - \sqrt{3})/3), C((2\sqrt{3} - 3)/9,$ $(21 - 2\sqrt{3})/9, (21 + \sqrt{3})/9)$.

**43.** $\alpha = 1 + \lambda$.

**44.** $(a^2 + 3b^2 - c^2)/4$. *Solution.* It is known that $\overrightarrow{CD} = (\overrightarrow{CA} + \overrightarrow{CB})/2$. Therefore, $\overrightarrow{CD} \cdot \overrightarrow{CA} = (\overrightarrow{CA} + \overrightarrow{CB}) \cdot \overrightarrow{CA}/2 = (\overrightarrow{CA}^2 + \overrightarrow{CB} \cdot \overrightarrow{CA})/2 = (b^2 + ab\cos(\widehat{\overrightarrow{CB}, \overrightarrow{CA}}))/2$. To find $\cos(\widehat{\overrightarrow{CB}, \overrightarrow{CA}})$, we use the cosine rule $c^2 = a^2 + b^2 - 2ab\cos(\overrightarrow{CA}, \overrightarrow{CB})$, whence we get $\cos(\widehat{\overrightarrow{CA}, \overrightarrow{CB}}) = (a^2 + b^2 - c^2)/2ab$. Consequently, $\overrightarrow{CD} \cdot \overrightarrow{CA} = (b^2 + (a^2 + b^2 - c^2)/2)/2 = (a^2 + 3b^2 - c^2)/4$.

**45.** *Proof.* We assume that the sum of the vectors which connect the centre of a regular $n$-gon (a triangle in particular) with its vertices is a certain vector $\mathbf{x}$. Then we turn the $n$-gon about its centre through the angle of $2\pi/n$; then the whole figure, all the vectors considered on it, including the vector $\mathbf{x}$, will turn through the angle of $2\pi/n$. Assume that $\mathbf{x}$ occupies now the position $\mathbf{x}^*$. But upon the rotation through the angle of $2\pi/n$, the $n$-gon we consider coincides with the initial $n$-gon and, consequently, all the vectors drawn from the centre to the vertices remain the same in their totality, and that means that their sum i.e. the vector $\mathbf{x}$, remains the same but two vectors, $\mathbf{x}$ and $\mathbf{x}^*$, which have turned through the angle of $2\pi/n$ relative to each other, can coincide only if $\mathbf{x} = \mathbf{x}^* = 0$.

**46.** $\overrightarrow{AB} + \overrightarrow{CD} = -(\overrightarrow{DC} + 2\overrightarrow{CQ})$.

**47.** $\pi/3$ *Solution.* We designate $(\widehat{\mathbf{a}, \mathbf{b}}) = \alpha$. Then, $0 = \mathbf{c} \cdot \mathbf{d} = (\mathbf{a} + 2\mathbf{b}) \cdot (5\mathbf{a} - 4\mathbf{b}) = 5\mathbf{a}^2 + 10\mathbf{ab} - 4\mathbf{ab} - 8\mathbf{b}^2 =$ $5 + 10\cos\alpha - 4\cos\alpha - 8 = 6\cos\alpha - 3$, whence it follows that $\cos\alpha = 1/2$ and $\alpha = \pi/3$.

**48.** $(a^2 + 3b^2 - c^2)/2$. **49.** $(3a^2 + b^2 - c^2)/2$.

**50.** $\pm\left(\mathbf{q}-\left(\dfrac{\mathbf{pq}}{\mathbf{p}^2}\right)\mathbf{p}\right)$. *Solution.* We assume for definiteness that the vector which coincides with the altitude is directed from the vertex to the side $\mathbf{p}$. Then (see the figure) $\vec{BE} = \vec{BA} + \vec{AE}$ (if $\angle BAD$ is an obtuse angle, then the point $E$ lies outside the segment $[AD]$). The vector $\vec{AE}$ is collinear with the vector $\mathbf{p}$ and, consequently, $\vec{AE} = k\mathbf{p}$ (the sign of $k$ defines the position of the point $E$), $\vec{BA} = -\mathbf{q}$. Then the scalar product of the vectors $\vec{BE}$ and $\vec{AD}$ is equal to zero, i.e.

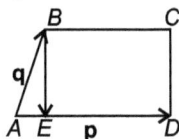

$$0 = \vec{BE} \cdot \vec{AD} = (-\mathbf{q} + k\mathbf{p})\cdot \mathbf{p} = -\mathbf{pq} + k\mathbf{p}^2,$$

whence it follows that $k = (\mathbf{pq})/\mathbf{p}^2$ and, consequently,

$$\vec{BE} = -\mathbf{q}+\left(\dfrac{\mathbf{pq}}{\mathbf{p}^2}\right)\mathbf{p}.$$

If we considered the vector $\vec{EB}$, the signs would be opposite, which fact is registered above.

**51.** $3(a^2 + b^2 - c^2)/2$.    **52.** 0.    **53.** $\lambda = -1/5$.

**54.** $\vec{AC} = 2(\mathbf{a} + \mathbf{b})/3.\vec{BD} = 2(\mathbf{b} - \mathbf{a})$.    **55.** $\vec{NP} = \dfrac{3}{4}\vec{AB}-\dfrac{1}{2}\vec{AF}$.

**56.** $\cos\alpha = -1/\sqrt{10}$.    **57.** 0.    **59.** $c^2$.    **60.** 0.

**62.** $|AC|=1/\cos\alpha, |AA_1|= 2\sin\alpha, |CA_1|=|\cos 2\alpha|/\cos\alpha$. *Solution.* It follows from the hypothesis that there are two possibilities of the location of the point $A$ : on the positive or on the negative semi–axis of ordinates. For definiteness we consider the case when the point $A$ lies on the negative semi–axis of ordinates (see the figure). The other position is symmetric about the $Ox$ axis and, therefore, the lengths of the sides of the triangle $ACA_1$ are the same in both cases. Let us find the coordinates of the points $A$ and $A_1(x, y)$. From the triangle $AOC$, with due account of the fact that $|OC|=1$ and $\angle OCA = \alpha$, we find that $|OA| = \tan\alpha$ and, consequently, $A(-\tan\alpha, 0)$. The length of $|AC| =1/\cos\alpha$. We complete the triangle $OAC$ to obtain a rectangle. Since in the rectangle $|AB|=1$, $\angle ABO = \alpha$, it follows that $|AK|= \sin\alpha$. The length of $|AA_1| = 2\sin\alpha$. The point $A_1$ lies in the fourth quadrant since $\alpha > \pi/4$.

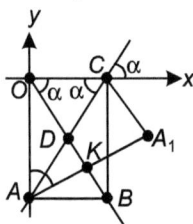

$x = \text{Proj}_{Ox}\,\vec{OA_1} = \text{Proj}_{Ox}(\vec{OA}+ \vec{AA_1}) = 0+|AA_1|\sin\alpha$

$= 2\sin\alpha \cdot \sin\alpha = 2\sin^2\alpha =1-\cos 2\alpha$.

$y = \text{Proj}_{Oy}\,\vec{OA_1} = \text{Proj}_{Oy}(\vec{OA} + \vec{AA_1}) = -\tan\alpha +|AA_1|\cos\alpha$

$= -\tan\alpha + 2\sin\alpha\cos\alpha = \tan\alpha \cdot \cos 2\alpha$.

Thus, $A_1(1 - \cos 2\alpha, \tan\alpha\cos 2\alpha)$. Hence

$$|CA_1|= \sqrt{\cos^2 2\alpha + \tan^2\alpha\cos^2 2\alpha} =|\cos 2\alpha|/\cos\alpha.$$

**63.** $1:\sin^2\alpha$. **64.** 29/2. **65.** 20. **66.** $D(2,9), S =14$ or $D(7,0), S = 22.5$. **67.** 5. **68.** $\lambda = 1/4$. **69.** $12b^2$. **70.** $\arccos\sqrt{2/5}$. **71.** 4. **72.** 26. **73.** 30. **74.** $2\sqrt{41}$. **75.** 10. **76.** 5. **77.** $-148$. **78.** $-6$. **79.** 3. **80.** $y = x^2 - 4x + 7$.

## Sec. 3.2

**2.** We use the fact that by connecting the midpoints of the successive sides of a convex quadrangle we get a parallelogram whose area is equal to half the area of the initial quadrangle.

**3.** Prove that for any convex quadrangle we can construct a circle which touches its any three sides or their extensions. Then, draw a tangent to the circle parallel to the fourth side.Compare the perimeters of the constructed and given quadrangles and use the theorem on tangents to a circle drawn from the same point.

**4.** Prove that a perpendicular drawn to the midpoint of the base of a trapezoid is, in that case, the axis of symmetry of the trapezoid.

**5.** Use the fact that a perpendicular drawn to the midpoint of the base of a trapezoid is, in that case, the axis of symmetry of the circle and the trapezoid.

**6.** Calculate the length of the inscribed circle, designating the radius of the initial circle as $R$.

**7.** Assume that $a$ and $b$ are the lengths of the legs, $c$ is the length of the hypotenuse of the right triangle, $r$ and $R$ are the radii of the inscribed and the circumscribed circle. Prove that $2R = c, 2r = a + b - c$,using the theorem on the equality of the lengths of the tangents drawn to the circle from the same point.

**8.** We have $|a| = 2|b|\sin 10°$ . Substitute it into the equation and use the formula for the sine of a triple angle.

**9.** Use the formula for calculating the length of a median in terms of the lengths of the sides of a triangle and the property that the point of intersection divides the medians in the ratio 1:2.

**10.** Use the fact that the radius of a circle drawn to the point of tangency is perpendicular to the tangent and the fact that the straight line which connects the centres of the tangential circles passes through the point of tangency.

**12.** Prove that the sum of the areas of the indicated triangles does not vary when the point $O$ is displaced parallel to a side of the parallelogram.It follows that this sum is always equal to half the area of the parallelogram.

**13.** Rotate the triangle $ABK$ through the angle of $90°$ so that the point $B$ coincides with the point $D$ (a rotation about the point $A$). Show that the side $(BK)$ passes along the side $(CD)$ and that $\triangle AMK'$ is an isosceles triangle ($K'$ is the point into which the point $K$ passes).

**14.** Use the formula for calculating the area of a triangle in terms of the radius of the inscribed circle and half the perimeter.

**15.** Prove that the centre of the inscribed circle coincides with the centre of the given circle, and the midpoints of the segments of the sides which lie in the interior of the given circle are the points of tangency.

**16.** Use the sine rule.

**17.** Prove that the segment which connects the midpoints of the adjacent sides of the convex quadrangle is parallel to the corresponding diagonal, and the length of the segment is equal to half the length of the diagonal.

**18.** Assume that $a, b$, and $c$ are the lengths of the sides of the triangle, $m_a$ is the median drawn to the side $a$. Prove that $b - 0.5_a < m_a < (b + c) / 2$. Write similar inequalities for the medians drawn to the other sides.

**19.** Assume that $A, B$, and $C$ are vertices of $\triangle ABC$, $O$ is the centre of the inscribed and the circumscribed circle. Show that $\angle OAC = \angle OCA = \angle OAB = \angle OBA$.

**20.** Show that the straight line is the axis of symmetry of the trapezoid.

**21.** Assume that $O$ is the point of intersection of the nonparallel sides of the trapezoid, $O_1$ and $O_2$ are the midpoints of the bases. Show that the points $O, O_1$, and $O_2$ lie on the same straight line. Use the fact that $[OO_1]$ and $[OO_2]$ are the medians of similar triangles.

**22.** Prove that the line of centres is the axis of symmetry of two circles.

**23.** Use the hint to the preceding problem.

**24.** Prove that the sum of these distances is equal to the length of the altitude drawn to a lateral side of the triangle.

**25.** Assume that $\triangle ABC$ is a given regular triangle, $O$ is the centre of symmetry of $\triangle ABC$, $|OA| = R$ is the radius of the circumscribed circle, $\alpha$ is the angle between $(OA)$ and the given straight line. Prove that the required sum of squares is
$$R^2\left[\sin^2\alpha + \sin^2\left(\frac{\pi}{3}-\alpha\right) + \sin^2\left(\frac{\pi}{3}+\alpha\right)\right] = \frac{3}{2}R^2.$$

**26.** Connect the centres of the squares with the vertices of the parallelogram and consider the resulting triangles. It follows from the equality of these triangles that the sides of the quadrangle in question are equal and mutually perpendicular.

**27.** Use the fact that the required distance is also equal to $|h - (R + r)|$, where $h$ is the altitude drawn to the base. Express $R$ and $r$ in terms of $h$ and the magnitude of the base angle and verify the corresponding equality.

**28.** Assume that $R$ is the radius of the circle circumscribed about $\triangle ABC$, the point $O$ lies on the circle (say, between the vertices $B$ and $C$), $\angle OAC = \alpha$. Then, using the sine rule, we get $|OC| = 2R\sin\alpha, |OB| = 2R\sin\left(\frac{\pi}{3}-\alpha\right)$ and $|OA| = 2R\sin\left(\frac{2\pi}{3}-\alpha\right)$. Hence, $|OA| = |OC| + |OB|$.

**29.** Prove the equality of the respective angles of the trapezoids.

**30.** Assume that $ABCD$ is the given trapezoid ($[BC]\|[AD]$). Prove that $\triangle ABC$ and $\triangle BCD$ are isosceles triangles.

**31.** Assume that $ABCD$ is the given trapezoid, $|AD| = a$, and $|BC| = b$. Draw a straight line through the point $C$ parallel to the diagonal $[BD]$. Assume that $D'$ is the point of intersection of that straight line and $(AD)$. Calculate the length of the midline of $\triangle ACD'$ and find the distance between the midpoints of the diagonals of the trapezoid.

**32.** Assume that $ABCD$ is the given trapezoid, $[MN]$ is a line segment which passes through the point of intersection of the diagonals, i.e. the point $O$. Use the similarity of the triangles $ABD$ and $MBO; ACD$ and $OCN$.

**33.** Assume that $\triangle ABC$ is the given triangle. The direct statement is well known. To prove the converse statement, we consider the point $D \in [AC]$ such that $|AD|:|AB| = |CD|:|CB|$ and use the sine rule. We obtain $\sin ABD = \dfrac{|AD|}{|AB|}\sin ADB$, $\sin CBD$, $= \dfrac{|DC|}{|BC|} \cdot \sin CDB$. Since $\sin ADB = \sin CDB$, we have $\sin ABD = \sin CBD$ and this equality yields the equality of $\angle ADB$ and $\angle CDB$.

**34.** Assume that $\triangle ABC$ is the given triangle; $m, \beta$, and $h$ are the median, the bisector, and the altitude drawn to the side $[AC]$; $M, N$, and $K$ are the points of intersection of the median, the bisector, and the altitude with $[AC]$ (or its extension). Assume $|AB| \geq |AC|$. Prove that $\angle BAC \leq \angle BMC \leq \angle BNC \leq \angle BKC = \pi/2$. Hence, results the required statement.

**38.** Use the result of problem 21.

**39.** Use the fact that the bisectors of the adjacent angles of a parallelogram are mutually perpendicular, and the bisectors of the opposite angles are parallel.

**40.** Assume that $ABCD$ is the given trapezoid, $O$ is the point of intersection of the diagonals, $|AO| = |OC|$. Prove that $\triangle AOD = \triangle COB$.

**41.** Here is a brief solution. Assume that $ABCD$ is the given quadrangle, $M$ and $N$ are the midpoints of the sides $[AB]$ and $[DC]$ respectively. Let us draw a straight

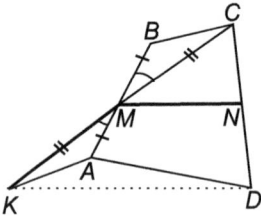

line $(MC)$ through the points $M$ and $C$ and lay off a segment $[MK]$ on it such that $|MK| = |MC|$. We connect the point $K$ with the points $A$ and $D$ (see the figure). Then, $\triangle AMK = \triangle BMC \Rightarrow (KA) \| (BC)$ and $|KA| = |BC|$. But $[MN]$ is the median of $\triangle KCD \Rightarrow |MN| = |KD|/2 \leq (|BC| + |AD|)/2$, the equality sign is possible if and only if the points $K, A$, and $D$ lie on the same straight line. But then $(BC) \| (KA) \| (AD)$, i.e. $ABCD$ is a trapezoid.

42. Assume that $\triangle ABC$ is the given triangle, $[AN]$ and $[CM]$ are the medians drawn to the sides $[BC]$ and $[BA]$ respectively, $O$ is the point of intersection of the medians. Prove that $\triangle AOM = \triangle CON$.

43. Assume that $\triangle ABC$ is the given triangle, $[AN]$ and $[CM]$ are the altitudes drawn to the sides $[BC]$ and $[BA]$. Prove that $\triangle ANC = \triangle CMA$.

44. As an illustration, we carry out a solution of the problem by means of the method of coordinates (see the end of Sec. 3.2). Assume that $O$ and $O_1$ are the centres of symmetry of the figure $\phi$. We draw a straight line through the points $O$ and $O_y$ and choose a scale such that $|OO_1| = 1$. We draw the $O_y$ axis through the point $O$ at right angles to $(OO_1)$ (see the figure). Assume that $M \in \Phi$ and has coordinates $(x_0, y_0)$ in the system of coordinates we have constructed. Let us consider the point $M_1(2 - x_0, -y_0)$ which is symmetric with respect to the point $M$ about $O_1$, $M_2 (x_0 - 2, y_0)$ which is symmetric with respect to $M_1$ about $O$, and then $M_3 (4 - x_0, -y_0)$ which is symmetric with respect to $M_2$ about $O_1$, and so on. All these points $M_k$ belong to the figure $\Phi$. On the other hand, the distance $|OM_{2k-1}| = |OM_{2k}|$ is equal to $\sqrt{(x_0 - 2k)^2 + y_0^2}$ i.e. it increases indefinitely with an increase in $k$. Consequently, the figure cannot be bounded.

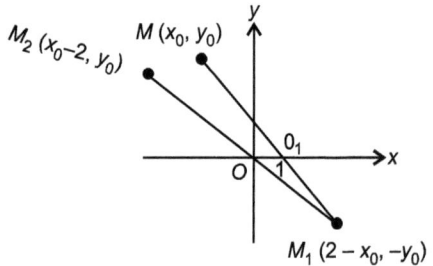

45. Prove that the axis of symmetry either cuts two opposite sides of the convex quadrangle at their midpoints, and in that case the quadrangle is an isosceles trapezoid, or the axis of symmetry passes through two opposite vertices and divides the quadrangle into two symmetric triangles.

47. Circumscribe a circle about the triangle, extend the bisector and the midperpendicular to the respective side till it intersects the circumscribed circle. Prove that they meet at one point. Next use the fact that the altitude and the respective midperpendicular are parallel

48. Assume that $\widehat{AB}$ and $\widehat{CB}$ are arcs between the points $A$ and $D$, $C$ and $B$. Prove that the sum of the lengths of the arcs is $\pi R$, i.e. half the length of the circle. Next connect the points $A, B, C$, and $D$ with the centre $O$ of the circle and use $\triangle AOD$ and $\triangle COB$ and the cosine rule to find the required value.

49. Prove that the area of the required part of the triangle is equal to half the area of the whole triangle minus the area of the $1/6$ th part of the circle.

50. Prove that $\triangle CAE \sim \triangle CPA$ and $\angle CAE + \angle CEA = 45°$.

52. Prove that the sum of the lengths of the segments is equal to double the length of the altitude drawn to the base of the triangle.

53. First prove that the diagonals of the square pass through the centre of symmetry of the rhombus. Draw the diagonals and drop perpendiculars from the centre of the square to the sides of the rhombus. Use the equality of the resulting right triangles.

54. Use the results of problems 27 and 19, having first proved the following statement. Assume that $\triangle ABC$ is inscribed in a circle of radius $R$. We also assume that the base $[AC]$ of the triangle is fixed and the vertex $B$ is displaced along the arc of the circle. Then, the radius of the circle inscribed in $\triangle ABC$ is the greatest if $\triangle ABC$ is an isosceles triangle.

55. Here is a solution based on the calculation of the maximum of the function. Assume that the magnitude of the angle $C$ in $\triangle ABC$ is fixed : $\angle C = \alpha$, and the angles $A$ and $B$ vary. We set $\angle A = x$; then $\angle B = \pi - \alpha - x$. Let us consider the function $f(x) = \cos A + \cos B + \cos C = \cos x + \cos\alpha + \cos(\pi - \alpha - x)$. We test this function for maximum-minimum for $x \in (0, \pi)$, and find that the maximum is attained for $x = (\pi - \alpha)/2$, i.e. in the case when $\triangle ABC$ is an isosceles triangle ($\angle A = \angle B$). Assume now that $\triangle ABC$ is an isosceles triangle, $\angle A = \angle B = x$. Then, $\angle C = \pi - 2x$. We consider the function $f(x) = \cos A + \cos B + \cos C = 2\cos x + \cos(\pi - 2x)$. Let us test this function for maximum for $x \in (0, \pi/2)$. We find that $f(x)$ attains its maximum for $x = \pi/3$, i.e. in the case of an equilateral triangle. Thus, the sum of the cosines of the angles of the triangle attains its greatest value for an equilateral triangle and is equal to 3/2 in that case.

56. To prove the first and the third relation, use the formula for calculating the area of a triangle $S = \dfrac{ab}{2} \sin\alpha$, where $\alpha$ is the magnitude of the angle between the sides of the triangle. To prove the second relation, construct a quadrangle of the same size as the given quadrangle with the same lengths of the sides but such that the sides of the lengths $a$ and $c, b$ and $d$ are adjacent.

57. Draw a diameter $(CD)$ through the point $M$ and prove that $|AM|.$ $|MB| = |CM| \cdot |MD|$.

58. *Solution.* We choose a system of rectangular coordinates taking the point of intersection of the given straight lines as the origin of the system and the straight

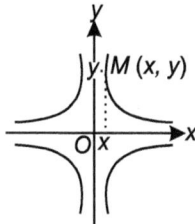

lines themselves as the coordinate axes, choosing positive directions on them and indicating scale units (see the figure). Then, the distances from any point $M$ with coordinates $(x, y)$ to the given straight lines are $|x|$ and $|y|$ respectively. Consequently, the point $M(x, y)$ belongs to the required locus of points if and only if $|x| + |y| = \dfrac{1}{|x|} + \dfrac{1}{|y|}$. Then $O(0, 0)$ cannot belong to the required set and, therefore, $|x|+|y| \neq 0$ and, consequently, the equation we have written is equivalent to the equation $|x||y|=1$. The graph of the last equation is the union of the graphs of the function $y = 1/x$ and $y = -1/x$, i.e. two hyperbolas. We have thus constructed the required set of points.

59. *Solution.* We again use the method of coordinates. Assume that the given points are $O$ and $A$. We consider the rectangular system of coordinates $Oxy$ whose origin coincides with the point $O$ and the $Ox$ axis contains the segment $[OA]$. We can assume that the point

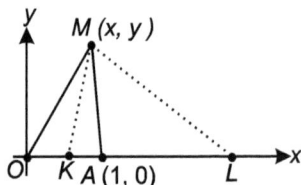

$A$ has the coordinates $(1, 0)$ (see the figure). Using the formula for calculating the distance between two points on a plane with given coordinates, we obtains $|OM| = \sqrt{x^2 + y^2}$, $|AM| = \sqrt{(x-1)^2 + y^2}$, where $M(x,y)$ is any point of the plane. Thus, the point $M(x,y)$ belongs to the required locus of points if and only if $\sqrt{x^2 + y^2}:\sqrt{(x-1)^2 + y^2} = m:n$, i.e.

$$n^2(x^2 + y^2) = m^2[(x-1)^2 + y^2].$$

We transform the relation obtained :

$$(n^2 - m^2)x^2 + 2m^2 x + (n^2 - m^2)y^2 = m^2.$$

If $n = m$, then we get $x = 1/2$, i.e. the required locus of points is a straight line which passes through the midpoint of the segment $[OA]$ at right angles to that segment. Thus, for $m = n$ we get the theorem on midperpendicular. Assume now that $m \neq n$, we can assume that $m > n > 0$. We carry out the following transformation :

$$x^2 - 2\frac{m^2}{m^2 - n^2}x + y^2 = -\frac{m^2}{m^2 - n^2},$$

$$\left(x - \frac{m^2}{m^2 - n^2}\right)^2 + y^2 = \frac{m^4}{(m^2 - n^2)^2} - \frac{m^2}{m^2 - n^2},$$

$$\left(x - \frac{m^2}{m^2 - n^2}\right)^2 + y^2 = \left(\frac{mn}{m^2 - n^2}\right)^2.$$

The last equation defines a circle with centre at the point $N$ $(m^2/(m^2 - n^2),0)$ of radius $mn/(m^2 - n^2)$. We have thus shown that for $m = n$ (i.e. when the distances from the required point to the two given points are equal) the required locus of points is a midperpendicular drawn to the segment with the ends at the two given points, and for $m \neq n$ it is a certain circle with centre on the straight line which passes through the given points. We must indicate the close connection between the constructed set of points $M$ with the interior and exterior bisectors of $\triangle OMA$ (see the figure). Assume that $K \in [OA]$ and $L \notin [OA]$ are the points of intersection of the given circle and the straight line $(OA)$. Then, $[MK]$ is the bisector of the interior angle $OMK$ of the triangle $OMA$ since the straight line $(MK)$ divides the side $[OA]$ into parts which are proportional to the sides $[OM]$ and $[AM]$ (see problem 33 of this section). Since, the straight line $(ML)$ is perpendicular to the bisector $(MK)$, it follows that $(ML)$ is the bisector of the exterior vertex angle $M$ of the triangle $OMA$ (see problem 51 of this section).

60. The required sets of points are described by the equations

(1) $|x||y|=||x|-|y||$.  (2) $\left|\dfrac{1}{|x|} - \dfrac{1}{|y|}\right| = ||x|-|y||$.

Construct the corresponding sets on a plane.

**61.** A parabola with the symmetry axis passing through the given point $F$ is at right angles to the given straight line $l$. The vertex of the parabola lies on the axis of symmetry at equal distances from the given straight line and the given point. The branches of the parabola recede from the given straight line. If we assume the axis of the parabola to be the $Oy$ axis, the distance from $l$ towards $F$ as the positive direction, the vertex to be the origin of coordinates $O$, the straight line passing through the vertex at right angles to $Oy$ as the $Ox$ axis, and designate the distance from the given point to the given straight line as $p$, then the equation of the parabola assumes the form $x^2 = 2py$.

**62.** We designate the midpoint of the segment $[AB]$ as $O$ and the given constant as $a^2$. Then two straight lines which are perpendicular to the straight line $(AB)$ and lie at the distance $a^2 / 2 |AB|$ from the point $O$ are the required locus of points.

**63.** Use the result of problem 24 of this section.

**64.** Assume that $A$ is a point which lies on the side of the angle at the given distance from the other side. Draw a straight line through the point $A$ parallel to the bisector. Then a part of the straight line, lying in the interior of the angle, satisfies the conditions of the problem. Consider all possible cases.

**65.** The straight lines which pass through the point of intersection of the straight lines $(AB)$ and $(CD)$, the distances of whose points to $(AB)$ and $(CD)$ are reciprocal to the ratio $|AB|:|CD|$ (excluding the point of intersection of the straight lines).

**66.** The circle constructed on the segment, which connects the given point with the centre of the circle, as a diameter.

**67.** A part of the circle with centre at the vertex of the given right angle and radius $a/2$.

**68.** A circle which is concentric with the given circle and whose radius is $R = r / \sin(\alpha/2)$, where $r$ is the radius of the given circle.

## Sec. 3.4

**1.** 50 cm. **2.** $\pi / 6$. **3.** 12 cm, 16 cm. **4.** 9.5 cm. **5.** $\dfrac{a \cos(\alpha / 2)}{\sin((\pi + 3\alpha)/4)}$. **6.** $2\sqrt{S} \tan(\alpha / 2)$.

**7.** 14 cm. **8.** $ab\sqrt{2} / (a + b)$. **9.** 85.

**10.** $7(\sqrt{7} - 1) / 9\sqrt{2}$ *Solution.* We designate the lengths of the sides $[AB],[BC]$, and $[CA]$ as $c, a$ and $b$ respectively. Then, by the hypothesis, $a / R = 2, b / R = 3 / 2$, i.e. $a = 2R$, $b = 3R / 2$. On the other hand, $a = 2R \sin A$ and, consequently, $\sin A = 1$, i.e. the angle $A$ is a right angle. In the right triangle $ABC$ we find the third side $|AB| = c = \sqrt{a^2 - b^2} = \sqrt{4R^2 - 9R^2 / 4} = R\sqrt{7} / 2$. The bisectors (see the figure).

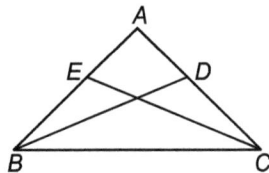

$$\beta_b = |BD| = \sqrt{|AB|^2 + |AD|^2},$$

$$\beta_c = |CE| = \sqrt{|AC|^2 + |AE|^2},$$

$$|AD| = |AC| \cdot \frac{|AB|}{|AB| + |BC|} = \frac{bc}{a + c},$$

$$|AE| = |AB| \cdot \frac{|AC|}{|AC| + |BC|} = \frac{bc}{a + b}.$$

Then
$$|AD| = \frac{3R \cdot R\sqrt{7}}{2 \cdot 2 \cdot (2R + R\sqrt{7}/2)} = \frac{3R\sqrt{7}}{2(4+\sqrt{7})},$$

$$|AE| = \frac{3R \cdot R\sqrt{7}}{2 \cdot 2 \cdot (2R + 3R/2)} = \frac{3R\sqrt{7}}{14},$$

$$\beta b = \sqrt{\frac{7R^2}{4} + \frac{9 \cdot 7 \cdot R^2}{4(4+\sqrt{7})^2}} = \sqrt{\frac{7R^2(16 + 8\sqrt{7} + 7 + 9)}{4(4+\sqrt{7})^2}}$$

$$= \sqrt{\frac{7R^2(32 + 8\sqrt{7})}{4(4+\sqrt{7})^2}} = \frac{\sqrt{7}R}{2}\sqrt{\frac{8(4+\sqrt{7})}{(4+\sqrt{7})^2}} = \frac{\sqrt{7}R\sqrt{2}}{\sqrt{4+\sqrt{7}}}$$

$$= \frac{\sqrt{7}R \cdot 2}{\sqrt{8+2\sqrt{7}}} = \frac{2R\sqrt{7}}{\sqrt{7+2\sqrt{7}+1}} = \frac{2R\sqrt{7}}{\sqrt{7}+1}$$

$$\beta_c = \sqrt{\frac{9R^2}{4} + \frac{9 \cdot 7R^2}{4 \cdot 7^2}} = \frac{3R}{2}\sqrt{1 + \frac{1}{7}} = \frac{3R \cdot 2\sqrt{2}}{2\sqrt{7}} = \frac{3R\sqrt{2}}{\sqrt{7}}.$$

The required ratio is

$$\frac{\beta_b}{\beta_c} = \frac{2R\sqrt{7} \cdot \sqrt{7}}{(\sqrt{7}+1)3R \cdot \sqrt{2}} = \frac{14R(\sqrt{7}-1)}{(7-1)3R\sqrt{2}} = \frac{7(\sqrt{7}-1)}{9\sqrt{2}}.$$

**11.** $\sqrt{3}(\sqrt{5}+1)/4$. **12.** $\pi/6, \pi/3, \pi/2$. **13.** $\arccos(13/14)$. **14.** 25. **15.** $2\arctan(1/3)$. **16.** $\arccos(1/\sqrt{10})$. **17.** 15 cm, 5 cm. **18.** $2\sqrt{2}\sqrt{\sqrt{5}+1}$ **19.** $(9\sqrt{3}/4)$ cm$^2$. **20.** 14 cm, 8 cm.

**21.** 1. *Solution.* Assume that the sides of the parallelogram are $a$ and $b$, the smaller diagonal is $d_1$, and the larger diagonal is $d_2$. Then, $2a^2 + 2b^2 = d_1^2 + d_2^2$, but by the hypothesis $d_2^2 = 3d_1^2$ and, therefore, $a^2 + b^2 = 2d_1^2$. By the cosine rule

$$d_1^2 = a^2 + b^2 - 2ab\cos 60° = 2d_1^2 - ab, \text{ i.e. } ab = d_1^2.$$

Consequently, $a^2 + b^2 + 2ab = 2d_1^2 + 2d_1^2 = 4d_1^2$, or $(a+b)^2 = 4d_1^2$, i.e. $a + b = 2d_1$. We have got a system

$$\begin{cases} a + b = 2d_1, \\ ab = d_1^2. \end{cases}$$

whose solution is $a = b = d_1$, i.e. $a/b = 1$.

**22.** 96 cm$^2$. **23.** $|a-b|/2$. **24.** $l^2\sin\alpha\cos\alpha$. **25.** $0.5\arcsin(4/5), \pi - 0.5\arcsin(4/5)$.

**26.** $2ab/|a-b|$. *Solution.* Assume $|AD| > |BC|$ (see the figure); we assume for definiteness that $AD = a$ and $BC = b$. Then, $\frac{|LC|}{|AL|} = \frac{b}{a} \Rightarrow \frac{|LC|}{|AL|+|LC|} = \frac{b}{a+b} \Rightarrow$

$$\frac{|LC|}{|AC|} = \frac{b}{a+b}; \frac{|KC|}{|KD|} = \frac{b}{a} \Rightarrow \frac{|KC|}{|KD|-|KC|}$$

$$= \frac{b}{a-b} \Rightarrow \frac{|KC|}{|CD|} = \frac{b}{a-b} \Rightarrow \frac{|KC|}{|CD|} = \frac{|NC|}{|AC|} = \frac{b}{a-b}$$

Multiplying the proportions $\frac{|AC|}{|LC|} =$

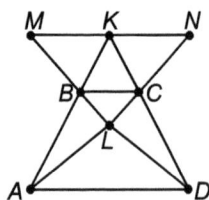

$\dfrac{a+b}{b}$ and $\dfrac{|NC|}{|AC|} = \dfrac{b}{a-b}$ ,we obtain

$\dfrac{|NC|}{|LC|} = \dfrac{a+b}{a-b} \Rightarrow \dfrac{|LC|+|NC|}{|LC|} = \dfrac{2a}{a-b} \Rightarrow \dfrac{|LN|}{|LC|} = \dfrac{2a}{a-b}$ but

$\dfrac{|LN|}{|LC|} = \dfrac{|MN|}{|BC|} = \dfrac{2a}{a-b}$ ,whence we get $|MN| = \dfrac{2ab}{a-b}$. If we had $a < b$, then they

would change parts in the derivation and we would get $|MN| = 2ab / (b-a)$.

**27.** $\arctan (2/3)$. **28.** $\sqrt{5/2}$ cm. **29.** $\pi/6$ and $\pi/3$. **30.** $\sqrt{7}$ m. **31.** $\sqrt{m^2 + n^2}/2$.

**32.** $21\sqrt{13}$ cm. **33.** $\sqrt{91}$ m. **34.** $1/\sqrt{13}$. **35.** 2 and 6. **36.** $R^2 \cos (\alpha/2)(1 + \sin(\alpha/2))$.

**37.** $\tan (\varphi/2) \sin 2\varphi$. **38.** 4 cm and 12.5 cm. **39.** $(\sqrt{15} + \sqrt{35})$ cm. **40.** $(3 - \sqrt{7})/2$.

**41.** $b/(2\cos(\alpha/2))$ . **42.** $r/8$ . **43.** $2\pi R/3$. **44.** $2\sqrt{R r} 2R$, $\sqrt{r/(r+R)}$,

$2r\sqrt{R/(r+R)}$. **45.** $Rr/(\sqrt{R} + \sqrt{r})^2$, $Rr/(\sqrt{R} - \sqrt{r})^2$ $(R > r)$. **46.** $r(r+R)/R$, $r+R$.

**47.** $r_1 r_2 / 2(r_1 + r_2)$. **48.** $\sqrt{3}$. **49.** Two times. **50.** $2R^2(1 - \sin(\alpha/2))^2 / \sin\alpha$. **51.** $2Rr + r^2$.

**52.** 15/8. **53.** $\dfrac{2}{a+b}\sqrt{p(p-a)(p-b)(p-c)}$, where $p = (a+b+c)/2$. **54.** $2(3 + 2\sqrt{3})$.

**55.** $9r^2/2$. **56.** $(S/\sin\alpha)^{1/2}$. **57.** $2R^2 \sin^3\alpha$. **58.** $100\pi/9$ cm$^2$. **59.** $a$. **60.** $3\sqrt{3}r^2$.

**61.** $\sqrt{24 - 6\sqrt{3}}/2$. **62.** $2a/\sqrt{5}$. **63.** 9 cm, 12 cm, and 15 cm. **64.** 240 cm$^2$.

**65.** $20\pi$ cm . **66.** $0.5 \arccos (3 - 8S)(1/4 < S < 3/8)$. **67.** Assume that $0 < \beta < \alpha$ $\leq \pi$; $H/(2\sin((\alpha+\beta)/4)\sin((\alpha+\beta)/4))$ and $H^2 \cot((\alpha-\beta)/4)$ if the bases of the trapezoid are on the same side of the centre of the circle, $H/(2\cos((\alpha-\beta)4)\cos((\alpha+\beta)/4))$ and $H^2 \tan((\alpha+\beta)/4)$ if the bases of the trapezoid are on different sides of the centre of the circle. **68.** 1 cm and 17 cm. **69.** $\dfrac{1}{8}$

$b\dfrac{\sin\alpha(5 - 4\cos\beta)}{\sin\beta \sin(\alpha + \beta)}$. **70.** $k\dfrac{\sin\beta (5 - 2\sqrt{6}\cos\alpha)}{6\sin\alpha \sin(\alpha + \beta)}$. **71.** 2 cm and 14 cm. **72.** 4 cm *Solution.*

Since $\dfrac{S_1}{S_2} = \dfrac{7}{13}$, it follows that $\dfrac{b+5}{5+a} = \dfrac{7}{13}$, i.e. $7a - 13b = 30$. On the other hand, $a + b = 10$.

Consequently, $a = 8$ and $b = 2$. The lateral side $l = \dfrac{a+b}{2} = 5$. Therefore, the altitude

$H^2 = l^2 - \left(\dfrac{a-b}{2}\right)^2 = 25 - 9 = 16$, i.e. $H = 4$ cm. **73.** $3\sqrt{3}$. **74.** $|BB_1| = \sqrt{53}/2, \angle B =$ $\arccos (11\sqrt{6}/30)$. **75.** $S = 216$ cm$^2$. **76.** $20\sqrt{3}$ cm$^2$. **77.** $|AB| = 2\sqrt{46}/3$. **78.** $\sqrt{10}$ cm.

**79.** $\sqrt{65}/2$. **80.** $P = 4ab/(a+b)$. **81.** $\tan\alpha = \sqrt{3}/7$, where $\alpha$ is the smallest angle of the triangle. **82.** 6 m and 6/5 m$^2$. **83.** $\sin\alpha = (1 + \sqrt{73})/12$, where $\alpha$ is a base angle of the triangle. **84.** 18 cm, 24 cm, and 30 cm. **85.** $\sqrt{b^2 + c^2 \pm 6bc}/5$. **86.** $\arccos (4/5)$ or $\pi - \arccos (4/5)$. **87.** $|EF| = 9\sqrt{2/7}$ cm. **88.** $6\pi$ cm. **89.** $40\sqrt{2}/11$ cm. **90.** 9 cm. **91.** $16\sqrt{7}/11$ cm$^2$. **92.** 20 cm and 5 cm. **93.** 45/2 cm$^2$. **94.** $8/\sqrt{3}$ cm. **95.** $|CA| = 39$ cm and $|CB| = 26$ cm. **96.** The lateral sides are $2R(2 + \sqrt{3})/\sqrt{3}$, the base is $2R(2 + \sqrt{3})$. **97.** $|OK| = 4/3$ cm.

**98.** $8\sqrt{5}$ cm and $4\sqrt{5}$ cm. **99.** $R_1 - r, R_2 - r, r(R_1 - r)(R_2 - r) / (R_1R_2 - r^2)$.**100.** $1 : 4$ $(3 - 2\sqrt{2})$. **101.** $2(\sqrt{2} - 1)$ cm. **102.** $a / 2$ and $a / 2$. **103.** $32$ cm$^2$. **104.** $4$ cm. **105.** $8 / \sqrt{3}$ cm$^2$. **106.** $a/2$ and $h/2$. **107.** $h = H / 2$, where $h$ and $H$ are the altitudes of the parallelogram and the triangle respectively. **108.** $2a$. **109.** $3R / 2$. **110.** $(a - b)^3$ $/16(a + b), a > b$. **111.** A square with the side $R\sqrt{2}, S_{max} = 2R^2$. **112.** A square with the side equal to 3 dm. **113.** 12 cm and $3\sqrt{3}$ cm. **114.** 12 cm and 9 cm. **115.** 9 cm and 12 cm. **116.** 9 cm and 7.5 cm. **117.** In a circle of radius $7\sqrt{2}$ cm. **118.** $c / \sqrt{2}$ and $c / \sqrt{2}$.

## Sec. 3.5

1. 7. *Hint.* Consider the case when two vertices lie on either side of the plane, and the case when one and three vertices lie on different sides of the plane.

2. If the straight lines do not pass through the same point, then show that they lie in the plane passing through three points of their intersection.

3. Consider the tetrahedron *ABCD* and show that the vertex *D* is projected into the point of intersection of the altitudes of the triangle *ABC*. Use the theorem on three perpendiculars.

4. Draw a plane through [*AB*] parallel to [*CD*]. Draw a straight line through the point *A* in that plane parallel to [*CD*] and lay off a segment [*AD'*] equal to [*CD*] on it. Consider the triangle *D'DB* and show that |*MN*| is equal to the length of the median of triangle *D'DB*, and hence obtain the required inequality.

5. Assume that $l$ is an oblique line, and $l_1, l_2$, and $l_3$ are the three given straight lines. Prove that the oblique line $l$ is perpendicular to the bisectors of the angles between the lines $l_1$ and $l_2$, and the lines $l_1$, and $l_3$. Hence, it follows that $l$ is perpendicular to the plane in which $l_1, l_2$, and $l_3$ lie.

6. Complete the triangular pyramid to obtain a triangular prism with the bases lying in the given planes. Prove that the volume of the prism does not vary upon the translation of the segments, and the volume of the given triangular pyramid is equal to one-third of the volume of the constructed prism.

8. Consider the section of the tetrahedral angle by planes which are perpendicular to the line of intersection of the planes passing through its opposite edges.

9. No. *Hint.* Consider a trihedral angle whose two plane angles are right angles and the third angle is smaller than $\pi / 3$.

10. Prove that if $\alpha$ is the magnitude of the plane angle at the vertex of the given trihedral angle, then the magnitude of the dihedral angle which does not adjoin it is greater than $\alpha$.

11. 0, 1, 3, and 6.

16. Here is a solution based on the theorem relating the area of the base and those of the lateral faces. Assume that all the dihedral angles of the pyramid are equal to $\alpha$; $S_1, S_2, S_3$, and $S_4$ are the areas of the faces. We obtain

$$\begin{cases} S_1 = \cos\alpha \ (S_2 + S_3 + S_4), \\ S_2 = \cos\alpha \ (S_1 + S_3 + S_4), \\ S_3 = \cos\alpha \ (S_1 + S_2 + S_4), \\ S_4 = \cos\alpha \ (S_1 + S_2 + S_3). \end{cases}$$

Hence $S_1 = S_2 = S_3 = S_4$ and $\cos\alpha = 1 / 3$. In addition, in this case the altitudes of the pyramid are projected into the centres of the circles inscribed in the faces of the pyramid. We can easily find from this that the apothems of the lateral faces are equal (if we take one of the faces to be the base). But then the lengths of the sides of the base are equal. Repeating these arguments for each face, we get an equality of all the edges of the pyramid.

17. Here is a solution. We designate the vertices of the pyramid by the digits 1, 2, 3, and 4; $a_{12}, a_{13}, a_{14}, a_{23}, a_{24}$, and $a_{34}$ are the lengths of the edges which connect the respective vertices; $R_1, R_2, R_3$, and $R_4$ are the radii of the spheres with centres at the corresponding vertices. Then the solution of the problem reduces to the solubility of the system

$$\begin{cases} R_1 + R_2 = a_{12}, \\ R_1 + R_3 = a_{13}, \\ R_1 + R_4 = a_{14}, \\ R_2 + R_3 = a_{23}, \\ R_2 + R_4 = a_{24}, \\ R_3 + R_4 = a_{34} \end{cases}$$

under the condition that $a_{12} + a_{34} = a_{13} + a_{24} = a_{14} + a_{23} = d$, $d$ being given. Then the system we have written is equivalent to the following system :

$$\begin{cases} R_1 + R_2 = a_{12}, \\ R_1 + R_3 = a_{13}, \\ R_1 + R_4 = a_{14}, \\ R_1 + R_2 + R_3 + R_4 = d. \end{cases}$$

Hence we find

$$\begin{cases} R_1 = (a_{12} + a_{13} + a_{14} - d)/2 > 0, \\ R_2 = (a_{12} + d - a_{13} - a_{14})/2 > 0, \\ R_3 = (a_{13} + d - a_{12} - a_{14})/2 > 0, \\ R_4 = (a_{14} + d - a_{12} - a_{13})/2 > 0, \end{cases}$$

i.e. we have proved that spheres of that kind exist and indicated their radii. (Prove that $R_i > 0$ for $i = 1, 2, 3$, and 4 using the relation between the lengths of the sides of the triangle.)

18. Draw a plane through the altitude of the pyramid and a lateral edge and consider the line of intersection of that plane and the base of the pyramid.

19. No. Construct the corresponding example.

20. Prove that this quantity is equal to $3H$, where $H$ is the altitude of the pyramid. For that purpose draw perpendiculars from the point $P$ onto the sides of the base and prove that the sum of their lengths is constant and equal to the length $h$ of the altitude of the base, and the required quantity is equal to $h \tan \alpha$, where $\alpha$ is the magnitude of the dihedral angle at the base of the pyramid.

21. A triangle, a square, and a hexagon.

24. It exists. Consider, for example, a polyhedron which results when one oblique parallelopiped is placed onto the other parallelopiped of the same size so that their bases are superimposed.

26. Assume that $M$ is the given point, $l_1$ and $l_2$ are tangents to the sphere, and $K$ and $M$ are the points of tangency of $l_1$ and $l_2$ with the sphere. Draw a plane through the points $M, K$, and $N$ and use the theorem on tangents to a circle.

29. Assume that $O$ is the centre of the sphere circumscribed about the pyramid. Consider four pyramids whose vertices coincide with the point $O$ and, whose bases are the faces of the initial tetrahedron. Prove that they are equal to one another. It will follow that the point $O$ is equidistant from the planes of the

faces, i.e. it is, at the same time, the centre of the sphere inscribed in the tetrahedron.

32. Assume that $H_k$ is the length of the altitude of the pyramid, drawn to the face of area $S_k$, and the escribed sphere of radius $R_k$ touches that face. Prove that $\dfrac{1}{R_k} = \dfrac{1}{r} - \dfrac{2}{H_k}$. Next make use of the equality $H_k S_k = r(S_1 + S_2 + S_3 + S_4) = 3V$.

Hence we obtain

$$\frac{1}{R_1} + \frac{1}{R_2} + \frac{1}{R_3} + \frac{1}{R_4} = \frac{1}{r}\left(1 - \frac{2S_1}{S_1 + S_2 + S_3 + S_4}\right)$$
$$+ \frac{1}{r}\left(1 - \frac{2S_2}{S_1 + S_2 + S_3 + S_4}\right) + \frac{1}{r}\left(1 - \frac{2S_3}{S_1 + S_2 + S_3 + S_4}\right)$$
$$+ \frac{1}{r}\left(1 - \frac{2S_4}{S_1 + S_2 + S_3 + S_4}\right) = \frac{2}{r}.$$

33. To prove the sufficiency, show that the centre of the circumscribed sphere coincides with the point of intersection of the perpendicular drawn to the base from the centre of the circle circumscribed about the base and the plane, which is perpendicular to the segment connecting the vertex of the pyramid with any vertex of the base, drawn through its midpoint.

34. Assume that $M, N$, and $K$ are the points of tangency of the sphere and the faces of the trihedral angle. Consider the pyramid $SMNK$. Prove that $|SM| = |SN| = |SK|$. Hence it follows that the altitude of the pyramid passes through the centre of the circle circumscribed about the base. Using this fact, show that the extension of the altitude passes through the point $O$, i.e. through the centre of the given sphere.

36. Use the theorem on the equality of the lengths of the tangents drawn from the same point to the given sphere (see problem 26).

39. Prove that the indicated angle is equal in magnitude to the dihedral angle formed by the planes $\alpha$ and $\beta$.

40. Assume that the points $M, N, K$, and $L$ are the midpoints of the sides $[SB], [SC], [AC]$, and $[AB]$ respectively, and $O$ is the midpoint of $[SA]$. Prove that the quadrangle $MNKL$ is a rectangle and $\Delta SAC$ and $\Delta SAB$ are equilateral triangles. Then prove that the point $O$ is equidistant from the points $M, N, K, L, S$, and $A$. Hence you will get that $O$ is the centre of the given sphere.

41. Consider the sections of the given sphere by the plane of the base and the plane passing through the midpoints of the lateral edges of the pyramid. Since the ring resulting in the section of the sphere by the base of the pyramid touches the sides of the base at their midpoints, the base of the pyramid is an equilateral triangle. Consider the straight line which passes through the centres of the rings resulting in the sections and prove that it is perpendicular to the base. Then, show that the lengths of the lateral edges of the pyramid are equal to one another.

46. Assume that $2\alpha$ is the magnitude of the angle at the base of the axial section of the given cone and $l$ is the length of the generatrix of the truncated cone. Then $R_1 = r \cot \alpha, R_2 = r \tan \alpha$, and $l = R_1 + R_2$, where $r$ is the radius of the inscribed sphere, $R_1$ and $R_2$ are the radii of the bases of the given cone. Using the formula for the area of the lateral surface of a truncated cone, we obtain

$$S_{\text{lat of trunc cone}} = 2\pi (R_1 + R_2)l / 2 = \pi r^2 (\tan \alpha + \cot \alpha)^2.$$

Next, since $\tan \alpha + \cot \alpha \geq 2 (0 < \alpha < \pi / 2)$, it follows that $S_{\text{lat of trunc cone}} \geq 4\pi r^2 = S$, where $S$ is the surface area of the inscribed sphere.

## Sec. 3.6

1. $2\pi / 3$. *Hint.* Draw a straight line $L_2' \| L_2$ through the point $C \in L_1$, and a straight line $L_2'' \| (CD)$ through the point $B \in L_2$. Assume that $B'$ is the point of intersection of $L_2'$ and $L_2''$. Consider the triangle $B'CA$.

2. $\sqrt{m^2 - 4l^2 \sin^2(\alpha / 2)}$ or $\sqrt{m^2 - 4l^2 \cos^2(\alpha / 2)}$ depending on the mutual positions of the points $A$ and $B$ on the straight lines $L_1$ and $L_2$.

3. 29.6. *Hint.* Use the theorem on three perpendiculars.

4. (1) $(d + b\cos(\pi / 5))/(1 + \cos(\pi / 5))$ if the points $A, B,$ and $C$ lie on the same side of the plane $\gamma$;

   (2) $|d - b\cos(\pi / 5)|/(1 + \cos(\pi / 5))$ if the points $A$ and $B$ lie on the same side of $\gamma$, and the point $C$ lies on the other side.

   (3) $b\cos(\pi / 5)/(1 + \cos(\pi / 5))$ if $A$ and $B$ lie on different sides of $\gamma$.

5. (1) $|c - b|$ if $B$ and $C$ lie on the same side of the plane $\beta$;
   (2) $b + c$ if $B$ and $C$ lie on different sides of the plane $\beta$.

6 Into three equal parts of length $\sqrt{3} / 3$ each. 7. $3a^2 / 2$.

8. $\text{arccot } \sqrt{\cot^2\alpha + \cot^2\beta}$. 9. $\text{ancsin } \tan (\alpha / 2)(0 < \alpha < \pi / 2)$. 10. $2\sqrt{2}b^2$ $\times \sin 2\beta \sin((2\alpha + \pi) / 4)$. 11. $b^3 / \sqrt{2}$. 12. $|BD_1| = 5\sqrt{3}$ cm. 13. $2l\sin\alpha \times \sqrt{2S + l^2 \cos^2\alpha}$.

14. $V = h^3(\cot\alpha \cot\beta) / 2, S = 2h^2 \sqrt{\cot^2\alpha + \cot^2\beta}$. 15. $2\sqrt{S_1^2 + S_2^2}$. 16. $l^3 \sin\beta$ $\cos^2\beta \tan(\alpha / 2)2$. 17. $S^{3/2} \sin\alpha\sqrt{\sqrt{3}\cos\alpha} / 3$. 18. $Q^{3/2} \sin\varphi\sqrt{\sqrt{3}\cos\varphi}$. 19. $a^3(\sin\alpha \sin\beta \cos^2\beta) / 2$. 20. $a^3(\sin 2\alpha \tan\beta) / 2$. 21. $l^3(\sin 2\alpha \tan^2\beta) / 4$. 22. $c^3 / 32$. 23. $\sqrt{3}$ $(V \cot\alpha)^{2/3} \sec\alpha$. 24. $4 + 6\sqrt{3}$. 25. $4p^2 \sin 2\alpha / (1 + \sin\alpha + \cos\alpha)$. 26. $l^3(\sin 2\alpha\cos\alpha) / 4$. 27. $7b^2 / (8\cos\alpha)$, 2 $\arctan (\cos\alpha)$. 28. $a^3\sqrt{\cos 2\alpha} / \sin\alpha$. 29. $6\sqrt{21}$ cm. 30. $d^3$ $(\tan\beta \cot (\beta / 2))/2$. 31. $a^3(\tan\alpha \tan(\alpha / 2))/8$. 32. $3d^3(\sin 2\alpha \sin\alpha)/16$. 33. $a^2b$ $\sqrt{\sin^2\alpha - \cos^2\beta} / 2$.The problem has a solution if $\sin\alpha > \cos\beta$. 34. $\sqrt{2S \cos\alpha / \sqrt{3}}$ $/ \cos(\alpha / 2)$. 35. $a^3\sqrt{3\cot^2(\alpha / 2) - 1} / 24(\alpha < 2\pi / 3)$. 36. $\arctan ((1 \pm \sqrt{1 - 8\tan^2\alpha})$ $/2\tan\alpha)$. The problem has a solution for $\tan^2\alpha \le 1/8$. 37.$125\sqrt{3} / 9$ cm$^3$.

38. $l^3(\sin 2\alpha \cos\alpha)\sqrt{3} / 8$. 39. $d^3 / (3\sin(\alpha / 2)(3 - 4\sin^2(\alpha / 2)))$. 40. $a^2\sqrt{3}$ $/(27 \cos\alpha)$. 41. $32\sqrt{133} / 27$ cm$^2$. 42. $b^3(\cos^3\beta \tan\alpha) / (6\sin\beta)$. 43. $(12 + 13\sqrt{3}) / 2$ cm$^2$. 44. $c^3(\sin 2\alpha\tan\beta) / 24$. 45. $25/4 (1 + \tan 2 + \sqrt{1 + 2\tan^2\alpha})$. 46. $S \cot\beta \times \sqrt{2S \sin\alpha}$ $/(6\sin\alpha \cos(\alpha /2))$. 47. $a^2(1 + \cos\beta)/(4\cos\beta \tan(\alpha / 2))$. 48. $(2/3)R^3 \sin\alpha \times$ $\sin\beta \sin(\alpha + \beta) \tan\varphi$. 49. $a^3 \cos(\alpha / 2) / (12\sqrt{4\sin^2(\alpha / 2) - 1})$. 50. $\sqrt{3} l^3 \tan^2\alpha \times (4$ $+ \tan^2\alpha)^{-3/2}$. 51. $6r^2\left(1 + \sin\dfrac{\alpha}{2}\right)^2 \text{cosec}\alpha$. 52. $a^3 / 24$. 53. $2 \cdot 3^{-1/6}\sqrt[3]{V}$. 54. $H^3\sqrt{3} / 8$.

55. $\sqrt{3} \dfrac{1 + \cos\alpha}{\cos\alpha} (3V \cot\alpha)^{2/3}$. 56. $(a^3 \tan\varphi)/12, a^2 \times \sqrt{3(1 + 4\tan^2\varphi)} / 4$. 57. $4d^3 /$ $(9\sqrt{3}\sin\alpha \sin 2\alpha)$. 58. $3aH / 16$. 59. $2a^{2\sqrt{11}} / 49$. 60. $1/7$. 61. $\sqrt{3}a^2 / (48\cos\alpha)$, $(a^3\tan\alpha) / 48$. 62. $H^2\sin\alpha/(4\cos\alpha - 2)$. 63. $R^3\sin^3\alpha\sqrt{3\cot^2(\alpha / 2) - 1} / 3$. 64. $3\pi a^2 /$ $(1 + 2\cos 2\alpha)$. 65. $b^3 \sin(\alpha/2)\times \sqrt{1 + 2\cos\alpha}/6(0 < \alpha < 2\pi / 3)$. 66. $a^3(\sin(\alpha / 2)$ $\tan\beta) / 6$. 67. $\sqrt{3}a^3 \tan\alpha / (1 - \tan\alpha)^3$. 68. $b^3 / 6$. 69. $\pi / 2$. 70. $84\sqrt{2}$ cm$^2$.

71. $h^3(\tan^2\alpha + \tan^2\beta)^{3/2} / (12\tan\alpha \tan\beta)$. *Solution.* Since all the edges are equal, their projections onto the base are also equal; consequently, the vertex $S$ (see the figure) is projected into the centre $D$ of the circle circumscribed about the triangle $ABC$. The triangle is right-angled and therefore, $D$ is the midpoint of the

hypotenuse. Assume that $|SK| \perp [BC]$ and $[SL] \perp [AC]$, $\angle SKD = \alpha$, $\angle SLD = \beta$, $|DK| = |AC|/2$, and $|DL| = |BC|/2$. Then, $H = |SD| = \dfrac{|AC|\tan\alpha}{2}$

$= \dfrac{|BC|\tan\beta}{2}$, we set $|AC| = k\tan\beta$ and $|BC| = k\tan\alpha$, where $k$ is a certain coefficient which we shall find later on.

Then, $H = (k\tan\alpha \times \tan\beta)/2$. The required volume is $V = \dfrac{H}{3}\dfrac{|AC|\cdot|BC|}{2}$

$= \dfrac{k^3 \tan^2\alpha \tan^2\beta}{12}$. The area of the triangle is $S = \dfrac{|AB|h}{2} = \dfrac{|AC|\cdot|BC|}{2}$.

This relation yields the value of $k$. Indeed, $|AB| = \sqrt{|AC|^2 + |BC|^2} = k\sqrt{\tan^2\alpha + \tan^2\beta}$

and, therefore, $kh\sqrt{\tan^2\alpha + \tan^2\beta}$

$= k^2 \tan\alpha \tan\beta$, whence $k = h\sqrt{\tan^2\alpha + \tan^2\beta}\,/\tan\alpha\tan\beta$. Then the volume of the pyramid is

$$V = \dfrac{h^3(\sqrt{\tan^2\alpha + \tan^2\beta}\,)^3}{\tan^3\alpha \tan^3\beta} \cdot \dfrac{\tan^2\alpha \tan^2\beta}{12} = \dfrac{h^3(\sqrt{\tan^2\alpha + \tan^2\beta}\,)^3}{12\tan\alpha \tan\beta}.$$

72. $\dfrac{r^2}{6\sin^2(\varphi/2)}\cot^2\left(\dfrac{\pi - \varphi}{4}\right)\sqrt{a^2\sin^2\varphi - r^2\cot^2\left(\dfrac{\pi - \varphi}{4}\right)}$.

73. $\dfrac{2R^3\tan\beta\sin^2\alpha\cos^3(\alpha/2)}{3(1 + \sin(\alpha/2))} \equiv \dfrac{2}{3}R^3\tan\beta\sin^2\alpha\cos^2(\alpha/2)\tan\dfrac{\pi - \alpha}{4}$.

74. $b^3(\sin(\alpha/2)\cot\beta)/6$. 75. $r\sqrt{1 + \cos^4(\alpha/2)\tan^2\beta}$. 76. $m^3\cot\alpha\sqrt{3 - \cot^2\alpha}$

$/24(\pi/6 \le \alpha)$.

77. We designate $\alpha = \arcsin(2S(\sqrt{2} - 1)/l^2)$. Then if $\pi/4 < \alpha \le \pi/3$, then the required angles are $\alpha - (\pi/4), \alpha, \alpha + (\pi/4)$; if $\pi/3 < \alpha < \pi/2$, then the required angles are $\alpha - (\pi/4), \alpha, \alpha + (\pi/4)$ or $(3\pi/4) - \alpha, \pi - \alpha, (5\pi/4) - \alpha$.

78. $a^3\sqrt{\cos\alpha}/(12\sin(\alpha/2))$. Solution. Assume that $|AB| = |CD| = a$,

$\angle ADB = \angle ACB = \angle DAC = \angle DBC = \alpha$ (see the figure). We draw a plane through $[DC]$ at right angles to the side $[AB]$, it cuts the side at a point $K$ such that $|AK| = |KB|$. Then, the volume of the pyramid is equal to $1/3\,|AB|S_{CDK}$. From the right triangle $ADK$ we find that $|DK| = |AK|\cot(\alpha/2) = a/(2\tan(\alpha/12))$.

From the right triangle $DKM$ we find that

$|KM|^2 = |KD|^2 - |DM|^2 = \dfrac{a^2}{4\tan^2(\alpha/2)} - \dfrac{a^2}{4}$

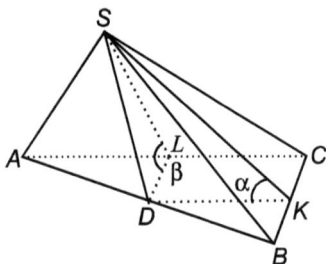

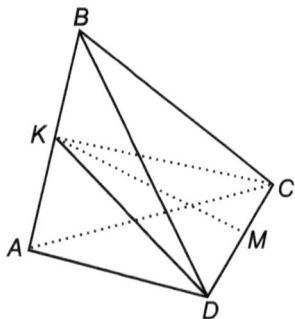

$$= \frac{a^2(\cos^2(\alpha/2) - \sin^2(\alpha/2))}{4\sin^2(\alpha/2)}$$

$$= \frac{a^2\cos\alpha}{4\sin^2(\alpha/2)} \Rightarrow |KM|$$

$$= \frac{a\sqrt{\cos\alpha}}{2\sin(\alpha/2)}(0 < \alpha < \pi/2).$$

Then we have

$$S_{CDK} = \frac{1}{2}|DC| \cdot |KM| = \frac{a^2\sqrt{\cos\alpha}}{4\sin(\alpha/2)},$$

and the required volume of the pyramid is equal to $a^3\sqrt{\cos\alpha}/(12\sin(\alpha/2))(0 < \alpha < \pi/2).$

**79.** $[3a^4 - 16(p^2 - Q^2)]:[3a^4 + 16(P^2 - Q^2)].$  **80.** (1) $2\sqrt[4]{\left|\frac{p+1}{p-1}\right|}|S^2 - T^2|$ for $p \neq 1$;

(2) for $p = 1$ the length of the hypotenuse can assume any value from the interval $(0, 2\sqrt{2}\,S)$ (in that case $S = T$).

**81.** $\dfrac{abc}{ab + bc + ac + \sqrt{(ab)^2 + (ac)^2 + (bc)^2}}.$

**82.** $\dfrac{1}{3}\sqrt{S_0} \cdot \sqrt{1 - \left(\dfrac{S_0}{S}\right)^2}\sqrt[4]{S(S - 2S_1)(S - 2S_2)(S - 2S_3)},$

where $S = S_1 + S_2 + S_3$.

**83.** $5\sqrt{3}/6.$ **84.** 7. **85.** 32 cm$^3$.

**86.** $(a^3\tan\alpha)/6.$  **87.** arccos $(\sqrt{2}\sin(\varphi/2)).$

**88.** 2 arccos $(1/\sqrt{1 - \cos\alpha})(\alpha > \pi/2).$

*Solution.* We draw $[BM]$ and $[DM]$ (see the figure) which are perpendicular to the edge $[SC]$. Then, $\angle BMD = \alpha$. The required angle $BSC = \beta$. We designate $|BC| = a$ and $|SB| = l$; then $|BN| = a\sqrt{2}/2$. Next, we calculate sin $(\beta/2)$;

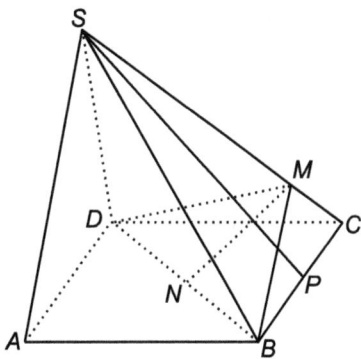

$$\sin(\beta/2) = \frac{|PB|}{|SB|} = \frac{a}{2l} = \frac{|BM|}{l} \cdot \frac{|BN|}{|BM|}$$

$$\times \frac{a}{2|BN|} = \sin\beta \cdot \sin\frac{\alpha}{2} \cdot \frac{1}{\sqrt{2}}.$$

Thus, $\sin(\beta/2) = \sqrt{2}\sin(\beta/2)\cos(\beta/2)\sin(\alpha/2),$

whence it follows that

$$\cos(\beta/2) = 1/(\sqrt{2}\sin(\alpha/2)) = 1/\sqrt{1 - \cos\alpha}\,(\alpha > \pi/2),$$

or $\beta = 2\arccos(1/\sqrt{1 - \cos\alpha})\,(\alpha > \pi/2).$

**89.** $l^3(\cot\alpha\cot\beta)/3(1 + \cot^2\alpha + \cot^2\beta)^{3/2}.$

**90.** $H\cos\varphi/(1 + \cos\varphi) \equiv H\cot\varphi\tan(\varphi/2).$ **91.** $a^2\sqrt{3}/6.$

**92.** $-4h^3\cos^2\alpha\,\cos2\alpha\,/\,3(\pi/4\le\alpha<\pi/2)$. For $0<\alpha<\pi/4$ the indicated plane does not cut the pyramid. **93.** $a^2\sin^2\alpha\,\cos(\alpha/2)\,/\,\sin^2(3\alpha/2)$.

**94.** $\sqrt{2}/12$. *Hint.* Prove that the point $M$ coincides with the centre of the base of the pyramid, and the point $N$ with the centre of the circle circumscribed about the face $BSC$.

**95.** $26\,\mathrm{m}^2$. **96.** $a^3(\sin^2\alpha\,\tan\beta)\,/\,3, a^2\sin\alpha(1+\sin\beta)\,/\,\cos\beta$. **97.** $a^3(\sin^2\alpha\times\tan\beta)\,/\,6$.

**98.** $2R^3(\sin\alpha\,\tan\beta\,\sin^3 2\beta)\,/\,3$. **99.** $4r^3\cot^3(\alpha/2)\sqrt{-\cos2\alpha}\,/\,(3\cos\alpha)$

$\equiv 4r^3\cot^3(\alpha/2)\sqrt{\tan^2\alpha-1}\,/\,3(\pi/4<\alpha<\pi/2)$. **100.** $2H^2\dfrac{\sqrt{1+\sin^2\alpha}}{\sin\alpha}\times\sqrt{2\cot^2\beta}$

$-\cot^2\alpha$. **101.** $6(7-4\sqrt{3})m^2\tan\alpha$. **102.** $4\sqrt{3}\,m^2\cos\alpha\,\cos^2(\alpha/2)$. **103.** $10\sqrt{19}$ cm$^2$.

**104.** $(\sqrt{S_1}+\sqrt{S_2})^2/4$. **105.** $\sqrt{(S\cos\alpha)\,/\,2(1+\sin\alpha-\cos\alpha)}$. **106.** $m^3(\sin^2\alpha\,\cos\alpha)$ $/6\sin^6(\alpha/2)$. **107.** $0.5\cdot h^2\tan\alpha\,/\,\cos\alpha$. **108.** $2\arcsin\,((\sqrt{10}-\sqrt{2})/4)$. **109.** $12(2+\sqrt{3})$ cm$^2$. **110.** $2a^2\sin\beta\,\sin^2(\gamma/2)\,/\,\cos\gamma$. **111.** $d\cot(\alpha/2)\times\sin^2(\beta/2)\,/\,\cos\beta$. **112.** $384\sqrt{10}/169$ cm$^2$. **113.** $a(1+\cos^2\alpha)\,/\,2\sin2\alpha$. **114.** $(a/6)\times$ $\sqrt{(\sqrt{3}-\tan(\alpha/2))\,/\,(\sqrt{3}+\tan(\alpha/2))}$. **115.** $H(\sqrt{4\tan^2\alpha+1}-1)\,/\,4\tan\alpha$. **116.** $(\pi a^3\tan\alpha)\,/\,9\sqrt{3}$. **117.** $l\cos^2\alpha\sqrt{3+\cos^2 2\alpha}\,/\,2\sin\alpha$. **118.** $2\pi a^3(3+\cos^2\alpha)\,/\,3\sin^2\alpha$. **119.** $2V$ $(\cos(\alpha/2)\sin\alpha)\,/\,\pi$. **120.** arccot 2. **121.** $\pi a^3\cot\beta\,/\,(24\sin^3(\alpha/2))$, $\pi a^2(1+\sin\beta)\,/\,(4\sin\beta\,\sin^2(\alpha/2))$. **122.** $3S/2$. **123.** $\pi a^3\times(\cos(\beta/2)\tan\alpha)\,/\,(24\sin^3(\beta/2)), \pi a^2\sqrt{1+\cos^2(\beta/2)\tan^2\alpha}\,/\,(4\sin^2(\beta/2))$. **124.** $R^2\sqrt{3}\,/\,(4\cos\alpha)$.

**125.** $15\pi(5+2\sqrt{3})\,/\,\sqrt{37+10\sqrt{3}}$. **126.** $\sqrt{(2S/\sin2\alpha)^3/(8\cos^3\alpha)}$.

**127.** $2\arctan\sqrt{1/3}$. *Solution.* Assume that $ABCD$ is the axial section of the cylinder and $ASD$ is the axial section of the cone. We designate the magnitude of the angle between the axis of the cone and its generatrix as $\alpha, 0<\alpha<\pi/2$, and the length of the radius of the common base of the cone and the cylinder as $r$. The total surface area of the cylinder is

$$S_{\mathrm{cyl}}=2\pi r^2+2\pi r\cdot r\cot\alpha=2\pi r^2(1+\cot\alpha),$$

and the total surface area of the cone is

$S_{\mathrm{cone}}=\pi r^2+\pi r\cdot r\,/\,\sin\alpha=\pi r^2(1+\sin\alpha)\,/\,\sin\alpha$.

By the hypothesis $S_{\mathrm{cyl}}:S_{\mathrm{cone}}=7:4$, i.e.

$$\frac{2(1+\cot\alpha)\sin\alpha}{1+\sin\alpha}=\frac{7}{4}.$$

Since $1+\sin\alpha\ne0$ for $0<\alpha<\pi/2$, we can multiply the given equation by $1+\sin\alpha$ and thus reduce it to an equation $\sin\alpha+8\cos\alpha=7$. Furthermore, taking into account that $\sin\alpha=2\sin(\alpha/2)\cos(\alpha/2),\cos\alpha=\cos^2(\alpha/2)-\sin^2(\alpha/2)$, and $1=\cos^2(\alpha/2)+\sin^2(\alpha/2)$, we reduce the given equation to the form

$$15\sin^2(\alpha/2)-2\sin(\alpha/2)\cos(\alpha/2)-\cos^2(\alpha/2)=0.$$

Dividing this equation term–by–term by $\cos^2(\alpha/2)\ne0$, we obtain an equation

$$15\tan^2(\alpha/2)-2\tan(\alpha/2)-1=0,$$

which has two roots $\tan(\alpha/2)=1/3$ and $\tan(\alpha/2)=-1/5$. Since $\alpha>0$ by the hypothesis, we must reject the second solution. Thus

$$\tan(\alpha/2)=1/3\Rightarrow\alpha=2\arctan(1/3).$$

**128.** $8Q \tan \varphi \cos^6(\varphi/2)\sqrt{Q/\pi}$. **129.** $\arctan((2 \pm \sqrt{2})/4)$. **130.** $2\pi R^2 \sin\alpha \times \sin(\alpha/2)$. **131.** $\arctan(4/(4-\sqrt{6}))$. **132.** $\pi R^2 \cot^2(\alpha/2)(1+\cos\alpha)/\cos\alpha$, $\pi R^3 (\cot^3(\alpha/2) \tan\alpha)/3$. **133.** $24\pi R^2 \sin^4\alpha/(2+\tan\alpha)^2$. **134.** $-\pi r^3 \tan 2\alpha/8\cos^6\alpha$. **135.** 10 cm. **136.** $\pi aS/\sin\alpha (0 < \alpha < \pi)$. **137.** $\pi S/\sin(\pi m/(m+n))(m \le n)$. **138.** $h(\sqrt{1+(n/m)}-1)^{1/2}$. **139.** $\arctan(\sin(\beta/2)\cot\varphi)$.

**140.** $\arcsin\left|\dfrac{2\sin^2(\beta/2)-\cos^2\varphi}{\cos\varphi}\right|$. *Hint.* Draw a straight line through the point $B$ in the plane $\alpha$, parallel to the straight line ($OA$), and calculate the angle between that line and the plane of the base of the other cone.

**141.** $7\pi l^3(\sin(\alpha/2)\sin\alpha)/54$. **142.** 4 cm. **143.** $2R\sqrt{1+2\cos\alpha}/\sqrt{3}$. **144.** $r(\sqrt{6}-2)/2, r(\sqrt{6}+2)/2$. **145.** $a(2-\sqrt{3}), a(2+\sqrt{3})$. **146.** $2\arccos((1+\sqrt{17})/8)$. **147.** $0.5a\sqrt[3]{\cot(\alpha/2)/\sin^2(\alpha/4)}$. **148.** $\dfrac{\pi R^3}{3}\left[\cos^4\dfrac{\alpha}{2}\operatorname{cosec}\dfrac{\alpha}{2} - \left(1-\sin\dfrac{\alpha}{2}\right)^2\left(2+\sin\dfrac{\alpha}{2}\right)\right]$. **149.** $\pi a^3(1+\cos\alpha)^3\cot^2\alpha/3$. **150.** $Q(1+\sin\alpha)/\sin\alpha)^3\tan^2\alpha$. **151.** $2\arcsin((3+\sqrt{3})/6)$ or $2\arcsin((3-\sqrt{3})/6)$. **152.** $\arccos(1/\sqrt[3]{2})$.

**153.** The ratio of the surface area of the sphere to that of the cube is $\pi/6$, and the ratio of their volumes is $\pi/6$.

**154.** $\dfrac{1}{4}Q\cot^2\left(\dfrac{\pi-\alpha}{4}\right)\dfrac{1+\sin(\alpha/2)}{\sin(\alpha/2)} \equiv \dfrac{Q(1+\sin(\alpha/2))^3}{2\cos(\alpha/2)\sin\alpha}$. **155.** $\sin^2\alpha(1+\operatorname{cosec}(\alpha/2))/4$. **156.** $5r/\sqrt{3}$. **157.** $\pi r^2(\sqrt{r^2+(d+r)^2}+r)^3/3(d+r)$, $\pi r^2(\sqrt{r^2+(d-r)^2}+r)^3/3(d-r)^2$. **158.** $4/3 \sin\alpha\cos(\alpha/2)$. **159.** $\dfrac{1}{\sqrt{\pi}}\left(\dfrac{S\cos\alpha}{1+\cos\alpha}\right)^{3/2}\tan\alpha$. **160.** $\dfrac{\pi}{4}r^3\dfrac{m}{n\sin\alpha}\left(1\pm\sqrt{1-\dfrac{2m}{n\cos\alpha}}\right)$. (the parameters $m, n$, and $\alpha$ must satisfy the condition $n\cos\alpha \ge 2m$).

**161.** $3\pi(l^2\sin^2 2\beta)/4, \pi(l^3\sin^3 2\beta)/12$. **162.** $S^{3/2}\sin 2\alpha\sin\varphi\sqrt{\cos\varphi}/(3\pi^{3/2})$.

**163.** $S\cot^2(\alpha/2)/(\pi\cos\alpha)$. **164.** $\pi a^3\sqrt{6}/1728$. **165.** $2h(2r-h)$. **166.** $\dfrac{4a^2\cos\alpha}{\sin^2\alpha}$ $(1+\cos\alpha)^2 \equiv 4a^2\cos\alpha\cot^2(\alpha/2), \dfrac{4}{3}a^3\cot^2\alpha(1+\cos\alpha)^3$. **167.** $a^2/\cos^4(\alpha/2)$. **168.** $ab\tan\alpha\sqrt{a^2+b^2}/12$. **169.** $\pi R^2H/12$. **170.** $r^3(1+\tan\varphi)[1+\cot((\pi-4\varphi)/4)]/6$. **171.** $\pi l^2\cos\alpha/(1+3\cos^2\alpha)$. **172.** $32r^3\sqrt{\cos\alpha}/(3\sin(\alpha/2))$. **173.** $\dfrac{\pi a^3}{4}\dfrac{\tan\alpha\cos^3\beta}{\sqrt{\cos^3(\alpha+\beta)\cos^3(\alpha-\beta)}} \equiv \dfrac{\pi a^3}{4}\dfrac{\cos^2\alpha\sin\alpha}{\sqrt{(\cos^2\alpha-\sin^2\beta)^3}}$. **174.** $\pi d^3(\cos^2(\alpha/2)\cot(\alpha/2)\tan\beta)/8$. **175.** 29/36 or 1. **176.** $1/2\sqrt{2}$. *Hint.* Prove that the sphere touches all the edges of the tetrahedron at their midpoints and its centre coincides with the centre of the sphere circumscribed about the tetrahedron.

**177.** $\sqrt{5}/3$. **178.** $a(3\sqrt{2}+4)/4$. **179.** For $\varphi = \operatorname{arccot}\sqrt{2}$ the volume of the cone is $2\pi l^3/9\sqrt{3}$. **180.** $R\sqrt{2}/3$. **181.** $Q\tan(\alpha/2)\tan\beta\sqrt{2Q\cot\alpha}/3$ and $\alpha = \pi/3$. *Solution.* Assume that in the given pyramid $\angle ABC = \pi/2$, $SBC \perp ABC$, and $\angle BAC = \alpha$, the dihedral angles at the sides $[AB]$ and $[AC]$ are equal to $\beta$, the respective plane dihedral angles shown in the figure are $\angle SBC = \angle SED = \beta$, with $[SD]$ being the altitude of the pyramid, $[SE] \perp [AC]$, $[DE] \perp [AC]$ and $[SB] \perp [AB]$. Note that $|BD| = |SD|\cot\beta = |DE|$ and $|SB| = |SD|/\sin\beta = |SE|$. It is easy to show that $|AB| = |AE|$ and $\angle BAD = \angle DAE = \alpha/2$. Since, $|BC| = |AB| \times \tan\alpha$, the area of the triangle $ABC$ is

$$Q = \frac{1}{2}|AB|\cdot|BC| = \frac{1}{2}|AB|^2\tan\alpha,$$

whence

$$|AB| = \sqrt{2Q\cot\alpha}.$$

Consequently, the altitude of the pyramid $|SD| = |BD|\tan\beta = |AB|\tan(\alpha/2)\tan\beta = \tan(\alpha/2)\tan\beta\sqrt{2Q}\cot\alpha$. Then the volume of the pyramid is

$$V = Q\tan(\alpha/2)\tan\beta\sqrt{2Q\cot\alpha}/3,$$

where, by the sense of the problem, restrictions are imposed on the angles $\alpha$ and $\beta: 0 < \alpha < \pi/2$ and $0 < \beta < \pi/2$. Let us find the greatest value of the function $V(\alpha)$; for that purpose, it is sufficient to find the greatest value of the function

$$f(\alpha) = \tan(\alpha/2)\sqrt{\cot\alpha}, 0 < \alpha < \pi/2.$$

Then, we have

$$f'(\alpha) = \frac{\sqrt{\cot\alpha}}{2\cos^2(\alpha/2)} - \frac{\tan(\alpha/2)}{2\sqrt{\cot\alpha}\sin^2\alpha}$$

$$= \frac{\sin\alpha(2\cos\alpha - 1)}{4\cos^2(\alpha/2)\sin^2\alpha\sqrt{\cot\alpha}}.$$

On the interval $(0, \pi/2)$ the derivative $f'(\alpha)$ is equal to zero if and only if $\cos\alpha = 1/2$, i.e. for $\alpha = \pi/3$. If $\alpha \in (0, \pi/3)$, then $f'(\alpha) > 0$, and if $\alpha \in (\pi/3, \pi/2)$, then $f'(\alpha) < 0$. Consequently, for $\alpha = \pi/3$ the function $f(\alpha)$, and, consequently, $V(\alpha)$ as well, assume the greatest value.

**182.** 2. **183.** $R/\sqrt{3}$.

**184.** The altitude of the cylinder must be equal to the radius of the circle of the base and to $\sqrt[3]{V/\pi}$. **185.** $4V\tan^3(\alpha/2)/\tan\alpha, \alpha = \arccos(1/3)$.

**186.** The length of the radius of the base of the cone is $3r/2, 9\pi r^2 h/4$.

**187.** (1) $(3a^2h)/4\sqrt{a^2 + 3h^2}$ for $h > a/\sqrt{6}$;

   (2) $(a/2)\sqrt{h^2 + (a^2/12)}$ for $0 < h \leq a/\sqrt{6}$.

**188.** $b^3(\tan\alpha\sin 2\varphi\cos\varphi)/24$. The volume of the pyramid assumes the greatest value for $\varphi = \operatorname{arccot}\sqrt{2}$.

**189.** (1) $2rR^3/(R^2 - r^2)$ for $r < R \leq (1 + \sqrt{2})r$;

   (2) $R^2((R^2 + r^2)/(R^2 - r^2))^2/2$ for $R > (1 + \sqrt{2})r$.

**190.** (1) $(H^2\cot\beta)/\sqrt{2}$ for $\arctan(\sqrt{2}/2) \leq \beta < \pi/2$;

   (2) $H^2(1 + \sin^2\beta)/(4\sin^2\beta)$ for $0 < \beta < \arctan(\sqrt{2}/2)$.

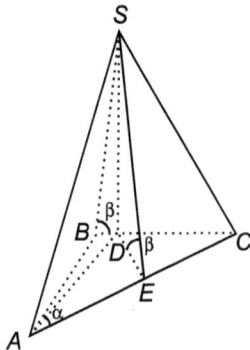

191. $\tan(\alpha/2)$. 192. $\tan(\alpha/2)$. 193. $\dfrac{4\pi a^3}{3}\dfrac{\cos^4(\alpha/2)}{\cos^2\alpha}\sin\dfrac{\alpha}{2}$. 194. $\pi a^3$ $\sin^2\alpha\sin(\alpha+\beta)\operatorname{cosec}\beta$. 195. $\pi a^3(\sin 2\alpha\sec^2 2\alpha)/6$. 196. $\pi(16S^2$ $+6Sh^2\cot^2\alpha+h^4\cot^4\alpha)/24h$. 197. $\sqrt{6V}\sin\alpha\cot\beta\,\tan(\beta/2)/2,\beta=\arccos(1/3)$. 198. $\pi R^3/(3\cos^2(\alpha/2)\sin(\alpha/2)),\alpha=2\operatorname{arccot}\sqrt{2}$.

# Part-4

## Sec. 4.1

**III. 3.** $D(y)=(-\infty,3-\sqrt{3})\cup(3+\sqrt{3},+\infty)$. **4.** $T=2\pi/3$.

**V. 2.** $C_3^1\cdot C_{10}^6+C_3^2\cdot C_{10}^5+C_3^3\cdot C_{10}^4=1596$. **3.** $4(\sqrt{2}-1)$.

**VI. 2.** $\{\pi n/2|n\in\mathbf{Z}\}$. **3.** $\sqrt{117}$.

**X. 3.** $1/2$.

**XI. 3.** $(-\infty,0)\cup(2,+\infty)$. **4.** $y_{\text{least}}=y(1/16)=-1/8;y_{\text{gr}}=y(4)=6$.

**XII. 3.** $\{\pi n/5\,|n\in\mathbf{Z}\}$. **4.** $y_{\text{least}}=y(2)=-1;y_{\text{gr}}=y(3)=3$.

**XIII. 3.** The function decreases on the interval $(-\infty,3/2)$ and increases on the interval $(3/2,+\infty)$ **4.** $(1,+\infty)$.

**XIV. 2.** $[-\sqrt{2},-1)\cup(1,\sqrt{2}]$.

**XV. 3.** $\{2\}$. **4.** $1/(a+b)$.

**XVI. 3.** $(-1,7/3)$.

**XVII. 3.** $(5-3\sqrt{2},5+3\sqrt{2})$. **4.** 1. **5A.** $y_{\text{gr}}=y(1)=y(6)=7/3$.
$y_{\text{least}}=y(\sqrt{6})=2\sqrt{6}/3$. **B.** $\{(\pi(2n+1))/4|n\in\mathbf{Z}\}$.

**XVIII. 3.** $y'=-\dfrac{3C\tan^2 x}{\sin^2 x}$ **XIX. 3.** $4\sqrt{5}$. **XX. 3.** $3\pi/4$. **XXI. 3.** $0$.

**XXII. 3.** $\left\{\pi n,\dfrac{6m+(-1)^m}{6}\pi\,|n,m\in\mathbf{Z}\right\}$.

**XXIII. 4.** $\left\{\dfrac{2n+1}{2}\pi,\dfrac{6m+(-1)^{m+1}}{6}\pi\,|m\in\mathbf{Z}\right\}$

## Sec. 4.2

**2.** Yes, it can. **5.** *Hint.* Use the fact that $n^4+4=(n^2+2n+2)(n^2-2n+2)$.

**3.** Yes, it can if one of them is the number 2.

**4.** *Solution.* Since, we can represent any odd number in the form $2m+1$, where $m\in Z$, the truth of the statement follows from the equation $2m+1=(m+1)^2-m^2$. **5.** 1692. **6.** (a) $\sin 3>\cos 3$; (b) $\sqrt[3]{3}>\sqrt{2}$ (since $(\sqrt[3]{3})^6=9,(\sqrt{2})^6=8)$. **7.** $A>B$.

**8.** *Hint.* Since $(\sqrt{2}+\sqrt{3})^2=5+2\sqrt{6}$, it follows that to prove the statement it is sufficient to prove that $\sqrt{6}$ is an irrational number.

**9.** *Hint.* Compare the area of a circle of unit radius with that of a regular inscribed 12–gon and the area of a circumscribed square.

**10.** $-1$ for $x<1;1$ for $x>1$.

**11.** $2/(1-a)$. **12.** $a^{511/512}$.

**13.** $n(n+1)/2$ for $x=1$; $(x-(n+1)x^{n+1}+nx^{n+2})/(1-x)^2$ for $x \neq 1$. *Solution.* For $x=1$ the given sum is equal to $1+2+3+\ldots+n = n(n+1)/2$. For $x \neq 1$ the sum

$$x + 2x^2 + 3x^3 + \ldots + nx^n$$
$$= x(1 + 2x + 3x^2 + \ldots + nx^{n-1}) = x(x + x^2 + x^3 + \ldots + x^n)'$$
$$= x\left(\frac{x-x^{n+1}}{1-x}\right)' = x\frac{1-(n+1)x^n + nx^{n+1}}{(1-x)^2}.$$

**14.** *Solution.* Since, $1 - \dfrac{1}{k^2} = \dfrac{k-1}{k} \cdot \dfrac{k+1}{k}, k \in \mathbf{Z}$, it follows that

$$\left(1-\frac{1}{4}\right)\left(1-\frac{1}{9}\right)\left(1-\frac{1}{16}\right)\ldots\left(1-\frac{1}{n^2}\right)$$
$$= \frac{1}{2} \cdot \frac{3}{2} \cdot \frac{2}{3} \cdot \frac{4}{3} \cdot \frac{3}{4} \cdot \frac{5}{4} \ldots \frac{n-1}{n} \cdot \frac{n+1}{n} = \frac{1}{2} \cdot \frac{n+1}{n} = \frac{n+1}{2n}.$$

**15.** *Solution.* Since the equality $\dfrac{1}{k(k+1)} = \dfrac{1}{k} - \dfrac{1}{k+1}$ holds for any $k \neq 0$,

we have $\dfrac{1}{1 \cdot 2} + \dfrac{1}{2 \cdot 3} + \dfrac{1}{3 \cdot 4} + \ldots + \dfrac{1}{n(n+1)}$

$$= \frac{1}{1} - \frac{1}{2} + \frac{1}{2} - \frac{1}{3} + \frac{1}{3} - \frac{1}{4} + \ldots + \frac{1}{n} - \frac{1}{n+1} = 1 - \frac{1}{n+1} = \frac{n}{n+1}.$$

**16.** *Hint.* Prove the equality by induction.

**17.** $3 \cdot 3 \cdot P_3 = 54$.      **18.** 3970.      **19.** $C_9^5 = 126$.

**20.** $n(n-3)/2$. *Hint.* The number of line segments connecting the vertices of a convex $n$-gon is equal to $C_n^2$, $C_n^2 - n$ out of which are diagonals (the other segments are the sides of the polygon). There are other methods of solving the problem. Here is one of them. Each vertex is connected by diagonals with $n-3$ vertices. Since, there are $n$ vertices, we get $n(n-3)$; but we have counted each diagonal twice (a diagonal connects two vertices of the polygon and we have considered it to emanate from the first and from the second vertex). Therefore, there are $n(n-3)/2$ different diagonals.

**21.** Not more than $C_{m+n+k}^3 - C_m^3 - C_n^3 - C_k^3$.

**22.** *Hint.* Multiply the left–hand and right–hand sides of the equations $(1+x)^n = C_n^0 + C_n^1 x + \ldots + C_n^n x^n$, $(x+1)^n = C_n^0 x^n + C_n^1 x^{n-1} + \ldots + C_n^n$ and equate the coefficients in $x^n$ of the left–hand and right–hand sides of the resulting equation to each other.

**23.**

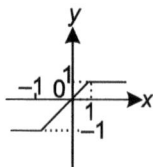

**24.**

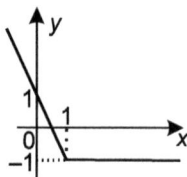

25.

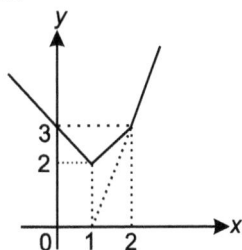

26.

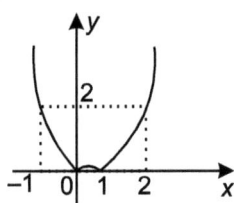

27.

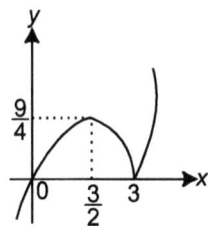

28.

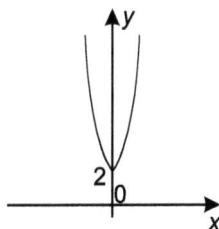

29.

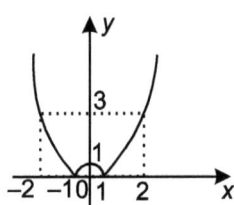

30.

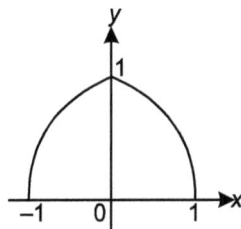

31.

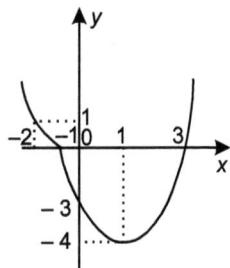

32.

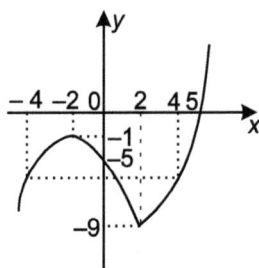

33.

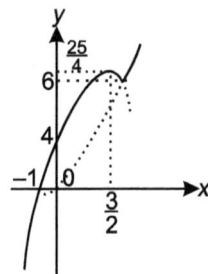

**34.**

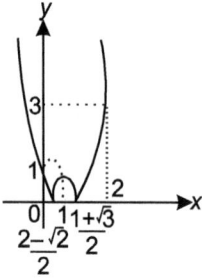

**35.**

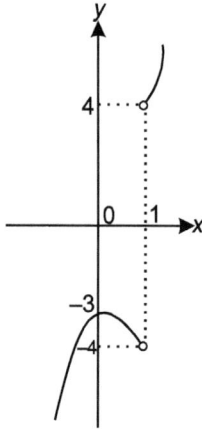

**36.**

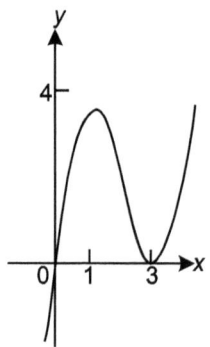

**37.**

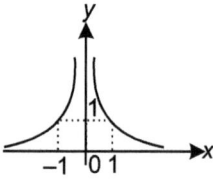

**38.**

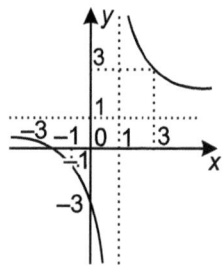

**39.**

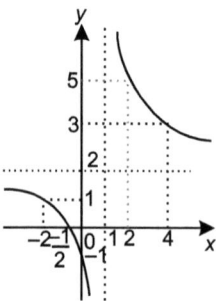

**40.**

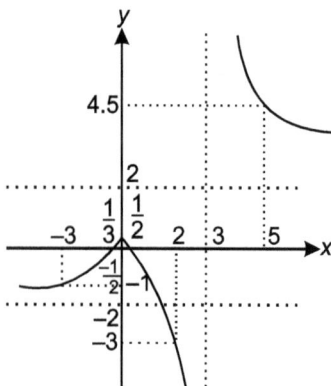

**41.**

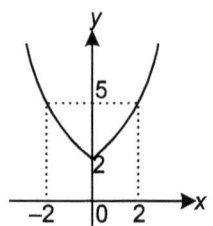

**42.**

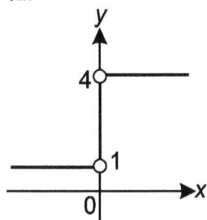

**43.**

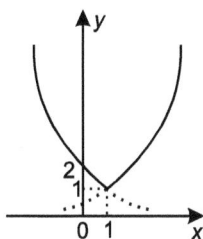

**44.**

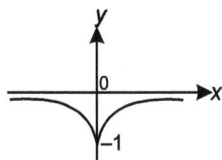

**45.**

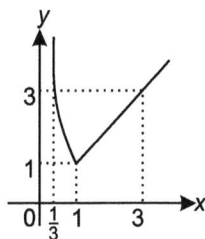

**46.**

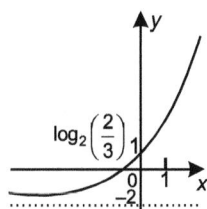

**47.**

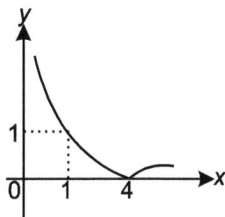

**48.**

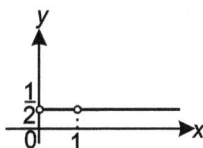

**49.**

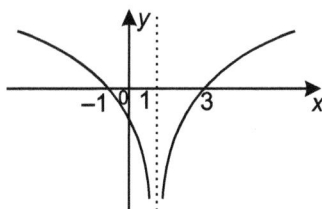

**50.**

51.

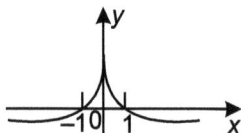

52.

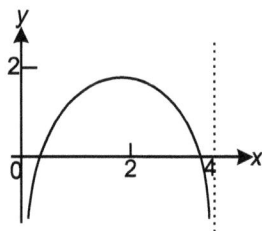

53.

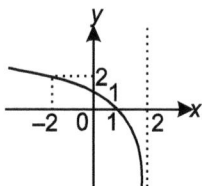

54.

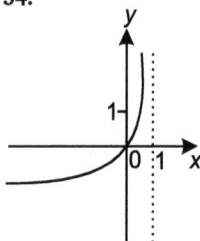

55.

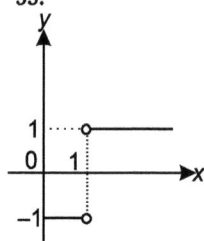

56.

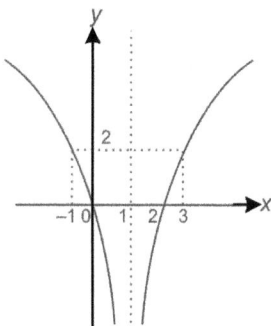

57.

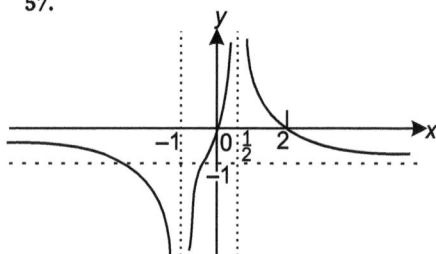

58.

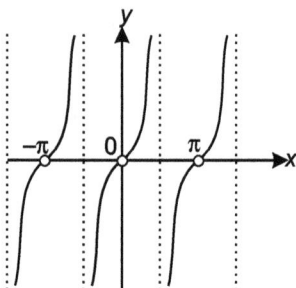

**59.**

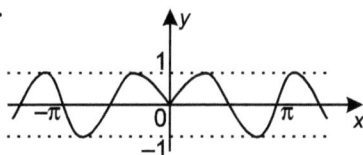

**60.**

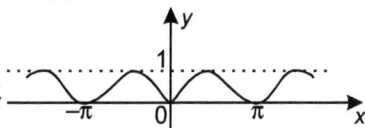

**61.**

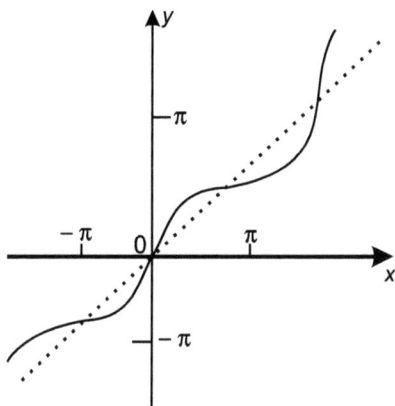

**62.**

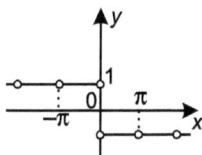

**63.**

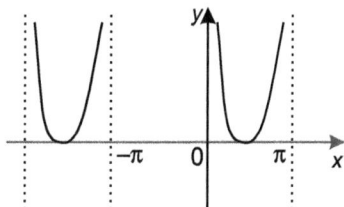

**64.**

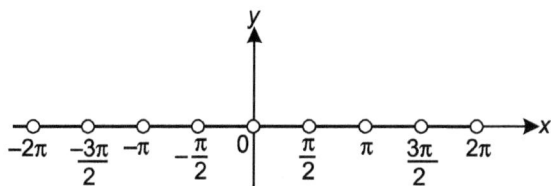

**65.**

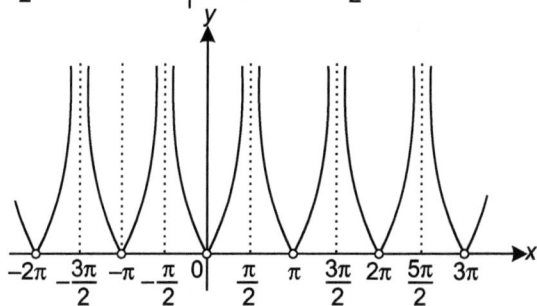

**66.**

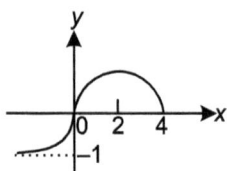

**67.**

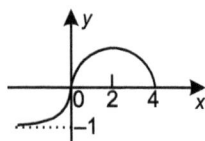

**68.**

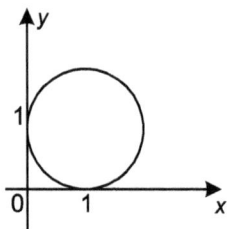

**69.**

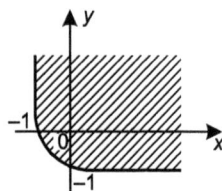

**70.**

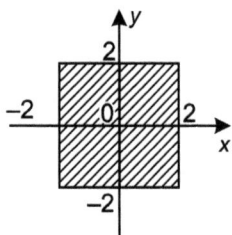

**71.**

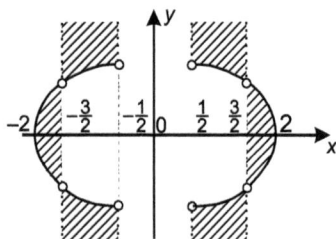

**72.**

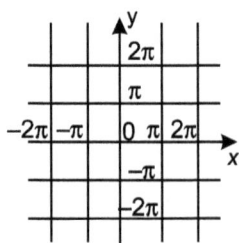

**73.**

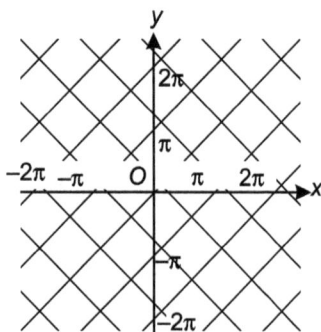

**74.** $y_{\text{least}} = 2$ for all $x \in [1,3]$. *Hint.* Consider the function for $x < 1$; for $1 \le x \le 3$; for $x > 3$.

**75.** $\{1 - \sqrt{2}\}$. *Hint.* For $x \ge 0$ the given equation is equivalent to the system.
$$\begin{cases} x^2 + 2x + 1 = 0, \\ x \ge 0, \end{cases}$$
and for $x < 0$ it is equivalent to the system
$$\begin{cases} x^2 - 2x - 1 = 0, \\ x < 0. \end{cases}$$

**76.** $z_{\text{least}} = 1$ for $x = 0$ and $y = -1$. The truth of the statement follows from the identity
$$z = x^2 + 2xy + 3y^2 + 2x + 6y + 4 \equiv (x + y + 1)^2 + 2(y + 1)^2 + 1.$$

**77.** $(2,7)$.      **78.** $(-4,-1)\cup(1,4)$.

**79.** $(-\infty, 2 - \sqrt{5})\cup(2 - \sqrt{3}, 2 + \sqrt{3})\cup(2 + \sqrt{5}, +\infty)$.

**80.** $(-\infty, -3/2]\cup[-1,1]\cup[3/2, +\infty)$.

**82.** *Hint.* It follows from the hypothesis that $c, a - b + c, 4a - 2b + c$ are integers and therefore $c, a - b, 2a, 2b, a + b$ are integers. Taking the conclusion into account, consider separately the case of an even $x$ and the case of an odd $x$.

**84.** $a = 1$. *Hint.* It follows from the hypothesis that $x_1 + x_2 = 5/a, x_1 x_2 = 6/a$, and $x_1 = 2x_2/3$. Solving this system, we get $a = 1$.

**85.** $a > 0$. *Solution.* By the hypothesis
$$a - b + c > -4, \tag{1}$$
$$a + b + c < 0 \Rightarrow -a - b - c > 0, \tag{2}$$
$$9a + 3b + c > 5. \tag{3}$$
Adding (1) and (2) together, we obtain
$$-2b > -4. \tag{4}$$
Adding (2), (3), and (4) together, we get $8a > 1 \Rightarrow a > 0$.

**86.** $\{3 - 2\sqrt{5}, 3, 3 + 2\sqrt{5}\}$. *Solution.* We rewrite the equation as follows :
$$((x^2 - 6x + 9) - 9)^2 - 2(x - 3)^2 = 81$$
or as    $((x - 3)^2 - 9)^2 - 2(x - 3)^2 = 81,$
i.e.    $(x - 3)^4 - 20(x - 3)^2 = 0$
or    $(x - 3)^2(x - 3 + \sqrt{20})(x - 3 - \sqrt{20}) = 0,$
whence it follows that
$$x_1 = 3 - 2\sqrt{5}, x_2 = 3, x_3 = 3 + 2\sqrt{5}.$$

**87.** $\{3/2, 2/3\}$. *Hint.* Reduce the equation to the form
$$6\left(x^2 + \frac{1}{x^2}\right) - 13\left(x + \frac{1}{x}\right) + 12 = 0,$$
set $x + \dfrac{1}{x} = t$, and take into account that $x^2 + \dfrac{1}{x^2} = t^2 - 2$.

**88.** $7 - x$. *Hint. 1st method.* Group together $x^3 + 2x^2 - 2x + 5$
$= (x^3 - x) + (2x^2 - 2) - x + 7 = (x^2 - 1)(x + 2) - x + 7.$

*2nd method.* Taking into account that the remainder of the division of the polynomial by $x^2 - 1$ has the form $ax + b$, use the Bezout theorem.

**89.** No roots if $a < -15\sqrt[3]{10}/16$; one root if $a = -15\sqrt[3]{10}/16$; two roots if $a > -15\sqrt[3]{10}/16$. *Solution.* Let us consider the function $y = x^4 - 5x - 2a$. The values of $x$ for which $y = 0$ are roots of the given equation. The graph of the given function has two symmetric branches drawn from the vertex in the direction of the increase of $y$. Thus at the vertex $(x_0, y_0)$ the function assumes its least value. Let us find $x_0$ and $y_0$. For that purpose we find $y' = 4x^3 - 5$ and equate the derivative to zero. Then $4x_0^3 - 5 = 0, x_0^3 = 5/4, x_0 = \sqrt[3]{10}/2$ and, consequently,

$$y_0 = x_0(x_0^3 - 5) - 2a = -\frac{15\sqrt[3]{10}}{8} - 2a.$$

Consequently, if $y_0 > 0$, i.e. $a < -15\sqrt[3]{10}/16$, then the vertex lies above the $Ox$ axis and the graph of the function does not cut the $Ox$ axis; therefore, the given equation has no roots. If $y_0 = 0$, i.e. $a = -15\sqrt[3]{10}/16$, then the graph of the function touches the $Ox$ axis and the given equation has one root; $x_0 = \sqrt[3]{10}/2$. If $a > -15\sqrt[3]{10}/16$, then the vertex lies below the $Ox$ axis and the graph of the function cuts the $Ox$ axis at two different points ; consequently, the given equation has two roots.

**90.** One root for $a > -3$; two roots for $a = -3$; three roots for $a < -3$. *Hint.* Consider the function $y = x^3 + ax + 2$. If the function has no extrema, then there is one root. If the function possesses a maximum and a minimum, and $y > 0$ at the point of maximum and $y < 0$ at the point of minimum, then there are three roots and so on.

**92.** $\{13/12\}$. **93.** $(-\infty, 0)$. **94.** $(-1, 0)$. **95.** $(-4, -3)$. **96.** $(-6, -2) \cup (0, 3)$.

**97.** $(-\infty, -1) \cup (-1, 1/2)$. *Hint.* When solving the problem, it is necessary to take into account that the given inequality is equivalent to the collection of two inequalities : $(2x - 4)/(x + 1) > 2$ and $(2x - 4)/(x + 1) < -2$.

**98.** $[3, 4) \cup (4, +\infty)$. **99.** $(-1, 0] \cup \{1\}$. **100.** $\{4\}$. **101.** $\{0\}$.

**102.** $[-2, 0) \cup [\sqrt{2 - \sqrt{3}}, \sqrt{2 + \sqrt{3}}]$. *Solution.* The domain of definition of the inequality is the set of numbers $[-2, 0) \cup (0, 2]$. All $x \in X_1 = [-2, 0)$ are solutions of the inequality since the left–hand side of the inequality is nonnegative and the right–hand side is negative for these values of $x$. If $x \in (0, 2]$, then both sides of the inequality are nonnegative and, therefore, in that set of numbers the given inequality is equivalent to the system of inequalities

$$\begin{cases} 0 < x \le 2, \\ 4 - x^2 \ge 1/x^2, \end{cases} \text{or} \begin{cases} 0 < x \le 2, \\ x^4 - 4x^2 + 1 \le 0; \end{cases}$$

whence it follows that

$$\begin{cases} 0 < x \le 2, \\ (x + \sqrt{2 + \sqrt{3}})(x + \sqrt{2 - \sqrt{3}})(x - \sqrt{2 - \sqrt{3}})(x - \sqrt{2 + \sqrt{3}}) < 0. \end{cases}$$

The solution of the system is a set of numbers $X_2 = [\sqrt{2 - \sqrt{3}} \sqrt{2 + \sqrt{3}}]$. The solution of the given inequality is a set $X_1 \cup X_2$.

**103.** $[-3/2, 1]$. *Solution.* The domain of definition of the inequality is the set of numbers $[-3/2, 1]$. Both sides of the inequality are positive for all $x$ from the domain of definition and, therefore, both sides of the inequality can be squared.

After the requisite transformations, we get an inequality
$$2\sqrt{1-x}\sqrt{2x+3} < 21-x,$$
whose both sides are nonnegative in the domain of definition. Rationalizing the inequality, we get an inequality $9x^2 - 38x + 429 > 0$, which is satisfied for all $x \in \mathbf{R}$. Consequently, the given inequality holds true in the whole domain of definition. *Remark.* In the domain of definition $\sqrt{1-x} \le \sqrt{1+3/2} < 2$, $\sqrt{2x+3} \le \sqrt{5} < 3$ and, therefore, $\sqrt{1-x} + \sqrt{2x+3} < 5$ for all $x$ belonging to the domain of definition.

**104.** $(-\infty, 2)$. *Hint.* The given inequality is equivalent to the collection of two systems of inequalities $\begin{cases} x+1 < 0, \\ 11-x \ge 0 \end{cases}$ and $\begin{cases} 11-x \ge 0, \\ x+1 \ge 0, \\ (x+1)^2 < 11-x. \end{cases}$

**105.** $[-\sqrt{2}/4, 0) \cup (0, 1/3)$. *Solution.* The domain of definition of the inequality is the set of numbers $[-\sqrt{2}/4, 0) \cup (0, \sqrt{2}/4]$. Since $1 - \sqrt{1-8x^2} \ge 0$, it follows that the initial inequality holds true for all $x \in [-\sqrt{2}/4, 0)$. Assume that $x \in (0, \sqrt{2}/4]$, then the initial inequality is equivalent to the inequality
$$1 - 2x < \sqrt{1-8x^2}, \tag{1}$$
with $1 - 2x > 0$ for the values of $x$ being considered. Squaring both sides of inequality (1), we get, after a transformation, an inequality $3x^2 - x < 0$, or $3x - 1 < 0$, i.e. $x < 1/3$. Consequently, the inequality holds true for all $x \in [-\sqrt{2}/4, 0) \cup (0, 1/3)$.

**106.** $[a/2, +\infty)$ for $a < 0$; $[-a/2, +\infty)$ for $a \ge 0$. *Solution.* If $x \ge 0$, then the inequality holds true. For $x < 0$ the given inequality is equivalent to $(\sqrt{5x^2+a^2})^2 \ge (-3x)^2$, or to $x^2 \le a^2/4$, or to $|x| \le |a|/2$. Thus the inequality also holds true for all $-|a|/2 \le x < 0$. Combining the results obtained, we get the intervals indicated above.

**107.** $\{(c, 1-c)\}$ for $0 \le c \le 1$; $\{(c, -1-c)\}$ for $-1 \le c \le 0$. *Hint.* Solve the system by graphical means.

**108.** $\varnothing$. **109.** $\{(11/5, 1/5)\}$ for $a = 3$; $\varnothing$ for $a \ne 3$. *Solution.* The solution of the system of the first two equations is a pair of numbers $x = 11/5, y = 1/5$. This pair of numbers is a solution of the third equation of the given system only for $a = 3$.

**110.** $\{(-1, 2)\}$ for $a = -2$; $\varnothing$ for $a \ne 2$.

**111.** $\{(1, 8), (8, 1)\}$.　　　**112.** $\{(5, 7)\}$.

**113.** $\varnothing$ if $|a| < \sqrt{2}/2$, or $|a| > 1$; four if $|a| = 1$; $\sqrt{2}/2$, or $|a| = 1$; eight if $\sqrt{2}/2 < |a| < 1$. *Hint.* Solve the system by graphical means. Take into account that $x^2 + y^2 = a^2$ for $a \ne 0$ is an equation of a circle of radius $|a|$; for $a = 0$ the equation is satisfied by the coordinates of the point $O(0,0)$; for $|a| < 0$ the equation has no solutions. The equation $|x| + |y| = 1$ is satisfied by the coordinates of the points of the sides of the square with vertices at the points $(-1, 0)$, $(0, 1)$, $(1, 0)$ and $(0, -1)$.

**114.** $a = \pm 1$ and $a = \pm 2/\sqrt{5}$. *Solution.* From the second equation of the system, we find that $1 + xy = -(1 + x + y)$. Removing $1 + xy$ from the first equation of the system, we get an equation
$$(a+1)x + (a-1)y = -a. \tag{1}$$

For $a = 1$, we have $x = -1/2$ and $y = -3$; for $a = -1$, we have $y = -1/2$ and $x = -3$; for $a \neq \pm 1$ we find the expression for $y$ from equation (1) :

$$y = \frac{a}{1-a} + \frac{a+1}{1-a}x.$$

Removing $y$ from the second equation of the given system, we find the following equation to define $x$ :

$$x^2 + \frac{a+2}{a+1}x - \frac{a-2}{a+1} = 0;$$

the discriminant of this equation is equal to zero for $a = \pm 2 / \sqrt{5}$. For each of these values of $a$ the system of equations also has a unique solution.

**115.** $-1/2$. **116.** (a) $\sqrt{5}$; (b) 1. **117.** $-18$. **118.** 16. **119.** 4. **120.** 0. **121.** Positive.

**122.** Negative.

**123.** $\dfrac{a + 2b - 1}{1 - a}$. *Solution.* $\log_5 9.8 = \dfrac{\log 9.8}{\log 5} = \dfrac{\log(98/10)}{\log(10/2)}$

$$= \frac{\log 2 + 2\log 7 - \log 10}{\log 10 - \log 2} = \frac{a + 2b - 1}{1 - a}.$$

**124.** $(5 - b)/(2ab + 2a - 4b + 2)$. *Solution.* Since $\log_{20} 50 = b$, it follows that

$$\frac{\log(100/2)}{\log 20} = b \Rightarrow \frac{2 - \log 2}{1 + \log 2} = b \Rightarrow \log 2 = \frac{2 - b}{1 + b} \qquad (1)$$

Since $\log 15 = a$, we have

$\log 3 + \log 5 = a \Rightarrow \log 3 + \log(10/2) = a \Rightarrow \log 3 - \log 2 = a - 1$
$\Rightarrow \log 3 = \log 2 + a - 1$. Taking (1) into account, we find that

$$\log 3 = \frac{2 - b}{1 + b} + a - 1. \qquad (2)$$

Taking (1) and (2) into consideration, we obtain

$$\log_9 40 = \frac{1 + 2\log 2}{2\log 3} = \frac{5 - b}{2ab + 2a - 4b + 2}.$$

**125.** $D(y) = (-\infty, 0]$. **126.** $D(y) = [2, 3) \cup (3, 4]$. **127.** $D(f) = (1, 3]$.
**128.** $D(y) = (0, 1]$. **129.** $D(f) = (5, +\infty)$.
**130.** $D(y) = (-\infty, -3]$. *Hint.* To find the domain of the given function, we must solve the inequality $\sqrt{x^2 - 5x - 24} > x + 2$ which is equivalent to the collection of two systems of inequalities

$$\begin{cases} x^2 - 5x - 24 \geq 0, \\ x + 2 < 0 \end{cases} \text{and} \begin{cases} x^2 - 5x + 24 \geq 0, \\ x + 2 \geq 0, \\ x^2 - 5x - 24 > (x + 2)^2. \end{cases}$$

**131.** $D(y) = \{(2\pi n, \pi(2n + 1)) | n \in \mathbf{Z}\}$. **132.** $\{1/2\}$. **133.** $\{6\}$. **134.** $\{1/3\}$.
**135.** $\{4 - \sqrt{11}\}$. **136.** $\{16\}$. **137.** $\{-1, 3\}$. **138.** $\{2\}$. **139.** $\{-9/10\}$. **140.** $\{5\}$.
**141.** $\{-2, 3\}$. *Hint.* Multiply the equation by $3^{-2x-12}$ and set $3^{x^2-x-6} = t$. Obtain a quadratic equation to define $t$.
**142.** $\{\pi(2n + 1)/2 | n \in \mathbf{Z}\}$. *Hint.* Reduce the equation to the form $2^{\sin^2 x} + 8 \cdot 2^{-\sin^2 x} = 6$ and set $2^{\sin^2 x} = t$. Obtain a quadratic equation to define $t$.

**143.** $\{-7/5, 61\}$. *Hint.* Set $\log_5(2x+3) = t$ to simplify the calculations.

**144.** $\{2\}$. *Hint.* Set $2^x = t$ to simplify the calculations.

**145.** $\{2\}$. *Hint.* Assuming $|\log_8 x| < 1$, write the left–hand side of the equation in the form $(\log_8 x)/(1 - \log_8 x)$.

**146.** $\{2\pi k \mid k \in \mathbf{Z}\}$.      **147.** $\{\pi(8n+3)/4 \mid n \in \mathbf{Z}\}$.

**148.** $\varnothing$. *Solution.* The given equation is meaningless if $\cos 2x = 0$, i.e. if $x = \pi(2k+1)/4, k \in \mathbf{Z}$. Note that

$$\frac{\sin^2\left(x - \dfrac{\pi}{4}\right)}{\cos 2x} = \frac{1 - \cos\left(2x - \dfrac{\pi}{4}\right)}{2\cos 2x} = \frac{1 - \sin 2x}{2\cos 2x} = \frac{(\sin x - \cos x)^2}{-2(\sin^2 x - \cos^2 x)}$$

$$= \frac{\sin x - \cos x}{-2(\sin x + \cos x)} = -\frac{\tan x - 1}{2(1 + \tan x)}$$

$$= -\frac{1}{2}\tan\left(x - \frac{\pi}{4}\right).$$

In this case the initial equation has the form $2^{\tan\left(x-\frac{\pi}{4}\right)} = 1$. The solutions of the last equation are $x = \pi(4n+1)/4, n \in \mathbf{Z}$, but all these numbers are excluded from the consideration (they are obtained for $k = 2n$). Consequently, the given equation has no solutions.

**149.** $\{2\pi n + \arctan((\sqrt{17}-1)/4) \mid n \in \mathbf{Z}\} \equiv \{2\pi n + 0.5\arctan 4 \mid n \in \mathbf{Z}\}$.

**150.** $\{a\}$ for $a \in (0,1) \cup (1, +\infty); \varnothing$ for $a \in (-\infty, 0] \cup \{1\}$.

**151.** $\{10, 100\}$. *Solution.* Applying the fundamental logarithmic identity, we write the given equation in the form $10^{\log x\,(\log x - 3)} = 10^{-2} \Rightarrow \log^2 x - 3\log x + 2 = 0 \Rightarrow \log x = 1, \log x = 2; x_1 = 10$ and $x_2 = 100$. *Remark.* We can solve the equation by taking logarithms of both sides of the equation to the base 10, etc.

**152.** $\{1, 100\}$.

**153.** $\{1/11\}$. *Hint.* Take logarithms of both sides of the equation to the base $x$.

**154.** $\{100\}$. *Hint.* When solving the problem, bear in mind that $a^{\log_b c} = c^{\log_b a}$ for $a > 0$, $b > 0, c > 0$, and $b \neq 1$. Indeed, applying the fundamental logarithmic identity, we obtain $a^{\log_b c} = (b^{\log_b a})^{\log_b c} = (b^{\log_b c})^{\log_b a} = c^{\log_b a}$.

**155.** (a) For all $a \in (-\infty, 0]$; (b) for all $a \in (0, +\infty)$.

**156.** One root for $a = 0$; two roots for $a = 1$; four roots for $0 < a < 1; \varnothing$ for $a < 0$ and $a > 1$. *Hint.* Solve the problem by graphical means: construct the graphs of the functions $y = x^2 e^{2-|x|}$ and $y = 4a$ and determine how many common points the graphs of these functions possess for different values of $a$.

**157.** $\{(1, 2), (2,1)\}$. **158.** $\{(-3, -9), (3, 9)\}$. **159.** $\{(-1/4, -9/14), (5/4, -3/4)\}$. *Hint.* Taking the first equation of the system $(3x + y)^{x-y} = 9$ into account, transform the second equation of the system as follows :

$$324^{1/(x-y)} = 2(3x+y)^2 \Rightarrow 324 = 2^{x-y} \cdot (3x+y)^{2(x-y)} \Rightarrow 324$$
$$= 81 \cdot 2^{x-y} \Rightarrow x - y = 2.$$

Obtain the system of equations $\begin{cases} x - y = 2, \\ (3x+y)^2 = 9. \end{cases}$ to define $x$ and $y$.

**160.** $\{25, 36\}$. **161.** $[0, +\infty)$. **162.** $(\log_{3/2} 6, +\infty)$.

Part-4 • Hints and Answers

**163.** $(\log_2((5+\sqrt{29})/2),+\infty)$. **164.** $(-\infty,-1)$. **165.** $(0,2)$. **166.** $[-15/4,+\infty)$.
**167.** $(3/2,2)$. **168.** $(1/3,1)$. **169.** $(-\infty,0)\cup(0,8)$. **170.** $(1/9,9)$.
**171.** $(0,1)\cup(3,+\infty)$. *Solution.* $\dfrac{1}{\log_3 x}-1<0,$ or $\dfrac{1-\log_3 x}{\log_3 x}<0,$ or

$$\dfrac{\log_3 x-1}{\log_3 x}>0 \Leftrightarrow \begin{bmatrix}\log_3 x<0,\\ \log_3 x>1,\end{bmatrix} \Leftrightarrow \begin{bmatrix}0<x<1,\\ x>3.\end{bmatrix}$$

**172.** $(-8,0)$. *Solution.* $\log_{1/2}\log_3(1-x)>-1 \Leftrightarrow 0<\log_3(1-x)$
$<(1/2)^{-1} \Leftrightarrow 1<1-x<9 \Leftrightarrow -1>x-1>-9 \Leftrightarrow -8<x<0.$

**173.** $(1-\sqrt{5},-1)\cup(3,1+\sqrt{5})$. *Solution.* $(1/2)^{\log_2(x^2-2x-3)}>1$

$$\Leftrightarrow \log_2(x^2-2x-3)<0 \Leftrightarrow \begin{cases}x^2-2x-3>0,\\ x^2-2x-3<1,\end{cases}$$

$$\Leftrightarrow \begin{cases}(x+1)(x-3)>0,\\ (x-1+\sqrt{5})(x-1-\sqrt{5})<0,\end{cases} \Leftrightarrow \begin{cases}1-\sqrt{5}<x<-1,\\ 3<x<1+\sqrt{5}.\end{cases}$$

**174.** $[-4,-3)\cup(3,4]$. **175.** $(2^{-\sqrt{2}},1/2)\cup(1,2^{\sqrt{2}})$. *Hint.* Transform the logarithms to the base 2, designate $\log_2 x = t$, and solve the resulting inequality by the method of intervals. **176.** $(0,1)$.

**177.** $(4,+\infty)$. *Hint.* Since the base of the logarithm exceeds unity in the domain of definition of the inequality, the given inequality is equivalent to the system of inequalities

$$\begin{cases}x-4>0,\\ x-4<(x-3)^2.\end{cases}$$

**178.** $(3,4)\cup(5,+\infty)$. *Hint.* When taking antilogarithms, consider two cases: (1)$0<x-3<1;(2)x-3>1.$

**179.** $(-2,-1)\cup(-1,0)\cup(0,1)\cup(2,+\infty)$. *Hint.* When taking antilogarithms, consider two cases: (1)$0<x^2<1;(2)x^2>1.$

**180.** $[7,+\infty)$.
**181.** $(0,(3-\sqrt{5})/10)\cup((3+\sqrt{5})/10,4/5)\cup(1,+\infty)$. *Solution.* Let us transform the inequality as follows :

$$\log_{x+0.2}2<\log_x 4$$

$$\Leftrightarrow \dfrac{1}{\log_2(x+0.2)}<\dfrac{2}{\log_2 x} \Leftrightarrow \dfrac{\log_2 x-2\log_2(x+0.2)}{\log_2 x\cdot\log_2(x+0.2)}<0$$

$$\Leftrightarrow \log_2\dfrac{x}{(x+0.2)^2}\cdot\log_2 x\cdot\log_2(x+0.2)<0; \qquad (1)$$

$\log_2 x$ changes sign when $x$ passes through the point $x_1=1;\log_2(x+0\cdot2)$ changes sign when $x$ passes through the point $x_2=4/5;\log_2\dfrac{x}{(x+0.2)^2}$ changes sign when $x$ passes through the points $x_3=(3-\sqrt{5})/10$ and $x_4=(3+\sqrt{5})/10$. There are no other points where the left-hand side of inequality (1) would change sign upon an increase of $x$. Marking these points on the $Ox$ axis and taking into account that the inequality is not defined for $x\leq 0$ and the left-hand side of it is negative for $x>1$, we can easily find the set of solutions by the method of intervals.

**182.** $(-3/2,-1)\cup(-1,0)\cup(0,3)$. *Hint.* When taking antilogarithms, consider two cases: (1) $0<2x+3<1;(2)2x+3>1.$

**183.** (1, 3). *Hint.* Since $x > 1$, the given inequality is equivalent to the system of inequalities $(x + 3)/(x - 1) > x$ and $x > 1$.

**184.** $\{((2n-1)\,\pi,\,2\pi n)\,|n \in \mathbf{Z}\}$. *Hint.* The given inequality is equivalent to the inequality $2^{-x} - 100\sin x > 2^{-x} \Leftrightarrow -100\sin x > 0 \Leftrightarrow \sin x < 0$.

**185.** $(-1, -\sqrt{1-a}) \cup (\sqrt{1-a}, 1)$ for $0 < a < 1; \varnothing$ for the other values of $a$.

**186.** (1, 2) $\cup$ (3, $+\infty$). *Hint.* The inequality is defined for $x \in (0,2) \cup (2, +\infty)$. In this domain the inequality is equivalent to the inequality

$$\log_4 \frac{x + 2}{x^2} \cdot \log_4 |x - 2| < 0.$$

On the left–hand side of the last inequality the function changes sign when $x$ passes through the points $x_1 = 1, x_2 = 2$, and $x^3 = 3$. We can solve the inequality by the method of intervals, with due account of the domain of definition.

**187.** (0, arcsin $((\sqrt{5} - 1)/2))$. *Hint.* The given system of inequalities is equivalent to the system

$$\begin{cases} \cos x > \tan x, \\ \tan x > 0, \\ 0 \le x \le \pi, \end{cases} \Leftrightarrow \begin{cases} \cos^2 x > \sin x, \\ 0 < x < \pi/2, \end{cases} \Leftrightarrow \begin{cases} \sin^2 x + \sin x - 1 < 0, \\ 0 < x < \pi/2. \end{cases}$$

**188.** $|a| > 1/e$. **189.** $\sin 1980°$. **190.** $\tan 1$. **191.** $\sin 2$. **192.** $-7\pi/6$.

**193.** $-7/25$. *Solution.* $\sin$ (arcsin (3/5) $-$ arccos (3/5)) $= \sin$ (arcsin (3/5)) $\times$ cos

(arccos (3/5)) $-$ cos (arcsin (3/5)) sin (arccos (3/5)) $= \dfrac{3}{5} \cdot \dfrac{3}{5} - \sqrt{1 - \dfrac{9}{25}}$

$\times \sqrt{1 - \dfrac{9}{25}} = -\dfrac{7}{25}$.

**194.** $3\pi/4$. *Hint.* Prove that arcsin (1/3) + arccos (1/3) = $\pi/2$.

**195.** $3\pi/4$. *Hint.* Prove that arcsin (1/3) + arccos (1/3) = $\pi/2$.

**196.** $\pi/2$. *Solution.* arctan $(1/\sqrt{2})$ = arccot $\sqrt{2}$, arctan $\sqrt{2}$ + arccot $\sqrt{2}$ = $\pi/2$ since arctan $x$ + arccot $x$ = $\pi/2$ for any $x \in \mathbf{R}$.

**197.** $4/5$. *Solution.* sin(2 arctan 2) = 2 sin (arctan 2) cos (arctan 2)

$$= 2 \cdot \frac{2}{\sqrt{1 + 4}} \cdot \frac{1}{\sqrt{1 + 4}} = \frac{4}{5}.$$

**198.** $5/8$.

**199.** $1/\sqrt{5}$. *Solution.* $\sin\left(\dfrac{1}{2} \arctan \dfrac{4}{3}\right) \sqrt{\dfrac{1 - \cos(\arctan(4/3))}{2}}$

$$= \sqrt{\frac{1}{2} - \frac{1}{2} \cdot \frac{1}{\sqrt{1 + (4/3)^2}}} = \sqrt{\frac{1}{2} - \frac{3}{2 \cdot 5}} = \sqrt{\frac{2}{10}} = \frac{1}{\sqrt{5}}.$$

**200.** $\pi - 3$. *Solution.* arcsin (sin 3) = arcsin (sin ($\pi - 3$)) = $\pi - 3$.

**201.** $6x/7$. *Solution.* $\arccos\left(\cos\dfrac{8\pi}{7}\right) = \arccos\left(\cos\left(2\pi - \dfrac{8\pi}{7}\right)\right)$ = arccos

$\left(\cos\dfrac{6\pi}{7}\right) = \dfrac{6\pi}{7}.$

**202.** $\pi/7$. *Solution.* $\arctan\left(\tan\dfrac{8\pi}{7}\right) = \arctan\left(\tan\left(\dfrac{8\pi}{7} - \pi\right)\right) = \dfrac{\pi}{7}.$

**203.** 24/25. *Solution.* $\left(\sin\dfrac{\alpha}{2}+\cos\dfrac{\alpha}{2}\right)^2=\dfrac{49}{25}\Rightarrow\sin\alpha+1=\dfrac{49}{25}$.

**204.** $b^2-1$ if $|b|\le\sqrt2;\varnothing$ if $|b|>\sqrt2$ .

**205.** $-7/25$. *Solution.* $\sin 2\alpha=\dfrac{2\sin\alpha\cos\alpha}{\sin^2\alpha+\cos^2\alpha}=\dfrac{2\cot\alpha}{1+\cot^2\alpha}=-\dfrac{14}{50}=-\dfrac{7}{25}$.

**206.** 7/9. *Solution.* $\dfrac{\cos^4\alpha+\sin^3\alpha\cos\alpha}{\sin^2 2\alpha}=\dfrac{\cos^4\alpha+\sin^3\alpha\cos\alpha}{4\sin^2\alpha\cos^2\alpha}$.

$$=\dfrac{1+\tan^3\alpha}{4\tan^2\alpha}=\dfrac{1+27}{4\cdot9}=\dfrac{7}{9}.$$

**207.** 10/11. *Solution.* $\dfrac{\sin\alpha}{\sin^3\alpha+3\cos^3\alpha}=\dfrac{\tan\alpha\cdot\dfrac{1}{\cos^2\alpha}}{\tan^3\alpha+3}$

$=\dfrac{\tan\alpha(1+\tan^2\alpha)}{\tan^3\alpha+3}=\dfrac{10}{11}$.

**208.** $a^2-2$ if $|a|\ge 2;\varnothing$ if $|a|<2$ .

**209.** $-3$ or $-1/3$. *Hint.* Take into account that
$$\sin\alpha=\dfrac{2\sin(\alpha/2)\cos(\alpha/2)}{\sin^2(\alpha/2)+\cos^2(\alpha/2)}=\dfrac{2\cot(\alpha/2)}{1+\cot^2(\alpha/2)}.$$

**210.** 2.

**211.** $\sqrt{10}-3$ . *Solution.* Since $0<\alpha<\pi/4$, it follows that $\cos2\alpha=\sqrt{1-(0.6)^2}=0.8$;
$$\cos\alpha=\sqrt{\dfrac{1+\cos2\alpha}{2}}=\dfrac{3}{\sqrt{10}};\tan\dfrac{\alpha}{2}=\sqrt{\dfrac{1-\cos\alpha}{1+\cos\alpha}}=\sqrt{\dfrac{\sqrt{10}-3}{\sqrt{10}+3}}$$
$$=\sqrt{(\sqrt{10}-3)^2}=\sqrt{10}-3.$$

**212.** $-33$ . **213.** $\sin\alpha=-12/13,\cos\alpha=-5/13,\cot\alpha=5/12$ .

**214.** $\sin2\alpha=120/169$ and $\sin(\alpha/2)=5/\sqrt{26}$.

**215.** $1/2\sin^2(\alpha/2)$.

**216.** *Hint.* Transform the expression $\tan(\alpha+\beta)=\tan(\pi/4)$.

**217.** *Proof.* $\sin\alpha\sin\beta-\cos\gamma=\sin\alpha\sin\beta-\cos(\pi-(\alpha+\beta))=\sin\alpha\times\sin\beta$.
$+\cos(\alpha+\beta)=\sin\alpha\sin\beta+\cos\alpha\cos\beta-\sin\alpha\sin\beta=\cos\alpha\cos\beta$.

**218.** $\sin 2\alpha$.

**219.** *Hint.* Use the formulas $\cos^2\beta=\dfrac{1+\cos2\beta}{2}$ and $\sin^2\beta=\dfrac{1-\cos2\beta}{2}$.

**220.** *Proof.* $2(\sin^6\alpha+\cos^6\alpha)-3(\sin^4\alpha+\cos^4\alpha)+1=2(\sin^6\alpha+3\sin^4\alpha\cos^2\alpha+3$
$\sin^2\alpha\cos^4\alpha+\cos^6\alpha)-2\cdot3\sin^2\alpha\cos^2\alpha(\sin^2\alpha+\cos^2\alpha)$
$-3(\sin^4\alpha+2\sin^2\alpha\cos^2\alpha+\cos^4\alpha)+3\cdot2\sin^2\alpha\cos^2\alpha+1=2(\sin^2\alpha+$
$\cos^2\alpha)^3-6\sin^2\alpha\cos^2\alpha-3(\sin^2\alpha+\cos^2\alpha)^2+6\sin^2\alpha\cos^2\alpha+1$
$$=2-3+1=0.$$

**223.** *Solution.* $\dfrac{(1+\cos x)(1+\cos2x)}{(1+\sin x)(1-\cos2x)}=\dfrac{(1+\cos x)2\cos^2 x}{(1+\sin x)2\sin^2 x}$
$=\dfrac{(1+\cos x)(1-\sin x)(1+\sin x)}{(1+\sin x)(1-\cos x)(1+\cos x)}=\dfrac{1-\sin x}{1-\cos x}$.

**224.** $\{\pi n,\pi(2k+1)/8|n,k\in\mathbf{Z}\}$.

**225.** $\{\pi(2n+1)/2|n\in\mathbf{Z}\}$.

**226.** $\{2\pi n, 2\pi(3k+1)/3 | n, k \in \mathbf{Z}\}$. *Solution.* *1st method.* $\sqrt{3}\sin x = 1 - \cos x$

$\Leftrightarrow 2\sqrt{3}\sin(x/2)\cos(x/2) = 2\sin^2(x/2) \Leftrightarrow 2\sin(x/2)(\sin(x/2)$

$-\sqrt{3}\cos(x/2)) = 0 \Rightarrow \sin(x/2) = 0, x = 2\pi n; \tan(x/2) = \sqrt{3},$

$x = 2\pi(3k+1)/3, n, k \in \mathbf{Z}.$

*2nd method.* $\dfrac{\sqrt{3}}{2}\sin x + \dfrac{1}{2}\cos x = \dfrac{1}{2} \Leftrightarrow \sin\dfrac{\pi}{3}\sin x + \cos\dfrac{\pi}{3}\cos x = \dfrac{1}{2} \Leftrightarrow$

$\cos\left(x - \dfrac{\pi}{3}\right) = \dfrac{1}{2} \Leftrightarrow x - \dfrac{\pi}{3} = \pm\dfrac{\pi}{3} + 2\pi n, n \in \mathbf{Z}.$

**227.** $\{\pi n, \pi(6k \pm 1)/3 | n, k \in \mathbf{Z}\}$.

**228.** $\{\pi(2n+1)/4, \pi(6k \pm 1)/3 | k, n \in \mathbf{Z}\}$.

**229.** $\{\pi n | n \in \mathbf{Z}\}$. *Solution.* $\tan^2 33x = \cos 2x - 1 \Leftrightarrow \tan^2 33x + 2\sin^2 x$

$= 0 \Leftrightarrow \begin{cases} \sin x = 0, \\ \tan 33x = 0, \end{cases} \Rightarrow x = \pi n, n \in \mathbf{Z}.$

**230.** $\{\pi(6n + (-1)^n)/6 | n \in \mathbf{Z}\}$.

**231.** $\{\pi n / 7, \pi(4k+3)/2 | n, k \in \mathbf{Z}\}$. *Solution.* $\sin 3x + \cos 4x - 4\sin 7x$

$= \cos 10x + \sin 17x \Leftrightarrow (\sin 3x - \sin 17x) + (\cos 4x - \cos 10x) - 4\sin 7x$

$= 0 \Leftrightarrow -2\sin 7x \cos 10x + 2\sin 3x \sin 7x - 4\sin 7x$

$= 0 \Leftrightarrow -2\sin 7x \times (\cos 10x - \sin 3x + 2) = 0,$

whence $\sin 7x = 0, x = \pi n / 7, n \in \mathbf{Z}$; or

$\cos 10x - \sin 3x + 2 = 0 \Leftrightarrow \begin{cases} \cos 10x = -1, \\ \sin 3x = 1. \end{cases} \qquad (1)$

Let us solve system (1). We first solve the second equation of system (1) :

$\sin 3x = 1, 3x = \pi(4m+1)/2, x = \pi(4m+1)/6.$

From these numbers, we choose those which satisfy the first equation of system (1),

$\cos(10\pi(4m+1)/6) = \cos(5\pi(4m+1)/3) = -1. \qquad (2)$

Equality (2) holds for $m = 2 + 3k$. Consequently, system (1) is satisfied by the numbers $x = \pi(4k+3)/2, k \in \mathbf{Z}$. Combining the results obtained, we find that $x \in \{\pi n / 7, \pi(4k+3)/2 | n, k \in \mathbf{Z}\}$.

**232.** $\{\pi(2n+1)/8 | n \in \mathbf{Z}\}$. *Solution.* Applying the formulas $\cos^2\alpha = (1 + \cos 2\alpha)/2$ and $\sin^2\alpha = (1 - \cos 2\alpha)/2$, we have

$$\left(\frac{1 - \cos 2x}{2}\right)^2 + \left(\frac{1 + \cos 2x}{2}\right)^2 = \cos^2 2x + \frac{1}{4}$$

$\Leftrightarrow 2\cos^2 2x = 1 \Leftrightarrow 1 + \cos 4x = 1 \Rightarrow \cos 4x = 0, x = \pi(2n+1)/8, n \in \mathbf{Z}.$

**233.** $\varnothing$.

**234.** $\{(-\pi(2k+1)/2, -\pi k] \cup [\pi k, \pi(2k+1)/2) | k \in \mathbf{N}\}$.

**235.** It does not have a solution. *Solution.* $4\sin 2x + \cos x = 5$ is equivalent to the system of equations, $\begin{cases} \sin 2x = 1, \\ \cos x = 1. \end{cases}$

The equation $\cos x = 1$ is satisfied by the numbers $2\pi k, k \in \mathbf{Z}$. None of these numbers satisfies the first equation of the system.

**236.** $\{2\pi n, 2\pi k - 2 \arctan (3/5) | n, k \in \mathbf{Z}\}$. *Solution.* $5\cos x - 3\sin x = 5$

$\Leftrightarrow 10\sin^2(x/2) + 6\sin(x/2)\cos(x/2) = 0 \Leftrightarrow 2\sin(x/2)(5\sin(x/2) +$

$3\cos(x/2)) = 0 \Rightarrow \sin(x/2) = 0, \ x = 2\pi n, n \in \mathbf{Z};$
$5\sin(x/2) + 3\cos(x/2) = 0 \Leftrightarrow \tan(x/2) = -3/5$
$x = -2\arctan(3/5) + 2\pi k, k \in \mathbf{Z}.$
Combining the results obtained, we find the required solution.

**237.** For irrational values of $a$. *Hint.* Consider separately the equation for the rational values of $a$ and the irrational values of $a$.

**238.** Three. *Hint.* Determine the number of points of intersection of the graphs of the functions $y = \log_{3\pi/2} x$ and $y = \cos x$.

**239.** $\{(\pi(24k+1)/12, \pi(24k+17)/12) \,|\, k \in \mathbf{Z}\}$. *Remark.* The given inequality is equivalent to the inequality $\sin(x - (\pi/4)) > -1/2$.

**240.** $\{(\pi(12n-5)/6, \pi(4n+1)/2 \,|\, n \in \mathbf{Z}\}$. *Hint.* The given inequality is equivalent to the collection of two inequalities $\sin(x - (\pi/3)) > 2$, which has no solution and $\sin(x - (\pi/3)) < 1/2$, which is satisfied by $-(7\pi/6) + 2\pi n < x - (\pi/4) < (\pi/6) + 2\pi n, n \in \mathbf{Z}.$

**241.** $f_{\max}(x) = 5.$

**242.** *Proof.* By the hypothesis
$$b^2 - a^2 = c^2 - b^2. \qquad (1)$$
We must prove that
$$\frac{1}{c+a} - \frac{1}{b+c} = \frac{1}{a+b} - \frac{1}{c+a}, \qquad (2)$$
or $\dfrac{b-a}{(c+a)(b+c)} - \dfrac{c-b}{(a+b)(a+c)} = 0,$
or $\dfrac{(b^2-a^2)-(c^2-b^2)}{(a+c)(b+c)(a+b)} = 0. \qquad (3)$

Equation (1) yields the validity of equation (3) and, consequently, of equation (2).

**243.** $3069/256$. **244.** Yes, it is. *Solution.* For any $n \in \mathbf{N}$, we have
$$a_{n+1} - a_n = \frac{(n+1)^2 - (n+1)+1}{(n+1)^2+(n+1)+1} - \frac{n^2-n+1}{n^2+n+1} = \frac{n^2+n+1}{n^2+3n+3} - \frac{n^2-n+1}{n^2+n+1}$$
$$= \frac{2(n^2+n-1)}{(n^2+n+1)(n^2+3n+3)} > 0.$$

**245.** $-1/9$. *Solution.* $\lim\limits_{n\to\infty} \dfrac{n^2}{1-9n^2} = \lim\limits_{n\to\infty} \dfrac{1}{\frac{1}{n^2}-9} = \dfrac{1}{\lim\limits_{n\to\infty}\left(\frac{1}{n^2}-9\right)} = -\dfrac{1}{9}.$

**246.** $0.$ **247.** $1/4.$

**248.** $1$. *Solution.* $\lim\limits_{n\to\infty} \dfrac{\frac{1}{2}+1+\frac{3}{2}+...+\frac{n}{2}}{\frac{n^2}{4}+n+3} = \lim\limits_{n\to\infty} \dfrac{2(1+2+3+...+n)}{n^2+4n+12}$
$$= \lim\limits_{n\to\infty} \frac{n(n+1)}{n^2+4n+12} = \lim\limits_{n\to\infty} \frac{1+\frac{1}{n}}{1+\frac{4}{n}+\frac{12}{n^2}} = 1.$$

**249.** 1/2.    **250.** 1. *Solution.* $\lim\limits_{n\to\infty} \dfrac{2^n + 5^n}{3^n + 5^n} = \lim\limits_{n\to\infty} \dfrac{(2/5)^n + 1}{(3/5)^n + 1} = \dfrac{0+1}{0+1} = 1.$

**251.** 5.

**252.** 1. *Solution.* $\lim\limits_{x\to3} \dfrac{x^2 - 5x + 6}{x - 3} = \lim\limits_{x\to3} \dfrac{(x-3)(x-2)}{x-3} = \lim\limits_{x\to3}(x-2) = 1.$

**253.** 6/5.    **254.** 1/4.    **255.** −7    **256.** 13/5 .

**257.** 2 . *Solution.* $\lim\limits_{x\to-3} \dfrac{x+3}{\sqrt{x+4}-1} = \lim\limits_{x\to-3} \dfrac{(x+3)(\sqrt{x+4}+1)}{(x+4)-1}$

$= \lim\limits_{x\to-3} (\sqrt{x+4}+1) = 2.$

**258.** $\sqrt{6}/6.$    **259.** −1/56.

**260.** 4. *Solution.* $\lim\limits_{x\to-1} \dfrac{\sqrt{2x+3}-1}{\sqrt{5+x}-2} = \lim\limits_{x\to-1} \dfrac{((2x+3)-1)(\sqrt{5+x}+2)}{((5+x)-4)(\sqrt{2x+3}+1)}$

$= 2\lim\limits_{x\to-1} \dfrac{\sqrt{5+x}+2}{\sqrt{2x+3}+1} = 2\cdot\dfrac{4}{2} = 4.$

**261.** $-\sqrt{3}/24.$

**262.** 3. *Solution.* We set $\sqrt[6]{x} = t, t \to 2$ as $x \to 64.$ Then we have

$$\lim\limits_{x\to64} \dfrac{\sqrt{x}-8}{\sqrt[3]{x}-4} = \lim\limits_{t\to2} \dfrac{t^3-8}{t^2-4}$$

$$= \lim\limits_{t\to2} \dfrac{(t-2)(t^2+2t+4)}{(t-2)(t+2)} = 3 .$$

**263.** $3/2 .$

**264.** 2/5. *Solution.* $\lim\limits_{x\to0} \dfrac{\cos4x - \cos6x}{\sin^2 5x} = \lim\limits_{x\to0} \dfrac{2\sin5x\sin x}{\sin^2 5x}$

$= 2\lim\limits_{x\to0} \dfrac{\sin x}{x} \times \lim\limits_{x\to0} \dfrac{5x}{\sin5x}\cdot\dfrac{1}{5} = \dfrac{2}{5}.$

**265.** 2/3. *Solution.* $\lim\limits_{x\to0} \dfrac{\sqrt[3]{1+\sin x} - \sqrt[3]{1-\sin x}}{x}$

$= \lim\limits_{x\to0} \dfrac{(1+\sin x) - (1-\sin x)}{x(\sqrt[3]{(1+\sin x)^2} + \sqrt[3]{1+\sin x}\,\sqrt[3]{1-\sin x} + \sqrt[3]{(1-\sin x)^2})}$

$= 2\lim\limits_{x\to0} \dfrac{\sin x}{x} \lim\limits_{x\to0} \dfrac{1}{\sqrt[3]{(1+\sin x)^2} + \sqrt[3]{1-\sin^2 x} + \sqrt[3]{(1-\sin x)^2}} = \dfrac{2}{3}.$

**266.** 8. *Solution.* We set $2^x = t, t \to 4$ as $x \to 2.$ Then

$$\lim\limits_{x\to2} \dfrac{2^x + 2^{3-x} - 6}{\sqrt{2^{-x}} - 2^{1-x}} = \lim\limits_{t\to4} \dfrac{t + \dfrac{8}{t} - 6}{\dfrac{1}{\sqrt{t}} - \dfrac{2}{t}}$$

$$= \lim\limits_{t\to4} \dfrac{t^2 - 6t + 8}{\sqrt{t} - 2}$$

$$= \lim\limits_{t\to4} \dfrac{(\sqrt{t} + 2)(t-2)(t-4)}{t-4}$$

$$= \lim\limits_{t\to4} (\sqrt{t} + 2)(t-2) = 8.$$

**267.** $a(a + \sqrt{3}).$ *Hint.* The lengths of the successive segments of the infinite polygonal line we have constructed form a decreasing geometric progression. By the length of an infinite polygonal line, we mean the sum of the progression.

**268.** (a) $-\dfrac{3\cot^2 x}{\sin^2 x}$; (b) $\dfrac{\cos\sqrt{x}}{2\sqrt{x}}$. **269.** $\dfrac{4x-3}{(2x^2-3x+1)\ln 2}$. **270.** $6x^2\tan x + \dfrac{2x^3-5}{\cos^2 x}$.

**271.** (a) $\dfrac{1}{2x\sqrt{\ln x}}$; (b) $\dfrac{\cos 2x}{\sqrt{\sin 2x}}$; (c) $6(\sin 2x + 8)^2\cos 2x$. **272.** $-1$. **273.** 40. **274.** 1/8.

**275.** $-2$. **276.** The function increase on the intervals $(-\infty,-1)$ and $(1, +\infty)$ and decreases on the intervals $(-1, 0)$ and $(0, 1)$.

**277.** The function increases on the interval $[-\infty,-3/4)$ and decreases on the interval $(-3/4,1/4)$. *Remark.* Do not forget to take the domain of the function into account.

**278.** The function increases on the interval $(0, 2)$ and decreases on the intervals $(-\infty, 0)$ and $(2, +\infty)$.

**279.** The function increases on the interval $(e, +\infty)$ and decreases on the intervals $(0,1)$ and $(1,e)$. *Remark.* Do not forget to take the domain of the function into account.

**280.** The function increases on the interval $(-1/\sqrt[3]{2}, +\infty)$ and decreases on the interval $(-\infty,-1/\sqrt[3]{2})$.

**282.** *Proof.* $f'(x) = 6(x^8 - x^5 + x^2 - x + 1) = 6x^2\left(x^6 - x^3 + \dfrac{1}{4}\right) + 3x^2 +$

$6\left(\dfrac{x^2}{4} - x + 1\right) = 6x^2\left(x^3 - \dfrac{1}{2}\right)^2 + 3x^2 + 6\left(\dfrac{x}{2} - 1\right)^2 > 0$ for all $x \in \mathbf{R}$.

**284.** *Solution.* $f'(x) = (a^2 - 1)x^2 + 2(a-1)x + 2$. The discriminant of the quadratic trinomial $D = 4(a-1)^2 - 8(a^2-1) = -4(a-1)(a+3)$ is negative for all $a \in (-\infty,-3)\cup(1, +\infty)$. Since the coefficient in $x^2$ is positive for all these values of $a$, the quadratic trinomial is positive for all $x \in \mathbf{R}$. For $a = 1$ we have $f'(x) = 2 > 0$. For $a = -1$, we have $f'(x) = -4x + 2$, $f'(x) \le 0$ for $x \ge 1/2$. For $-1 < a < 1$ the coefficient in $x^2$ is smaller than zero and, therefore, for each of these values of $a$ there are values of $x$ for which $f'(x) < 0$. For $a = -3$ we have $f'(x) = 2(2x-1)^2 = 0$ only for $x = 1/2$. For $-3 < a < -1$ the quadratic trinomial has two distinct real roots $x_1$ and $x_2$. If $x_1 < x_2$, then $f'(x) < 0$ for all $x \in (x_1, x_2)$. It follows from the aforesaid that $f'(x) > 0$ for all $x \in \mathbf{R}$ for any $a \in (-\infty,-3]\cup[1, +\infty)$ and only for these values of $a$.

**285.** For all $a \ge 0$. *Remark.* $f'(x) = e^{-x}(e^x + a)(2e^x + 1) > 0$ for all $a \ge 0$.

**286.** For all $a \in [1, +\infty)$. *Solution.* $y' = \cos x - 2a\cos 2x - \cos 3x + 2a = 4a\sin^2 x + 2\sin x\sin 2x = 4a\sin^2 x + 4\cos x\sin^2 x = 4\sin^2 x(a + \cos x)$. For $a \ge 1$ we have $y' > 0$ for all $x$, except for the points $x = \pi n, n \in \mathbf{Z}$, at which $y' = 0$. Therefore, the function increases for all $a \ge 1$.

**287.** For all $a \in (-\infty, -3]$. **288.** $x_1 = -2$; $x_2 = 0$; $x_3 = 2$. **289.** $x_1 = -5$; $x_2 = 1$.

**290.** $x = 1/2$ is a point of minimum; $y(1/2) = (1 - 4\ln 2)/4$.

**291.** $a \in (-1, 3)$. *Solution.* $y' = 3(x^2 - 2ax + a^2 - 1) = 3(x - a + 1)\times(x - a - 1)$. For the roots of the trinomial to belong to the interval $(-2, 4)$, it is necessary and sufficient that the inequalities $a - 1 > -2$ and $a + 1 < 4$ are satisfied and, consequently, that $a \in (-1, 3)$. **292.** $f_{gr} = 5$; $f_{least} = 11/4$. **293.** $f_{gr} = f(1) = 5$; $f_{least} = f(1/2) = 3$.

**294.** $f_{gr} = f(100) = 10 - 2\sqrt{10}$; $f_{least} = f(1) = -1$. **295.** $f_{gr} = f(-5) = e^{48}$; $f_{least}$

$= f(2) = 1/e$. **296.** $y_{gr} = y(4) = 6$; $y_{least} = y(1/16) = -1/8$. **297.** For all $\cup(3, 29/7)$ $a \in (-\infty, -3)$. *Solution.* $f'(x) = 3(x^2 + 2(a-7)x + a^2 - 9)$. For the function $f(x)$ to have a maximum at the point $x_1$, it is necessary that the function $f'(x)$ should change sign from plus to minus when $x$ passes through the point $x_1$ in the order of increase. Therefore, the quadratic trinomial

$$x^2 + 2(a-7)x + a^2 - 9 \tag{1}$$

must have two real roots $x_1$ and $x_2$. If $x_1 < x_2$, then $x_1$ is a point of maximum since $f'(x) > 0$ for $x < x_1$ and for $x_1 < x < x_2$ the derivative $f'(x) < 0$. Since, (by the hypothesis) the point of maximum must be positive, both roots of the quadratic trinomial (1) must be positive (with $x_1 < x_2$). That occurs if and only if

$$\begin{cases} (a-7)^2 - (a^2-9) > 0, \\ a - 7 < 0, \\ a^2 - 9 > 0. \end{cases} \tag{2}$$

Solving system (2), we obtain $a \in (-\infty, -3) \cup (3, 29/7)$.

**298.** $a = 2$. *Remark.* It follows from the hypothesis that $f'(x) = 6(x - 2a)(x - a)$. The roots $x_1$ and $x_2$ of the trinomial $x^2 - 3x + 2a^2$ must be real and distinct, and if $x_1 < x_2$, then $x_1$ is a point of maximum and $x_2$ is a point of minimum.

**299.** For all $a \in (1, +\infty)$. *Remark.* It follows from the hypothesis that

$$f(x) = ax^2 + 2(a+2)x + a - 1.$$

We must consider three cases : $a = 0$, $a > 0$, and $a < 0$. For $a > 0$ and for $a < 0$ we must consider the values of $a$ for which both roots are real, and if in that case $x_1 < x_2$, then for $a > 0$ the minimum is at the point $x_2$ and for $a < 0$ the minimum is at the point $x_1$.

**300.** $a \in (-3\sqrt{3}, -2\sqrt{3}) \cup (2\sqrt{3}, 3\sqrt{3})$. *Solution.* Let us solve the inequality

$$\frac{x^2+x+2}{x^2+5x+6} \le 0, \text{ or } \frac{x^2+x+2}{(x+3)(x+2)} \le 0.$$

Since $x^2 + x + 2 = \left(x + \dfrac{1}{2}\right)^2 + \dfrac{7}{4} > 0$ for $x \in R$, the inequality is satisfied if $(x + 2)$ and $(x + 3)$ have different signs, i.e. $x + 2 < 0$ and $x + 3 > 0$, or $x < -2$ and $x > -3$. Consequently, the inequality holds true for all $x \in (-3, -2)$. Let us find the critical points of the function $y = 1 + a^2x - x^3$. We find $y' = a^2 - 3x^2$ and equate it to zero. We obtain two critical points $x_1 = -|a|\sqrt{3}$ and $x_2 = |a|/\sqrt{3}$. On the intervals $(-\infty, -|a|/\sqrt{3})$ and $(|a|/\sqrt{3}, +\infty)$ the derivative $y'$ is negative and on the interval $(-|a|/\sqrt{3}, |a|/\sqrt{3})$ it is positive; consequently, $x_1 = -|a|/\sqrt{3}$ is a point of minimum of the function and it must satisfy the inequality

$$-3 < -|a|/\sqrt{3} < -2, \text{ or } 2\sqrt{3} < |a| < 3\sqrt{3}.$$

Thus the hypothesis is satisfied by all $a \in (-3\sqrt{3}, -2\sqrt{3}) \cup (2\sqrt{3}, 3\sqrt{3})$.

**301.** $x = \pi/4$.

**302.** $-1/2$. *Solution.* Assume that $x$ is the required number. We must find the point of minimum of the function $f(x) = x + x^2$. Since $f'(x) = 2x + 1$, it follows that $x = -1/2$ is a point of minimum of the function $f(x)$.

**303.** 1.

**304.** A triangle with sides $a, a, a\sqrt{2}$ and area $a^2/2$. *Solution.* Assume that $b$ is the length of the base of the triangle and $S$ is its area. Since $S = (b/4) \times \sqrt{4a^2 - b^2} = S(b)$, we have $S' = (2a^2 - b^2)/2\sqrt{4a^2 - b^2}$, $S'(b) = 0 \Rightarrow b = a\sqrt{2}$, $S_{max} = a^2/2$.

**305.** $\pi / 3$. *Solution.* Assume that $l$ is the length of a lateral side, $a$ is the length of the base, $h$ is the altitude and $S$ is the area of the triangle. Then $l = 2R \times \cos (\alpha / 2), h = 2R \cos^2 (\alpha / 2) = R (1 + \cos\alpha), a / 2 = 2R \cos$ $(\alpha / 2) \sin (\alpha / 2 = R \times \sin\alpha$, and $S = ah / 2 = R^2 (1 + \cos\alpha)\sin\alpha$. Let us find the maximum of the function $S'(\alpha)$. We have $S(0) = S(\pi) = 0$ and $S(\alpha) > 0$ for $0 < \alpha < \pi$. Furthermore, $S'(\alpha) = R^2 (\cos\alpha + \cos^2 \alpha - \sin^2\alpha)$ $= R^2(\cos\alpha + \cos 2\alpha) = 2R^2 \cos (\alpha / 2)\cos (3\alpha / 2)$. The interval $[0, \pi]$ contains the critical points $x = \pi$ and $x = \pi / 3$; excluding the extraneous solutions from the consideration, we find that $\alpha = \pi / 3$.

**306.** 2. **307.** (1,1). **308.** $\pi / 3$. **309.** $y = -x - 2 + 2 \ln 2$. **310.** $y = \dfrac{1}{2}$
$+ \sqrt{3} \times \left( x - \dfrac{\pi}{12} \right)$. **311.** $y = 12x - 16$. **312.** $y = x / e$. **313.** $y = 2$. **314.** $y = -4$. **315.**
$y = -4x, y = 4x - 8$. **316.** $y = -5x + 2$.

**317.** *Solution.* Assume that $x_0$ is the abscissa of the point of tangency of the straight line and the hyperbola. The ordinate of the point of tangency is $y_0 = a^2 / x_0$, and the equation of the tangent is

$$y = \frac{a^2}{x_0} - \frac{a^2}{x_0^2} (x - x_0).$$

Let us find the coordinates of the points of intersection of the tangent and the axes of coordinates. We set $x = x_1 = 0$ and get $y_1 = 2a^2 / x_0, M_1 (0, 2a^2 / x_0)$. Setting $y = y_2 = 0$, we get $x_2 = 2x_0, M_2(2x_0, 0)$. The area of the right triangle $M_1 O M_2$ is
$$\frac{1}{2}|OM_1| \cdot |OM_2| = \frac{1}{2} \frac{2a^2}{|x_0|} \cdot 2|x_0| = 2a^2.$$

**319.** $x_1 = 0$ and $x_2 = 3$.

**320.** Yes, it does. *Solution.* Let us find the coordinates of the points of intersection of the straight line and the hyperbola. For that purpose we solve the equation
$$\frac{1}{x} = \frac{4 - x}{4} \Leftrightarrow (x - 2)^2 = 0 \Rightarrow x_0 = 2, y_0 = 1 / 2.$$
Since in that case the straight line and the hyperbola have only one point in common, the straight line touches the hyperbola.

**321.** [1/4, 3/2]. **322.** (a) Even; (b) and (c) odd. **323.** $2^{x^2}, 2^{2x}$.

**324.** $f(x) = \dfrac{7}{6} x^2 - \dfrac{13}{6} x + 1$. *Hint.* Write the function in the form $f(x) = ax^2 + bx + c$.

**325.** $x = -\sqrt{y + 1}, y \in [-1, + \infty)$.

**326.** $A (1/2, 3/4)$. *Solution.* Assume that $x_0$ and $y_0$ are the coordinates of the point $A$. Since $y'(x_0) = 2 - 2x_0$, the equation of the tangent to the parabola at the point $A$ is $y = x_0^2 + (2 - 2x_0)x$. The coordinates of the points of intersection of the tangent and the straight lines $x = 0$ and $x = 1$ are $x_1 = 0$, $y_1 = x_0^2$; $x_2 = 1, y_2 = x_0^2 - 2x_0 + 2$. The area of the trapezoid is
$$S = \frac{y_1 + y_2}{2} \cdot 1 = x_0^2 - x_0 + 1.$$
In that case $S' = 2x_0 - 1$. Note that $S' < 0$ for $x_0 < 1 / 2; S' > 0$ for $x_0 > 1 / 2$ and, consequently, the area $S$ has a minimum at the point $x_0 = 1 / 2$. Thus $x_0 = 1 / 2$ and $y_0 = 3 / 4$.

**327.** $y = (5 - 2 \cot 3x) / 3$. **328.** $F(x) = (2 \sin 4x + 7) / 8$.

**329.** $x = 2$. Solution. $F(x) = -\cos\pi x + x^2 - 4x + C$, $F(1) = -\cos\pi + 1 - 4 + C = 3 \Rightarrow$
$C = 5$. Let us solve the equation
$F(x) = 0 \Leftrightarrow -\cos\pi x + x^2 - 4x + 5 = 0 \Leftrightarrow (x-2)^2 + 2\sin^2(\pi x / 2)$
$= 0 \Leftrightarrow \begin{cases} x - 2 = 0, \\ \sin(\pi x / 2) = 0, \end{cases} \Rightarrow x = 2.$

**330.** $x - \cos x + C$.      **331.** 1.      **332.** $4\sqrt{2} / 3$.    **333.** $3 / 8$.

**334.** $2\ln(e + 1)$.

**335.** $\pi / 2$. Hint. Use the formula $\cos^2 x = (1 + \cos 2x) / 2$.

**336.** $1/2$. Hint. Use the formula $\sin x \cos x = (\sin 2x) / 2$.

**337.** 2. Hint. $\int_{-\pi/2}^{\pi/2} |\sin x| \, dx = 2\int_0^{\pi/2} \sin x \, dx$.

**338.** 0. Hint. Use the formula $\sin\alpha \cos\beta = (\sin(\alpha - \beta) + \sin(\alpha + \beta)) / 2$.

**339.** $125 / 3$.      **340.** $50 / 3$.      **341.** $243 / 4$.

**342.** $5/12$.      **343.** $(4 - \pi) / 2$.      **344.** $(4 - \pi) / 4$.

**345.** $64 / 3$.      **346.** $7 / 3$.    **347.** $(4 + \pi) / 2$.

**348.** $44 / 27$. Hint. Since the figure bounded by curve $y = -3x^2 - |x| + 2$ and the straight line $y = 0$ is symmetric about the $Oy$ axis, we must calculate the area of the figure lying in the first quadrant and bounded by the curve $y = -3x^2 - x + 2$ and the straight lines $x = 0$ and $y = 0$, and then double the result obtained.

**349.** $8 / 3$. Hint. Since the figure is symmetric about the $Ox$ and $Oy$ axes, it is sufficient to calculate the area of the figure lying in the first quadrant and bounded by the curve $y = 1 - x^2$ and the straight lines $x = 0$ and $y = 0$, and multiply the result obtained by 4.

**350.** No, it is not, since the area of the figure is equal to $e - \dfrac{1}{e} > 2.7 - \dfrac{1}{2.7} > 2$.

**351.** $2 / 3$.

**352.** $y = 2x / 3$. Solution. The area of the curvilinear triangle is
$$S = \int_0^1 (2x - x^2) \, dx = 2 / 3.$$
The required straight line $y = kx$ cuts off from the curvilinear triangle a right triangle whose area is $S / 2 = 1 / 3$ and the base is equal to 1 and, consequently, its other leg, the altitude of the triangle, is equal to $2 / 3$ and, therefore, $k = 2 / 3$.

**353.** $k = 0$. The area is equal to $20\sqrt{5} / 3$. Hint. Construct the graphs of the functions $y = x^2 - 3$ and $y = kx + 2$. Use geometric means to prove that the area of the figure bounded by the parabola $y = x^2 - 3$ and the straight line $y = 2$ is the least. In that case the area is
$$S = 2\int_0^{\sqrt{5}} (2 - (x^2 - 3)) \, dx = 2\int_0^{\sqrt{5}} (5 - x^2) \, dx = \frac{20\sqrt{5}}{3}.$$

**354.** $3\pi^2 / 8$. Solution.
$$V = \pi\int_{-\pi/2}^{\pi/2} \cos^4 x \, dx = 2\pi\int_0^{\pi/2} \cos^4 x \, dx = 2\pi\int_0^{\pi/2}\left(\frac{1 + \cos 2x}{2}\right)^2 dx$$
$$= \frac{\pi}{2}\int_0^{\pi/2}\left(1 + 2\cos 2x + \frac{1 + \cos 4x}{2}\right) dx = \frac{3\pi^2}{8}.$$

**355.** $a_1 = 1/2$ and $a_2 = 2$. **356.** For all $a \in [-2, \ 4]$. **357.** $6\sqrt{2}$. **358.** $k \in \varnothing$.
**359.** $\sqrt{14}/2$. *Hint.* Since $\mathbf{a} \parallel \mathbf{b}$, it follows that $\mathbf{a} = \lambda\mathbf{b} = (3\lambda, -2\lambda, \lambda)$. **360.** $\pi$
$-\arccos(11/\sqrt{406})$. **361.** $m = -1/2$. **362.** $a = \sqrt{2/3}$.
   **363.** $\mathbf{a} = (5, 7/2, -4)$. *Hint.* Setting $\mathbf{a} = (x, y, z)$, we get a system of equations
$x + 2y + 3z = 0, -2x + 4y + z = 0$, and $\underline{x - 2y + z = -6}$ to detion $x, y,$ and $z$.
   **364.** $|\mathbf{a} + \mathbf{b}| = 15$ and $|\mathbf{a} - \mathbf{b}| = \sqrt{593}$, **365.** *Hint.* Transform the expression
$(\mathbf{a} + \mathbf{b} + \mathbf{c})^2 = 0$, **373.** Yes. it can.

   **374.** *Hint.* Prove that if $a$ is the length of a side of the triangle inscribed in a circle of
   radius $R$, lying opposite the angle $\alpha$, then $a = 2R \sin\alpha$, and use that result.

   **375.** A rhombus when the diagonals of the initial quadrangle are perpendicular. A
   square when they are, in addition, equal to each other.
   **378.** *Hint.* Make use of vector algebra. **381.** $(\sqrt{2} - 1):1$. **382.** $\pi/3; 2\pi/3$.

   **383.** $3, 5, 7; 4, 5, 6; 5, 5, 5$. **384.** $36°, 36°, 108°$. **385.** $\arccos(4/5)$. *Hint.* Make use of
vector algebra. **386.** $h^2 \cot(\alpha/2)$. *Hint.* Assume that $a$ and $b$ are the lengths of the bases
of the trapezoid. Make a drawing and prove that $(a + b)/2 = h \cot(\alpha/2)$.
   **387.** $2(\sqrt{2} - 1) a^2$. **388.** The perimeter of a square. **389.** Yes, there is. **390.** $18\sqrt{2}$ cm .
**391.** 2 radians. **392.** By 125 per cent. **393.** $abc\sqrt{-\cos 2\alpha}$ ; $\pi/4 < \alpha < 3\pi/4$.

   **394.** $\sqrt{S}\,(\cot(\alpha/2) - 1)/2$. **395.** $\sqrt{(4H^3 + 3V)/4H}$. **396.** $\arccos(\tan(\alpha/2))$
$= \arcsin(\sqrt{\cos\alpha}/\cos(\alpha/2))$. **397.** $90°$. *Hint.* Apply the theorem on three
perpendiculars. **398.** $7(\sqrt{2} - 1)a^3/3$. **399.** $R(H - h)/H, 0 < h < H$.

   **400.** $\dfrac{3\sqrt{3}\,\pi}{2\sin^2(\alpha/2)(1 + 2\cos\alpha)^2}$. **401.** For all $a \in (-1, 1)$ and $b \in (-1, 1)$. **402.** 943/81.

*Solution.* Since $x_1 + x_2 = 5/3$ and $x_1 x_2 = -1/3$, it follows that
$(x_1^4 + x_2^4) = (x_1^2 + x_2^2)^2 - 2x_1^2 x_2^2 = ((x_1 + x_2)^2 - 2x_1 x_2)^2 - 2 \cdot \dfrac{1}{9}.$

$$= \left(\left(\frac{5}{3}\right)^2 + \frac{2}{3}\right)^2 - \frac{2}{9} = \frac{943}{81}.$$

   **403.** $\{1, 11\}$. *Solution.* Let us consider two cases : (1) $x \ge 0$; (2) $x < 0$.
   (1) $x \ge 0$. We can write the given equation in the form $3^x + 1 - (3^x - 1) =$
   $2\log_5 |6 - x| \Leftrightarrow 1 = \log_5 |6 - x| \Leftrightarrow |x - 6| = 5 \Rightarrow x_1 = 1, x_2 = 11$.
   (2) $x < 0$. The given equation can be written in the form
   $3^x + 1 + (3^x - 1) = 2\log_5(6 - x) \Leftrightarrow 3^x = \log_5(6 - x).$
   This equation has no roots for $x < 0$ since $3^x < 1$ for $x < 0$, and
   $\log_5(6 - x) > 1$.
   **404.** $\{1/10, 2, 1000\}$. **405.** $\{-2\pi/3, \pi/2, -2\pi(3n \mp 1)/3, \pi(6k \pm 1)/3 \mid$
$n, k \in \boldsymbol{N}\}$. *Solution.* Let us consider three cases (1) $x < 0$; (2) $0 \le x \le \pi$; (3) $x > \pi$.
   (1) $x < 0$. We can write the equation in the form
   $$2\pi \cos x = -x - (\pi - x) \Leftrightarrow \cos x = -\frac{1}{2}$$
   $$\Rightarrow x = \pm \frac{2\pi}{3} - 2\pi n, \ n \in \boldsymbol{Z}.$$
   The condition $x < 0$ is satisfied by the numbers $-2\pi(3n \pm 1)/3, n \in \boldsymbol{N}$, and
   $-2\pi/3$.
   (2) $0 \le x \le \pi$. We can write the equation in the form
   $2\pi \cos x = x - (\pi - x) = 2x - \pi$

On the interval $[0, \pi]$ the graphs of the functions $y = 2\pi \cos x$ and $y = 2x - \pi$ meet at only one point with the abscissa $x = \pi / 2$. Consequently, on the interval $[0, \pi]$ the equation has only one root $x = \pi / 2$.

(3) $x > \pi$. We can write the equation in the form

$$2\pi \cos x = x - (x - \pi) = \pi \Leftrightarrow \cos x = \frac{1}{2} \Rightarrow x = \pm \frac{\pi}{3} + 2\pi k, \ \ k \in \mathbf{Z}.$$

The condition $x > \pi$ is satisfied by the numbers $\pm \dfrac{\pi}{3} + 2\pi k, \ k \in \mathbf{N}$.

**406.** $\{41\}$. *Hint.* Write $x (1 - \log 5) = x (\log 10 - \log 5) = x \log 2 = \log 2^x$.

**407.** $\{\pi (2k + 1) / 6, \ 2\pi n \pm \arccos ((-1 \pm \sqrt{5}) / 4) \mid k, \ n \in \mathbf{Z}\}$. *Hint.* Carry out the following transformations :
$\cos x + \cos 2x + \cos 3x + \cos 4x + \cos 5x = 2 \cos 3x \cos 2x + 2 \cos 3x \cos x +$
$\cos 3x = \cos 3x (2 \cos 2x + \cos x + 1) = \cos 3x (4 \cos^2 x + \cos x - 1)$.

**408.** $\{2\}$. **409.** $\{\pi (6k + 1) / 3 \mid k \in \mathbf{Z}\}$. *Hint.* The given equation is equivalent to the system of equations.

$$\begin{cases} \cos 3x = -1, \\ \cos\left(2x - \dfrac{7\pi}{6}\right) = -1. \end{cases}$$

**410.** 4. **411.** 2. *Hint.* Construct the graph of the function $y = x^4 + x^3 - 10$ and determine the number of its points of intersection with the $Ox$ axis.

**412.** 4. *Hint.* Construct the graphs of the functions $y = \mid 2 - \mid x \mid \mid$ and $y = 3^{-|x|}$ and determine the number of their points of intersection.

**413.** 4. *Hint.* Construct the graphs of the functions $y = x^2 - 2x - 3$ and $y = \log_2 \mid 1 - x \mid$ and determine the number of their points of intersection.

**414.** $a < -3$. *Hint.* For the equation to have three roots, it is necessary that the function $y = x^3 + ax + 2$ should have a minimum and a maximum; if $x_1$ is a point of maximum and $x_2$ is a point of minimum of the function, then $x_1 > x_2, y (x_1) > 0$, and $y (x_2) < 0$.

**415.** $0 < a < 4 / e^2$. **416.** $0 < a < 1 / e$. **417.** $\{-\pi, \ 0, \ \pi\}$. **418.** $-24 / 25$.

**419.** $(0, 3] \cup [4, 5]$. *Hint.* The given inequality is equivalent to the system of inequalities $x > 0, \ 25 - x^2 \geq 0$, and $25 - x^2 \leq 144 / x^2$.

**420.** $[-2, \ -1) \cup (-1, 0) \cup (0, 1) \cup (1, 2]$. *Hint.* For all $x \neq 0$ and $\mid x \mid \neq 1$ we have $(1 + x^2) / (2 \mid x \mid) > 1$. **421.** $(1 / 4, 3)$. **422.** $[-2, 2)$. *Hint.* The inequality is equivalent to the collection of two systems of inequalities : (1) $6 + x - x^2 \geq 0, \ 4x - 2 < 0$; (2) $6 + x - x^2 \geq 0, \ 4x - 2 \geq 0, \ 9 (6 + x - x^2) > (4x - 2)^2$. **423.** $(-\infty, \ -5) \cup (1, + \infty)$.

**424.** 2. **425.** $2 \sin 2$. **426.** On $(-\infty, 1 / 3)$ it increases, on $(1 / 3, + \infty)$ it decreases, and $x = 1 / 3$ is a point of maximum. **427.** The intervals of decrease are $(0, 1)$ and $(1, e), x = e$ is a point of minimum, and $(e, + \infty)$ is the interval of increase.

**428.** *Hint.* Since $y' = 12x (x^2 - x + 1) + a$, we must prove that the equation $12x (x^2 - x + 1) + a = 0$ has only one root $x_1$, and the derivative $y'$ changes sign when $x$ passes through $x_1$. If $a = 0$, then there is one root : $x = 0$. For $a \neq 0$ the value $x = 0$ is not a root. Represent the equation in the form $x^2 - x + 1 = a / (12x)$ and prove that the graphs of the functions $y = a / (12x)$ and $y = x^2 - x + 1$ meet at one point. **429.** $\pm \sqrt{2}$. **430.** $a = 0$. **431.** $\pi / 2$. **432.** 4.5. **433.** 15. **434.** $1 / 3$. **435.** $\pi^2 / 2$. **437.** $M (0, 0, 2 / 3)$ and $N (1 / 3, 1 / 3, 2 / 3)$.

*Hint.* The vector $\overrightarrow{MN}$ is parallel to the bisector of the first coordinate angle of the plane $xOy$. **438.** $\pi / 3$.

# Appendix

**The appendix contains material for reference.**

## 1. Algebra and Trigonometry

**Properties of a power.** Assume $a > 0$, $b > 0$, and $\alpha$, $\beta \in \mathbf{R}$. Then :

(1) $a^0 = 1$ (by definition);

(2) $a^\alpha \cdot a^\beta = a^{\alpha+\beta}$;

(3) $a^\alpha : a^\beta = a^{\alpha-\beta}$;

(4) $(a^\alpha)^\beta = a^{\alpha\cdot\beta}$;

(5) $a^\alpha \cdot b^\alpha = (a \cdot b)^\alpha$;

(6) $a^\alpha : b^\alpha = \left(\dfrac{a}{b}\right)^\alpha$.

**Properties of logarithms.** The *logarithm* of a positive number $x$ to the base $a$ ($a > 0$, $a \neq 1$) is a number which is equal to the power exponent to which $a$ must be raised to obtain $x$.

Designations : $\log_a x$; $\log_{10} x \equiv \log x$; $\log_e x \equiv \ln x$. By definition $a^{\log_a x} = x$,

(1) $\log_a (x_1 \cdot x_2) = \log_a |x_1| + \log_a |x_2|$, $(x_1 \cdot x_2) > 0$;

(2) $\log_a \dfrac{x_1}{x_2} = \log_a |x_1| - \log_a |x_2|$, $(x_1 \cdot x_2) > 0$;

(3) $\log_a x^p = p \log_a x$, $x > 0$;

$\log_a x^{2n} = 2n \log_a |x|$, $n \in \mathbf{N}$, $x \neq 0$;

(4) $\log_a x = \dfrac{\log_b x}{\log_b a}$, $x > 0$, $b > 0$, $b \neq 1$.

**Solution of the fundamental exponential and logarithmic equations.**

$a^x = b$ $(a > 0, a \neq 1)$          $\log_a x = b$

(1) $b \leq 0 \Rightarrow x \in \varnothing$,

(2) $b > 0 \Rightarrow x = \log_a b$.          $x = a^b$.

**Solution of the fundamental exponential and logarithmic inequalities.**

$a^x < b$ $(a > 0, \ a \neq 1)$          $\log_a x < b$

(1) $b \leq 0 \Rightarrow x \in \varnothing$,

(2) $b > 0$, $a > 1 \Rightarrow x < \log_a b$,

$b > 0$, $0 < a < 1 \Rightarrow x > \log_a b$.

(1) $a > 1 \Rightarrow 0 < x < a^b$,

(2) $0 < a < 1 \Rightarrow x > a^b$.

Inequalities with the signs $\geq$, $>$, and $\leq$ can be solved by analogy.

**Solution of the fundamental trigonometric equations.**

$$\sin x = a \qquad\qquad\qquad\qquad \cos x = a$$

(1) $|a| > 1 \Rightarrow x \in \varnothing,$          (1) $|a| > 1 \Rightarrow x \in \varnothing,$

(2) $|a| \le 1$                    (2) $|a| \le 1$

$\Rightarrow x = (-1)^m \arcsin a + \pi m, \quad m \in \mathbf{Z}.$    $\Rightarrow x = \pm \arccos a + 2\pi n, \; n \in \mathbf{Z}.$

$$\tan x = a \qquad\qquad\qquad\qquad \cot x = a$$

$x = \arctan a + \pi k, \; k \in \mathbf{Z}.$    $x = \operatorname{arccot} a + \pi k, \; k \in \mathbf{Z}.$

**Basic trigonometric formulas.**

(1) $\sin^2 \alpha + \cos^2 \alpha = 1.$

(2) $\sin(\alpha + \beta) = \sin\alpha \cos\beta + \cos\alpha \sin\beta.$

(3) $\sin(\alpha - \beta) = \sin\alpha \cos\beta - \cos\alpha \sin\beta.$

(4) $\cos(\alpha + \beta) = \cos\alpha \cos\beta - \sin\alpha \sin\beta.$

(5) $\cos(\alpha - \beta) = \cos\alpha \cos\beta + \sin\alpha \sin\beta.$

(6) $\tan(\alpha + \beta) = \dfrac{\tan\alpha + \tan\beta}{1 - \tan\alpha \tan\beta}, \; \alpha, \beta, \alpha + \beta \ne \dfrac{\pi}{2} + \pi n, \; n \in \mathbf{Z}.$

(7) $\sin 2\alpha = 2\sin\alpha \cos\alpha.$

(8) $\cos 2\alpha = \cos^2 \alpha - \sin^2 \alpha = 2\cos^2 \alpha - 1 = 1 - 2\sin^2 \alpha.$

(9) $\sin 3\alpha = 3\sin\alpha - 4\sin^3 \alpha.$

(10) $\cos 3\alpha = 4\cos^3 \alpha - 3\cos\alpha.$

(11) $\sin\alpha + \sin\beta = 2\sin\dfrac{\alpha + \beta}{2} \cos\dfrac{\alpha - \beta}{2}.$

(12) $\sin\alpha - \sin\beta = 2\sin\dfrac{\alpha - \beta}{2} \cos\dfrac{\alpha + \beta}{2}.$

(13) $\cos\alpha + \cos\beta = 2\cos\dfrac{\alpha + \beta}{2} \cos\dfrac{\alpha - \beta}{2}.$

(14) $\cos\alpha - \cos\beta = -2\sin\dfrac{\alpha - \beta}{2} \sin\dfrac{\alpha + \beta}{2}.$

(15) $\sin\alpha \cos\beta = \dfrac{1}{2}[\sin(\alpha - \beta) + \sin(\alpha + \beta)].$

(16) $\sin\alpha \sin\beta = \dfrac{1}{2}[\cos(\alpha - \beta) - \cos(\alpha + \beta)].$

(17) $\cos\alpha \cos\beta = \dfrac{1}{2}[\cos(\alpha - \beta) + \cos(\alpha + \beta)].$

also formulas on pp. (106-108).

**Recursion formulas ($n \in \mathbf{Z}$).**

(1) $\sin(\pi n + \alpha) = (-1)^n \sin\alpha, \; \sin(\pi n - \alpha) = (-1)^{n+1} \sin\alpha.$

(2) $\cos(\pi n \pm \alpha) = (-1)^n \cos\alpha.$

(3) $\sin\left(\dfrac{\pi}{2} + \pi n \pm \alpha\right) = (-1)^n \cos\alpha.$

(4) $\cos\left(\dfrac{\pi}{2} + \pi n + \alpha\right) = (-1)^{n+1} \sin\alpha, \; \cos\left(\dfrac{\pi}{2} + \pi n - \alpha\right) = (-1)^n \sin\alpha.$

**Inverse trigonometric function.** The *arc sine* of the number $x \in [-1, +1]$ (designated as arcsin $x$) is a number $y \in \left[-\dfrac{\pi}{2}, \dfrac{\pi}{2}\right]$ whose sine is equal to $x$.

     The *arc cosine* of the number $x \in [-1, 1]$ (designated as arccos $x$) is a number $y \in [0, \pi]$ whose cosine is equal to $x$.

Table of the Values of Trigonometric Functions

| $\alpha°$ | $\alpha$ (radians) | $\sin\alpha$ | $\tan\alpha$ | $\cot\alpha$ | $\cos\alpha$ | | |
|---|---|---|---|---|---|---|---|
| 0 | 0 | 0 | 0 | ∞ | 1 | 1.571 | 90 |
| 1 | 0.017 | 0.017 | 0.017 | 57.29 | 1.000 | 1.553 | 89 |
| 2 | 0.035 | 0.035 | 0.035 | 28.64 | 0.999 | 1.536 | 88 |
| 3 | 0.052 | 0.052 | 0.052 | 19.08 | 0.999 | 1.518 | 87 |
| 4 | 0.070 | 0.070 | 0.070 | 14.30 | 0.998 | 1.501 | 86 |
| 5 | 0.087 | 0.087 | 0.087 | 11.43 | 0.996 | 1.484 | 85 |
| 6 | 0.105 | 0.105 | 0.105 | 9.514 | 0.995 | 1.466 | 84 |
| 7 | 0.122 | 0.122 | 0.123 | 8.144 | 0.993 | 1.449 | 83 |
| 8 | 0.140 | 0.139 | 0.141 | 7.115 | 0.990 | 1.431 | 82 |
| 9 | 0.157 | 0.156 | 0.158 | 6.314 | 0.988 | 1.414 | 81 |
| 10 | 0.175 | 0.174 | 0.176 | 5.671 | 0.985 | 1.396 | 80 |
| 11 | 0.192 | 0.191 | 0.194 | 5.145 | 0.982 | 1.379 | 79 |
| 12 | 0.209 | 0.208 | 0.213 | 4.705 | 0.978 | 1.361 | 78 |
| 13 | 0.227 | 0.225 | 0.231 | 4.331 | 0.974 | 1.344 | 77 |
| 14 | 0.244 | 0.242 | 0.249 | 4.011 | 0.970 | 1.326 | 76 |
| 15 | 0.262 | 0.259 | 0.268 | 3.732 | 0.966 | 1.309 | 75 |
| 16 | 0.279 | 0.276 | 0.287 | 3.487 | 0.961 | 1.292 | 74 |
| 17 | 0.297 | 0.292 | 0.306 | 3.271 | 0.956 | 1.274 | 73 |
| 18 | 0.314 | 0.309 | 0.325 | 3.078 | 0.951 | 1.257 | 72 |
| 19 | 0.332 | 0.326 | 0.344 | 2.904 | 0.946 | 1.239 | 71 |
| 20 | 0.349 | 0.342 | 0.364 | 2.747 | 0.940 | 1.222 | 70 |
| 21 | 0.367 | 0.358 | 0.384 | 2.605 | 0.934 | 1.204 | 69 |
| 22 | 0.384 | 0.375 | 0.404 | 2.475 | 0.927 | 1.187 | 68 |
| 23 | 0.401 | 0.391 | 0.424 | 2.356 | 0.921 | 1.169 | 67 |
| 24 | 0.419 | 0.407 | 0.445 | 2.246 | 0.914 | 1.152 | 66 |
| 25 | 0.436 | 0.423 | 0.466 | 2.145 | 0.906 | 1.134 | 65 |
| 26 | 0.454 | 0.438 | 0.488 | 2.050 | 0.899 | 1.117 | 64 |
| 27 | 0.471 | 0.454 | 0.510 | 1.963 | 0.891 | 1.100 | 63 |
| 28 | 0.489 | 0.469 | 0.532 | 1.881 | 0.883 | 1.082 | 62 |
| 29 | 0.506 | 0.485 | 0.554 | 1.804 | 0.875 | 1.065 | 61 |
| 30 | 0.524 | 0.500 | 0.577 | 1.732 | 0.866 | 1.047 | 60 |
| 31 | 0.541 | 0.515 | 0.601 | 1.664 | 0.857 | 1.030 | 59 |
| 32 | 0.559 | 0.530 | 0.625 | 1.600 | 0.848 | 1.012 | 58 |
| 33 | 0.576 | 0.545 | 0.649 | 1.540 | 0.839 | 0.995 | 57 |
| 34 | 0.593 | 0.559 | 0.675 | 1.483 | 0.829 | 0.977 | 56 |
| 35 | 0.611 | 0.574 | 0.700 | 1.428 | 0.819 | 0.960 | 55 |
| 36 | 0.628 | 0.588 | 0.727 | 1.326 | 0.809 | 0.942 | 54 |
| 37 | 0.646 | 0.602 | 0.754 | 1.327 | 0.799 | 0.925 | 53 |
| 38 | 0.663 | 0.616 | 0.781 | 1.280 | 0.788 | 0.908 | 52 |
| 39 | 0.681 | 0.629 | 0.810 | 1.235 | 0.777 | 0.890 | 51 |
| 40 | 0.698 | 0.643 | 0.839 | 1.192 | 0.766 | 0.873 | 50 |
| 41 | 0.716 | 0.656 | 0.869 | 1.150 | 0.755 | 0.855 | 49 |
| 42 | 0.733 | 0.669 | 0.900 | 1.111 | 0.743 | 0.838 | 48 |
| 43 | 0.750 | 0.682 | 0.933 | 1.072 | 0.731 | 0.820 | 47 |
| 44 | 0.768 | 0.695 | 0.966 | 1.036 | 0.719 | 0.803 | 46 |
| 45 | 0.785 | 0.707 | 1.000 | 1.000 | 0.707 | 0.785 | 45 |
| | | $\cos\alpha$ | $\cot\alpha$ | $\tan\alpha$ | $\sin\alpha$ | $\alpha$ (radians) | $\alpha°$ |

The arc *tangent* of the number $x \in \mathbf{R}$ (designated as arctan $x$) is a number $y \in \left(-\dfrac{\pi}{2}, \dfrac{\pi}{2}\right)$ whose tangent is equal to $x$.

The basic identities :

(1) $\sin (\arcsin x) = x, \; x \in [-1, +1]$;

$\arcsin (\sin x) = x, \; x \in \left[-\dfrac{\pi}{2}, \dfrac{\pi}{2}\right]$;

(2) $\cos (\arccos x) = x, \; x \in [-1, +1]$;

$\arccos (\cos x) = x, x \in [0, \pi]$;

(3) $\tan (\arctan x) = x, x \in \mathbf{R}$;

$\arctan (\tan x) = x, \; x \in \left(-\dfrac{\pi}{2}, \dfrac{\pi}{2}\right)$;

(4) $\arcsin x + \arccos x = \dfrac{\pi}{2}, \qquad x \in [-1, +1]$.

## 2. Fundamentals of Analysis

**Sequences.** If there is a rule which associates a certain real number $a_n$ with every natural number $n$, then we say that we are given a *number sequence* $a_1, a_2, \ldots, a_n, \ldots$

In particular, if $a_{n+1} = a_n + d$, where $d$ is a fixed number, for all $n \in \mathbf{N}$, then the sequence is called an *arithmetic progression*. Now if $a_{n+1} = a_n \cdot q$, where $q$ is also fixed, then the sequence is called a *geometric progression* (see p. 140).

The number $a$ is called a *limit* of the sequence $(a_n)$ if for any number $\varepsilon > 0$ there is a number $\mathbf{N}$ (dependent on the number $\varepsilon$) such that the inequality $|a_n - a| < \varepsilon$ holds true for all the term of the sequence with number $n > \mathbf{N}$.

Designation : $a = \lim\limits_{n \to \infty} a_n$.

In particular, $\lim\limits_{n \to \infty} q^n = 0$ ( for $|q| < 1$),

$$\lim_{n \to \infty} \left(1 + \frac{1}{n}\right)^n = e \text{ (definition of the number } e\text{)}.$$

**The limit of a function.** Assume that the function $f(x)$ is defined on the set $D(f) \subseteq R$ and let the point $a$ be such that an infinite number of points of the set $D(f)$ lie in any of its neighbourhoods (*a point of cond cond ensation or a limit point of the set* $D(f)$.

The number $b$ is called the *limit* of the function $f(x)$ at the point $a$ if for any number $\varepsilon > 0$ there is a number $\delta > 0$ (dependent on $\varepsilon$) such that the inequality
$$|f(x) - b| < \varepsilon$$
holds true for all $x \in D(f)$ satisfying the condition $0 < |x - a| < \delta$.

Designation : $b = \lim\limits_{x \to a} f(x)$.

In particular, $\lim\limits_{x \to 0} \dfrac{\sin x}{x} = 1$.

The function $f(x)$ is said to be *continuous* at the point $a \in D(f)$ if $\lim\limits_{x \to a} f(x) = f(a)$.

The function is *continuous on the set* $X \subseteq D(f)$ if it is continuous at each point of the set.

The sum, the difference, and the product of two functions continuous on the same set, is also continuous on that set. If the denominator of the fraction does not vanish on a set, then the quotient of two functions, continuous on that set is also continuous.

a polynomial is continuous on the entire number axis, and a fractional rational function is continuous at all the points of the axis where the denominator is nonzero.

**Definition of a derivative.**

$$f'(x) = \lim_{h \to 0} \frac{f(x+h) - f(x)}{h}.$$

**Rules of differentiation.**

(1) $(f(x) + g(x))' = f'(x) + g'(x)$.

(2) $(Cf(x))' = Cf'(x) (C \text{ is constant})$.

(3) $(f(x) \cdot g(x))' = f'(x) g(x) + f(x) g'(x)$.

(4) $(f(x)/g(x))' = \dfrac{f'(x) g(x) - f(x) g'(x)}{g^2(x)}$.

**Derivatives of certain functions.**

(1) $(C)' = 0$.

(2) $(x^\alpha)' = \alpha x^{\alpha-1} (\alpha \in \mathbf{R})$.

(3) $(\sin x)' = \cos x$.

(4) $(\cos x)' = -\sin x$.

(5) $(\tan x)' = \dfrac{1}{\cos^2 x}$.

(6) $(\cot x)' = -\dfrac{1}{\sin^2 x}$.

(7) $(a^x)' a^x \ln a$,

(8) $(\log_a x)' = \dfrac{1}{x \ln a}$,

$(e^x)' = e^x$.

$(\ln x)' = \dfrac{1}{x}$.

(9) $(\arcsin x)' = \dfrac{1}{\sqrt{1-x^2}}$.

(10) $(\arccos x)' = -\dfrac{1}{\sqrt{1-x^2}}$.

(11) $(\arctan x)' = \dfrac{1}{1+x^2}$.

(12) $(\text{arccot } x)' = -\dfrac{1}{1+x^2}$.

**Geometrical sense of a derivative.** $f'(x)$ is an angular coefficient (slope) of a tangent to the graph of the function $y = f(x)$ at the point $(x, f(x))$.

The *equation of a tangent line* to the graph of the function $y = f(x)$ at the point $(x_0, f(x_0))$ is $y = f(x_0) + f'(x_0)(x - x_0)$.

If $f'(x) > 0 \ (< 0)$ on the interval $(a, b)$, then the function $f(x)$ increases (decreases) on that interval.

The point $x_0 \in D(f)$ is called a *point of maximum* (*minimum*) of the function $f(x)$ if the inequality $f(x) \le f(x_0) (f(x) \ge f(x_0))$ holds true for all $x \in D(f)$ from a certain neighbourhood of the point $x_0$. The points of the maximum and minimum are known as *points of extremum* of a function.

A *critical (stationary) point* of a function is a point belonging to the domain of definition of the function at which the derivative is equal to zero or does not exist.

**The necessary condition for an extremum (Fermat's theorem).** If the function $f(x)$, differentiable at the point $x_0$, has an extremum at that point, then $f'(x_0) = 0$.

**The sufficient condition for an extremum.** If the derivative of a function changes sign from plus to minus when $x$ increases and passes through the critical point $x_0$, then $x_0$ is a point of maximum ; if the derivative changes sign from minus to plus, then $x_0$ is a point of minimum of that function; if the derivative does not change sign, then $x_0$ is not a point of extremum.

**The integral.** The function $F(x)$ is called an *antiderivative* of $f(x)$ on an interval if $F'(x) = f(x)$ on that interval. We can write all the antiderivatives of the function $f(x)$ in the form $F(x) + C$, where $C$ are various constants.

If $F(x)$ and $G(x)$ are antiderivatives of the functions $f(x)$ and $g(x)$, then

$F(x)+G(x)$, $\alpha F(x), \frac{1}{k}F(kx+b)$ are antiderivatives of the functions $f(x)+$ $g(x), \alpha f(x), f(kx+b)$ respectively. The antiderivatives of certain elementary functions can be easily obtained from the tables of derivatives presented above.

The *integral* of the function $f(x)$ on the interval $[a, b]$ (designated as $\int_a^b f(x)\,dx$) is the limit of the integral sums $f(c_1)\Delta x_1 + f(c_2)\Delta x_2 + \dots + f(c_n)\Delta x_n$ provided that the length of the greatest one of the intervals $[x_{k-1}, x_k]$ tends to zero. Here we have $a = x_0 < x_1 < \dots < x_n = b, \Delta x_k = x_k - x_{k-1}, c_k \in [x_{k-1}, x_k]$.

**The Newton-Leibniz formula.** If $F(x)$ is an antiderivative of $f(x)$ on $[a, b]$, then
$$\int_a^b f(x)\,dx = F(b) - F(a).$$

**The area of a curvilinear trapezoid** with the base $[a, b]$ bounded above by the graph of a nonnegative function $f(x)$ is
$$S = \int_a^b f(x)\,dx.$$

**The volume of a body of revolution** resulting from the rotation of the curvilinear trapezoid $a \le x \le b, 0 \le y \le f(x)$ about the $Ox$ axis is
$$V = \pi \int_a^b f^2(x)\,dx.$$

**Vectors.** A *vector* is a directed line segment. The *length of a vector* is the length of the segment. The vectors are said to be *equal* if their lengths are equal and they have the same direction.

Designations of a vector : $\mathbf{a}$, $\vec{AB}$ (if $A$ is the origin and $B$ is the terminal point of the vector) ; $|\mathbf{a}|, \vec{AB}$ are the designations of the length of a vector. If $|\mathbf{a}| = 0$, then $\mathbf{a}$ is a zero vector.

For any pair of vectors $\mathbf{a}$ and $\mathbf{b}$ their *sum* $\mathbf{a}+\mathbf{b}$ and their *difference* $\mathbf{a}-\mathbf{b}$ are defined. Any vector $\mathbf{a}$ can be *multiplied* by any number $k \in R$.

The nonzero vectors are said to be *collinear* if they lie on parallel straight lines and *coplanar* if they are parallel to the same plane. A zero vector is said to be collinear with any vector; any three vectors among which there is a zero vector are said to be coplanar.

If $\mathbf{p}, \mathbf{q}$ and $\mathbf{r}$ are three noncoplanar vectors, then any vector $\mathbf{a}$ can be represented in the form $\mathbf{a} = x\mathbf{p}+ y\mathbf{q}+ z\mathbf{r}$. This representation is called a *resolution* of the vector $\mathbf{a}$ *into components with respect to the base vectors* $\mathbf{p}, \mathbf{q}$ and $\mathbf{r}$; the numbers $x, y$ and $z$ are the *coordinates* of the vector $\mathbf{a}$ in the basis $\mathbf{p}, \mathbf{q}$ and $\mathbf{r}$. The notation is $\mathbf{a} = \{x, y, z\}$.

If $\mathbf{a}_1 = \{x_1, y_1, z_1\}$ and $\mathbf{a}_2 = \{x_2, y_2, z_2\}$ in a given basis, then $\mathbf{a}_1 + \mathbf{a}_2 = \{x_1 + x_2, y_1 + y_2, z_1 + z_2\}$ and $k\mathbf{a}_1 = \{kx_1, ky_1, kz_1\}$ in that basis.

Let us lay off the vector $\mathbf{a} = \vec{OA}$ and $\mathbf{b} = \vec{OB}$ from an arbitrary point $O$. The angle between the nonzero vectors $\mathbf{a}$ and $\mathbf{b}$ is the angle between the rays $\vec{OA}$ and $\vec{OB}$. The designation is $(\mathbf{a}, \hat{\ }\mathbf{b})$.

The *scalar product* of the vectors **a** and **b** is defined by the equation

$$\mathbf{a} \cdot \mathbf{b} \equiv (\mathbf{a}, \mathbf{b}) = |\mathbf{a}|\,|\mathbf{b}|\cos(\widehat{\mathbf{a}, \mathbf{b}}).$$

If at least one of the vectors is equal to zero, then the scalar product is assumed to be equal to zero by definition. The scalar product of nonzero vectors is equal to zero if and only if the vectors are perpendicular.

If $\hat{\mathbf{i}}, \hat{\mathbf{j}}$ and $\hat{\mathbf{k}}$ is a basis of orthogonal vectors which are equal to unity in length (coordinate vectors), and $\mathbf{a} = \{x_1, y_1, z_1\}$, $\mathbf{b} = \{x_2, y_2, z_2\}$ in that basis, then

$$(\mathbf{a}, \mathbf{b}) = x_1 \cdot x_2 + y_1 \cdot y_2 + z_1 \cdot z_2;$$

$$|\mathbf{a}| = \sqrt{x_1^2 + y_1^2 + z_1^2};$$

$$\cos(\widehat{\mathbf{a}, \mathbf{b}}) = \frac{x_1 x_2 + y_1 y_2 + z_1 z_2}{\sqrt{x_1^2 + y_1^2 + z_1^2}\sqrt{x_2^2 + y_2^2 + z_2^2}}.$$

For the reference material in geometry see Part 3 of the book.

# ✳ arihant

# Popular Series for
# JEE (Main & Advanced)

All arihant books are available@ **www.arihantbooks.com**

## Classic Texts Series

## New Pattern JEE Books

## 37 Years' Chapterwise IIT JEE Solved

All arihant books are available@ **www.arihantbooks.com**

## IIT JEE Questions & Solutions (Yearwise)

| | | |
|---|---|---|
| C007 | 14 Years' Solved Papers IIT JEE Physics (Main & Advanced) | 560 |
| C008 | 14 Years' Unsolved Questions Papers | 290 |
| C009 | 14 Years' Objective Solved Papers IIT JEE (Main & Advanced) | 390 |

## Master Resource Books for JEE Main

| | | | |
|---|---|---|---|
| B063 | Master Resource Book for JEE Main Physics | *DB Singh* | 799 |
| B064 | Master Resource Book for JEE Main Chemistry | *Sanjay Sharma* | 799 |
| B065 | Master Resource Book for JEE Main Mathematics | *Prafull K Agarwal* | 799 |
| B031 | सम्पूर्ण स्टडी पैकेज JEE Main भौतिकी | *धर्मवीर सिंह* | 799 |
| B032 | सम्पूर्ण स्टडी पैकेज JEE Main रसायन विज्ञान | *अभय कुमार* | 799 |
| B033 | सम्पूर्ण स्टडी पैकेज JEE Main गणित | *मंजुल त्यागी* | 799 |

## Solved Papers & Mock Tests for JEE Main

| | | |
|---|---|---|
| C019 | 15 Years' JEE Main Solved Papers | 420 |
| C016 | 15 Years' JEE Main सॉल्वड पेपर्स | 420 |
| C102 | 16 Years' JEE Main Chapterwise Solutions Physics | 205 |
| C103 | 16 Years' JEE Main Chapterwise Solutions Chemistry | 205 |
| C104 | 16 Years' JEE Main Chapterwise Solutions Maths | 205 |

## 40 Days Revision Books for JEE Main

| | | | |
|---|---|---|---|
| C142 | 40 Days JEE Main Physics | *Saurabh A* | 380 |
| C143 | 40 Days JEE Main Chemistry | *Dr Praveen Kumar* | 380 |
| C144 | 40 Days JEE Main Mathematics | *Rajeev Manocha* | 380 |
| C127 | 40 Days JEE Main भौतिकी | *देवेन्द्र कुमार* | 370 |
| C126 | 40 Days JEE Main रसायन | *हंसराज मोदी* | 370 |
| C125 | 40 Days JEE Main गणित | *डॉ आरपी सिंह* | 370 |

## Objective Books for JEE Main & Advanced

| | | | |
|---|---|---|---|
| B122 | Objective Physics (Vol 1) | *DC Pandey* | 645 |
| B123 | Objective Physics (Vol 2) | *DC Pandey* | 645 |
| B121 | Objective Chemistry (Vol 1) | *Dr. RK Gupta* | 795 |
| B130 | Objective Chemistry (Vol 2) | *Dr. RK Gupta* | 795 |
| B048 | Objective Mathematics (Vol 1) | *Amit M Agarwal* | 780 |
| B053 | Objective Mathematics (Vol 2) | *Amit M Agarwal* | 780 |

BITSAT 2018
BITSAT Prep Guide
Code : C064    ₹810

JEE Main Prepguide 2018
(with CD)
Code : C211    ₹1295

All arihant books are available@ **www.arihantbooks.com**

# DPP Daily Practice Problems
## for JEE (Main & Advanced)

## Physics

## Chemistry

## Mathematics

# Solved & Mock Tests for Engineering Entrances

## Solved Papers & Mock Tests (2-Edge Series)

| | | |
|---|---|---|
| C084 | VIT Solved Papers & Mock Tests | 355 |
| C023 | BVP Engineering 2-Edge Solved Papers & Mock Tests | 370 |
| C136 | AMU Engineering 2-Edge Solved Papers & Mock Tests | 345 |
| C092 | Manipal Engineering 2-Edge Solved Papers & Mock Tests | 380 |

## Andhra Pradesh

| | | |
|---|---|---|
| C154 | 27 Years' Chapterwise EAMCET Physics | 300 |
| C155 | 27 Years' Chapterwise EAMCET Chemistry | 260 |
| C156 | 27 Years' Chapterwise EAMCET Mathematics | 340 |
| C061 | 17 Years' Solved Papers EAMCET Engineering | 410 |

## Bihar

| | | |
|---|---|---|
| C042 | BCECE Previous Years' Solved Papers | 330 |
| C043 | 12 Years' Solved Papers BCECE Mains Entrance Exam | 335 |

## Chhattisgarh Complete Success Packages

| | | |
|---|---|---|
| F022 | Chhattisgarh PET Complete Success Package | 795 |
| F028 | छत्तीसगढ़ PET सक्सेस पैकेज | 775 |
| F036 | Chhattisgarh PMT Complete Success Package | 790 |
| F037 | छत्तीसगढ़ PMT सक्सेस पैकेज | 790 |

## Solved Papers & Mock Tests

| | | |
|---|---|---|
| C105 | Chhattisgarh PET 2-Edge Solved Papers & Mock Tests | 350 |
| C137 | छत्तीसगढ़ PET 2-Edge मॉक टेस्ट सॉल्वड पेपर्स | 335 |
| C106 | Chhattisgarh PMT 2-Edge Mock Tests & Solved Papers | 355 |
| C138 | छत्तीसगढ़ PMT 2-Edge मॉक टेस्ट सॉल्वड पेपर्स | 335 |

## Delhi

| | | |
|---|---|---|
| C059 | GGSIPU Engineering Entrance Exam *2-Edge Solved Papers & Mock Tests* | 355 |

## Haryana/Jammu & Kashmir (Solved Papers & Mock Tests)

| | | |
|---|---|---|
| C081 | J&K CET Medical Entrance Exam | 385 |
| C091 | J&K CET Engineering Entrance Exam | 385 |

## Jharkhand (Solved Papers & Mock Tests)

| | | |
|---|---|---|
| C045 | 16 Years' Solved Papers JCECE Engineering Entrance Exam | 340 |

All arihant books are available@ **www.arihantbooks.com**

## Kerala (Solved Papers & Mock Tests)

| C076 | 16 Years' Solved Papers Kerala CEE Engineering Entrance Exam | 475 |

## Karnataka/Maharashtra (Solved Papers & Mock Tests)

| C032 | 16 Years' Solved Papers K-CET Engineering Entrance Exam | 375 |
| C107 | MHT-CET Engineering Entrance Exam | 310 |

## Uttar Pradesh (Complete Success Packages)

| F029 | UPTU/UPSEE संपूर्ण सक्सेस पैकेज 2018 | 775 |
| F024 | Complete Study Guide UPTU/UPSEE 2018 Physics | 390 |
| F025 | Study Guide Chemistry for SEE-GBT | 380 |
| F026 | Study Guide Mathematics for SEE-GBTU | 380 |

### Solved Papers

| C014 | 14 Years' सॉल्वड पेपर्स UPTU/UPSEE | 390 |
| C072 | 14 Years' Solved Papers UPTU/UPSEE | 405 |

## West Bengal

| C075 | WB JEE Engineering Solved Papers & Mock Tests | 415 |

## Science & Mathematics Olympiads

| B056 | Indian National Physics Olympiad | Saurabh A | 360 |
| B068 | Indian National Chemistry Olympiad | Dr.Praveen Kumar | 275 |
| B044 | Indian National Mathematics Olympiad | Rajeev Manocha | 390 |
| B043 | Indian National Biology Olympiad | Dr. RK Manglik | 275 |

# NCERT Exemplar Solutions

## Class XI

| F259 | NCERT Exemplar - Physics | 150 |
| F251 | NCERT Exemplar - Chemistry | 150 |
| F280 | NCERT Exemplar - Mathematics | 175 |
| F260 | NCERT Exemplar - Biology | 175 |

## Class XII

| F281 | NCERT Exemplar - Physics | 150 |
| F279 | NCERT Exemplar - Chemistry | 175 |
| F282 | NCERT Exemplar - Mathematics | 175 |
| F278 | NCERT Exemplar - Biology | 150 |

All arihant books are available@ **www.arihantbooks.com**

# *All in one*

## The Complete Study Resources for
## CBSE 11th & 12th

## Class XI

| F246 | All in One - Physics | 475 |
|------|---------------------|-----|
| F244 | All in One - Chemistry | 495 |
| F240 | All in One - Mathematics | 425 |
| F245 | All in One - Biology | 465 |
| F243 | All in One - Computer Science | 285 |
| F234 | All in One - English Core | 395 |

## Class XII

| F206 | All in One - Physics | 465 |
|------|---------------------|-----|
| F225 | All in One - Chemistry | 475 |
| F208 | All in One - Mathematics | 465 |
| F211 | All in One - Biology | 385 |
| F195 | All in One - Computer Science | 365 |
| F196 | All in One - English Core | 395 |

# Handbook Series

| C190 | Handbook Physics | 210 |
|------|-----------------|-----|
| C191 | Handbook Chemistry | 275 |
| C192 | Handbook Mathematics | 235 |
| C207 | Handbook Biology | 295 |

# Dictionaries

| C185 | Dictionary of Physics | 135 |
|------|----------------------|-----|
| C186 | Dictionary of Chemistry | 175 |
| C187 | Dictionary of Mathematics | 135 |
| C188 | Dictionary of Biology | 195 |

All arihant books are available@ **www.arihantbooks.com**

Ingram Content Group UK Ltd.
Milton Keynes UK
UKHW020720260623
424053UK00014B/692